Microwave Solid-State Circuits and Applications

WILEY SERIES IN MICROWAVE AND OPTICAL ENGINEERING

KAI CHANG, *Editor*
Texas A&M University

INTRODUCTION TO ELECTROMAGNETIC COMPATIBILITY
Clayton R. Paul

OPTICAL COMPUTING: AN INTRODUCTION
Mohammad A. Karim and Abdul Abad S. Awwal

COMPUTATIONAL METHODS FOR ELECTROMAGNETICS AND MICROWAVES
Richard C. Booton, Jr.

FIBER-OPTIC COMMUNICATION SYSTEMS
Covind P. Agrawal

ANTENNAS FOR RADAR AND COMMUNICATIONS: A POLARIMETRIC APPROACH
Harold Mott

OPTICAL SIGNAL PROCESSING, COMPUTING, AND NEURAL NETWORKS
Francis T.S. Yu and Suganda Jutamulia

MULTICONDUCTOR TRANSMISSION-LINE STRUCTURE: MODAL ANALYSIS TECHNIQUES
Jose A.M. Brandao Faria

MICROSTRIP CIRCUITS
Fred Gardiol

MICROWAVE DEVICES, CIRCUITS AND THEIR INTERACTION
Charles A. Lee and G. Conrad Dalman

MICROWAVE SOLID-STATE CIRCUITS AND APPLICATIONS
Kai Chang

Microwave Solid-State Circuits and Applications

KAI CHANG
Department of Electrical Engineering
Texas A&M University
College Station, Texas

A WILEY-INTERSCIENCE PUBLICATION
JOHN WILEY & SONS, INC.
NEW YORK / CHICHESTER / BRISBANE / TORONTO / SINGAPORE

This text is printed on acid-free paper.

Library of Congress Cataloging in Publication Data:

Chang, Kai, 1948–
Microwave solid-state circuits and applications / Kai Chang.
p. cm.—(Wiley series in microwave and optical engineering)
Includes bibliographical references and index.
ISBN 0-471-54044-7 (alk. paper)
1. Microwave devices. 2. Microwave circuits. 3. Semiconductors.
I. Title. II. Series.
TK7876.C44 1994
621.381'3—dc20 93-25582

Printed in the United State of America

10 9 8 7 6 5 4 3 2

To my parents and my family

Contents

Preface

Microwave technology has made rapid advances in the past four decades due to the development of new solid-state devices and circuits. These devices and circuits play a more and more important role in modern microwave engineering. Rectangular waveguides and tubes are still being used in numerous applications, but many of their uses have been replaced by microstrip and solid-state circuits. The objective of this book is to introduce students to modern microwave technologies and applications based on solid-state devices and circuits. Because the book is self-contained, it also serves as a reference book for microwave, antenna, optics, and solid-state engineers.

This book was written based on notes developed for a senior- or graduate-level course in microwave solid-state circuits at Texas A&M University. It is assumed that the reader has had previous courses in electromagnetics, transmission lines, and solid-state electronics. The book is organized into three major parts: (1) Review of fundamental principles of transmission lines and circuits, and semiconductor physics (Chapters 2, 3, and 4); (2) two-terminal solid-state devices, circuits, and applications (Chapters 5, 6, 8, 9, 10, and 11); and (3) three-terminal solid-state devices, circuits, and applications (Chapters 12, 13, 14, 15, and 16). Chapter 7 stands alone by introducing noise figures and some system parameters for receiver design. The instructor could cover this book in one semester with some sections assigned to students for self-study. The end-of-chapter problems will strengthen the reader's knowledge of the subject. The reference sections list the principal references used.

Throughout the book, the emphasis is on the basic operating principles and techniques incorporating the devices into circuit applications. Fundamental design equations are derived and practical examples are given whenever possible. Suggestions for further reading are given in some chapters.

I would like to thank all of my former students who used my notes in class for their helpful comments and suggestions. I would also like to thank Kenneth Hummer, Mingyi Li, and James McCleary for critical review of the manuscript. Sherry Orange has done an excellent job typing the manuscript. Finally, I wish to express my deep appreciation to my wife, Suh-jan, and my children, Peter and Nancy, for their constant encouragement and support.

CHAPTER 1

Introduction

The objective of this book is to introduce modern microwave technology and applications of solid-state devices and circuits. Microwave technology applies to the frequency range from 300 MHz to 300 GHz. The applications in this spectrum have been accelerating in the past four decades due to the advances of many microwave solid-state devices. This book presents the basic principles and applications of these devices and circuits.

1.1 HISTORY OF MICROWAVE SOLID-STATE DEVICES AND CIRCUITS

The first microwave semiconductor device was a crystal valve. During World War II, a great effort was directed toward the development of crystal detectors for radar applications. Silicon and germanium materials were used, and point-contact crystals were developed. Since the invention of the bipolar transistor in 1947, the applications of semiconductor devices have grown very rapidly. Devices using p–n junctions were widely used in the 1950s and 1960s, including varactors, mixers, detectors, parametric amplifiers, frequency converters, harmonic generators, and switches. Several other devices were also invented during this period. A tunnel diode with a negative-resistance effect was invented by Esaki in 1957. The device can be used for microwave generation, mixing, and detection. Read proposed an IMPATT diode in 1958, but it was not until 1965 that the structure was modified by Johnston, DeLoach, and Cohen to give microwave oscillation. Another negative-resistance device, the Gunn device, was discovered by J. B. Gunn, who in 1963 observed microwave oscillations in a homogeneous specimen of gallium arsenide at a critical electric field.

The first three-terminal device, the bipolar transistor, was invented in 1947. The microwave bipolar transistor has been refined and improved continuously since 1952 by reduction in the emitter stripe width and base-layer thickness. Its operating frequency is generally limited to 4 GHz. The GaAs MESFET (metal–semiconductor field-effect transistor) using a Schottky-barrier gate, first proposed by Mead in 1966, has become the dominant microwave solid-state device for many applications up to 100 GHz. The discovery in 1978 of two-dimensional electron gas (2-DEG) at the AlGaAs/GaAs heterojunction has led to great progress in the high electron mobility transistor (HEMT) or heterojunction FET (HFET). An operational frequency of up to 100 GHz has been demonstrated for this device. Another useful three-terminal device is the heterojunction bipolar transistor (HBT), which was first proposed by Kroemer in 1957 but became popular only recently. The HBT can operate at higher frequencies than the silicon bipolar transistor.

Early microwave solid-state circuits were built primarily in coaxial lines and waveguides. Since the introduction of the planar microstrip line in 1952, the use of microwave integrated circuits has become widespread. Recently, monolithic integrated circuits have been developed and made commercially available.

1.2 FREQUENCY SPECTRUMS

Microwaves and millimeter waves occupy the region of the electromagnetic spectrum from 300 MHz to 300 GHz, with a corresponding wavelength from 100 cm to 1 mm, as shown in Figure 1.1. Millimeter waves, which derive their name from the dimensions of the wavelengths (from 10 to 1 mm), can be classified as microwaves since millimeter-wave technology is quite similar to microwave technology. Below the microwave spectrum is the radio-frequency (RF) spectrum, and above the millimeter-wave spectrum are the submillimeter-wave, infrared, and optical spectrums. For convenience, microwave and millimeter-wave spectrums are further divided into many frequency bands. Table 1.1 shows the commonly used microwave band designations, and Table 1.2 shows the millimeter-wave bands. These band designations have not been accepted globally, and some other designations can be found in the literature.

As the frequency increases and the wavelength becomes shorter, the amount of energy in a signal that is absorbed as it goes through the atmosphere increases irregularly, as shown in Figure 1.2. In the low microwave frequency range, the absorption is small and the attenuation of transmission through the atmosphere is small. At millimeter-wave frequencies, the attenuation increases and the amount of absorption can increase further with the amount of moisture in the atmosphere (e.g., in the rain). There is an absorption peak at 60 GHz and "windows" or absorption minima occur at 35 and 94 GHz (as well as at higher frequencies). The windows at 35 and 94 GHz are of interest for radar applications, and 60 GHz is of interest

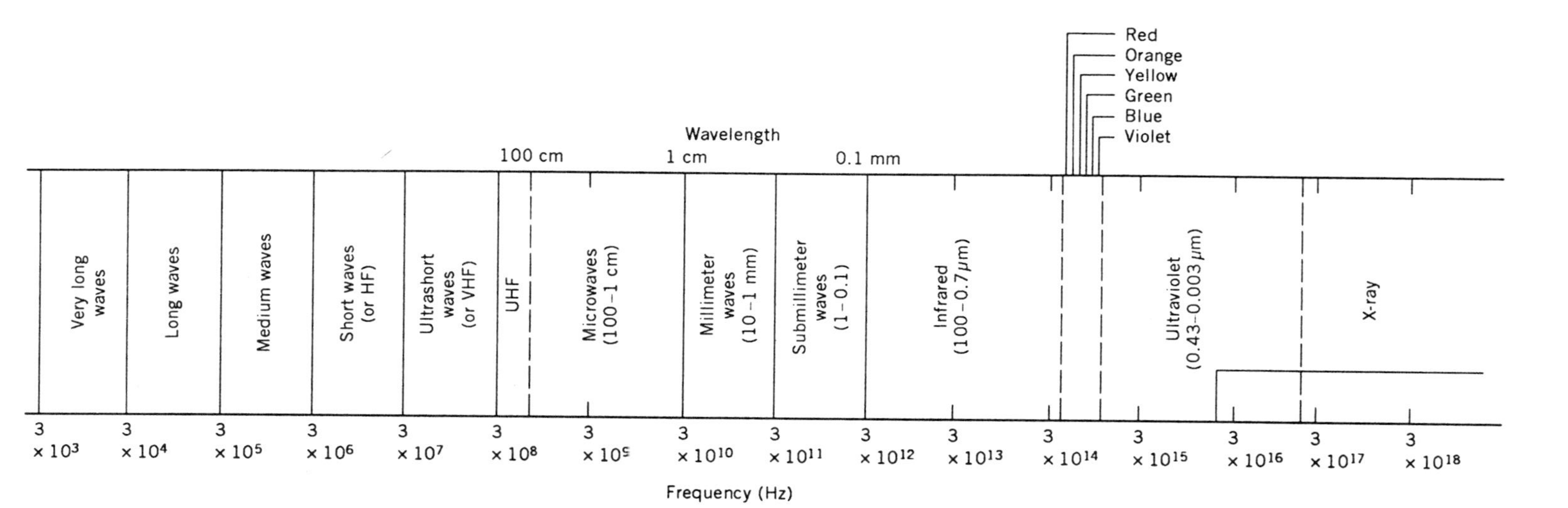

FIGURE 1.1 Electromagnetic spectrum.

TABLE 1.1 Microwave Band Designation

Designation	Frequency Range (GHz)
L-band	1–2
S-band	2–4
C-band	4–8
X-band	8–12
Ku-band	12–18
K-band	18–26
Ka-band	26–40

TABLE 1.2 Millimeter-Wave Band Designation

Designation	Frequency Range (GHz)
Q-band	33–50
U-band	40–60
V-band	50–75
E-band	60–90
W-band	75–110
D-band	110–170
G-band	140–220
Y-band	220–325

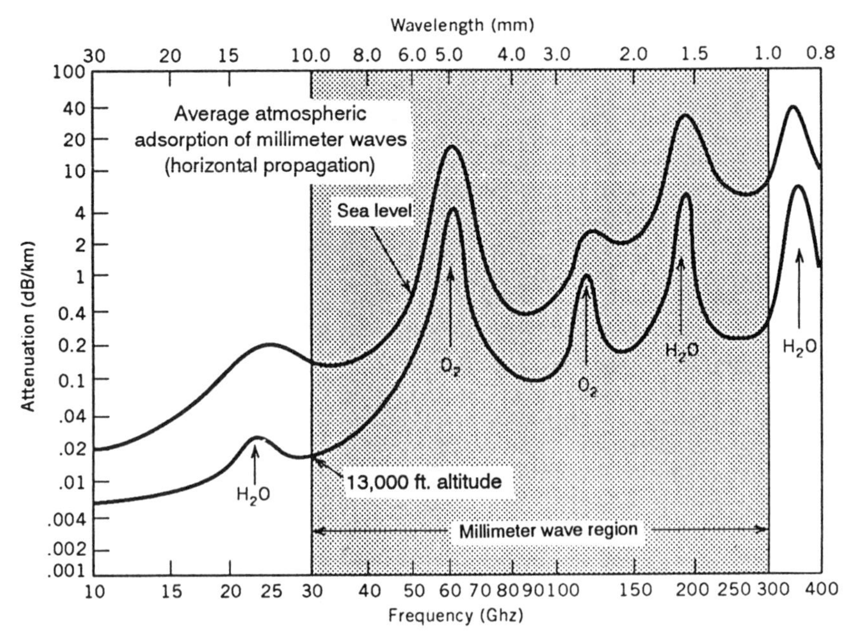

FIGURE 1.2 Absorption by the atmosphere.

for covert operations such as point-to-point short-distance communications on a battlefield or satellite-to-satellite communications in space.

1.3 MICROWAVE APPLICATIONS

During the past several decades, there has been a continuous trend to move to higher frequencies in system applications. The advantages of using higher frequencies are many: a less crowded spectrum, wider bandwidths, smaller antenna sizes, better resolution in radar applications, and less interference. The disadvantages are more expensive components, less mature technology, higher losses and lower power, and a heavy reliance on GaAs technology instead of silicon technology. Despite these disadvantages, operating at higher frequencies is the only choice in many applications simply because the low-frequency spectrum has been occupied.

Microwaves and millimeter waves have many military and commercial applications, as summarized in Table 1.3. It should be emphasized that this table is not meant to exhaust all applications. The two major applications are in communications and radar. The communication applications include satellite, space, long-distance telephone, marine, cellular telephone, mobile, aircraft, personal, data, wireless LAN, and vehicle, among others. Radar has applications for aircraft, weather, collision avoidance, imaging, air defense, ships, police, intrusion detection, smart weapons, and so on. The navigation systems are used for aircraft, ships, and vehicles. For example, the microwave landing system (MLS) is used in airports to guide aircraft during landing. The global positioning system (GPS) enables a person to find his or her exact location at any place on earth. In remote-sensing applications, many satellites are used to monitor the earth and atmosphere. These satellites constantly monitor weather, forests, oceans, soil moisture, snow thickness, icebergs, ozone, and other factors. Satellites can also be used for resource monitoring or exploration. In domestic and industrial use, microwave ovens, moisture sensors, automatic door openers, tank gauges, flow meters, fluid heating, and pest control are examples. Microwaves also have uses in medical applications for selective heating in cancer treatment, sterilization, and imaging. For high-speed computing and signal processing in gigabit logic circuits, connecting lines need to be treated as transmission lines, and microwave technology

TABLE 1.3 Major Microwave Applications

Communications	Digital circuits and high speed computing
Radar	Optical communications
Navigation	Surveillance
Remote sensing	Broadcast
Domestic and industrial applications	Plasma diagnosis
Medical applications	Astronomy and space exploration

TABLE 1.4 Commercial Applications of Microwaves

Category of Commercial Applications	Description
Agriculture	Crop protection from winter weather
	Germination
	Moisture detection
	Pesticides
	Soil treatment
Automobiles	Adaptive cruise control
	Airbag arming
	Antitheft radar or sensor
	Auto navigation aids and positioning
	Automatic headway control
	Automotive telecommunications
	Autonomous intelligent cruise control
	Blind-spot radar
	Collision-avoidance radar
	Collision-warning radar
	Current parking information
	Current traffic information
	Intelligent vehicle
	Near-obstacle detectors
	Optimum speed data
	Radar speed sensors
	Road-to-vehicle communications
	Vehicle identification
	Vehicle-to-vehicle communications
Broadcast	Direct broadcast satellite
	Downconverter chips for satellite TV
	High-definition TV
	Sky cable satellite
	Universal radio system
Communications	Millimeter-wave radios
	Personal communication systems
	Satellite-based data communications
	Satellite mobile communication systems
	Wireless local area networks
Domestic	Microwave clothes dryer
	Microwave heating
Highway	Automatic toll collection
	Buried-object sensors
	Highway traffic controls
	Highway traffic monitoring
	Intelligent highway
	Penetration radar for pavement
	Range and speed detection
	Road guidance

TABLE 1.4 (Continued)

Category of Commercial Applications	Description
	Structure inspection
	Truck position tracking
	Vehicle detection
Imaging	Aircraft traffic control and landing systems
	Hidden-weapon detection
	Medical imaging
	Obstacle detection and navigation
Industrial control	Chip defect detection
	Coal purification
	Hidden-object detection
	Moisture content measurements
	Pollutant detection
	Material property measurement
Medical	Balloon angioplasty
	Cautery
	Heart stimulation
	Hemorrhaging control
	Hyperthermia
	Microwave imaging
	Sterilization
	Thermography
Navigation/position information	Global positioning system
Office	Mail sorting
	Wireless local area networks
Pesticides	Slug and insect control
Power	Beamed-power propulsion
	Power transmission in space
Preservation	Food preservation
	Treated manuscript drying
Production control	Ceramic production assistance
	Etching system production
	Industrial drying
	Moisture control
	Rubber vulcanization assistance
Remote sensing	Earth monitoring
	Meteorology
	Pollution control
Security system	Intruder detection
	Surveillance

should be considered in the circuit designs. Optical communications using low-loss fiber normally have a microwave modulator in the transmitting side and a demodulator in the receiving end. The microwave signal acts as a modulating signal and the optical signal is the carrier. High-sensitivity EW (electronic warfare) receivers can be used to monitor signal traffic and to identify invaders in surveillance applications. In TV and radio broadcast, microwaves are used as the carriers for video and audio signals. The direct broadcast system (DBS) is designed to link satellites directly with home users. Now that the Cold War has ended, more commercial applications will be emerging. Some possibilities for such applications are given in Table 1.4.

1.4 ORGANIZATION OF THIS BOOK

The book is organized into 16 chapters. Chapters 2, 3, and 4 review some fundamental principles of transmission lines, waveguides, microwave circuit representations, and semiconductor physics. Chapters 5, 6, 8, 10, and 11 cover various two-terminal devices and circuits, including varactors, detectors, mixers, switches, Gunn devices, and IMPATTs. Chapters 7 and 9 give the operating theory and system design for receivers, oscillators, and amplifiers. Chapters 12 through 16 are devoted to three-terminal devices and their applications to low-noise amplifiers, power amplifiers, oscillators, mixers, switches, and phase shifters.

CHAPTER 2

Transmission Lines and Waveguides

2.1 INTRODUCTION

At low frequencies, two electronic components can be connected by using a wire or a line on a printed circuit board. At microwave frequencies the wire becomes lossy and radiates power. Special transmission lines and waveguides are required to connect microwave components and circuit elements. Many transmission lines and waveguides have been proposed and used in microwave and millimeter-wave frequencies. Figure 2.1 shows some of these lines, and Table 2.1 summarizes their properties. Among them, the rectangular waveguide, coaxial line, and microstrip line are the most commonly used. The choice of a suitable transmission medium for constructing microwave circuits, components, and subsystems is dictated by electrical and mechanical trade-offs. Electrical trade-offs involve such parameters as transmission-line loss, dispersion, higher-order modes, range of impedance levels, maximum operating frequency, and suitability for component and device implementation. Mechanical trade-offs include ease of fabrication, tolerance, and reliability.

In this chapter, transmission-line theory and impedance-matching techniques are reviewed. The most commonly used transmission lines and waveguides (rectangular waveguide, coaxial line, and microstrip line) are discussed.

2.2 TRANSMISSION-LINE EQUATION

Suppose that a transmission line is used to connect a source to a load as shown in Figure 2.2. At position x along the line, there exists a time-varying voltage $v(x,t)$ and current $i(x,t)$. For a small section between x and $x + \Delta x$, the equivalent circuit of this section Δx can be represented by the

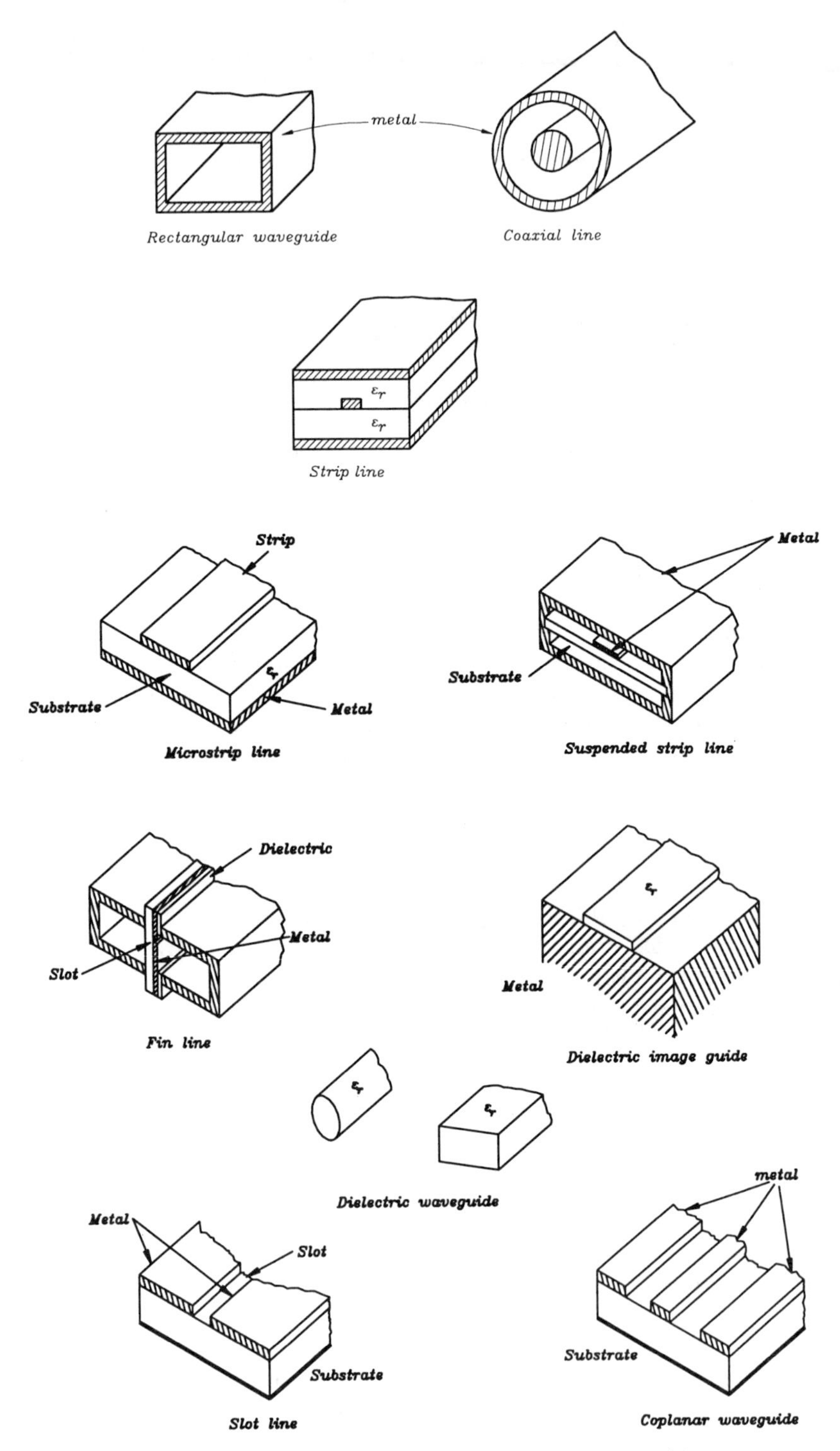

FIGURE 2.1 Various transmission lines and waveguides.

TABLE 2.1 Comparison of Guiding Media and Waveguides

Item	Useful Frequency (GHz)	Impedance Level (Ω)	Cross-Sectional Dimensions	Q Factor	Power Rating	Active Device Mounting	Potential for Low-Cost Production
Rectangular waveguide	< 300	100–500	Moderate to large	High	High	Easy	Poor
Coaxial line	< 50	10–100	Moderate	Moderate	Moderate	Fair	Poor
Strip line	< 10	10–100	Moderate	Low	Low	Fair	Good
Microstrip line	≤ 100	10–100	Small	Low	Low	Easy	Good
Suspended strip line	≤ 150	20–150	Small	Moderate	Low	Easy	Fair
Fin line	≤ 150	20–400	Moderate	Moderate	Low	Easy	Fair
Slot line	≤ 60	60–200	Small	Low	Low	Fair	Good
Coplanar waveguide	≤ 60	40–150	Small	Low	Low	Fair	Good
Image guide	< 300	20–30	Moderate	High	Low	Poor	Good
Dielectric guide	< 300	20–50	Moderate	High	Low	Poor	Fair

distributed elements of L, R, C, and G, which are the inductance, resistance, capacitance, and conductance per unit length. For a lossless line, $R = G = 0$. In most cases, R and G are small. This equivalent circuit can easily be understood by considering the coaxial line shown in Figure 2.3. L and R are due to the length and conductor losses of the outer and inner conductors. C and G are attributed to the separation and dielectric losses between the outer and inner conductors.

Now apply the Kirchhoff current and voltage laws to the equivalent circuit shown in Figure 2.2. We have

$$v(x+\Delta x,t) - v(x,t) = \Delta v(x,t) = -(R\Delta x)i(x,t) - (L\Delta x)\frac{\partial i(x,t)}{\partial t} \tag{2.1}$$

$$i(x+\Delta x,t) - i(x,t) = \Delta i(x,t) = -(G\Delta x)v(x+\Delta x,t) - (C\Delta x)\frac{\partial v(x+\Delta x,t)}{\partial t} \tag{2.2}$$

Divide Equations (2.1) and (2.2) by Δx and take the limit with Δx approaching 0. We have the following equations:

$$\frac{\partial v(x,t)}{\partial x} = -Ri(x,t) - L\frac{\partial i(x,t)}{\partial t} \tag{2.3}$$

$$\frac{\partial i(x,t)}{\partial x} = -Gv(x,t) - C\frac{\partial v(x,t)}{\partial t} \tag{2.4}$$

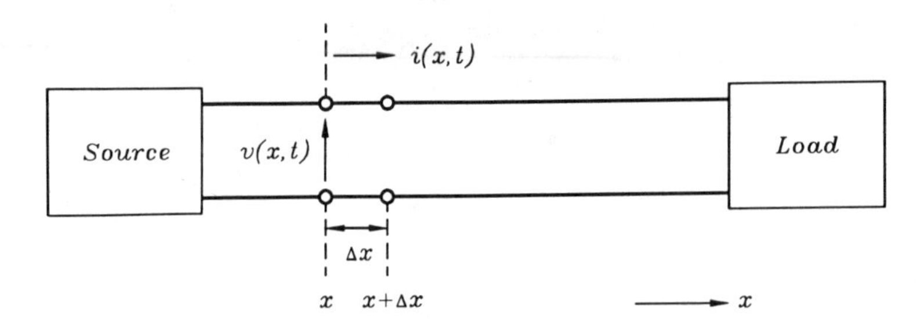

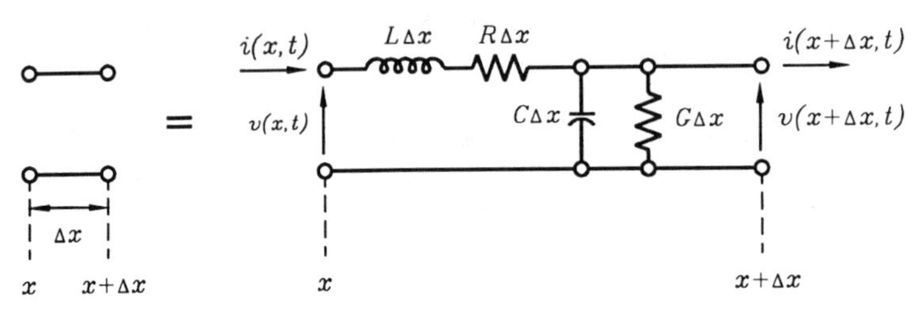

FIGURE 2.2 Transmission-line equivalent circuit.

Differentiating (2.3) with respect to x and (2.4) with respect to t gives

$$\frac{\partial^2 v(x,t)}{\partial x^2} = -R\frac{\partial i(x,t)}{\partial x} - L\frac{\partial^2 i(x,t)}{\partial x\,\partial t} \tag{2.5}$$

$$\frac{\partial^2 i(x,t)}{\partial t\,\partial x} = -G\frac{\partial v(x,t)}{\partial t} - C\frac{\partial^2 v(x,t)}{\partial t^2} \tag{2.6}$$

To eliminate $\partial i/\partial x$ and $\partial^2 i/\partial x\,\partial t$ in Equation (2.5), substitute (2.4) and (2.6)

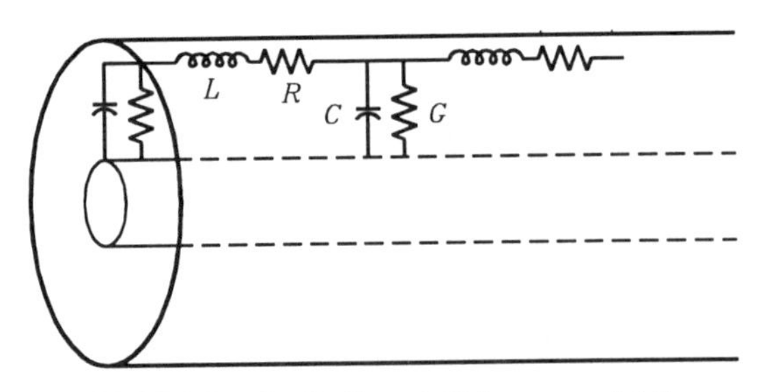

FIGURE 2.3 L, R, C, and G for a coaxial line.

into (2.5), resulting in

$$\frac{\partial^2 v(x,t)}{\partial x^2} - (RC + LG)\frac{\partial v(x,t)}{\partial t} - LC\frac{\partial^2 v(x,t)}{\partial t^2} - RGv(x,t) = 0 \tag{2.7}$$

Similarly,

$$\frac{\partial^2 i(x,t)}{\partial x^2} - (LG + RC)\frac{\partial i(x,t)}{\partial t} - LC\frac{\partial^2 i(x,t)}{\partial t^2} - RGi(x,t) = 0 \tag{2.8}$$

Equations (2.7) and (2.8) are called *transmission-line equations or telegrapher's equations in the time domain*. If only the steady-state sinusoidally time-varying solution is desired, phasor notation can be used to simplify these equations [1]. v and i can be expressed as

$$v(x,t) = \mathrm{Re}\left[V(x)e^{j\omega t}\right] \tag{2.9}$$

$$i(x,t) = \mathrm{Re}\left[I(x)e^{j\omega t}\right] \tag{2.10}$$

where Re is the real part and ω is the angular frequency equal to $2\pi f$. Equations (2.7) and (2.8) can be written as

$$\frac{d^2V(x)}{dx^2} - j\omega(RC + LG)V(x) - (RG - \omega^2 LC)V(x) = 0 \tag{2.11}$$

$$\frac{d^2I(x)}{dx^2} - j\omega(RC + LG)I(x) - (RG - \omega^2 LC)I(x) = 0 \tag{2.12}$$

Equations (2.11) and (2.12) are called *transmission-line equations in the frequency domain*. If we let $\gamma^2 = (R + j\omega L)(G + j\omega C)$, Equation (2.11) can be rewritten as

$$\frac{d^2V(x)}{dx^2} - \gamma^2 V(x) = 0 \tag{2.13}$$

Note that Equation (2.13) is a wave equation and γ is the wave propagation constant. The general solution to Equation (2.13) is

$$V(x) = V_+ e^{-\gamma x} + V_- e^{\gamma x} \tag{2.14}$$

The propagation constant γ is given by

$$\gamma = \left[(R + j\omega L)(G + j\omega C)\right]^{1/2} = \alpha + j\beta \tag{2.15}$$

where α is the attenuation constant in nepers/unit length and β is the phase constant in radians/unit length.

Equation (2.14) gives the solution for voltage along the transmission line. The voltage is the summation of a forward wave ($V_+ e^{-\gamma x}$) and a reflected wave ($V_- e^{\gamma x}$) propagating in the $+x$ and $-x$ directions, respectively. The current $I(x)$ can be found from Equation (2.3) in the frequency domain:

$$\frac{dV(x)}{dx} = -(R + j\omega L) I(x) \tag{2.16}$$

Therefore,

$$\begin{aligned} I(x) &= -\frac{1}{R + j\omega L} \frac{dV(x)}{dx} \\ &= \frac{\gamma}{R + j\omega L} (V_+ e^{-\gamma x} - V_- e^{\gamma x}) \\ &= I_+ e^{-\gamma x} - I_- e^{\gamma x} \end{aligned} \tag{2.17}$$

Hence

$$I_+ = \frac{\gamma}{R + j\omega L} V_+, \qquad I_- = \frac{\gamma}{R + j\omega L} V_-$$

The characteristic impedance of the line is defined by

$$Z_0 = \frac{V_+}{I_+} = \frac{V_-}{I_-} = \frac{R + j\omega L}{\gamma} = \left(\frac{R + j\omega L}{G + j\omega C} \right)^{1/2} \tag{2.18}$$

For a lossless line, $R = G = 0$, we have

$$\gamma = j\beta = j\omega\sqrt{LC} \tag{2.19}$$

$$Z_0 = \sqrt{\frac{L}{C}} \tag{2.20}$$

$$\text{phase velocity } v_p = \frac{\omega}{\beta} = f\lambda_g = \frac{1}{\sqrt{LC}} \tag{2.21}$$

where λ_g is the guided wavelength.

2.3 TERMINATED TRANSMISSION LINE

At high or microwave frequencies, the length of even a short transmission line plays an important role in circuit design. Figure 2.4(a) shows a transmission line with a length of l and a characteristic impedance of Z_0. If the line is

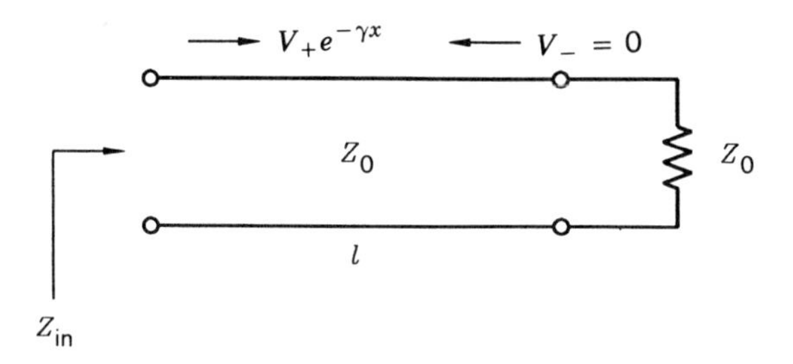

(a) A load Z_0 is connected to a transmission line with a characteristic impedance of Z_0

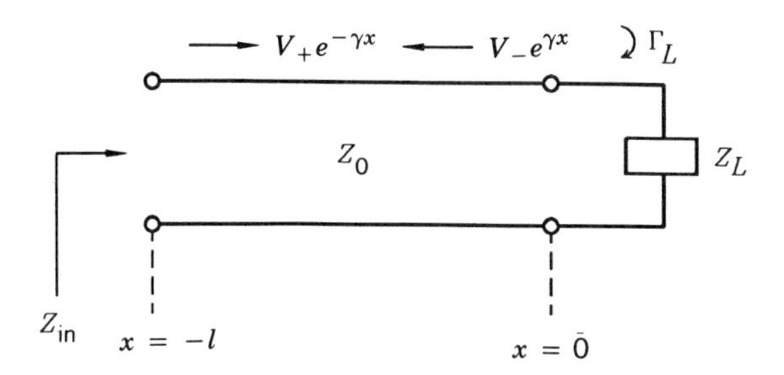

(b) A load Z_L is connected to a transmission line with a characteristic impedance of Z_0

FIGURE 2.4 Terminated transmission line.

terminated by a load Z_0, there is no reflection and the input impedance Z_{in} is always equal to Z_0 regardless of the length of the transmission line. If a load Z_L (Z_L could be real or complex) is connected to the line as shown in Figure 2.4(*b*) and Z_L is not equal to Z_0, there exists a reflected wave and the input impedance is no longer equal to Z_0. Instead, Z_{in} is a function of frequency (f), l, Z_L, and Z_0. Note that at low frequencies where the transmission-line length is small compared to a wavelength, $Z_{in} \approx Z_L$ regardless of l.

In Section 2.2 the voltage along the line was given by

$$V(x) = V_+e^{-\gamma x} + V_-e^{\gamma x} \tag{2.22}$$

The reflection coefficient $\Gamma(x)$ is defined as the ratio of the reflected voltage wave to the incident voltage wave anywhere along the line:

$$\Gamma(x) = \frac{\text{reflected } V(x)}{\text{incident } V(x)} = \frac{V_-e^{\gamma x}}{V_+e^{-\gamma x}} = \frac{V_-}{V_+}e^{2\gamma x}$$
$$= \Gamma_L e^{2\gamma x} \tag{2.23}$$

where

$$\Gamma_L = \frac{V_-}{V_+} = \Gamma(0)$$
$$= \text{reflection coefficient at the load} \tag{2.24}$$

Substituting Γ_L into Equation (2.22), we have

$$V(x) = V_+\left(e^{-\gamma x} + \frac{V_-}{V_+}e^{\gamma x}\right)$$
$$= V_+(e^{-\gamma x} + \Gamma_L e^{\gamma x}) \tag{2.25}$$

The current along the line is

$$I(x) = I_+ e^{-\gamma x} - I_- e^{\gamma x}$$
$$= \frac{V_+}{Z_0}(e^{-\gamma x} - \Gamma_L e^{\gamma x}) \tag{2.26}$$

The impedance along the line is given by

$$Z(x) = \frac{V(x)}{I(x)} = Z_0 \frac{e^{-\gamma x} + \Gamma_L e^{\gamma x}}{e^{-\gamma x} - \Gamma_L e^{\gamma x}} \tag{2.27}$$

At the position of the load, $x = 0$, Equation (2.27) reduces to $Z(x) = Z_L$. Therefore,

$$Z_L = Z_0 \frac{1 + \Gamma_L}{1 - \Gamma_L} \tag{2.28}$$

and

$$\Gamma_L = |\Gamma_L|e^{j\phi} = \frac{Z_L - Z_0}{Z_L + Z_0} \tag{2.29}$$

Substituting Equation (2.23) into (2.27) gives

$$Z(x) = Z_0 \frac{1 + \Gamma(x)}{1 - \Gamma(x)} \tag{2.30}$$

To find the input impedance, we set $x = -l$ in $Z(x)$. This gives

$$Z_{\text{in}} = Z(-l) = Z_0 \frac{e^{\gamma l} + \Gamma_L e^{-\gamma l}}{e^{\gamma l} - \Gamma_L e^{-\gamma l}} \tag{2.31}$$

Substituting Equation (2.29) into (2.31) leads to

$$Z_{\text{in}} = Z_0 \frac{(Z_L + Z_0)e^{\gamma l} + (Z_L - Z_0)e^{-\gamma l}}{(Z_L + Z_0)e^{\gamma l} - (Z_L - Z_0)e^{-\gamma l}} \tag{2.32}$$

Converting $e^{\gamma l}$ and $e^{-\gamma l}$ to hyperbolic functions, we have

$$Z_{\text{in}} = Z_0 \frac{Z_L + Z_0 \tanh \gamma l}{Z_0 + Z_L \tanh \gamma l} \tag{2.33}$$

For the lossless case $\gamma = j\beta$, Equation (2.33) becomes

$$\begin{aligned} Z_{\text{in}} &= Z_0 \frac{Z_L + jZ_0 \tan \beta l}{Z_0 + jZ_L \tan \beta l} \\ &= Z_{\text{in}}(l, f, Z_L, Z_0) \end{aligned} \tag{2.34}$$

Equation (2.34) is used to calculate the input impedance for a terminated transmission line.

Consider several special cases:

Case A.

$$Z_L = Z_0, \quad \text{matched load}$$

$$Z_{\text{in}} = Z_0$$

Case B.

$$Z_L = 0, \quad \text{shorted circuit}$$

$$Z_{\text{in},s} = jZ_0 \tan \beta l \tag{2.35}$$

Case C.

$$Z_L = \infty, \quad \text{open circuit}$$

$$Z_{\text{in},0} = -jZ_0 \cot \beta l \tag{2.36}$$

It is seen that any value of reactance or susceptance can be obtained by varying l for an open or shorted transmission line.

From Equations (2.35) and (2.36), we can calculate Z_0 by using the following equation:

$$Z_0 = \sqrt{Z_{\text{in},s} Z_{\text{in},0}}$$

The power transmitted and reflected can be expressed in terms of the incident power and the load reflection coefficient Γ_L as follows:

$$\text{incident power} = P_{\text{in}} = \frac{|V_+|^2}{Z_0} \tag{2.37}$$

$$\begin{aligned}\text{reflected power} = P_r &= \frac{|V_-|^2}{Z_0} \\ &= \frac{|V_+|^2|\Gamma_L|^2}{Z_0} = |\Gamma_L|^2 P_{\text{in}}\end{aligned} \tag{2.38}$$

$$\begin{aligned}\text{transmitted power} = P_t &= P_{\text{in}} - P_r \\ &= \left(1 - |\Gamma_L|^2\right)P_{\text{in}}\end{aligned} \tag{2.39}$$

A return loss in dB is defined as $-10\log(P_r/P_{\text{in}})$, which is equal to $-20\log|\Gamma_L|$.

2.4 VOLTAGE STANDING-WAVE RATIO

For a transmission line with a matched load, there is no reflection and the magnitude of the voltage along the line is equal to $|V_+|$. For a transmission line terminated with a load Z_L, a reflected wave exists and the incident and reflected waves interfere to produce a standing-wave pattern along the line. The voltage at point x along the line is given by

$$\begin{aligned}V(x) &= V_+ e^{-j\beta x} + V_- e^{j\beta x} \\ &= V_+ e^{-j\beta x}\left(1 + \Gamma_L e^{2j\beta x}\right)\end{aligned} \tag{2.40}$$

Here the line is assumed to be lossless. Substituting $\Gamma_L = |\Gamma_L|e^{j\phi}$ into Equation (2.40) gives the magnitude of $V(x)$ as

$$\begin{aligned}|V(x)| &= |V_+|\left|1 + |\Gamma_L|e^{j(2\beta x + \phi)}\right| \\ &= |V_+|\left|1 + |\Gamma_L|\cos(2\beta x + \phi) + j|\Gamma_L|\sin(2\beta x + \phi)\right| \\ &= |V_+|\left\{\left[1 + |\Gamma_L|\cos(2\beta x + \phi)\right]^2 + \left[|\Gamma_L|\sin(2\beta x + \phi)\right]^2\right\}^{1/2} \\ &= |V_+|\left[1 + |\Gamma_L|^2 + 2|\Gamma_L|\cos(2\beta x + \phi)\right]^{1/2} \\ &= |V_+|\left[\left(1 + |\Gamma_L|\right)^2 - 4|\Gamma_L|\sin^2\left(\beta x + \frac{\phi}{2}\right)\right]^{1/2}\end{aligned} \tag{2.41}$$

The identities used are $|a + jb| = \sqrt{a^2 + b^2}$ and $\cos 2A = 1 - 2\sin^2 A$.

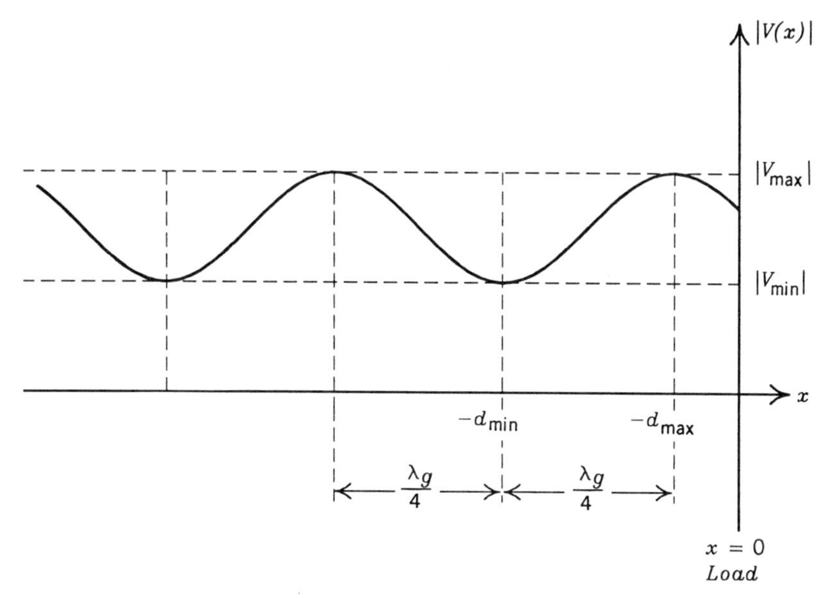

FIGURE 2.5 Pattern of voltage magnitude along the line.

Equation (2.41) shows that $|V(x)|$ oscillates between a maximum value of $|V_+|(1 + |\Gamma_L|)$ when $\sin(\beta x + \phi/2) = 0$ (or $\beta x + \phi/2 = n\pi$) and a minimum value $|V_+|(1 - |\Gamma_L|)$ when $\sin(\beta x + \phi/2) = \pm 1$ (or $\beta x + \phi/2 = m\pi - \pi/2$). Figure 2.5 shows the pattern, which repeats itself every $\lambda_g/2$.

The first maximum voltage can be found by setting $x = -d_{\max}$ and $n = 0$. We have

$$2\beta d_{\max} = \phi \tag{2.42}$$

The first minimum voltage, found by setting $x = -d_{\min}$ and $m = 0$, is given by

$$2\beta d_{\min} = \phi + \pi \tag{2.43}$$

The voltage standing-wave ratio (VSWR) is defined as the ratio of the maximum voltage to the minimum voltage. From (2.41)

$$\text{VSWR} = \frac{|V_{\max}|}{|V_{\min}|} = \frac{|V_+|(1 + |\Gamma_L|)}{|V_+|(1 - |\Gamma_L|)} = \frac{1 + |\Gamma_L|}{1 - |\Gamma_L|} \tag{2.44}$$

If the VSWR is known, $|\Gamma_L|$ can be found as

$$|\Gamma_L| = \frac{\text{VSWR} - 1}{\text{VSWR} + 1} = |\Gamma(x)| \tag{2.45}$$

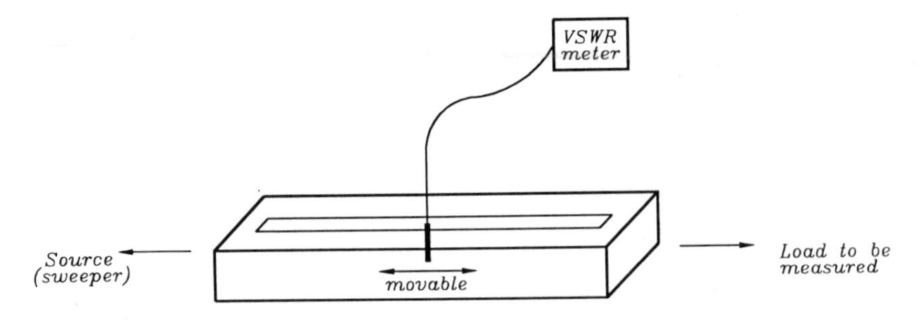

(a) Slotted line for VSWR and impedance measurement

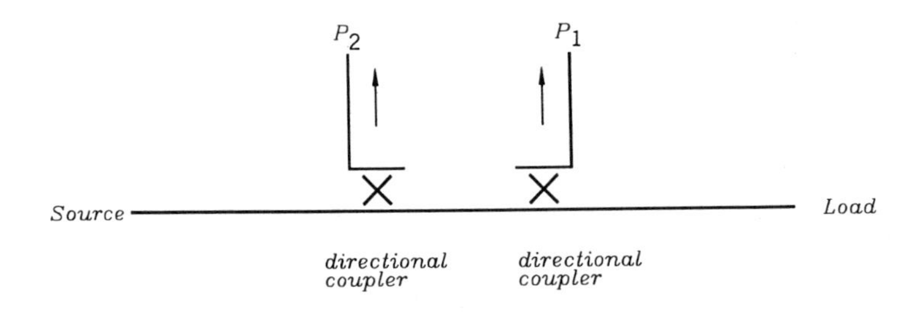

(b) Reflectometer for VSWR measurement; from P_1 and P_2, one can calculate the incident and reflected power

FIGURE 2.6 VSWR measurement.

VSWR is an important specification for all microwave components. For good matching, a low VSWR value close to 1 is generally required over the operating frequency bandwidth.

VSWR can be measured by a VSWR meter together with a slotted line as shown in Figure 2.6(a). By moving a probe in a slotted line, the VSWR meter shows deflections corresponding to the maximum and minimum voltage positions. The slotted line can also be used to measure d_{min} and λ_g, from which ϕ and β can be calculated. With the VSWR, ϕ, and β known, one can calculate Z_L from Equations (2.28) and (2.45). This was the method used to measure the load impedance before the network analyzer became available. With the modern automatic network analyzer, the VSWR and impedance information can be obtained directly.

VSWR can also be measured by using a reflectometer as shown in Figure 2.6(b). The reflectometer is used to measure the incident and reflected power, from which the reflection coefficient $|\Gamma_L|$ can be found from Equation (2.38). The VSWR can be calculated from $|\Gamma_L|$.

Example 2.1 Calculate the VSWR for a transmission line terminated by (a) a matched load, (b) a short, and (c) an open.

SOLUTION: (a) For a matched load, $Z_L = Z_0$.

$$\Gamma_L = \frac{Z_L - Z_0}{Z_L + Z_0} = 0$$

$$|\Gamma_L| = 0, \qquad \phi = 0$$

$$\text{VSWR} = \frac{1 + |\Gamma_L|}{1 - |\Gamma_L|} = 1$$

$$P_r = |\Gamma_L|^2 P_{\text{in}} = 0 \qquad \text{no reflection}$$

(b) For a shorted load, $Z_L = 0$.

$$\Gamma_L = \frac{Z_L - Z_0}{Z_L + Z_0} = -1 = e^{j180^\circ}$$

$$|\Gamma_L| = 1, \qquad \phi = 180^\circ$$

$$\text{VSWR} = \frac{1 + |\Gamma_L|}{1 - |\Gamma_L|} = \infty$$

$$P_r = |\Gamma_L|^2 P_{\text{in}} = P_{\text{in}} \qquad \text{total reflection}$$

(c) For an open load, $Z_L = \infty$.

$$\Gamma_L = \frac{Z_L - Z_0}{Z_L + Z_0} = \frac{1 - Z_0/Z_L}{1 + Z_0/Z_L} = 1$$

$$|\Gamma_L| = 1, \qquad \phi = 0^\circ$$

$$\text{VSWR} = \frac{1 + |\Gamma_L|}{1 - |\Gamma_L|} = \infty$$

$$P_r = |\Gamma_L|^2 P_{\text{in}} = P_{\text{in}} \qquad \text{total reflection}$$

Example 2.2 A load measured by an impedance measurement system shows a VSWR of 1.5 and a ϕ of 180° (where ϕ is the phase angle of Γ_L). The characteristic impedance for the measurement system is 50 Ω. Calculate the load impedance Z_L.

SOLUTION: VSWR = 1.5, $\phi = 180°$, and $Z_0 = 50\ \Omega$.

$$|\Gamma_L| = \frac{\text{VSWR} - 1}{\text{VSWR} + 1} = 0.2$$

$$\Gamma_L = |\Gamma_L| e^{j\phi} = 0.2 e^{j180°} = -0.2$$

$$Z_L = Z_0 \frac{1 + \Gamma_L}{1 - \Gamma_L} = 50 \frac{1 - 0.2}{1 + 0.2} = 33.3\ \Omega$$

Example 2.3 A load of $100 + j50\ \Omega$ is connected to a 50-Ω transmission line of length $0.3\lambda_g$. Calculate (a) the reflection coefficient at the load, (b) the VSWR, and (c) the input impedance.

SOLUTION: (a) $\Gamma_L = \dfrac{Z_L - Z_0}{Z_L + Z_0} = \dfrac{100 + j50 - 50}{100 + j50 + 50} = 0.447\ \angle 26.6°$

$|\Gamma_L| = 0.447 \qquad \phi = 26.6°$

(b)
$$\text{VSWR} = \frac{1 + |\Gamma_L|}{1 - |\Gamma_L|} = \frac{1 + 0.447}{1 - 0.447} = 2.62$$

(c)
$$Z_{in} = Z_0 \frac{Z_L + jZ_0 \tan \beta l}{Z_0 + jZ_L \tan \beta l}$$

$$\beta l = \frac{2\pi}{\lambda_g} \times 0.3\lambda_g = 1.885\ rad = 108°$$

$$Z_{in} = 50 \frac{(100 + j50) + j50 \tan 108°}{50 + j(100 + j50) \tan 108°}$$

$$= 19.53\ \angle 10.38°\ \Omega$$

2.5 SMITH CHART AND APPLICATIONS

The Smith chart was invented by P. H. Smith of Bell Laboratories in 1939. It is a graphical representation of the impedance transformation property of a length of transmission line. Although the impedance and reflection information can be obtained from equations in earlier sections, the calculations normally involve complex numbers, which can be complicated and time consuming. Use of the Smith chart avoids tedious computation. It also provides a graphical representation on the impedance locus as a function of frequency.

Define a normalized impedance $\bar{Z}(x)$ as

$$\bar{Z}(x) = \frac{Z(x)}{Z_0} = \bar{R}(x) + j\bar{X}(x) \tag{2.46}$$

The reflection coefficient $\Gamma(x)$ is given by

$$\Gamma(x) = \Gamma_r(x) + j\Gamma_i(x) \tag{2.47}$$

From Equation (2.30) we have

$$\bar{Z}(x) = \frac{Z(x)}{Z_0} = \frac{1 + \Gamma(x)}{1 - \Gamma(x)} \tag{2.48}$$

Therefore, substituting (2.46) and (2.47) into (2.48) gives

$$\bar{R}(x) + j\bar{X}(x) = \frac{1 + \Gamma_r + j\Gamma_i}{1 - \Gamma_r - j\Gamma_i} \tag{2.49}$$

By equating the real and imaginary parts of (2.49), two equations are generated:

$$\left(\Gamma_r - \frac{\bar{R}}{1 + \bar{R}}\right)^2 + \Gamma_i^2 = \left(\frac{1}{1 + \bar{R}}\right)^2 \tag{2.50}$$

$$(\Gamma_r - 1)^2 + \left(\Gamma_i - \frac{1}{\bar{X}}\right)^2 = \left(\frac{1}{\bar{X}}\right)^2 \tag{2.51}$$

In the Γ_r–Γ_i coordinate system, Equation (2.50) represents circles centered at $(\bar{R}/(1 + \bar{R}), 0)$ with radii of $1/(1 + \bar{R})$. These are called *constant-$\bar{R}$ circles*. Equation (2.51) represents circles centered at $(1, 1/\bar{X})$ with radii of $1/\bar{X}$. They are called *constant-$\bar{X}$ circles*. Figure 2.7 shows these circles in the Γ_r–Γ_i plane. A plot of these circles is called a *Smith chart*. On the Smith chart, any given value of $\Gamma(= |\Gamma|e^{j\phi})$ has the properties that a contour of constant $|\Gamma|$ is a circle centered at $(0, 0)$ with a radius of $|\Gamma|$. Hence motion along a lossless line gives a circular path on the Smith chart. From Equation (2.23) we know that for a lossless line,

$$\Gamma(x) = \Gamma_L e^{2\gamma x} = \Gamma_L e^{j2\beta x} \tag{2.52}$$

Hence given $\bar{Z}_L$, we can find $\bar{Z}_{\text{in}}$ at a distance $-l$ from the load by proceeding at an angle $2\beta l$ in the clockwise direction. Thus $\bar{Z}_{\text{in}}$ can be found graphically.

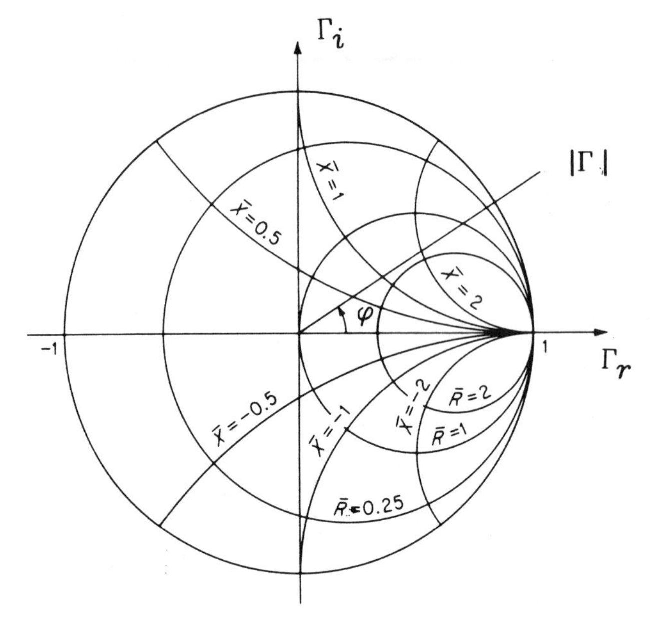

FIGURE 2.7 Constant $\bar{R}$ and $\bar{X}$ circles in the reflection-coefficient plane.

The Smith chart has the following features:

1. Impedance and admittance read from the chart are normalized values.
2. Moving away from the load (i.e., toward the generator) corresponds to moving in a clockwise direction.
3. A complete revolution around the chart is made in going a distance $l = \lambda_g/2$ along the transmission line.
4. The same chart can be used for reading admittance.
5. The center of the chart corresponds to an impedance-matched condition since $\Gamma(x) = 0$.

The Smith chart can be used to find (1) Γ_L from Z_L, and vice versa; (2) $\bar{Z}_{in}$ from $\bar{Z}_L$, and vice versa; (3) Z from Y, and vice versa; (4) VSWR; and (5) d_{min} and d_{max}. The Smith chart is also useful for designing an impedance-matching network, which is discussed in the next section.

Example 2.4 A load of $100 + j50\ \Omega$ is connected to a $50\text{-}\Omega$ transmission line. Use both equations and a Smith chart to find (a) Γ_L, (b) Z_{in} at $0.2\lambda_g$ away from the load, (c) Y_L, (d) VSWR, and (e) d_{max} and d_{min}.

SOLUTION: (a)

$$Z_L = 100 + j50$$

$$\bar{Z}_L = \frac{Z_L}{Z_0} = 2 + j$$

$$\Gamma_L = \frac{Z_L - Z_0}{Z_L + Z_0} = \frac{\bar{Z}_L - 1}{\bar{Z}_L + 1} = \frac{1 + j}{3 + j} = 0.4 + 0.2j$$

$$= 0.44 \angle 26°$$

From the Smith chart shown in Figure 2.8, $\bar{Z}_L$ is located by finding the intersection of two circles with $\bar{R} = 2$ and $\bar{X} = 1$. The distance from the center to $\bar{Z}_L$ is 0.44 and the angle can be read directly from the scale as 26°.*

(b)

$$Z_{\text{in}} = Z_0 \frac{Z_L + jZ_0 \tan \beta l}{Z_0 + jZ_L \tan \beta l}$$

$$l = 0.2\lambda_g \qquad \beta l = \frac{2\pi}{\lambda_g} \times 0.2\lambda_g = 0.4\pi$$

$$\tan \beta l = 3.07$$

$$\bar{Z}_{\text{in}} = \frac{2 + j + j3.07}{1 + j(2 + j) \times 3.07} = 0.5 - j0.5$$

From the Smith chart, we move $\bar{Z}_L$ a distance of $0.2\lambda_g$ on a constant-$|\Gamma|$ circle toward the generator to a point $\bar{Z}_{\text{in}}$. The value of $\bar{Z}_{\text{in}}$ can be read from the Smith chart as $0.5 - j0.5$ since $\bar{Z}_{\text{in}}$ is located at the intersection of two circles with $\bar{R} = 0.5$ and $\bar{X} = -0.5$. The distance of movement $(0.2\lambda_g)$ can be read from the outermost scale in the chart.

(c)

$$\bar{Z}_L = 2 + j$$

$$\bar{Y}_L = \frac{1}{\bar{Z}_L} = \frac{1}{2 + j} = 0.4 - j0.2$$

$$Y_L = \bar{Y}Y_0 = \frac{\bar{Y}}{Z_0} = \frac{0.4 - j0.2}{50} = 0.008 - j0.004$$

From the Smith chart, $\bar{Y}_L$ is found by moving $\bar{Z}_L$ a distance of $0.25\lambda_g$ toward the generator. Therefore, $\bar{Y}_L$ is located opposite $\bar{Z}_L$ along a line through the

*Note that the radius of the outermost circle is 1, and the distance from the center to $\bar{Z}_L$ is determined by a linear scale. A scale is normally provided under the commercially available Smith chart.

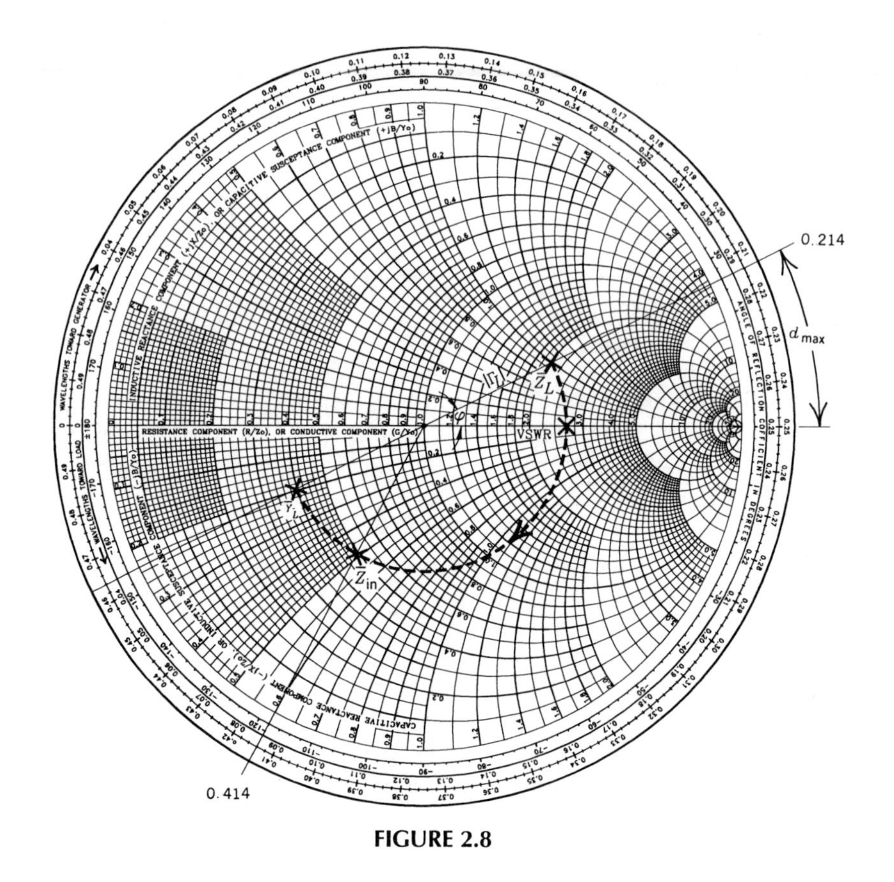

FIGURE 2.8

center of the Smith chart. This can be proved by the following:

$$\overline{Z}_{in}\left(l = \frac{\lambda_g}{4}\right) = \frac{Z_L + jZ_0 \tan \beta l}{Z_0 + jZ_L \tan \beta l}$$

$$= \frac{Z_0}{Z_L} = \frac{1}{\overline{Z}_L} = \overline{Y}_L$$

(d)

$$\Gamma_L = 0.44 \angle 26^\circ$$

$$|\Gamma_L| = 0.44$$

$$\text{VSWR} = \frac{1 + |\Gamma_L|}{1 - |\Gamma_L|} = \frac{1 + 0.44}{1 - 0.44} = 2.57$$

The VSWR can be found from the reading of $\overline{R}$ at the intersection of the constant-$|\Gamma|$ circle and the real axis. (This will be proved in Problem P2.12.)

(e) From Equation (2.42) we have

$$d_{\max} = \frac{\phi}{2\beta} = \frac{\phi}{4\pi}\lambda_g$$

Now $\phi = 26°$ and $d_{\max} = 0.036\lambda_g$. From the Smith chart, $d_{\max}$ is the distance obtained by moving from $\bar{Z}_L$ through a constant-$|\Gamma|$ circle to the real axis. This can be proved using Equation (2.41). The first $d_{\max}$ occurs when $\beta x + \phi/2 = 0$ or $2\beta x + \phi = 0$. Since $\Gamma(x) = \Gamma_L e^{2j\beta x} = |\Gamma_L| e^{j(2\beta x + \phi)}$, the first $d_{\max}$ occurs when $\Gamma(x)$ has zero phase (i.e., on the real axis). $d_{\min}$ is just $\lambda_g/4$ away from $d_{\max}$.

The Smith charts shown in Figures 2.7 and 2.8 are called the *Z chart* and the *Y chart*. One can read the normalized impedance or admittance directly from these charts. If we rotate the Y chart by 180°, we have a *rotated Y chart*, as shown in Figure 2.9(b). The combination of a Z chart and a rotated Y chart is called a *Z–Y chart*, shown in Figure 2.9(c). On a Z–Y chart, for any point A, one can read $\bar{Z}_A$ from the Z-chart and $\bar{Y}_A$ from the rotated Y chart. Therefore, this chart avoids the necessity of moving $\bar{Z}$ by $\lambda_g/4$ (i.e., 180°) to find $\bar{Y}$. The Z–Y chart is useful for impedance matching using lumped elements.

2.6 IMPEDANCE MATCHING

Impedance matching is important in the design of all microwave components, circuits, and subsystems. With an ideal matching case, $Z_{\text{in}} = Z_0$, VSWR $= 1$, $\Gamma(x) = 0$, $P_r = 0$, and the impedance locus is located at the center of the Smith chart. To accomplish this, a matching network is needed to transform a load impedance to match the transmission-line impedance. Figure 2.10 shows such a matching network placed between Z_L and the transmission line. Several methods can be used to achieve matching. In this section we discuss three methods: matching stubs (shunt or series), the quarter-wavelength transformer, and lumped elements. It should be emphasized that computer software such as SuperCompact or Touchstone can be used for practical matching network design. For broadband matching, multiple-section matching networks or tapered circuits are generally used [1].

A. Method 1: Use a Single-Stub Network

One can use a single- or double-stub network to accomplish impedance matching. Figure 2.11 shows a single-shunt-stub matching network that is used to match Z_L to Z_0. For a shunt stub, a Y chart is used for the design. A Z chart is more convenient for the series stub design. One needs to design l_1

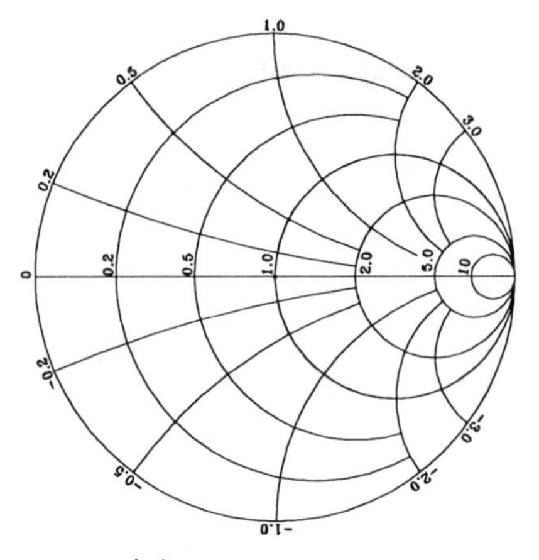

(a) Z or Y chart

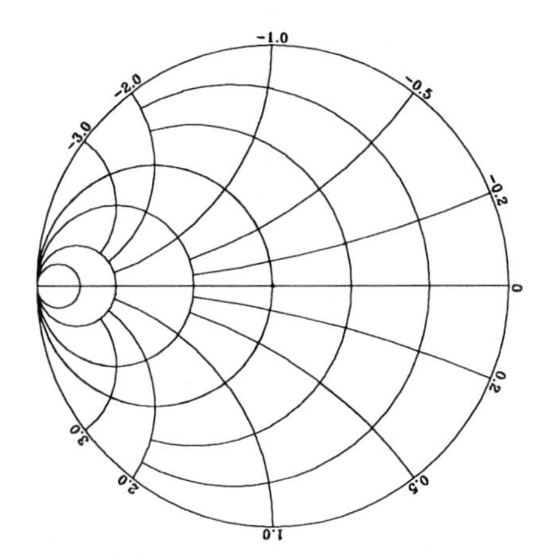

(b) Rotated Y chart

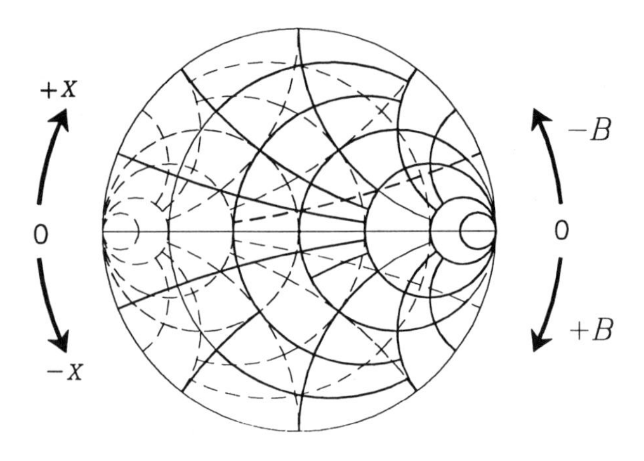

(c) Z–Y chart

FIGURE 2.9 Smith charts.

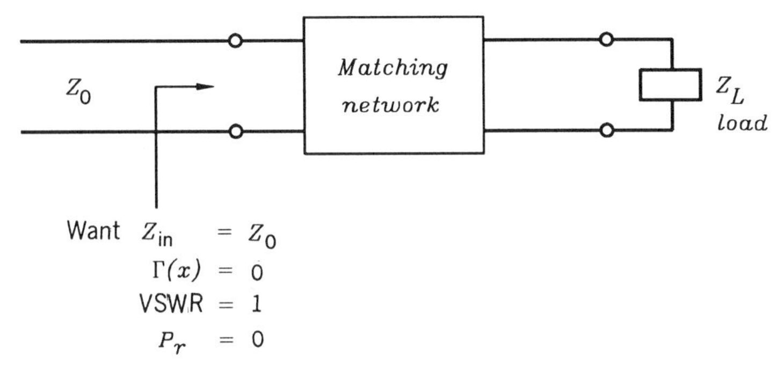

FIGURE 2.10 Impedance-matching network.

and l_2 such that $\overline{Y}_{in} = 1$. Although a shorted stub is employed in Figure 2.11, an open stub could also be used. To achieve the matching, l_1 is designed such that $\overline{Y}_A = 1 + jb$ and l_2 is designed such that $\overline{Y}_t = -jb$. The sum of $\overline{Y}_A$ and $\overline{Y}_t$ is then a matched condition:

$$\overline{Y}_{in} = \overline{Y}_t + \overline{Y}_A = 1 \tag{2.53}$$

The design procedure using a Smith chart for the circuit of Figure 2.11 is given below.

1. Plot $\overline{Y}_L$ from $\overline{Z}_L$ and use a Smith chart for admittance (i.e., a Y chart).
2. Draw a constant VSWR circle through $\overline{Y}_L$.

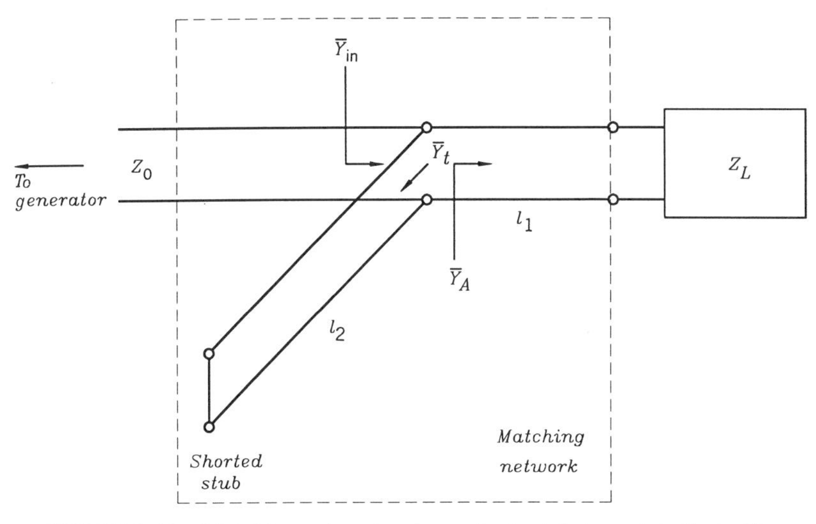

FIGURE 2.11 Matching network using a single stub to match Z_L to Z_0.

3. Rotate from $\overline{Y}_L$ toward the generator until it intersects with the circle of $\overline{R} = \overline{G} = 1$. The point of intersection gives $\overline{Y}_A = 1 + jb$ and the distance of rotation gives l_1 (two possible solutions).
4. Plot $\overline{Y}_t = -jb$ on a Smith chart.
5. Rotate from the short position ($\overline{Y} = \infty$) toward the generator until the point of $\overline{Y}_t$ is reached. The distance of rotation is l_2.

Example 2.5 In Figure 2.11, if $\overline{Z}_L = 0.4 - j0.4$, match Z_L to Z_0 and find the solution with the shortest l_1.

SOLUTION: From Figure 2.12,

$$\overline{Y}_L = 1.25 + j1.25$$

$$\overline{Y}_A = 1 - j1.15$$

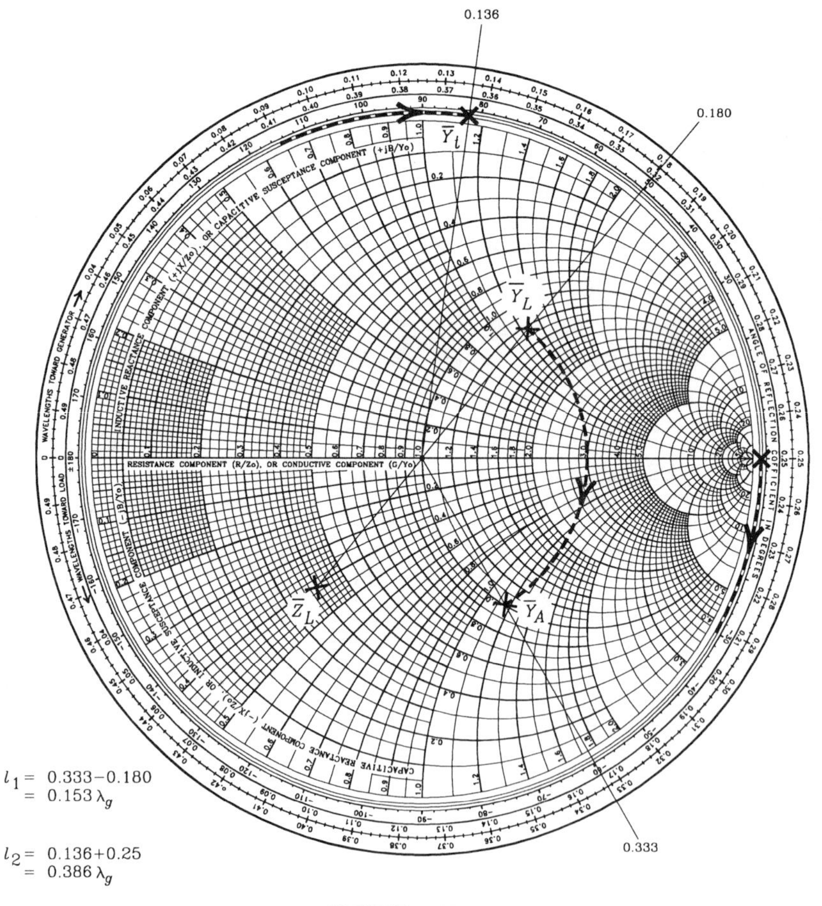

FIGURE 2.12

with

$$l_1 = 0.333\lambda_g - 0.18\lambda_g = 0.153\lambda_g \text{ and}$$

$$\bar{Y}_t = j1.15$$

with

$$l_2 = 0.136\lambda_g + 0.25\lambda_g = 0.386\lambda_g.$$

In many applications, one wants to match from Z_0 to Z_L. For example, in a FET amplifier design for maximum gain, one needs to match Z_0 to Z_s at the input port. Z_s is obtained from Γ_s, which is designed equal to S_{11}^*. S_{11} is an S parameter of the FET device (see Chapter 14).

Figure 2.13 shows the matching network to match Z_0 to Z_L. The procedure for using a Smith chart is as follows:

1. Locate $\bar{Y}_L$ on the Smith chart and use the chart for admittance.
2. Draw a constant-VSWR circle through $\bar{Y}_L$ which intersects with the $\bar{G} = 1$ circle at point A. $\bar{Y}_A = 1 + jb$.
3. Move from point A to $\bar{Y}_L$ toward the generator. The distance between $\bar{Y}_L$ and $\bar{Y}_A$ gives l_1.
4. From $\bar{Y}_t = jb$, find the distance l_2.

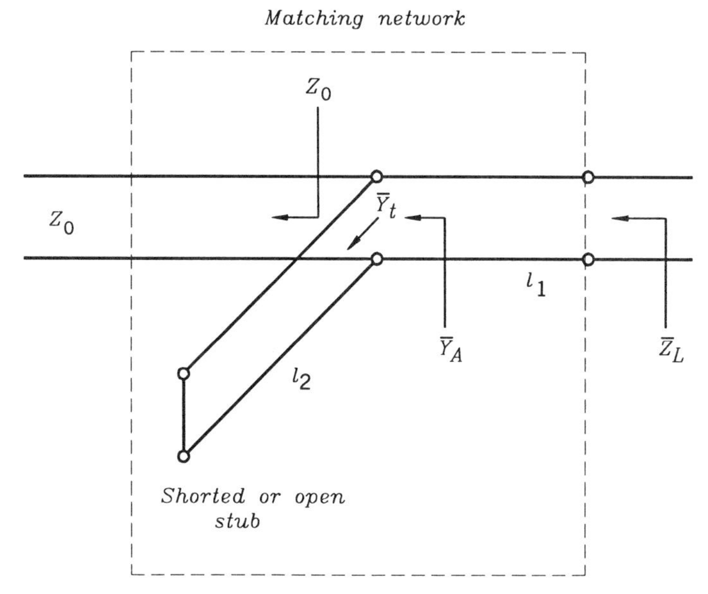

FIGURE 2.13 Matching network using a single short stub to match Z_0 to Z_L.

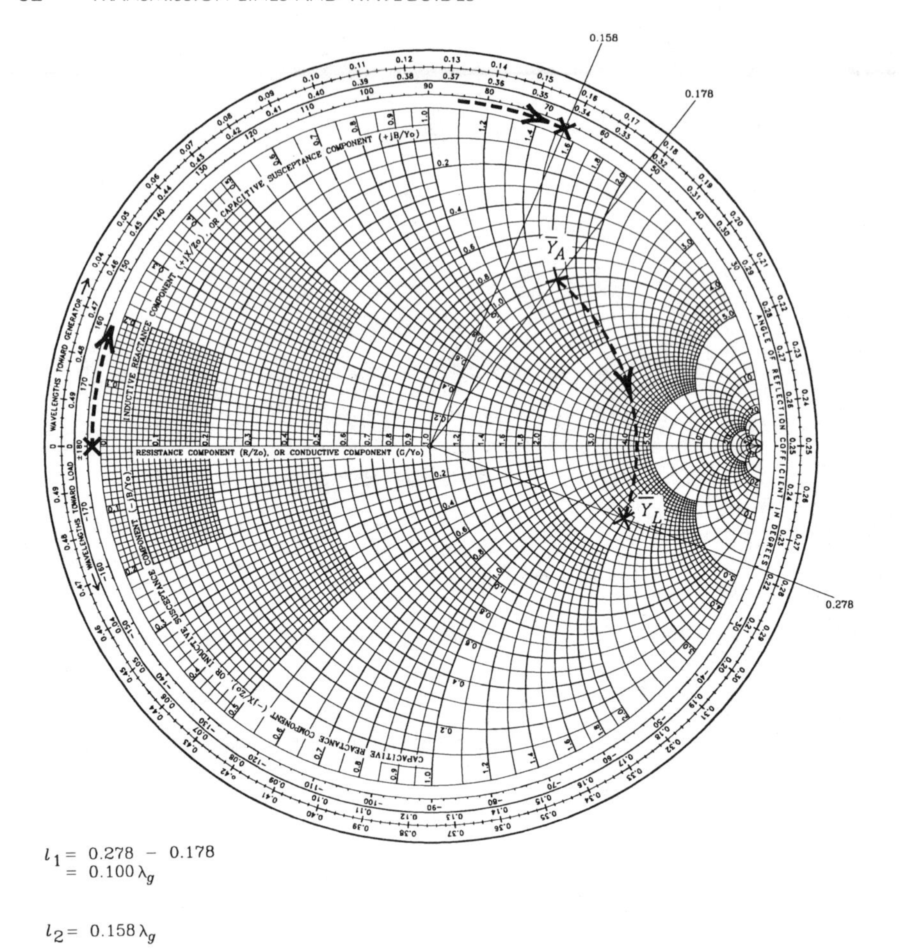

FIGURE 2.14

Example 2.6 In Figure 2.13, if $\bar{Z}_L = 0.25 + j0.18$, use an open stub to match Z_0 to Z_L and find l_1 and l_2.

SOLUTIONS: Plot $\bar{Y}_L$ in the Smith chart shown in Figure 2.14.

$$\bar{Y}_L = 2.8 - j1.9$$

From the Smith chart,

$$\bar{Y}_A = 1 + j1.55$$
$$jb = j1.55$$

with $l_1 = 0.278\lambda_g - 0.178\lambda_g = 0.1\lambda_g$ moving from $\bar{Y}_A$ to $\bar{Y}_L$ toward the generator. From $b = 1.55$, we find that $l_2 = 0.158\lambda_g$, moving from $\bar{Z} = \infty$ (or $\bar{Y} = 0$) to $j1.55$ toward the generator.

Another solution: $l_1 = 0.456\lambda_g$ and $l_2 = 0.34\lambda_g$.

B. Method 2: Use a Quarter-Wavelength Transformer

A quarter-wavelength transformer is a convenient way to accomplish impedance matching. For the resistive load shown in Figure 2.15, a quarter-wavelength transformer alone can achieve the matching. This can be seen from the following. To achieve impedance matching, we want $Z_{in} = Z_0$. For a line of length l_T and a characteristic impedance of Z_{0T}, the input impedance is

$$Z_{in} = Z_{0T}\frac{Z_L + jZ_{0T}\tan\beta l_T}{Z_{0T} + jZ_L\tan\beta l_T} \tag{2.54}$$

Since $l_T = \lambda_g/4$, $\beta l_T = \pi/2$, and $\tan\beta l_T = \infty$, we have

$$Z_{in} = \frac{Z_{0T}^2}{Z_L} = \frac{Z_{0T}^2}{R_L} = Z_0$$

Thus

$$Z_{0T} = \sqrt{Z_0 R_L} \tag{2.55}$$

Equation (2.55) implies that the characteristic impedance of the quarter-wavelength section should be equal to $\sqrt{Z_0 R_L}$ if a quarter-wavelength transformer is used to match R_L to Z_0. Since the electrical length is a function of frequency, the matching is good only for a narrow frequency range. The same argument applies to stub matching networks. To accomplish wideband matching, a multiple-section matching network is generally required.

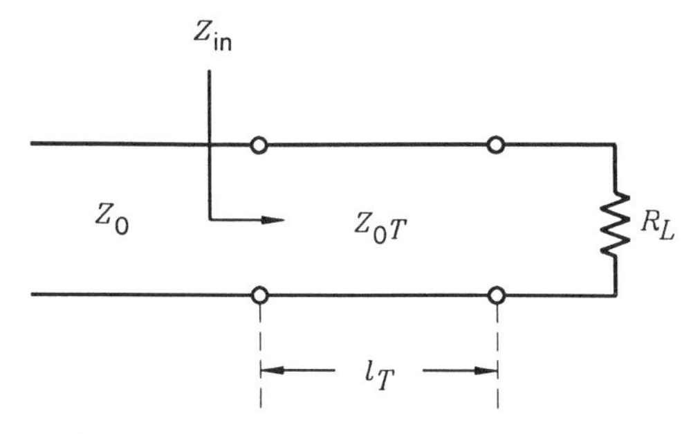

FIGURE 2.15 Using a quarter-wavelength transformer to match R_L to Z_0.

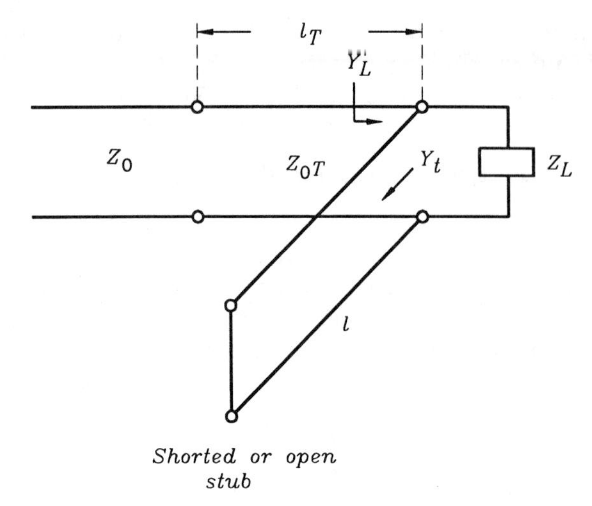

FIGURE 2.16 Matching network using a quarter-wavelength transformer together with a single-stub tuner.

For a complex load $Z_L = R_L + jX_L$, one needs to use a single-stub tuner to tune out the reactive part, then use a quarter-wavelength transformer to match the remaining real part of the load. Figure 2.16 shows this matching arrangement.

Example 2.7 In Figure 2.16, if $\overline{Y}_L = 1.25 + j1.25$, design l and Z_{0T} to match Z_L to 50 Ω at a frequency f_0 using a shorted stub tuner.

SOLUTION: $\overline{Y}_L = 1.25 + j1.25$. Design l such that $\overline{Y}_t = -j1.25$. Then $\overline{Y}'_L = \overline{Y}_L + \overline{Y}_t = 1.25$. From the Smith chart, $l = 0.108\lambda_g$.

$$\overline{Z}'_L = \frac{1}{\overline{Y}'_L} = 0.8 = \frac{Z'_L}{Z_0}$$

$$Z_0 = 50\ \Omega, \qquad Z'_L = 40\ \Omega$$

$$Z_{0T} = \sqrt{Z_0 Z'_L} = \sqrt{50 \times 40} = 44.7\ \Omega$$

$$l_T = \frac{\lambda_g}{4} + n\frac{\lambda_g}{2}, \qquad n = 1, 2, 3 \ldots$$

C. Method 3: Use Lumped Elements

At low microwave frequencies, lumped elements (L, C) can be used effectively to accomplish impedance matching. Lumped elements have the advantages of small size and wider bandwidth. At high frequencies, distributive circuits such as tuning stubs are better because of lower loss.

Many combinations of LC circuits can be used to match Z_L to Z_0 as shown in Figure 2.17. Adding a series reactance to a load produces motion along a constant-resistance circle in the Z-Smith chart. As shown in Figure 2.18, if the series reactance is an inductance, the motion is upward since the total reactance is increased. If it is a capacitance, the motion is downward. Adding a shunt reactance to a load produces a motion along a constant-conductance circle in the rotated Y chart. As shown in Figure 2.19, if the shunt element is an inductor, the motion is upward. If the shunt element is a capacitor, the motion is downward. Using a combination of constant-resis-

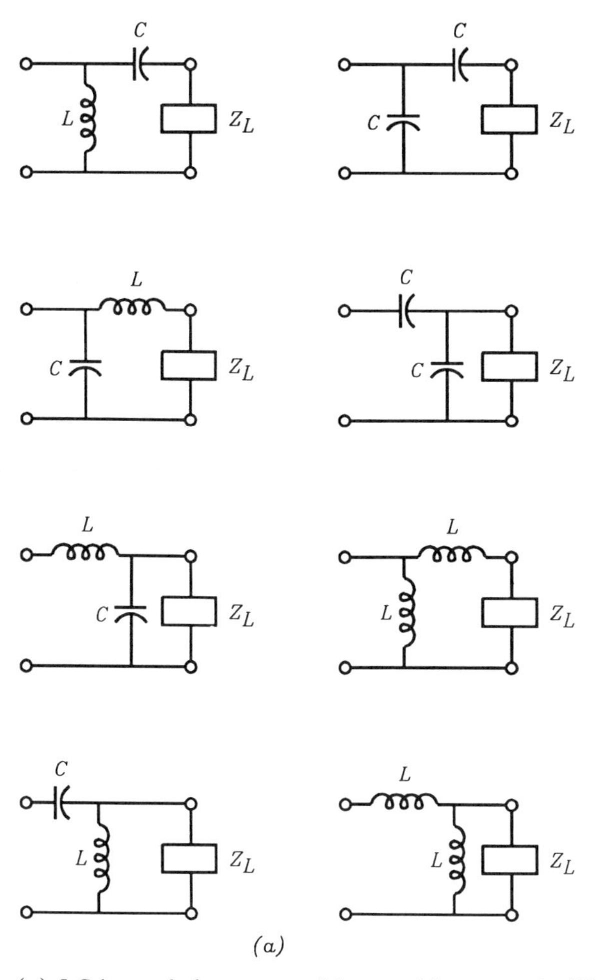

FIGURE 2.17 (a) LC lumped elements used for matching network; (b) some planar capacitor configurations; (c) some planar inductor configurations. [(b) and (c) From Ref. 2 with permission from Howard W. Sams.]

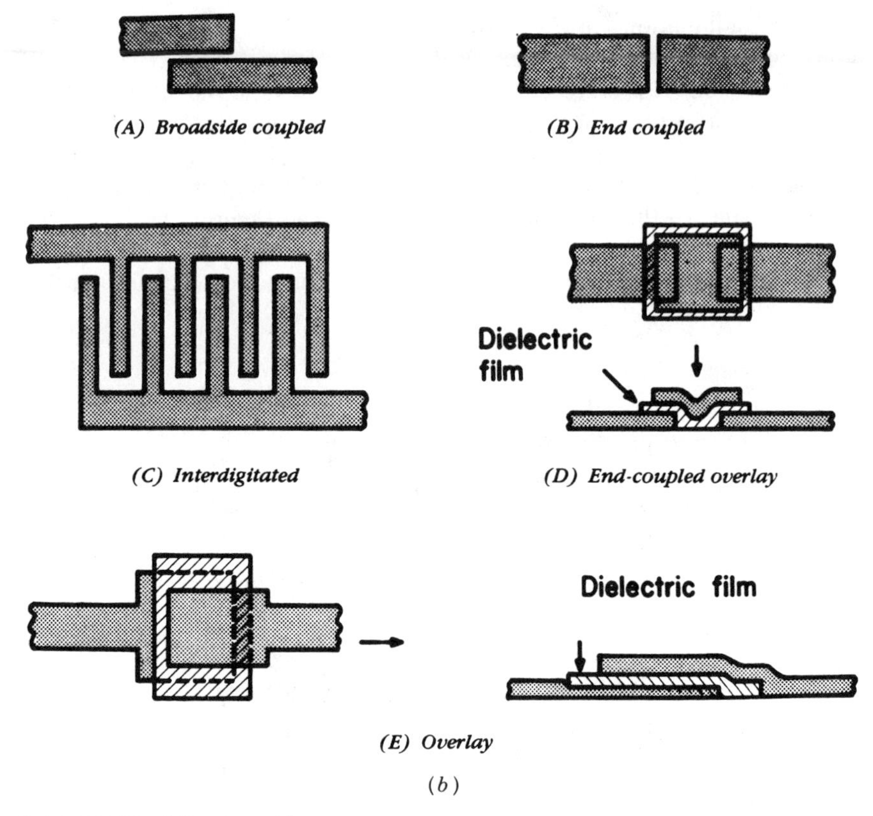

FIGURE 2.17 (Continued.)

tance circles and constant-conductance circles, one can move $\overline{Z}_L$ (or $\overline{Y}_L$) to the center of the Z–Y chart.

Consider a case of using a series L and a shunt C to match Z_L to Z_0, as shown in Figure 2.20. The procedure is as follows:

1. Locate A from $\overline{Z}_L$ in the Z chart.
2. Move A to B along the constant-resistance circle. B is located on the constant-conductance circle passing through the center (i.e., $\overline{G} = 1$ circle).
3. Move from B to C along the $\overline{G} = 1$ circle. C is the center of the chart.
4. From the Z chart and rotated Y chart, one can find L and C, respectively.

Example 2.8 In Figure 2.20, if $Z_L = 10 + j10\ \Omega$, design L, C to match Z_L to a 50-Ω line at 500 MHz.

(A) High-impedance line section

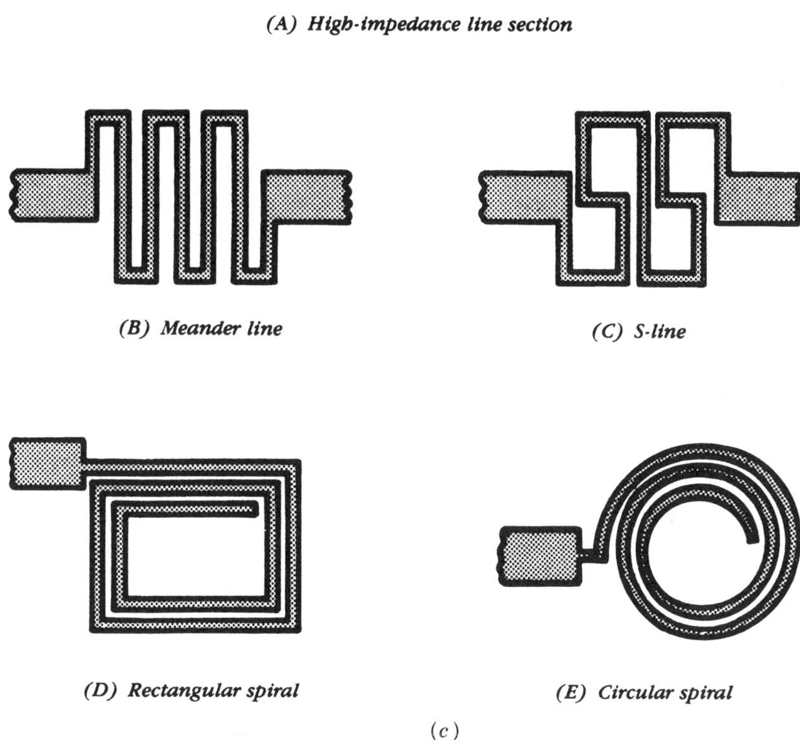

(B) Meander line

(C) S-line

(D) Rectangular spiral

(E) Circular spiral

(c)

FIGURE 2.17 (Continued.)

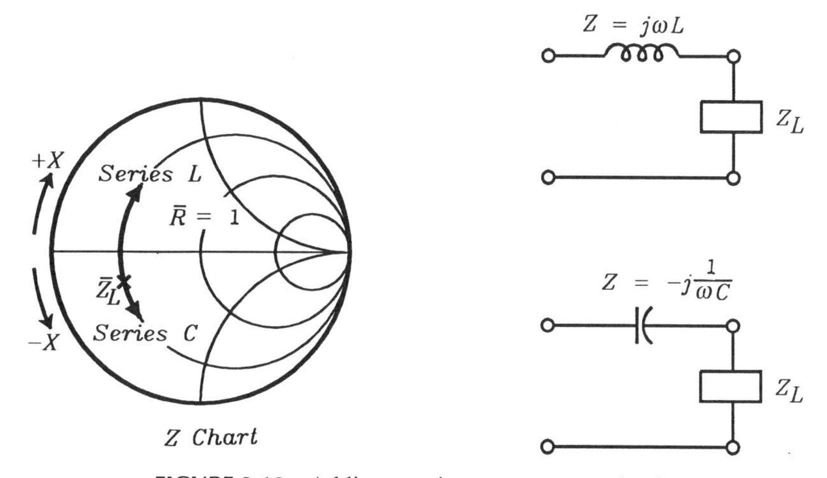

FIGURE 2.18 Adding a series reactance to a load.

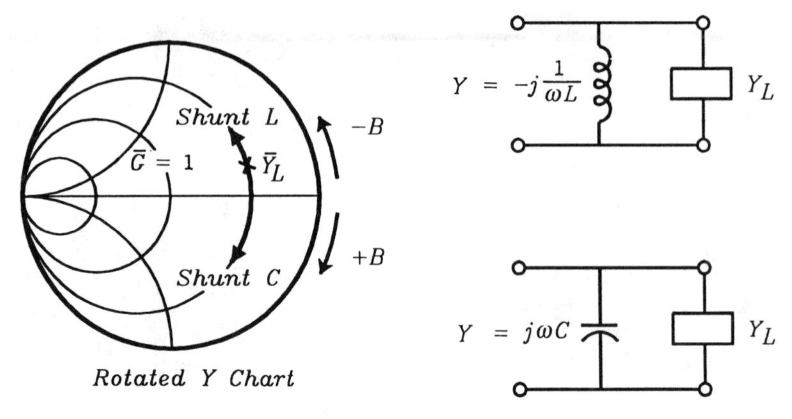

FIGURE 2.19 Adding a shunt reactance to a load.

SOLUTION:

$$\overline{Z}_L = \frac{Z_L}{Z_0} = 0.2 + j0.2$$

Locate point B on the Z–Y chart as shown in Figure 2.21.

$$\overline{Z}_B = 0.2 + j0.4$$

Using a series L, we have

$$\begin{aligned} j\omega L &= Z_0\left(j\overline{X}_B - j\overline{X}_L\right) \\ &= 50(j0.4 - j0.2) \\ \omega L &= 50 \times 0.2 = 10 = 2\pi f L = 2\pi \times 500 \times 10^6 \times L \\ L &= 3.18 \text{ nH} \end{aligned}$$

At point C, $\overline{Z}_c = 1$. From points B to C, we use a shunt C and the rotated Y chart.

$$\begin{aligned} \overline{Y}_c &= 1 \\ \overline{Y}_B &= 1 - j2 \\ j\omega C &= Y_0\left[\text{Im}\left(\overline{Y}_c\right) - \text{Im}\left(\overline{Y}_B\right)\right] \\ &= \frac{1}{50}[0 - (-j2)] \\ C &= \frac{1}{25\omega} = \frac{1}{25 \times 2\pi f} = 12.73 \text{ pF} \end{aligned}$$

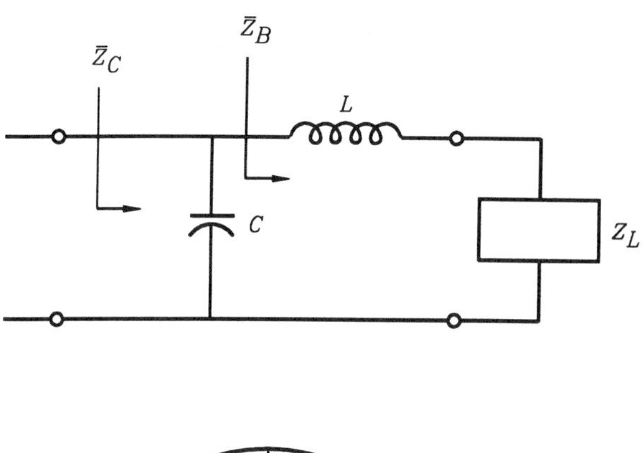

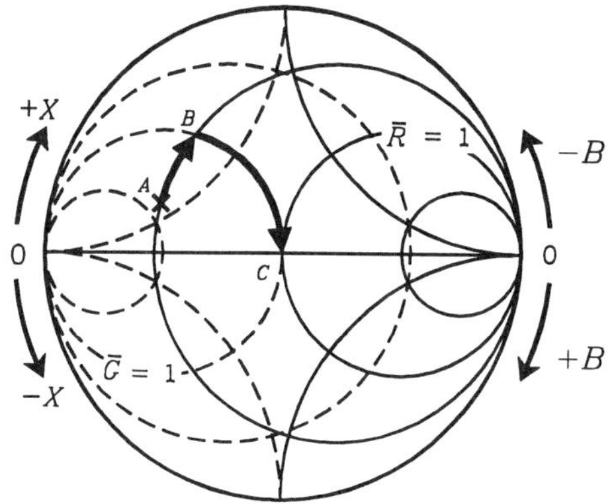

FIGURE 2.20 Using a series L and a shunt C to match Z_L to Z_0.

2.7 COAXIAL LINES

Coaxial line is commonly used for microwave and lower frequencies. Cable TV and video transmission cables are coaxial lines. At microwave frequencies, coaxial lines are used for interconnections, signal transmission, and measurements. Coaxial line normally operates in the TEM mode (i.e., no axial electric and magnetic field components). It has no cutoff frequency and can be used from dc to microwave or millimeter-wave frequencies. It can be made flexible. Presently, coaxial lines and connectors can be operated up to 50 GHz. The upper frequency limit is set by the excessive losses and the excitation of circular waveguide TE and TM modes. However, improved manufacturing of small coaxial lines will push the operating frequency even higher.

Since the coaxial line is operated in the TEM mode, the electric and magnetic fields can be found from the static case [1]. Figure 2.22 shows a coaxial transmission line. The electric potential of the coaxial line can be found by solving the Laplace equation in cylindrical coordinates.

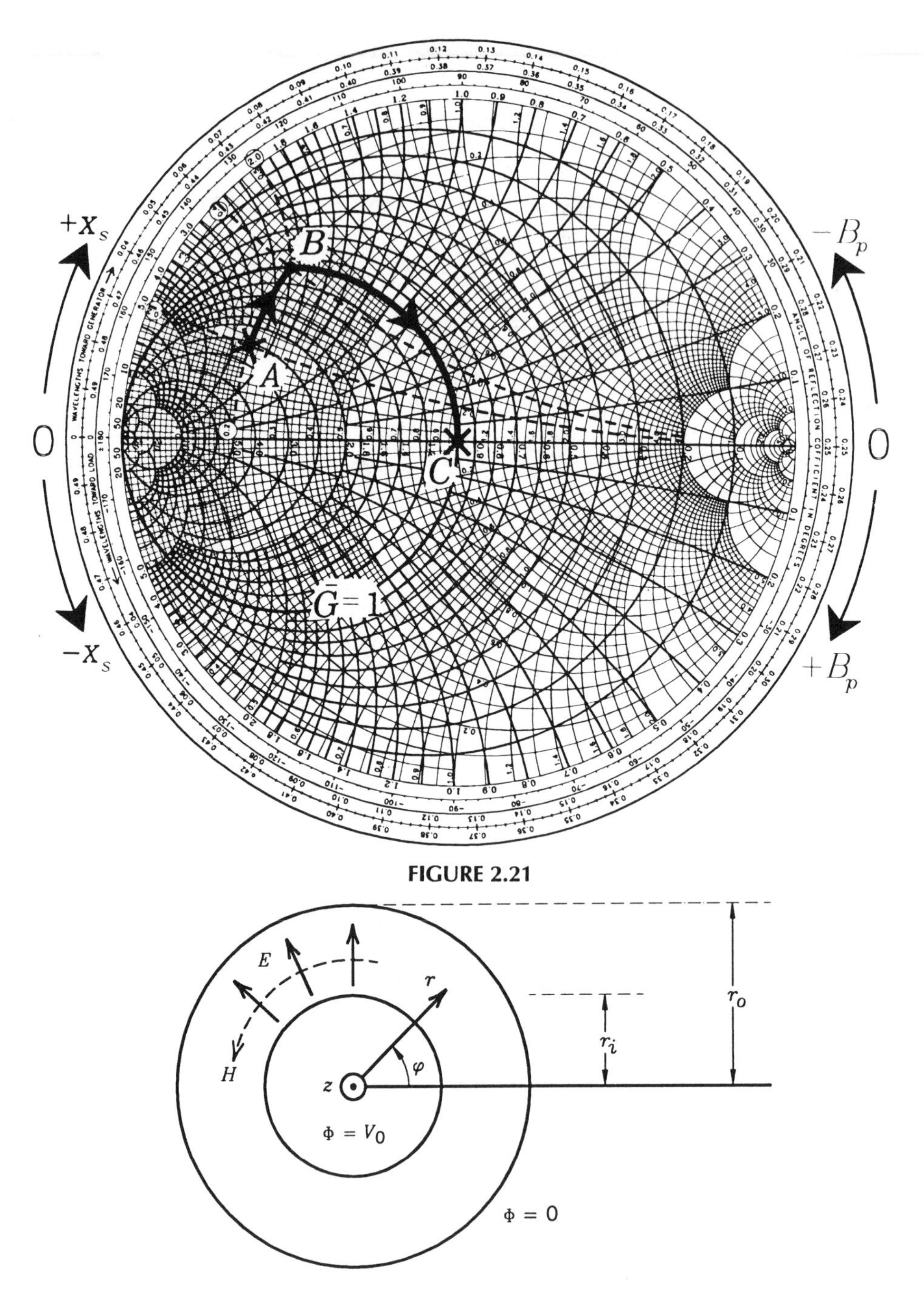

FIGURE 2.21

FIGURE 2.22 Coaxial line with inner and outer conductor radii of r_i and r_o.

The Laplace equation for the electric potential is

$$\nabla^2 \Phi = 0 \tag{2.56}$$

In cylindrical coordinates this equation can be written

$$\frac{1}{r}\left[\frac{\partial}{\partial r}\left(r\frac{\partial \Phi}{\partial r}\right) + \frac{\partial}{\partial \phi}\left(\frac{1}{r}\frac{\partial \Phi}{\partial \phi}\right) + \frac{\partial}{\partial z}\left(r\frac{\partial \Phi}{\partial z}\right)\right] = 0 \tag{2.57}$$

Since Φ is independent of z and ϕ due to symmetry, (2.57) is simplified to

$$\frac{1}{r}\frac{\partial}{\partial r}\left(r\frac{\partial \Phi}{\partial r}\right) = 0 \tag{2.58}$$

Integrating (2.58) once gives

$$r\frac{\partial \Phi}{\partial r} = A_1 \tag{2.59}$$

Integrating (2.59) gives

$$\Phi = A_1 \ln r + A_2 \tag{2.60}$$

The boundary conditions at the inner and outer conductors are:

$$\Phi = \begin{cases} 0 & \text{when } r = r_o \\ V_0 & \text{when } r = r_i \end{cases}$$

Using these boundary conditions in Equation (2.60), we have

$$A_1 \ln r_i + A_2 = V_0 \tag{2.61}$$

$$A_1 \ln r_o + A_2 = 0 \tag{2.62}$$

Solving these two equations gives

$$A_1 = \frac{V_0}{\ln(r_i/r_o)}$$

and

$$A_2 = -\frac{V_0}{\ln(r_i/r_o)} \ln r_o$$

The potential is then

$$\Phi(r) = V_0 \frac{\ln(r/r_o)}{\ln(r_i/r_o)} \tag{2.63}$$

The electric field can be derived by

$$\mathbf{E}(r) = -\nabla\Phi = -\hat{\mathbf{r}}\frac{\partial \Phi}{\partial r} \tag{2.64}$$

$$\mathbf{E}(r) = \frac{V_0}{\ln(r_o/r_i)} \frac{1}{r}\hat{\mathbf{r}} \tag{2.65}$$

This is the static solution. For a wave propagating in the positive z direction,

$$\mathbf{E}(r) = \frac{V_0}{\ln(r_o/r_i)} \frac{1}{r} e^{-j\beta z}\hat{\mathbf{r}} = E_r\hat{\mathbf{r}} \tag{2.66}$$

The magnetic field can be found from the Maxwell's equation using

$$\mathbf{H} = -\frac{1}{j\omega\mu}\nabla\times\mathbf{E} = -\frac{1}{j\omega\mu}\frac{1}{r}\begin{vmatrix} \hat{\mathbf{r}} & r\hat{\boldsymbol{\phi}} & \hat{\mathbf{z}} \\ \dfrac{\partial}{\partial r} & \dfrac{\partial}{\partial \phi} & \dfrac{\partial}{\partial z} \\ E_r & rE_\phi & E_z \end{vmatrix}$$

$$= \frac{V_0}{\eta\ln(r_o/r_i)} \frac{1}{r} e^{-j\beta z}\hat{\boldsymbol{\phi}} \tag{2.67}$$

where $\eta = \sqrt{\mu/\varepsilon}$ = wave impedance.

To find the characteristic impedance, we need to find the total surface current I_0 since $Z_0 = V_0/I_0$. The current density on the surface of the inner conductor ($r = r_i$) shown in Figure 2.23 is given by

$$\mathbf{J}_s = \hat{\mathbf{n}}\times\mathbf{H} = \hat{\mathbf{r}}\times\mathbf{H}$$

$$= \frac{V_0}{\sqrt{\mu/\varepsilon}\,\ln(r_o/r_i)} \frac{1}{r_i}\hat{\mathbf{z}}e^{-j\beta z} \tag{2.68}$$

The total current in the inner conductor is

$$I = \int_0^{2\pi} J_s\,dl = \int_0^{2\pi} J_s r_i\,d\phi$$

$$= \frac{2\pi V_0}{\ln(r_o/r_i)\sqrt{\mu/\varepsilon}} e^{-j\beta z} = I_0 e^{-j\beta z} \tag{2.69}$$

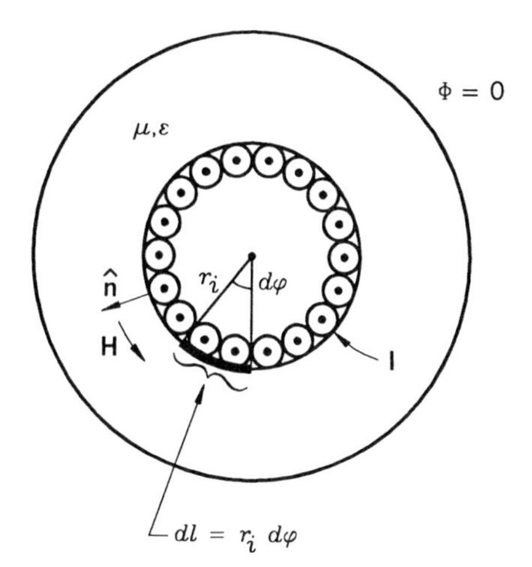

FIGURE 2.23 Inner conductor current calculation.

The characteristic impedance is written as

$$Z_0 = \frac{V}{I} = \frac{V_0 e^{-j\beta z}}{I_0 e^{-j\beta z}} = \frac{V_0}{I_0}$$

$$= \frac{\sqrt{\mu/\varepsilon}}{2\pi} \ln \frac{r_o}{r_i} \tag{2.70}$$

For an air-filled line, $\sqrt{\mu/\varepsilon} = \sqrt{\mu_0/\varepsilon_0} = 377\ \Omega$, and Equation (2.70) becomes

$$Z_0 = 60 \ln \frac{r_o}{r_i} \quad \Omega \tag{2.71}$$

The equations above are useful for characteristic impedance calculations.

Commonly used coaxial lines have a characteristic impedance of 50 Ω. The outer radii are 7, 3.5, and 2.4 mm operating up to a frequency of 18, 26, and 50 GHz, respectively. Some low-frequency coaxial lines have a characteristic impedance of 75 Ω.

2.8 RECTANGULAR WAVEGUIDE

Rectangular waveguide (Figure 2.24) is a metal pipe for guided-wave propagation. Rectangular waveguide is normally rigid, although some flexible waveguides are available. Precision machining is required, which makes waveguide expensive. Rectangular waveguide is operated in the TE_{10} mode. The operating frequency range for a particular size of waveguide is confined by the cutoff frequency at the lower end and the excitation of a higher-order

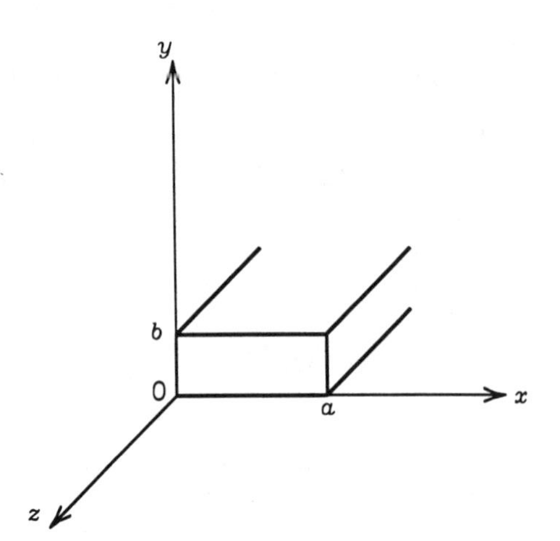

FIGURE 2.24 Rectangular waveguide.

mode at the upper end. Rectangular waveguide has been used from 1 to 300 GHz with many different bands, each band corresponding to a waveguide of certain dimensions (see Appendix B). For example, the X-band (8 to 12.4 GHz) waveguide has dimensions of 0.9 × 0.4 in, and the W-band (75 to 110 GHz) waveguide has dimensions of 0.1 × 0.05 in. Due to their large size, high cost, heavy weight, and difficulty in integration, at lower microwave frequencies, rectangular waveguides have been replaced by microstrip or coaxial lines except in very high power applications. In this section the operating principle of a rectangular waveguide is described briefly.

Consider the source-free interior region of rectangular waveguide. Maxwell's equations are written as

$$\varepsilon \nabla \cdot \mathbf{E} = 0 \tag{2.72}$$

$$\mu \nabla \cdot \mathbf{H} = 0 \tag{2.73}$$

$$\nabla \times \mathbf{E} = -j\omega\mu \mathbf{H} \tag{2.74}$$

$$\nabla \times \mathbf{H} = j\omega\varepsilon \mathbf{E} \tag{2.75}$$

From (2.74) to (2.75), we have

$$\nabla \times \nabla \times \mathbf{E} = -j\omega\mu \nabla \times \mathbf{H} = \omega^2 \mu\varepsilon \mathbf{E}$$

Using the identity $\nabla \times \nabla \times \mathbf{E} = \nabla(\nabla \cdot \mathbf{E}) - \nabla^2 \mathbf{E}$ and (2.72), we have

$$\nabla(\nabla \cdot \mathbf{E}) - \nabla^2 \mathbf{E} = \omega^2 \mu\varepsilon \mathbf{E} = -\nabla^2 \mathbf{E}$$

Therefore,

$$\nabla^2 \mathbf{E} + \omega^2 \mu\varepsilon \mathbf{E} = 0 \tag{2.76}$$

Similarly,

$$\nabla^2 \mathbf{H} + \omega^2 \mu \varepsilon \mathbf{H} = 0 \tag{2.77}$$

If $\mathbf{E} = E_x \hat{\mathbf{x}} + E_y \hat{\mathbf{y}} + E_z \hat{\mathbf{z}}$, Equation (2.76) can be written as three equations as follows:

$$\nabla^2 E_x + \omega^2 \mu \varepsilon E_x = 0$$

$$\nabla^2 E_y + \omega^2 \mu \varepsilon E_y = 0$$

$$\nabla^2 E_z + \omega^2 \mu \varepsilon E_z = 0$$

Considering only the z component gives

$$\nabla^2 E_z + \omega^2 \mu \varepsilon E_z = 0 \tag{2.78}$$

Similarly,

$$\nabla^2 H_z + \omega^2 \mu \varepsilon H_z = 0 \tag{2.79}$$

Assuming a wave propagating in the $+z$ direction with a propagation constant γ gives

$$E_z(x, y, z) = E_z(x, y) e^{-\gamma z} \tag{2.80}$$

$$H_z(x, y, z) = H_z(x, y) e^{-\gamma z} \tag{2.81}$$

Substituting (2.80) and (2.81) into (2.78) and (2.79) gives

$$\nabla_T^2 E_z(x, y) + (\omega^2 \mu \varepsilon + \gamma^2) E_z(x, y) = 0 \tag{2.82}$$

$$\nabla_T^2 H_z(x, y) + (\omega^2 \mu \varepsilon + \gamma^2) H_z(x, y) = 0 \tag{2.83}$$

Equations (2.82) and (2.83) are solved by using the boundary conditions. Once E_z and H_z are solved, E_x, E_y, H_x, and H_y can be found from Maxwell's equations since

$$\nabla \times \mathbf{E} = -j\omega\mu \mathbf{H} \tag{2.84}$$

$$\nabla \times \mathbf{H} = j\omega\varepsilon \mathbf{E} \tag{2.85}$$

From (2.84) and (2.85) we can derive

$$E_y = \frac{1}{\gamma^2 + \omega^2 \mu \varepsilon} \left(j\omega\mu \frac{\partial H_z}{\partial x} - \gamma \frac{\partial E_z}{\partial y} \right) \tag{2.86}$$

$$E_x = -\frac{1}{\gamma^2 + \omega^2 \mu \varepsilon} \left(j\omega\mu \frac{\partial H_z}{\partial y} + \gamma \frac{\partial E_z}{\partial x} \right) \tag{2.87}$$

$$H_y = -\frac{1}{\gamma^2 + \omega^2 \mu \varepsilon} \left(j\omega\varepsilon \frac{\partial E_z}{\partial x} + \gamma \frac{\partial H_z}{\partial y} \right) \tag{2.88}$$

$$H_x = \frac{1}{\gamma^2 + \omega^2 \mu \varepsilon} \left(j\omega\varepsilon \frac{\partial E_z}{\partial y} - \gamma \frac{\partial H_z}{\partial x} \right) \tag{2.89}$$

Two possible groups of modes can propagate in a rectangular waveguide. They are the transverse electric modes (TE) with $E_z = 0$ and $H_z \neq 0$ and the transverse magnetic modes (TM) with $H_z = 0$ and $E_z \neq 0$. Note that the TEM mode ($E_z = H_z = 0$) cannot propagate since no center conductor exists to support the current.

A. TE Modes

For the TE modes ($E_z = 0$), the following equation must be solved:

$$\nabla_T^2 H_z(x, y) + (\omega^2 \mu \varepsilon + \gamma^2) H_z(x, y) = 0 \tag{2.90}$$

To solve (2.90), the method of separation of variables is used. Let

$$H_z(x, y) = X(x)Y(y) \tag{2.91}$$

Equation (2.90) becomes

$$\frac{d^2 X(x)}{dx^2} Y(y) + \frac{d^2 Y(y)}{dy^2} X(x) + (\omega^2 \mu \varepsilon + \gamma^2) X(x) Y(y) = 0 \tag{2.92}$$

Dividing (2.92) by $X(x)Y(y)$ gives

$$\frac{1}{X(x)} \frac{d^2 X(x)}{dx^2} + \frac{1}{Y(y)} \frac{d^2 Y(y)}{dy^2} + (\omega^2 \mu \varepsilon + \gamma^2) = 0 \tag{2.93}$$

It is clear that

$$\frac{1}{X(x)}\frac{d^2X(x)}{dx^2} \quad \text{and} \quad \frac{1}{Y(y)}\frac{d^2Y(y)}{dy^2}$$

are constants independent of x and y. Let

$$\frac{1}{X(x)}\frac{d^2X(x)}{dx^2} = -k_x^2 \tag{2.94}$$

and

$$\frac{1}{Y(y)}\frac{d^2Y(y)}{dy^2} = -k_y^2 \tag{2.95}$$

Equations (2.94) and (2.95) become

$$\frac{d^2X(x)}{dx^2} + k_x^2X(x) = 0 \tag{2.96}$$

$$\frac{d^2Y(y)}{dy^2} + k_y^2Y(y) = 0 \tag{2.97}$$

From (2.93) we have

$$-k_x^2 - k_y^2 + \gamma^2 + \omega^2\mu\varepsilon = 0$$

Therefore,

$$\gamma = \sqrt{\left(k_x^2 + k_y^2\right) - \omega^2\mu\varepsilon} = \sqrt{k_c^2 - \omega^2\mu\varepsilon} \tag{2.98}$$

where $k_c = \sqrt{k_x^2 + k_y^2}$ is the cutoff wave number.

The general solutions to (2.96) and (2.97) are

$$X(x) = A\sin k_x x + B\cos k_x x$$

$$Y(y) = C\sin k_y y + D\cos k_y y$$

Thus we have

$$H_z(x, y) = (A\sin k_x x + B\cos k_x x)(C\sin k_y y + D\cos k_y y) \tag{2.99}$$

A, B, C, D, k_x, and k_y can be found from the boundary conditions, which require that the tangential electric fields vanish on the walls. Thus

$$E_x = 0 \quad \text{at} \quad y = 0 \quad \text{and} \quad b$$

and

$$E_y = 0 \quad \text{at} \quad x = 0 \quad \text{and} \quad a$$

From Equations (2.86) and (2.87), the boundary conditions are translated to

$$\frac{\partial H_z}{\partial y} = 0 \quad \text{at} \quad y = 0 \quad \text{and} \quad b$$

$$\frac{\partial H_z}{\partial x} = 0 \quad \text{at} \quad x = 0 \quad \text{and} \quad a$$

Now, the derivatives of H_z with respect to x and y are

$$\frac{\partial H_z(x,y)}{\partial x} = (Ak_x \cos k_x x - Bk_x \sin k_x x)(C \sin k_y y + D \cos k_y y)$$

$$\frac{\partial H_z(x,y)}{\partial y} = (A \sin k_x x + B \cos k_x x)(Ck_y \cos k_y y - Dk_y \sin k_y y)$$

Applying the boundary conditions leads to

$$\frac{\partial H_z}{\partial y} = 0 \quad \text{at} \quad y = 0 \qquad \text{requires} \quad C = 0$$

$$\frac{\partial H_z}{\partial y} = 0 \quad \text{at} \quad y = b \qquad \text{requires} \quad k_y b = m\pi, \quad m = 0, 1, 2, \ldots$$

$$\frac{\partial H_z}{\partial x} = 0 \quad \text{at} \quad x = 0 \qquad \text{requires} \quad A = 0$$

$$\frac{\partial H_z}{\partial x} = 0 \quad \text{at} \quad x = a \qquad \text{requires} \quad k_x a = n\pi, \quad n = 0, 1, 2, \ldots$$

The solution is then

$$\begin{aligned} H_z(x,y) &= BD \cos k_x x \cos k_y y \\ &= A_{nm} \cos \frac{n\pi x}{a} \cos \frac{m\pi y}{b} \end{aligned} \tag{2.100}$$

Note that $n = 0, 1, 2, \ldots$ and $m = 0, 1, 2, \ldots$, but n and m cannot both equal 0. Otherwise, all the other field components vanish. A_{nm} is the amplitude constant. The other field components (H_x, H_y, E_x, E_y) can be found from H_z using Equations (2.86) through (2.89).

The propagation constant is [from (2.98)] given by

$$\gamma_{nm} = \sqrt{\left(\frac{n\pi}{a}\right)^2 + \left(\frac{m\pi}{b}\right)^2 - \omega^2\mu\varepsilon} \tag{2.101}$$

The cutoff wave number can be found by

$$k_{c,nm}^2 = \left(\frac{n\pi}{a}\right)^2 + \left(\frac{m\pi}{b}\right)^2 = \left(2\pi f_c\sqrt{\mu\varepsilon}\right)^2 \tag{2.102}$$

The cutoff frequency is defined as the frequency which makes $\gamma_{nm} = 0$. Thus

$$\begin{aligned} f_{c,nm} &= \frac{1}{2\pi\sqrt{\mu\varepsilon}}\sqrt{\left(\frac{n\pi}{a}\right)^2 + \left(\frac{m\pi}{b}\right)^2} \\ &= \frac{c}{2\pi\sqrt{\mu_r\varepsilon_r}}\sqrt{\left(\frac{n\pi}{a}\right)^2 + \left(\frac{m\pi}{b}\right)^2} \end{aligned} \tag{2.103}$$

where $c = 1/\sqrt{\mu_0\varepsilon_0}$ is the speed of light in vacuum and $\mu = \mu_r\mu_0$, $\varepsilon = \varepsilon_r\varepsilon_0$.

If $f > f_{c,nm}$, the nmth mode will propagate and $\gamma_{nm} = j\beta_{nm}$.

$$\beta_{nm} = \sqrt{\omega^2\mu\varepsilon - \left(\frac{n\pi}{a}\right)^2 - \left(\frac{m\pi}{b}\right)^2} = \frac{2\pi}{\lambda_{g,nm}} \tag{2.104}$$

The guide wavelength λ_g is given by

$$\lambda_{g,nm} = \frac{\lambda_0}{\sqrt{\mu/\varepsilon}\sqrt{1 - (f_{c,nm}/f)^2}} \tag{2.105}$$

The wave impedance for the TE_{nm} mode is given by

$$Z_{\text{TE},nm} = \frac{E_x}{H_y} = -\frac{E_y}{H_x} = \frac{\omega\sqrt{\mu\varepsilon}}{\beta_{nm}}\eta \tag{2.106}$$

where $\eta = \sqrt{\mu/\varepsilon}$. If $f < f_{c,nm}$, from Equation (2.101), γ_{nm} is real, the wave does not propagate, and the wave is attenuated.

B. TE_{10} Mode

Rectangular waveguides normally operate in the TE_{10} mode, which is called the *dominant mode*. The field components for the TE_{10} mode are

$$H_z = A_{10} \cos \frac{\pi x}{a} e^{-j\beta_{10} z} \tag{2.107a}$$

$$H_x = \frac{j\beta_{10} a}{\pi} A_{10} \sin \frac{\pi x}{a} e^{-j\beta_{10} z} \tag{2.107b}$$

$$H_y = 0 \tag{2.107c}$$

$$E_x = 0 \tag{2.107d}$$

$$E_y = -\frac{j\omega\mu a}{\pi} A_{10} \sin \frac{\pi x}{a} e^{-j\beta_{10} z} \tag{2.107e}$$

$$E_z = 0 \tag{2.107f}$$

The propagation constant, guided wavelength, cutoff frequency, and cutoff wavelength for the TE_{10} mode are given by

$$\beta_{10} = \sqrt{\omega^2 \mu \varepsilon - \left(\frac{\pi}{a}\right)^2} \tag{2.108a}$$

$$\lambda_{g,10} = \frac{2\pi}{\sqrt{\omega^2 \mu \varepsilon - (\pi/a)^2}} \tag{2.108b}$$

$$f_{c,10} = \frac{1}{2a\sqrt{\mu\varepsilon}} \tag{2.108c}$$

$$\lambda_{c,10} = 2a \tag{2.108d}$$

For an X-band (8 to 12.4 GHz) waveguide with $a = 0.9$ in. and $b = 0.4$ in., the cutoff frequency is $f_{c,10} = 6.5$ GHz. The first higher-order mode is the TE_{20} mode, with a cutoff frequency $f_{c,20} = 13.1$ GHz. Operating between 8 and 12.4 GHz ensures that only one mode (i.e., the TE_{10} mode) propagates.

C. TM Modes

For TM modes, a similar procedure is followed as for TE modes. We solve the following equation for E_z:

$$\nabla_T^2 E_z(x, y) + (\omega^2 \mu \varepsilon + \gamma^2) E_z(x, y) = 0$$

By applying the boundary conditions that $E_z = 0$ at $x = 0$ and a and $y = 0$

and b, we have

$$E_z = B_{nm} \sin \frac{n\pi x}{a} \sin \frac{m\pi y}{b} \qquad n = 1, 2, 3, \ldots, \qquad m = 1, 2, 3, \ldots \quad (2.109)$$

The equations for γ_{nm}, β_{nm}, $f_{c,nm}$, and $\lambda_{g,nm}$ are the same as for TE_{nm} modes. The wave impedance is

$$Z_{TM,nm} = \frac{E_x}{H_y} = -\frac{E_y}{H_x} = \frac{\beta_{nm}}{\omega\sqrt{\mu\varepsilon}}\eta \quad (2.110)$$

Note that

$$Z_{TE,nm} Z_{TM,nm} = \eta^2 = \frac{\mu}{\varepsilon} \quad (2.111)$$

The other field components can be found from E_z.

2.9 MICROSTRIP LINES

The microstrip line is the most commonly used microwave integrated circuit transmission medium. The microstrip line has many advantages, such as low cost, small size, no critical machining, no cutoff frequency, ease of active-device integration, use of a photolithographic method for circuit production, good repeatability and reproducibility, ease of mass production, and compatibility to monolithic circuits. Monolithic circuits are microstrip circuits on a GaAs substrate with both active and passive devices on the same chip. Compared to a rectangular waveguide, the disadvantages of microstrip are its higher loss, lower power-handling capability, and greater temperature instability.

Figure 2.25 shows a schematic drawing of a microstrip transmission line. A conductor strip with width w is etched on top of a substrate with thickness h. Two types of substrates are generally used: soft substrates and hard sub-

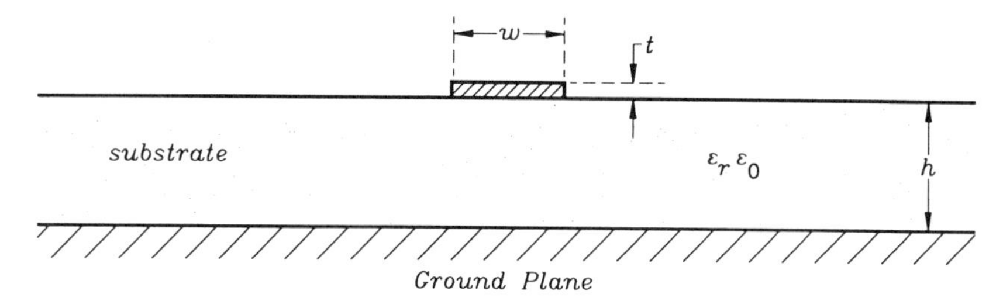

FIGURE 2.25 Microstrip transmission line.

strates. Soft substrates are flexible, cheap, and can be fabricated easily. However, they have higher thermal expansion coefficients. Some typical soft substrates are RT Duroid 5870 ($\varepsilon_r = 2.3$), RT Duroid 5880 ($\varepsilon_r = 2.2$), and RT Duroid 6010.5 ($\varepsilon_r = 10.5$). Hard substrates have better reliability and lower thermal expansion coefficients but are more expensive and nonflexible. Typical hard substrates are quartz ($\varepsilon_r = 3.8$), alumina ($\varepsilon_r = 9.7$), sapphire ($\varepsilon_r = 11.7$), and GaAs ($\varepsilon_r = 12.3$). Tables 2.2 and 2.3 provide the key design data for some common substrate materials.

The most important parameters in microstrip circuit design are w, h, and ε_r. The effects of strip thickness t and the conductivity σ are secondary. Since the structure is not uniform, it supports the quasi-TEM mode. For a symmetrical structure such as coaxial line or strip line, the TEM mode is supported. Analyses for microstrip lines can be complicated and are generally divided into two approaches:

1. Static or Quasi-TEM Analysis. This approach assumes a frequency of zero and solves the Laplace's equation ($\nabla^2\Phi = 0$). The characteristic

TABLE 2.2 Properties of Microwave Dielectric Substrates

Material	Relative Dielectric Constant	Loss Tangent at 10 GHz ($\tan\delta$)	Thermal Conductivity, K (W/cm/°C)	Dielectric Strength (kV/cm)
Sapphire	11.7	10^{-4}	0.4	4×10^3
Alumina	9.7	2×10^{-4}	0.3	4×10^3
Quartz (fused)	3.8	10^{-4}	0.01	10×10^3
Polystyrene	2.53	4.7×10^{-4}	0.0015	280
Beryllium oxide (BeO)	6.6	10^{-4}	2.5	—
GaAs ($\rho = 10^7\ \Omega \cdot$ cm)	12.3	16×10^{-4}	0.3	350
Si ($\rho = 10^3\ \Omega \cdot$ cm)	11.7	50×10^{-4}	0.9	300
3M 250 type GX	2.5	19×10^{-4}	0.0026	200
Keene DI-clad 527	2.5	19×10^{-4}	0.0026	200
RT Duroid 5870	2.35	12×10^{-4}	0.0026	200
3M Cu-clad 233	2.33	12×10^{-4}	0.0026	200
Keene DI-clad 870	2.33	12×10^{-4}	0.0026	200
RT Duroid 5880	2.20	9×10^{-4}	0.0026	200
3M Cu-clad 217	2.17	9×10^{-4}	0.0026	200
Keene DI-clad 880	2.20	9×10^{-4}	0.0026	200
RT Duroid 6010	10.5	15×10^{-4}	0.004	160
3M epsilam 10	10.2	15×10^{-4}	0.004	160
Keene DI-clad 810	10.2	15×10^{-4}	0.004	160
Air	1.0	0	0.00024	30

Source: Ref. 3 and E. C. Nichenke, "Microstrip Circuits," WMEC Class Notes, November, 1981.

TABLE 2.3 Typical Properties of *RT* / duroid Microwave Materials

Product Number, Type	RT/duroid 5880 Random Fiber			RT/duroid 5870 Random Fiber			RT/duroid 6006 Ceramic Filled			RT/duroid 6010 Ceramic Filled		
Dielectric constant, *Z* direction IPC-TM-650, 2.5.5.5	2.20			2.33			6.15			10.50		
Standard tolerance	±0.015			±0.02			±0.15			±0.25		
Isotropy ratio $X, Y/Z$	< 1.04			1.04			—			1.03		
Dissipation factor, IPC-TM-650, 2.5.5.5	0.0009			0.0012			0.0018			0.0021		
Specific gravity	2.2			2.2			2.7			2.9		
Specific heat (J/g/K)	0.96			0.96			0.97			1.00		
Thermal conductivity, *Z* direction, 23–100° C, Rogers T.R. 2721 (W/m/K)	0.26			0.26			0.48			0.41		
Direction of test	*X*	*Y*	*Z*	*X*	*Y*	*Z*	*X*	*Y*	*Z*	*X*	*Y*	*Z*
Tensile, ASTM D638, 23° C												
Modulus (MPa)	1070	860	—	1300	1280	—	510	627	—	931	559	—
Ultimate stress (MPa)	29	27	—	50	42	—	20	17	—	17	13	—
Ultimate strain (%)	6.0	4.9	—	9.8	9.8	—	12.5	5	—	12	10	—
Compression, ASTM D695, 23° C												
Modulus (MPa)	710	710	940	1210	1360	830	—	—	1069	—	—	2144
Ultimate stress (MPa)	27	28	52	30	37	54	—	—	54	—	—	47
Ultimate strain (%)	8.5	7.7	12.5	4.0	3.3	8.7	—	—	33	—	—	25
Linear thermal expansion ASTM D3386 mm/m total change from 35° C												
−50° C	−3.8	−5.7	−13.6	−3.1	−3.5	−8.3	−5.0	−4.8	−8.8	−2.0	−2.1	−2.6
0° C	−1.6	−2.7	−8.4	−1.2	−1.5	−5.0	−2.6	−2.5	−5.3	−1.0	−1.0	−1.4
150° C	2.3	3.2	28.4	1.8	2.2	22.0	3.5	1.4	12.0	2.2	2.2	1.7
250° C	3.8	5.5	69.5	3.4	4.0	58.9	9.0	5.0	32.1	3.7	3.8	4.3

Note: 1 MPa (megapascal) = 145 psi. ASTM, American Society for Testing and Materials, Philadelphia, Pa. RT/duroid is a registered trademark of Rogers Corporation for its microwave laminate. (Courtesy Rogers Corporation, Chandler, AZ)

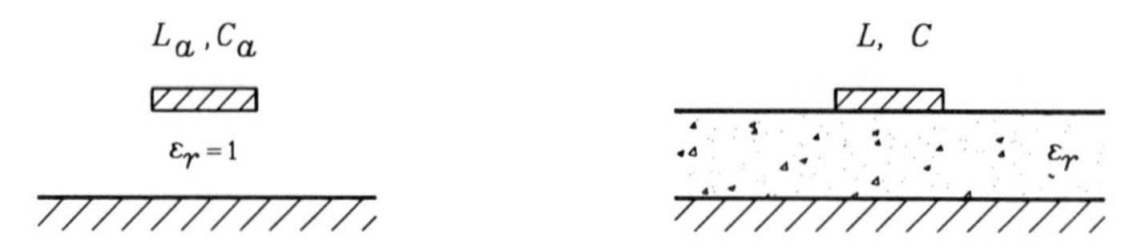

FIGURE 2.26 Microstrip with and without substrate.

impedance is independent of frequency. The results are accurate only for low microwave frequencies.

2. Full-Wave Analysis. This approach solves $\nabla^2\Phi + k^2\Phi = 0$. Therefore, the impedance is frequency dependent. The analysis provides good accuracy at millimeter-wave frequencies.

For simplicity, we limit our discussion to the static or quasi-TEM analysis. Recall that in Section 2.2, for a lossless line, the characteristic impedance is given by

$$Z_0 = \sqrt{\frac{L}{C}} \tag{2.112}$$

and the phase velocity by

$$v_p = f\lambda_g = \frac{2\pi f}{\beta} = \frac{\omega}{\omega\sqrt{LC}} = \frac{1}{\sqrt{LC}} \tag{2.113}$$

Consider a microstrip line without a substrate (i.e., an air line) with line inductance L_a and line capacitance C_a as shown in Figure 2.26. We have

$$L = L_a \tag{2.114}$$

but

$$C \neq C_a \tag{2.115}$$

where L_a and C_a are the inductance and capacitance per unit length of microstrip without the substrate (i.e., air substrate), and L and C are the inductance and capacitance per unit length of microstrip with the substrate. The phase velocity for an air line is

$$v_{pa} = \frac{1}{\sqrt{L_a C_a}} = c = \text{speed of light} \tag{2.116}$$

The characteristic impedance for microstrip with a substrate is

$$Z_0 = \sqrt{\frac{L}{C}} = \sqrt{\frac{L_a}{C}} = \sqrt{\frac{L_a C_a}{C C_a}} = \frac{1}{c}\frac{1}{\sqrt{C C_a}}$$

Therefore,

$$Z_0 = \frac{1}{c}\frac{1}{\sqrt{C C_a}} \quad \text{or} \quad \frac{1}{C v_p} \tag{2.117}$$

Thus Z_0 can be found if we can calculate the line capacitance with and without the substrate (i.e., C and C_a).

Let us define an effective dielectric constant ε_{eff} as

$$\varepsilon_{\text{eff}} = \frac{C}{C_a} \tag{2.118}$$

ε_{eff} is normally not equal to ε_r for a nonuniform structure. For a uniformly filled structure such as a strip line, coaxial line, or parallel plate, $\varepsilon_{\text{eff}} = \varepsilon_r$.

λ_g can be related to λ_0 and ε_{eff} by the following derivation:

$$\begin{aligned} v_p = f\lambda_g &= \frac{1}{\sqrt{LC}} = \frac{1}{\sqrt{L_a C}} \\ &= \frac{1}{\sqrt{L_a C_a}}\sqrt{\frac{C_a}{C}} = c\frac{1}{\sqrt{\varepsilon_{\text{eff}}}} \end{aligned} \tag{2.119}$$

Since $c = f\lambda_0$,

$$v_p = f\lambda_g = f\lambda_0 \frac{1}{\sqrt{\varepsilon_{\text{eff}}}}$$

and we have

$$\lambda_g = \frac{\lambda_0}{\sqrt{\varepsilon_{\text{eff}}}} \tag{2.120}$$

Let us consider two extreme cases for the effective dielectric constant. For the case $w \gg h$, shown in Figure 2.27, most fields are confined under the strip and the circuit acts like a parallel plate; thus $\varepsilon_{\text{eff}} \approx \varepsilon_r$. For the case $w \ll h$, where half of the fields are in the air and half in the dielectric, we have $\varepsilon_{\text{eff}} = \frac{1}{2}(1 + \varepsilon_r)$. For other cases

$$\tfrac{1}{2}(1 + \varepsilon_r) < \varepsilon_{\text{eff}} < \varepsilon_r \tag{2.121}$$

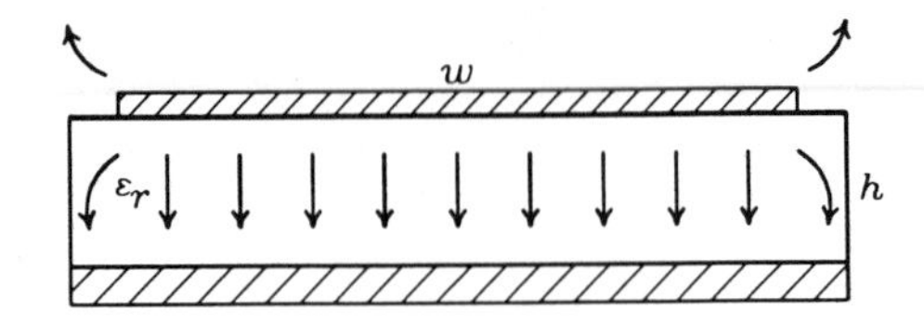

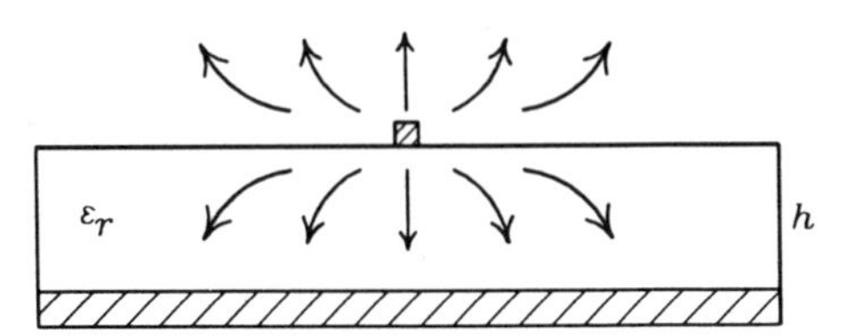

FIGURE 2.27 Two extreme cases.

A. Impedance and Effective Dielectric Constant Calculation

For a static analysis, the impedance can be calculated by

$$Z_0 = \frac{1}{c} \frac{1}{\sqrt{CC_a}} \tag{2.122}$$

C and C_a can be found by solving $\nabla^2 \Phi = 0$ for the cases with and without the substrate material. Many methods can be used to solve the Laplace's equation meeting the boundary conditions. Some examples are the conformal mapping method [4], finite-difference method [5, 6], and variational method [7, 8]. Computer programs are available for these calculations. In practice, closed-form equations obtained from curve fitting are convenient for quick calculations. Some examples of design equations are [3]

$$\varepsilon_{\text{eff}} = \frac{\varepsilon_r + 1}{2} + \frac{\varepsilon_r - 1}{2}\left[\left(1 + \frac{12h}{w}\right)^{-1/2} + 0.04\left(1 - \frac{w}{h}\right)^2\right] \tag{2.123}$$

and

$$Z_0 = 60(\varepsilon_{\text{eff}})^{-1/2} \ln\left(\frac{8h}{w} + \frac{0.25w}{h}\right) \quad \Omega \tag{2.124}$$

for $w/h \leq 1$.

$$\varepsilon_{\text{eff}} = \frac{\varepsilon_r + 1}{2} + \frac{\varepsilon_r - 1}{2}\left(1 + \frac{12h}{w}\right)^{-1/2} \tag{2.125}$$

and

$$Z_0 = \frac{120\pi(\varepsilon_{\text{eff}})^{-1/2}}{(w/h) + 1.393 + 0.667\ln(1.444 + w/h)} \quad \Omega \qquad (2.126)$$

for $w/h > 1$. Synthesis formulas (from Z_0 and ε_r to find w and h) are also available from the literature [9] (see Appendix D).

The equations for Z_0 and ε_{eff} above assume a strip thickness of $t = 0$. For a finite t, the electric field from the edge makes the line width w appear larger. The effective width is introduced as [3]

$$w_{\text{eff}} = \begin{cases} w + \dfrac{t}{\pi}\left(1 + \ln\dfrac{2h}{t}\right) & \text{for } \dfrac{w}{h} > \dfrac{1}{2\pi} \qquad (2.127) \\ w + \dfrac{t}{\pi}\left(1 + \ln\dfrac{4\pi w}{t}\right) & \text{for } \dfrac{w}{h} \le \dfrac{1}{2\pi} \qquad (2.128) \end{cases}$$

Note that if $t = 0$, $w_{\text{eff}} = w$. Otherwise, use w_{eff} to replace w in Equations (2.123) through (2.126).

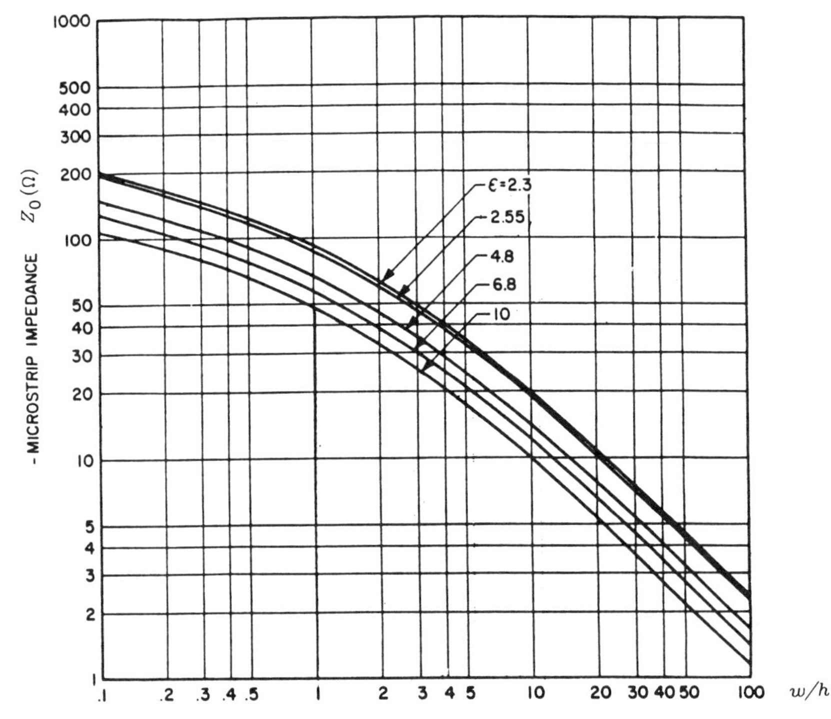

FIGURE 2.28 Microstrip characteristic impedance as a function of w/h and ε_r.

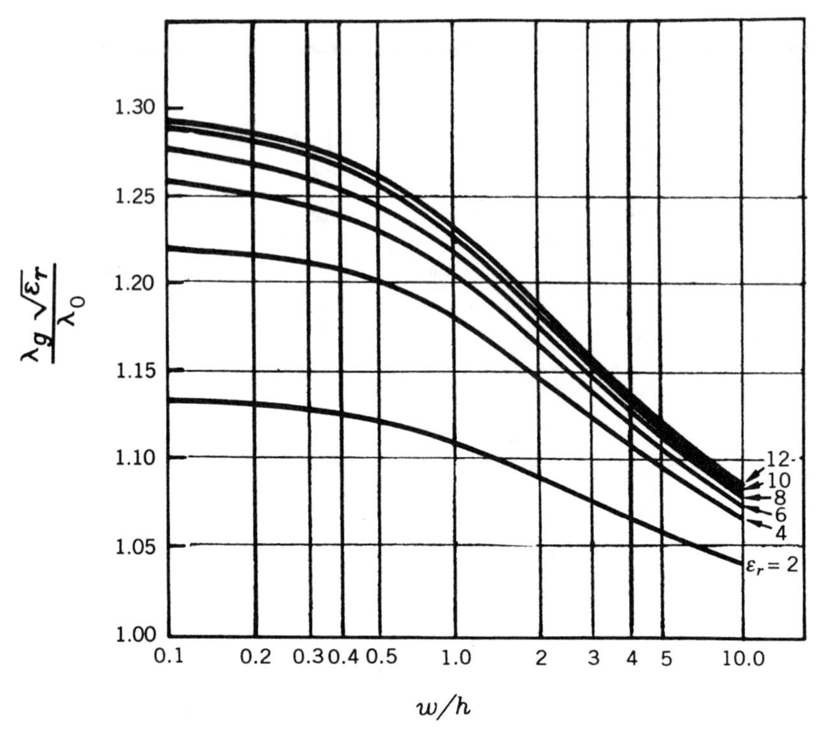

FIGURE 2.29 Microstrip guide wavelength as a function of w/h and ε_r.

For convenience, a graphical method can be used to estimate ε_{eff} and Z_0. Figures 2.28 and 2.29 show two charts that can be used to find Z_0 and ε_{eff} if w/h is known, and vice versa. An example is given below.

Example 2.9 For a microstrip line with $w/h = 1$ and $\varepsilon_r = 10$, find Z_0 and ε_{eff} from Figures 2.28 and 2.29.

SOLUTION: $w/h = 1$ and $\varepsilon_r = 10$. From Figure 2.28, $Z_0 = 50\ \Omega$; from Figure 2.29, $(\lambda_g/\lambda_0)\sqrt{\varepsilon_r} = 1.225$. Therefore,

$$\lambda_g = \frac{1.225}{\sqrt{10}}\lambda_0 = 0.38\lambda_0 = \frac{\lambda_0}{\sqrt{\varepsilon_{\text{eff}}}}$$

$$\varepsilon_{\text{eff}} = 6.66$$

B. Effects Due to an Enclosure or Shield

The microstrip line is often shielded to reduce the radiation loss and

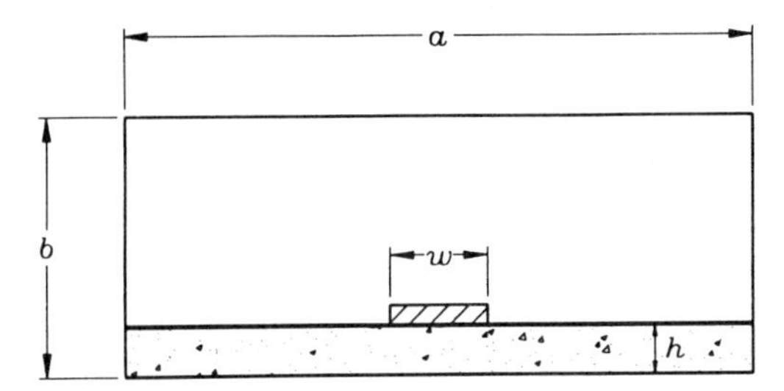

FIGURE 2.30 Shielded microstrip line.

crosstalk (Figure 2.30). The effects of the enclosure are negligible if

$$b > 5h$$
$$a > 5w$$

The reason is that there is little change in the field distribution with and without the enclosure if the conditions above are met. In other words, the enclosure does not change the field distribution appreciably.

C. Dispersion

In the previous discussion for Z_0 and ε_{eff}, we assumed a static case and that Z_0 and ε_{eff} are independent of frequency. In reality, a full-wave analysis is required and Z_0 and ε_{eff} are functions of frequency. This is especially true at high frequencies. The effect of frequency on Z_0 and ε_{eff} is called *dispersion*. The dispersion effect increases as the frequency, or the substrate thickness, or the dielectric constant is increased. For compensation, the effects of dispersion can be obtained from the nondispersive calculation. The frequency variations of $\varepsilon_{\text{eff}}(f)$ and $Z_0(f)$ are given [10, 11] as

$$\varepsilon_{\text{eff}}(f) = \left[\frac{\sqrt{\varepsilon_r} - \sqrt{\varepsilon_{\text{eff}}(0)}}{1 + 4F^{-1.5}} + \sqrt{\varepsilon_{\text{eff}}(0)}\right]^2 \tag{2.129}$$

where

$$F = \frac{4hf\sqrt{\varepsilon_r - 1}}{c}\left\{0.5 + \left[1 + 2\log\left(1 + \frac{w}{h}\right)\right]^2\right\}$$

and

$$Z_0(f) = Z_0(0)\frac{\varepsilon_{\text{eff}}(f) - 1}{\varepsilon_{\text{eff}}(0) - 1}\sqrt{\frac{\varepsilon_{\text{eff}}(0)}{\varepsilon_{\text{eff}}(f)}} \tag{2.130}$$

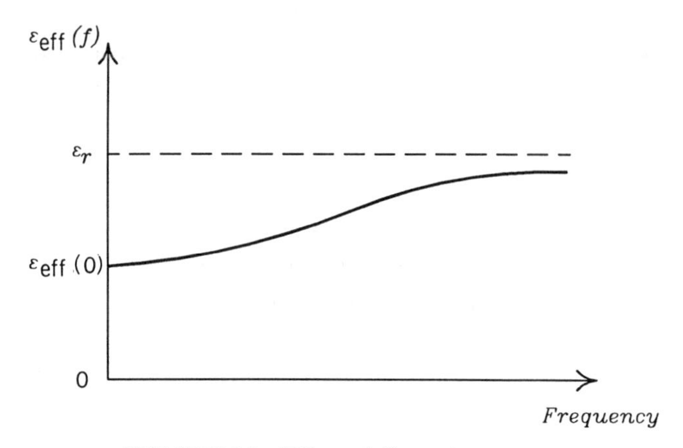

FIGURE 2.31 Effect of dispersion on ε_{eff}.

where $\varepsilon_{eff}(0)$ and $Z_0(0)$ are nondispersive results assuming that $f = 0$. These parameters are obtained using Equations (2.123) through (2.126) or a graphical method. Figure 2.31 shows how ε_{eff} varies with frequency.

D. Losses

A microstrip line has higher losses than those with a rectangular waveguide. The losses consist of three major components: conductor loss, dielectric loss, and radiation loss. Conductor loss is due to the current flow in the strip conductor and the ground plane shown in Figure 2.32. At microwave frequencies, conductor loss dominates and the other losses are small for most substrate materials except silicon. Perhaps the most useful expression for

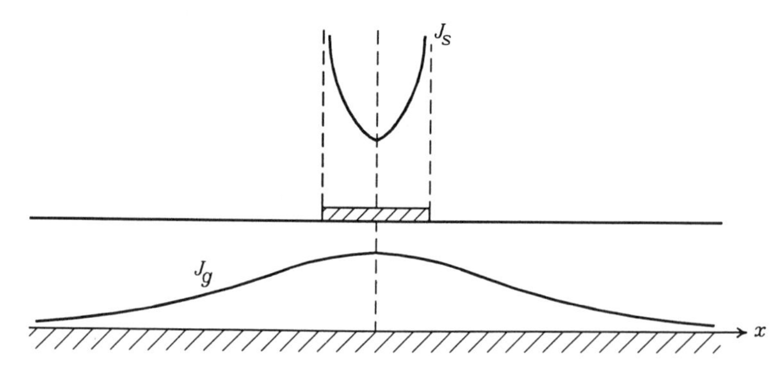

FIGURE 2.32 Current distribution on the strip and the ground plane.

conductor loss has been given by Hammerstad and Bekkadal [9, 12]:

$$\alpha_c = 0.072 \frac{\sqrt{f}}{wZ_0} \lambda_g \qquad \text{dB/guide wavelength} \tag{2.131}$$

where the frequency f is an GHz and Z_0 is in ohms.

The dielectric loss is due to the substrate and is related to the loss tangent of the substrate. The loss can be written as [9]

$$\alpha_d = 27.3 \frac{\varepsilon_r(\varepsilon_{\text{eff}} - 1)\tan\delta}{\varepsilon_{\text{eff}}(\varepsilon_r - 1)} \qquad \text{dB/guide wavelength} \tag{2.132}$$

The loss tangents for various materials are given in Table 2.2.

Radiation loss is due to discontinuities on the microstrip line. These discontinuities excite higher-order modes and radiate energy. Surface-wave propagation can also cause radiation loss.

The *quality factor* (Q), defined as the ratio of stored energy to the loss, is used to assess the loss. The total Q is expressed in terms of Q_c, Q_d, and Q_r, which account for conductor, dielectric, and radiation loss, respectively. The total Q is given by

$$\frac{1}{Q} = \frac{1}{Q_c} + \frac{1}{Q_d} + \frac{1}{Q_r} \tag{2.133}$$

Q is a strong function of frequency and the substrate thickness. Figure 2.33 shows the calculated Q factors as a function of substrate thickness for different substrate materials at various frequencies [13]. The optimum substrate thickness for a substrate at a particular frequency can be determined from this figure. For example, the optimum substrate thickness for a $\varepsilon_r = 13$ substrate at 64 GHz is about 0.1 mm, which gives the maximum value of the Q factor.

E. Spurious Modes

At high frequencies, spurious mode excitation can limit the maximum substrate thickness and linewidth used. Two types of spurious modes can exist in a microstrip line: surface-wave modes and transverse resonance modes.

Surface waves can be excited on a dielectric rod, dielectric-coated conducting wire, and dielectric sheet placed on a conducting plane. The field of a surface wave is confined inside the dielectric layer and decays exponentially in the direction normal to and away from the dielectric layer. If the phase velocity of the microstrip line is close to the phase velocity of a surface wave, strong coupling between waves is expected. The lowest frequency for this

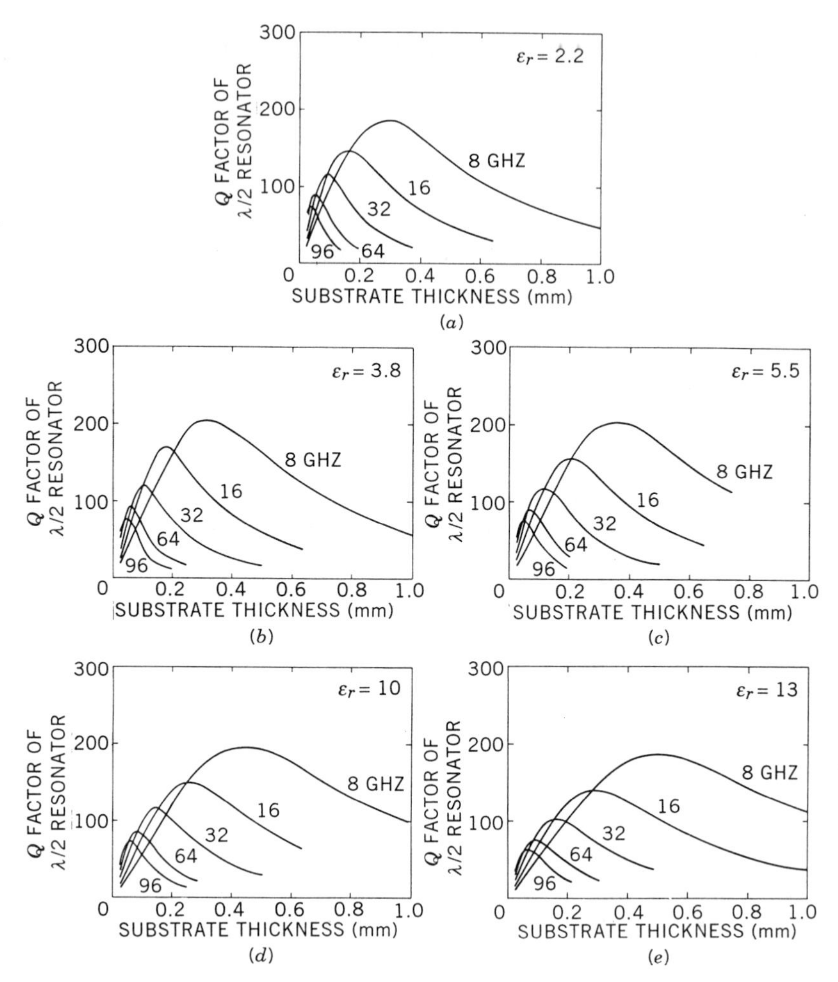

FIGURE 2.33 Q factor of $\lambda/2$ resonators against substrate thickness in millimeters for $Z_0 = 50\ \Omega$, for various frequencies. Strip metal gold 3 μm thick. Cases considered as follows: (*a*) $\varepsilon_r = 2.2$, $\tan\delta = 8 \times 10^{-4}$; (*b*) $\varepsilon_r = 3.8$, $\tan\delta = 1 \times 10^{-4}$; (*c*) $\varepsilon_r = 5.5$, $\tan\delta = 1 \times 10^{-4}$; (*d*) $\varepsilon_r = 10$, $\tan\delta = 1 \times 10^{-4}$; (*e*) $\varepsilon_r = 13$, $\tan\delta = 2 \times 10^{-4}$ up to 16 GHz; $\tan\delta = 6 \times 10^{-4}$ at 32 GHz; $\tan\delta = 1 \times 10^{-3}$ at 64 GHz; $\tan\delta = 1.5 \times 10^{-3}$ at 96 GHz. (From Ref. 13 with permission from IEEE.)

type of surface-wave coupling defines an upper frequency limit for microstrip operation.

Vendelin has indicated that the modal limitation in microstrip is due to the strong coupling between the microstrip quasi-TEM mode and the lowest-TM mode of the dielectric-coated conducting plane as shown in Figure 2.34 [14, 15]. From Collin [14], the propagating constants in the dielectric and air

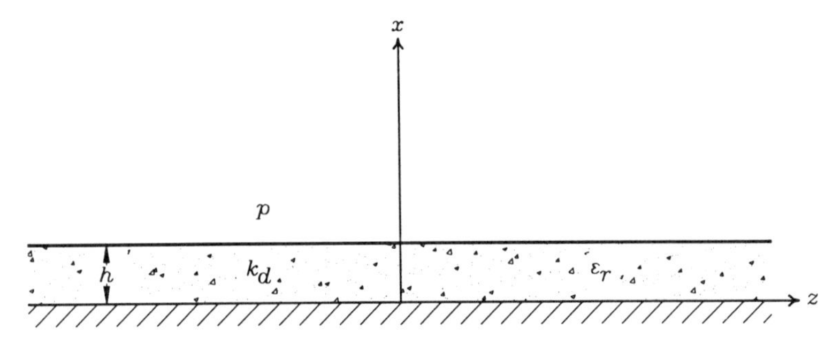

FIGURE 2.34 Dielectric-coated conducting plane.

are k_d and p, which can be found by solving the following two equations:

$$\varepsilon_r ph = k_d h \tan k_d h \tag{2.134}$$

$$k_d^2 + p^2 = (\varepsilon_r - 1)k_0^2 \tag{2.135}$$

where k_0 is the free-space propagation constant, which equals $\omega\sqrt{\mu_0 \varepsilon_o}$, and h is the substrate thickness. Equations (2.134) and (2.135) are obtained from the boundary conditions that the tangential electric and magnetic fields must be continuous for all values of z. The z components of electric field inside and outside the dielectric are

$$E_z = \begin{cases} A \sin(k_d x)e^{-j\beta z} & \text{for } h > x > 0 \quad (2.136) \\ A \sin(k_d h)e^{-p(x-h)-j\beta z} & \text{for } x > h \quad (2.137) \end{cases}$$

The propagating constant β is given by

$$\beta = \sqrt{k_0^2 + p^2} = \sqrt{\varepsilon_r k_0^2 - k_d^2} \tag{2.138}$$

and the phase velocity is

$$v_p = \frac{k_0}{\beta} c \tag{2.139}$$

The lowest order mode is the TM_0 mode, which has no cutoff frequency. Strong coupling could occur if the phase velocity of the microstrip quasi-TEM wave is close to the phase velocity of the surface wave. The result is given by

$$f_s = \frac{c}{2\pi h}\sqrt{\frac{2}{\varepsilon_r - 1}} \tan^{-1} \varepsilon_r \tag{2.140}$$

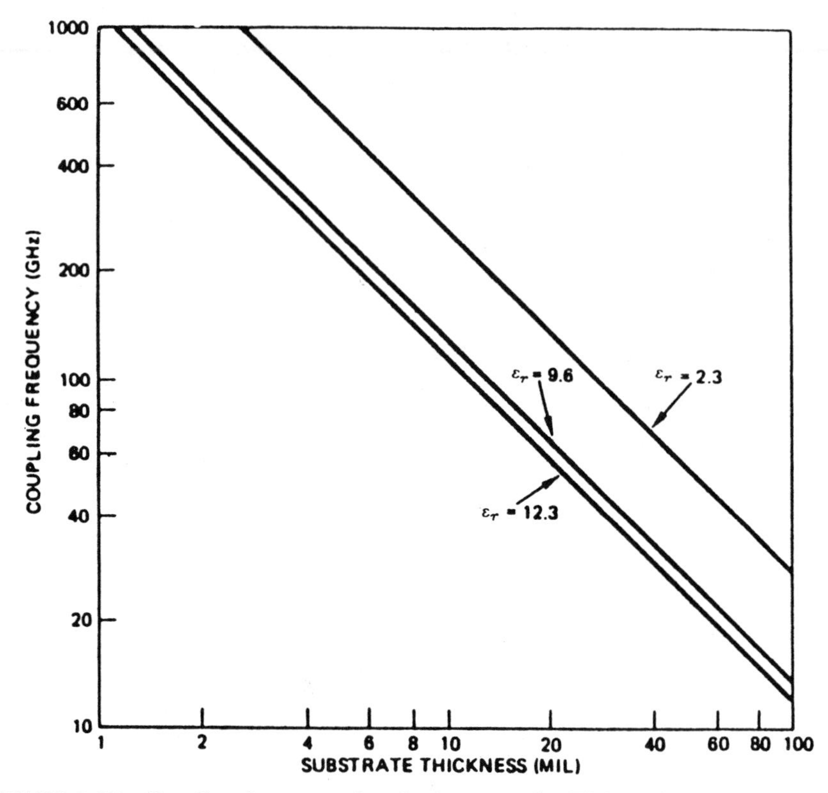

FIGURE 2.35 Coupling frequency for the lowest-order TM mode as a function of dielectric constant and substrate thickness for a microstrip line.

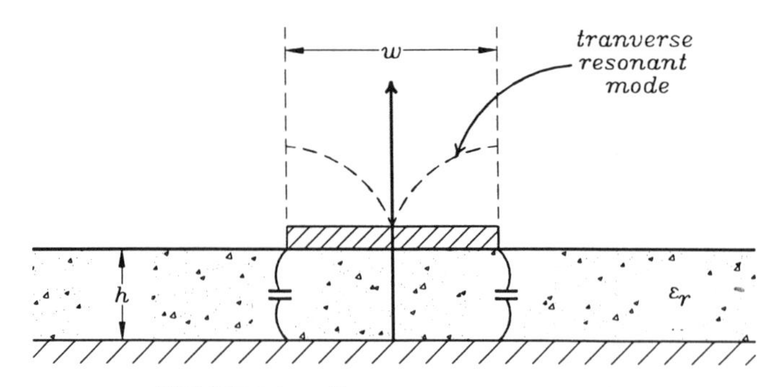

FIGURE 2.36 Transverse resonant mode.

where f_s is the lowest coupling frequency, which decreases as the substrate thickness and the dielectric constant are increased. A plot is given in Figure 2.35.

In sufficiently wide microstrip lines, a transverse TE mode may exist that will couple strongly to the microstrip-line mode. Considering a microstrip line with width w as shown in Figure 2.36, the first resonant mode in the transverse direction is

$$\frac{\lambda_T}{2} = w + 0.4h \tag{2.141}$$

where $0.4h$ accounts for the increase in electric length due to the fringe

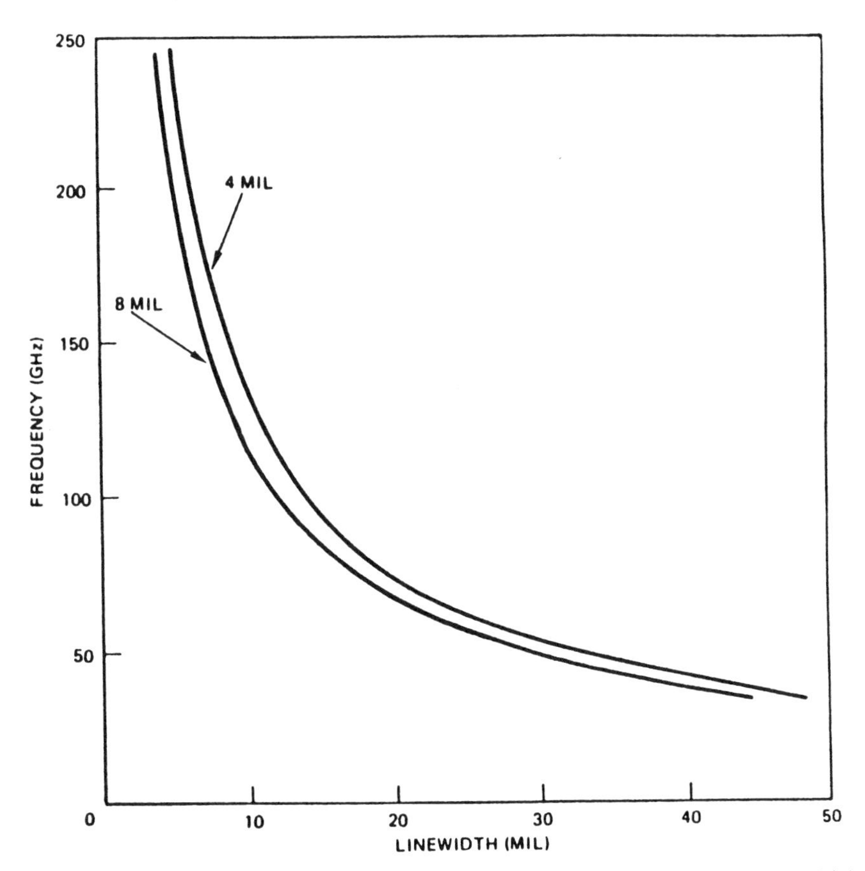

FIGURE 2.37 Transverse resonance frequency versus linewidth and substrate thickness for a microstrip line on GaAs substrate.

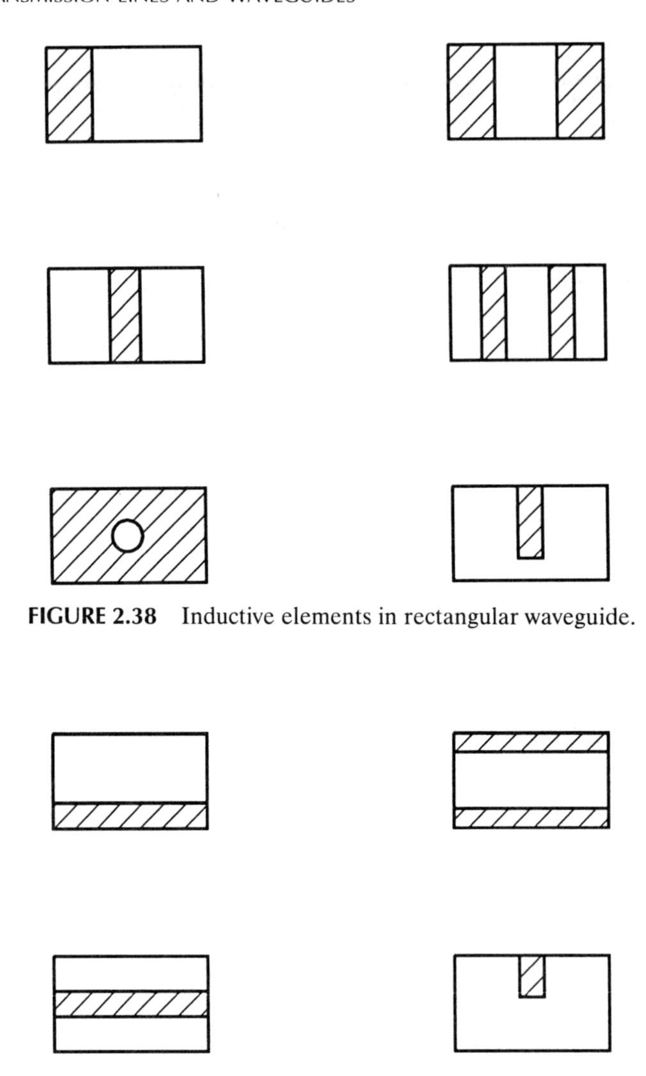

FIGURE 2.38 Inductive elements in rectangular waveguide.

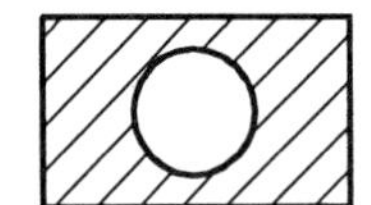

FIGURE 2.39 Capacitive elements in rectangular waveguide.

capacitance. λ_T, the wavelength of the transverse resonant mode, can be expressed by (see Problem P2.20)

$$\lambda_T = \frac{c}{\sqrt{\varepsilon_r}\, f_T} \tag{2.142}$$

where

$$f_T = \frac{c}{\sqrt{\varepsilon_r}\,\lambda_T} = \frac{c}{\sqrt{\varepsilon_r}} \frac{1}{2(w + 0.4h)} = \frac{30 \times 10^9}{2.54 \times 2(w + 0.4h)\sqrt{\varepsilon_r}}$$

$$= \frac{5.9}{\sqrt{\varepsilon_r}\,(w + 0.4h)} \quad \text{GHz} \tag{2.143}$$

where w and h in (2.143) are in inches and f_T is the transverse resonance frequency. Figure 2.37 plots f_T as a function of linewidth for two different substrate thicknesses of GaAs substrates. It can be seen that the maximum linewidth that can be used at 94 GHz for a 4-mil (or 0.004-in.) substrate thickness, without excitation of the transverse resonance mode, is about 16 mils. This linewidth is equivalent to a characteristic impedance of approximately 18 Ω.

Example 2.10 (a) Determine the optimum substrate thickness for a microstrip line with $\varepsilon_r = 2.2$ operating at 60 GHz.

(b) At what frequency will the first surface mode be excited for this thickness?

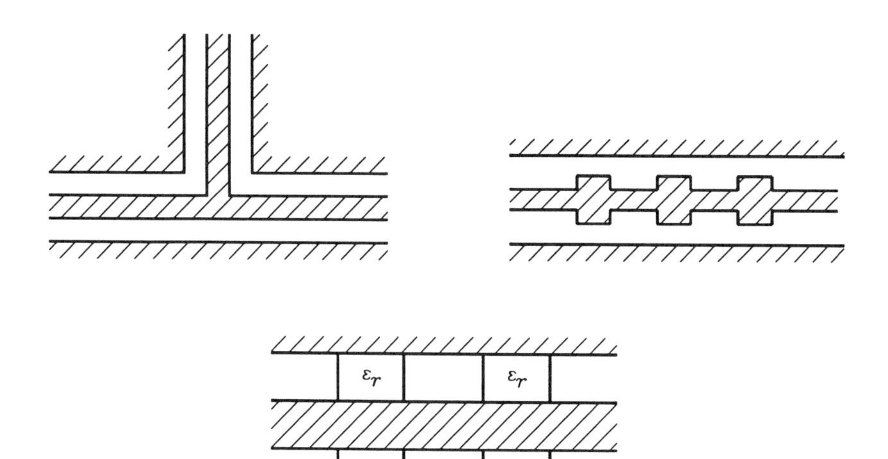

FIGURE 2.40 Coaxial discontinuities.

(c) What is the maximum width that the first transverse resonant mode will be excited?

(d) Calculate Z_0 corresponding to this width.

SOLUTION: (a) From Figure 2.33, h is estimated to be 0.055 mm (or 2.16 mils).

(b) $f_s = \dfrac{c}{2\pi h}\sqrt{\dfrac{2}{\varepsilon_r - 1}}\,\tan^{-1}\varepsilon_r = 1282$ GHz.

(c) $f_T = \dfrac{5.9}{\sqrt{\varepsilon_r}\,(w + 0.4h)} = 60$ GHz, $w = 65.4$ mils.

(d) From Figure 2.28, Z_0 is estimated to be about 8 Ω. This is the minimum impedance one can use at 60 GHz.

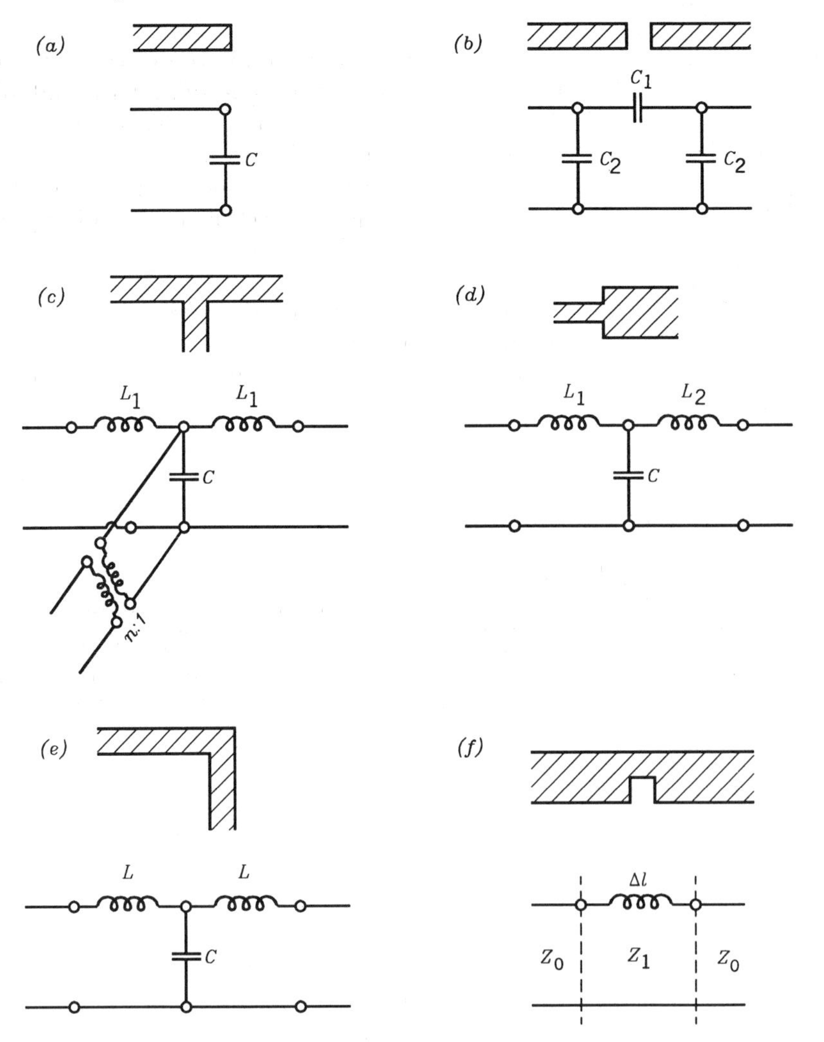

FIGURE 2.41 Microstrip discontinuities: (*a*) open end; (*b*) gap; (*c*) T junction; (*d*) step; (*e*) bend; (*f*) slit.

2.10 TRANSMISSION-LINE AND WAVEGUIDE DISCONTINUITIES

A straight, uninterrupted transmission line is considered continuous. Any bends, shorted or open circuits, gaps, width changes, external loadings, and transitions are discontinuities in the transmission line. Sometimes, discontinuities are undesirable and cause reflection and mismatch. In other cases, discontinuities can be used to accomplish impedance matching, resonance, filtering, and many other functions.

In a rectangular waveguide, inductive susceptances can be obtained by using the discontinuities shown in Figure 2.38 and the capacitive susceptances shown in Figure 2.39. The circular aperture could be an inductive or a capacitive element, depending on the size and frequency. A small aperture is inductive and a big aperture is capacitive. A nontouching post or strip could be inductive or capacitive, depending on the depth and frequency. Normally, a deep strip is more inductive. Analyses of these elements appear in many papers and books. One example is the book by Collin [14].

In coaxial lines, commonly used discontinuities for circuit construction are stubs, changes of center conductor diameter, changes of dielectric filling, and so on. Figure 2.40 shows some of these discontinuities.

Discontinuities in microstrip lines are easily obtained. Some examples shown in Figure 2.41 are open end, gap, stub or T-junction, step width change, bend, and transverse slit. The equivalent circuits of these discontinuities can be represented by L and C elements. Closed-form formulas obtained from curve fitting are available for these discontinuities [9]. For accurate design, numerical methods are required to solve the equivalent circuits.

PROBLEMS

P2.1 A lossless transmission line is connected to a load with $Z_L = 0$. The characteristic impedance of the line is 50 Ω.

(a) Plot $|V(x)|$ as a function of x.

(b) Calculate Z_{in} at $\lambda_g/4$ away from the load.

(c) Calculate Z_{in} at $\lambda_g/2$ away from the load.

(d) What is Γ_L?

(e) What is VSWR?

(f) What are the transmitted power and reflected power as a percentage of the incident power P_{in}?

P2.2 Repeat Problem P2.1 for an open load, $Z_L = \infty$.

P2.3 Repeat Problem P2.1 for a load with $Z_L = 40\ \Omega$.

P2.4 A load $Z_L = 5 + j25\ \Omega$ is connected to a 50-Ω lossless transmission line. Use the Smith chart to determine

(a) the input impedance Z_{in} at a distance $0.1\lambda_g$ from the load,
(b) the VSWR on the line,
(c) the reflection coefficient at the load, and
(d) the reflection coefficient at a distance $0.1\lambda_g$ from the load.

P2.5 Consider the lossless circuit shown in Figure P2.5.
(a) What is the load impedance, $\overline{Z}_L$?
(b) What is the distance from the load to the first voltage minimum?
(c) What length of open-circuit line could be used to replace $\overline{Z}_L$?

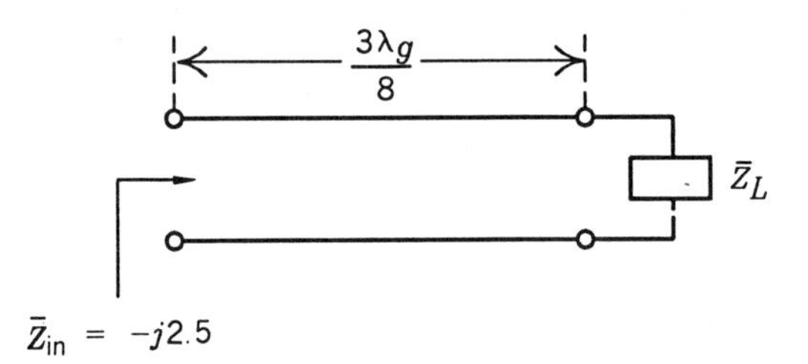

FIGURE P2.5

P2.6 Determine l_1 and l_2 for a single, open stub matching circuit to match a load impedance of $100 + j150\ \Omega$ to a lossless line with a characteristic impedance of 50 Ω (Figure P2.6).

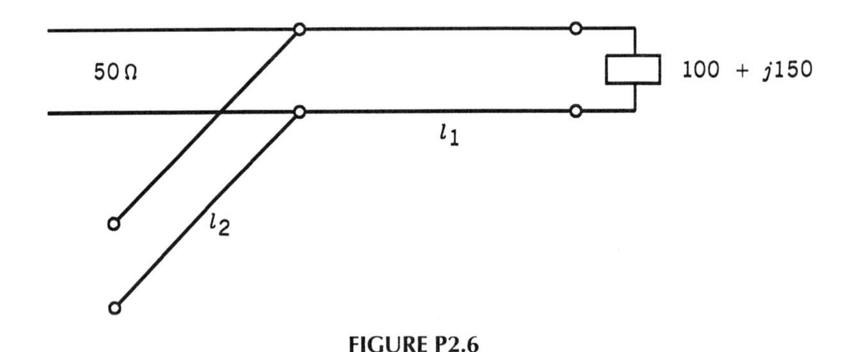

FIGURE P2.6

P2.7 In Example 2.8, design a series C and shunt L circuit to match the load $Z_L = 10 + j10\ \Omega$ to a 50-Ω lossless line at 500 MHz. (Figure P2.7).

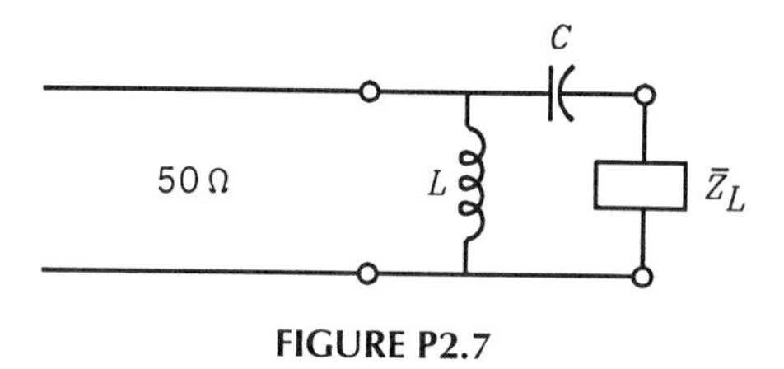

FIGURE P2.7

P2.8 A quarter-wavelength microstrip transformer with a characteristic impedance of Z_{0T} and a section of 50-Ω microstrip line are used to match a load $Z_L(50 + j50\ \Omega)$ to a 50-Ω microstrip line as shown in Figure P2.8. The operating frequency is 1 GHz and $Z_{0T} < 50\ \Omega$. The circuit is built on RT Duroid 6010 ($\varepsilon_r = 10$) substrate with a thickness of $h = 0.030$ in. Find w_1, w_2, l_1, and l_2 in inches using a Smith chart and the graphical method.

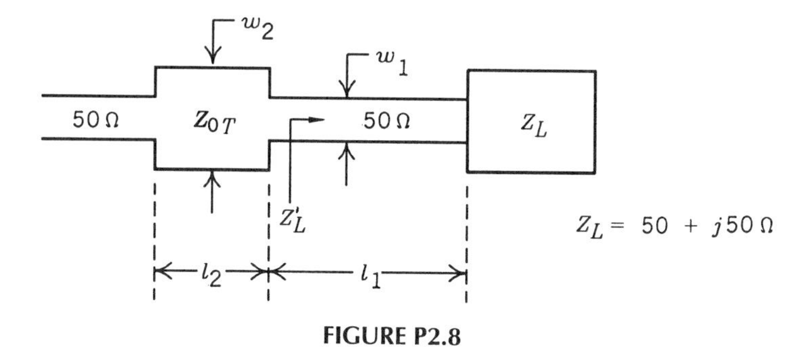

FIGURE P2.8

P2.9 For a circuit shown in Figure P2.9, determine Z_1 in terms of Z_0 to achieve impedance matching.

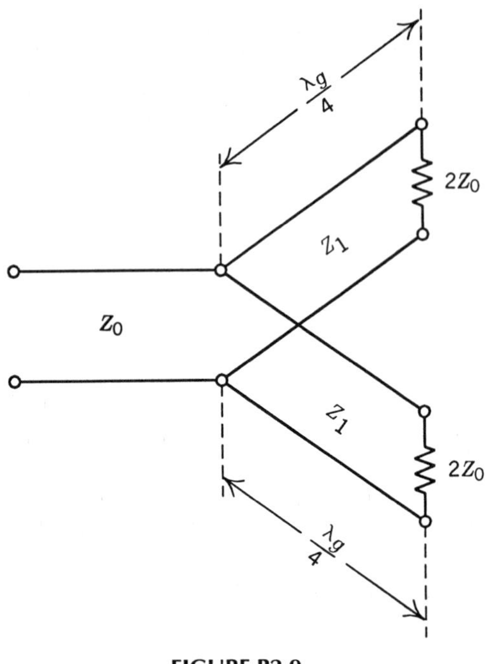

FIGURE P2.9

P2.10 Using a Smith chart, find w, l_1, and l_2 in inches for a shunt open stub to match a load of $Z_L = 100 + j50\ \Omega$. The microstrip line has a characteristic impedance of 50 Ω operating at 10 GHz. The substrate has $\varepsilon_r = 2.3$ and $h = 0.050$ in. The circuit is shown in Figure P2.10. (Select the solution with the smaller l_1 value.)

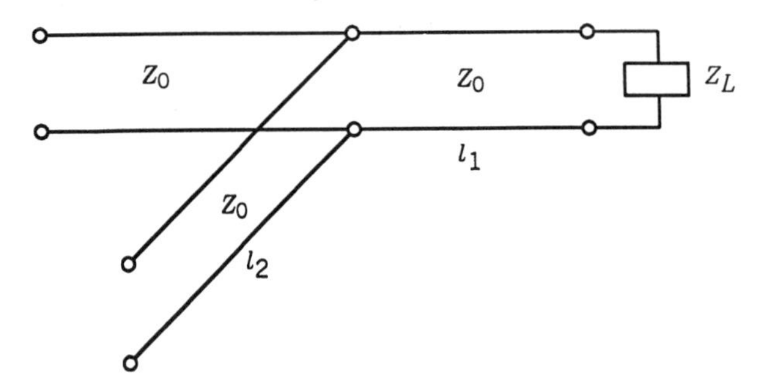

FIGURE P2.10

P2.11 A coaxial line with an inner conductor diameter of 0.5 cm and an outer conductor diameter of 1 cm is filled with a dielectric material of $\mu_r = 1$ and $\varepsilon_r = 10$. At 10 GHz, calculate the characteristic impedance in ohms and the guide wavelength in centimeters.

P2.12 Prove that the VSWR can be found from reading $\overline{R}$ at the intersection of the constant $|\Gamma|$ circle and the real axis.

P2.13 An air-filled rectangular waveguide with inside dimensions of 3.5×7 cm operates in the dominant mode.

(a) Find the cutoff frequency of the dominant mode.

(b) What is the cutoff frequency of the first higher-order mode?

(c) What is the guided wavelength at 3.5 GHz?

P2.14 A rectangular waveguide is operated in its dominant mode at 10 GHz, at this frequency the wavelength in the waveguide is 20% greater than the wavelength in free space. What is the cutoff frequency?

P2.15 A waveguide with dimensions of $a = 0.28$ in., $b = 0.14$ in. is filled with air. Identify which modes can propagate at the following frequencies: **(a)** 20 GHz; **(b)** 30 GHz; **(c)** 40 GHz; **(d)** 50 GHz.

P2.16 Design L and C to match Y_L to a 50-Ω transmission line at 1 GHz (Figure P2.16). $Y_L = (8 - j12) \times 10^{-3}$ S. Express C in pF and L in nH.

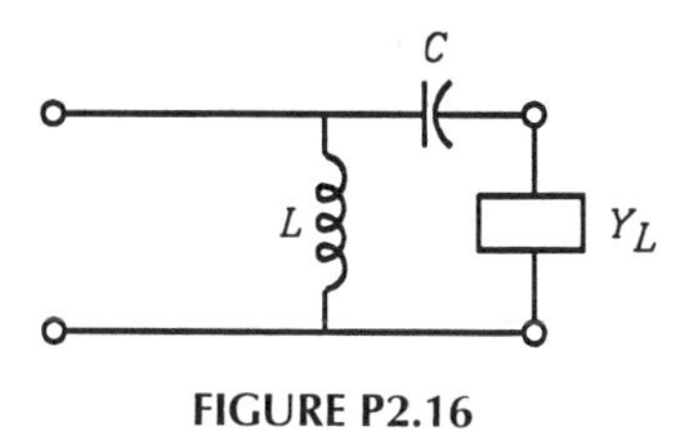

FIGURE P2.16

P2.17 A microstrip line of width equal to 0.090 in. is built on a substrate with a dielectric constant of 2.3 and a thickness of 0.030 in. The operating frequency is 10 GHz. Calculate the characteristic impedance and guide wavelength **(a)** using a graphical method and **(b)** using equations.

P2.18 Repeat Example 2.10 with $\varepsilon_r = 10$.

P2.19 The discontinuity of a microstrip open end can be represented by a fringe capacitor C_f. The equivalent effect can be represented by an extra length l_{e0}. Prove

$$l_{e0} = \frac{1}{\beta} \cot^{-1} \frac{1}{\omega C_f Z_0}$$

where ω is the angular frequency, β the propagation constant, and Z_0 the characteristic impedance.

P2.20 In Equation (2.142), explain why ε_r is used instead of ε_{eff}.

P2.21 Describe the procedure used for selecting the optimum substrate thickness. What are the maximum and minimum strip width that can be used?

REFERENCES

1. R. E. Collin, *Foundations for Microwave Engineering*, 2nd ed., McGraw-Hill, New York, 1992.
2. R. A. Pucel, "Technology and Design Considerations of Monolithic Microwave Integrated Circuits," in *Gallium Arsenide Technology*, D. K. Ferry, Editor, Howard W. Sams, Indianapolis, Ind., 1985, pp. 189–248.
3. E. A. Wolff and R. Kaul, *Microwave Engineering and Systems Applications*, Wiley, New York, 1988.
4. H. A. Wheeler, "Transmission-Line Properties of Parallel Strips Separated by a Dielectric Sheet," *IEEE Transactions on Microwave Theory and Techniques*, Vol. MTT-13, March 1965, pp. 172–185.
5. H. E. Brenner, "Numerical Solution of TEM-Line Problems Involving Inhomogeneous Media," *IEEE Transactions on Microwave Theory and Techniques*,Vol. MTT-15, August 1967, pp. 485–487.
6. H. E. Stinehelfer, "An Accurate Calculation of Uniform Microstrip Transmission Lines," *IEEE Transactions on Electron Devices*, Vol. ED-15, July 1968, pp. 501–506.
7. E. Yamashita and R. Mittra, "Variational Method for the Analysis of Microstrip Lines," *IEEE Transactions on Microwave Theory and Techniques*, Vol. MTT-16, August 1968, pp. 529–535.
8. B. Bhat and S. K. Koul, "Unified Approach to Solve a Class of Strip and Microstrip-like Transmission Lines," *IEEE Transactions on Microwave Theory and Techniques*, Vol. MTT-30, May 1982, pp. 679–686.
9. T. C. Edwards, *Foundations for Microstrip Circuit Design*, 2nd ed., Wiley, New York, 1991, Chaps. 3, 4, and 5.
10. E. Yamashita, K. Atsuki, and T. Ueda, "An Approximate Dispersion Formula of Microstrip Lines for Computer-Aided Design of Microwave Integrated Circuits," *IEEE Transactions on Microwave Theory and Techniques*, Vol. MTT-27, December 1979, pp. 1036–1038.
11. P. Bhartia and I. J. Bahl, *Millimeter Wave Engineering and Applications*, Wiley, New York, 1984, Chap. 6.
12. E. D. Hammerstad and F. Bekkadal, "A Microstrip Handbook," *ELAB Report*, STF 44A74169, University of Trondheim, Norway, 1975.
13. A. Gopinath, "Maximum *Q*-Factor of Microstrip Resonators," *IEEE Transactions on Microwave Theory and Techniques*, Vol. MTT-29, February 1981, pp. 128–131.
14. R. E. Collin, *Field Theory of Guided Waves*, McGraw-Hill, New York, 1st edition 1960; IEEE Press, New York, 2nd edition 1991.
15. G. D. Vendelin, "Limitations on Stripline *Q*," *Microwave Journal*, May 1970, pp. 63–69.

CHAPTER 3

S Parameters and Circuit Representations

3.1 INTRODUCTION

For simple circuit analysis, one can use the Smith chart and transmission-line equations given in Chapter 2. For more complicated cascaded circuits, it is more convenient to represent each circuit element in matrix form (e.g., *ABCD* matrix) and calculate the overall performance by cascading the matrices. Computer software can be used to accomplish the calculation.

Microwave circuits and components can be classified as one-, two-, three-, four-, or N-port networks. Some examples of one-, two-, three-, and four-port networks are shown in Figure 3.1. A majority of circuits are two-port networks.

3.2 CIRCUIT REPRESENTATIONS OF TWO-PORT NETWORKS

A two-port network can be represented by $z, y, h, ABCD$ (transmission) parameters and many others. Figure 3.2 shows a two-port network with the input voltage and current v_1 and i_1, and output voltage and current v_2 and i_2. The z, y, h, and $ABCD$ parameters are defined by the following equations:

$$v_1 = z_{11}i_1 - z_{12}i_2 \tag{3.1a}$$

$$v_2 = z_{21}i_1 - z_{22}i_2 \tag{3.1b}$$

$$[z] = \text{impedance matrix} = \begin{bmatrix} z_{11} & z_{12} \\ z_{21} & z_{22} \end{bmatrix} \tag{3.1c}$$

$$i_1 = y_{11}v_1 + y_{12}v_2 \tag{3.2a}$$

$$-i_2 = y_{21}v_1 + y_{22}v_2 \tag{3.2b}$$

$$[y] = \text{admittance matrix} = \begin{bmatrix} y_{11} & y_{12} \\ y_{21} & y_{22} \end{bmatrix} \tag{3.2c}$$

$$v_1 = h_{11}i_1 + h_{12}v_2 \tag{3.3a}$$

$$-i_2 = h_{21}i_1 + h_{22}v_2 \tag{3.3b}$$

$$[h] = \text{hybrid matrix} = \begin{bmatrix} h_{11} & h_{12} \\ h_{21} & h_{22} \end{bmatrix} \tag{3.3c}$$

$$v_1 = Av_2 + Bi_2 \tag{3.4a}$$

$$i_1 = Cv_2 + Di_2 \tag{3.4b}$$

$$[ABCD] = \text{transmission matrix} = \begin{bmatrix} A & B \\ C & D \end{bmatrix} \tag{3.4c}$$

The z parameters are in ohms relating (v_1, v_2) to (i_1, i_2). The y parameters are in siemens relating (i_1, i_2) to (v_1, v_2). The h parameters relate (v_1, i_2) to (i_1, v_2). h_{11} is in ohms and h_{22} in siemens, but h_{12} and h_{21} are unitless; therefore, they are termed *hybrid parameters*.* The *ABCD* matrix relates input (v_1, i_1) to output (v_2, i_2). Using the matrix, Equations (3.4a) and (3.4b) can be written as

$$\begin{bmatrix} v_1 \\ i_1 \end{bmatrix} = \begin{bmatrix} A & B \\ C & D \end{bmatrix} \begin{bmatrix} v_2 \\ i_2 \end{bmatrix} \tag{3.5}$$

The *ABCD* matrix is very useful for microwave circuit computation because it relates the output to the input. As shown in Figure 3.3, the *ABCD* matrix of a cascaded circuit can be obtained by multiplication of elements working from left to right. The overall matrix is

$$\begin{bmatrix} A_t & B_t \\ C_t & D_t \end{bmatrix} = \begin{bmatrix} A_1 & B_1 \\ C_1 & D_1 \end{bmatrix} \begin{bmatrix} A_2 & B_2 \\ C_2 & D_2 \end{bmatrix} \cdots \begin{bmatrix} A_n & B_n \\ C_n & D_n \end{bmatrix} \tag{3.6}$$

$$\begin{bmatrix} v_1 \\ i_1 \end{bmatrix} = \begin{bmatrix} A_t & B_t \\ C_t & D_t \end{bmatrix} \begin{bmatrix} v_{n+1} \\ i_{n+1} \end{bmatrix} \tag{3.7}$$

*In the definition of z, y, and h parameters, $-i_2$ is used instead of i_2 in Equations (3.1a), (3.1b), (3.2b), and (3.3b) because by convention the current is flowing into port 2 in the z, y, and h definition.

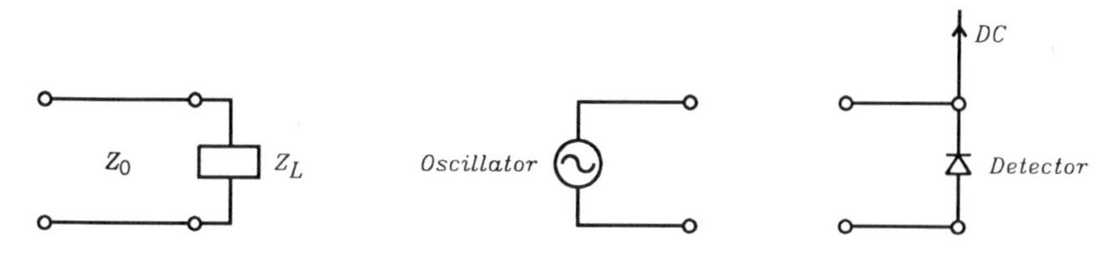

(a) One Port Circuits

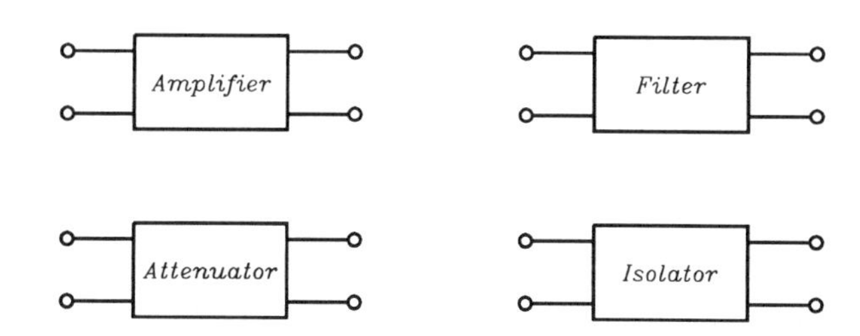

(b) Two Port Circuits

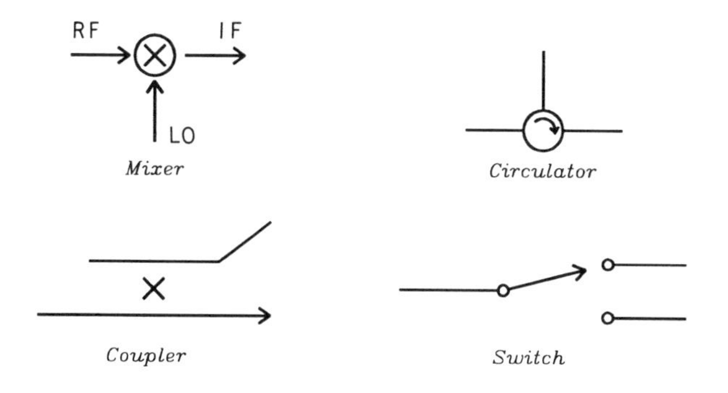

(c) Three and Four Port Circuits

FIGURE 3.1 One-, two-, three-, and four-port networks.

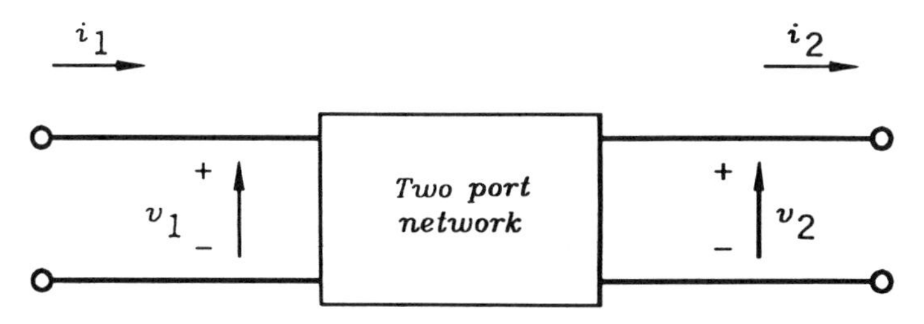

FIGURE 3.2 Two-port network.

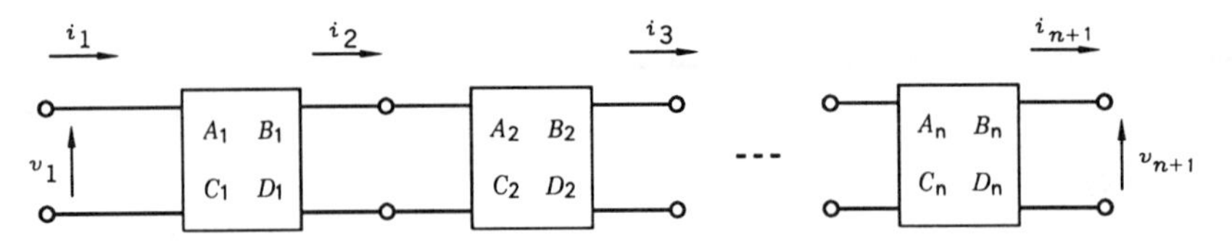

FIGURE 3.3 Cascaded circuits.

Therefore, if we represent each individual element by an *ABCD* matrix and use Equation (3.6), the entire circuit will be represented by one *ABCD* matrix. As will be shown, the *ABCD* matrix can be related to scattering parameters which can be measured directly in microwave frequencies. z, y, h, and *ABCD* parameters, on the other hand, are difficult to measure.

3.3 EXAMPLES OF *ABCD*-MATRIX REPRESENTATION OF ELEMENTS

The *ABCD* matrix has some special properties [1, 2]:

1. If the circuit is symmetric, $A = D$.
2. If the circuit is reciprocal, $AD - BC = 1$.
3. If the circuit is lossless, A and D are real and B and C are imaginary.

Any two-port element can be represented by an *ABCD* matrix. Some examples are given below. Note that the *ABCD* matrices are required to satisfy the properties listed above.

A. Series Impedance

The series impedance shown in Figure 3.4 is governed by the following equations:

$$v_1 = v_2 + Zi_2 = Av_2 + Bi_2 \tag{3.8a}$$

$$i_1 = i_2 = 0 + i_2 = Cv_2 + Di_2 \tag{3.8b}$$

The *ABCD* matrix is

$$\begin{bmatrix} A & B \\ C & D \end{bmatrix} = \begin{bmatrix} 1 & Z \\ 0 & 1 \end{bmatrix} \tag{3.8c}$$

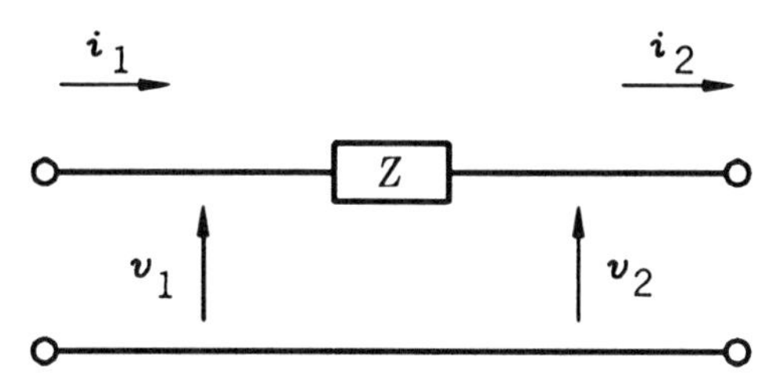

FIGURE 3.4 Series impedance.

B. Shunt Admittance

The *ABCD* matrix for a shunt admittance shown in Figure 3.5 can be found from the following equations:

$$v_1 = v_2 = v_2 + 0 = Av_2 + Bi_2 \tag{3.9a}$$

$$i_1 = Yv_2 + i_2 = Cv_2 + Di_2 \tag{3.9b}$$

The *ABCD* matrix is

$$\begin{bmatrix} A & B \\ C & D \end{bmatrix} = \begin{bmatrix} 1 & 0 \\ Y & 1 \end{bmatrix} \tag{3.9c}$$

C. Series Impedance Cascaded with a Shunt Admittance

The *ABCD* matrix for the cascaded circuit shown in Figure 3.6 can be found from the following equation:

$$\begin{aligned} \begin{bmatrix} A_t & B_t \\ C_t & D_t \end{bmatrix} &= \begin{bmatrix} A_1 & B_1 \\ C_1 & D_1 \end{bmatrix} \begin{bmatrix} A_2 & B_2 \\ C_2 & D_2 \end{bmatrix} \\ &= \begin{bmatrix} 1 & Z \\ 0 & 1 \end{bmatrix} \begin{bmatrix} 1 & 0 \\ Y & 1 \end{bmatrix} = \begin{bmatrix} 1 + ZY & Z \\ Y & 1 \end{bmatrix} \end{aligned} \tag{3.10}$$

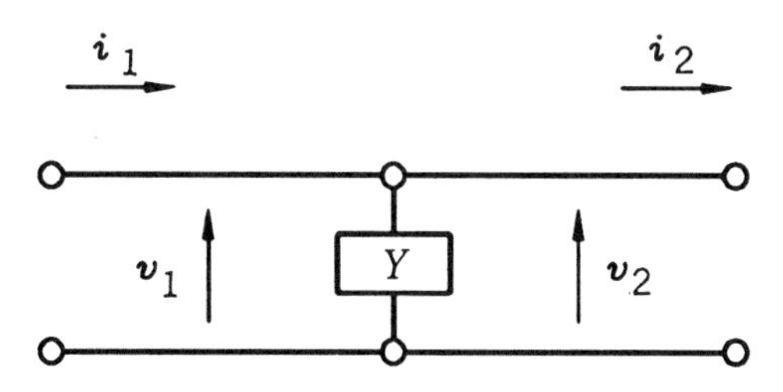

FIGURE 3.5 Shunt admittance.

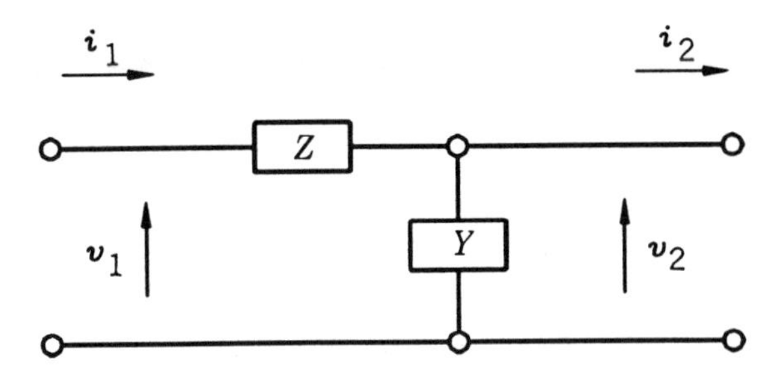

FIGURE 3.6 Series impedance cascaded with shunt admittance.

D. Transformer

The transformer shown in Figure 3.7 can be represented by the following equations:

$$v_1 = nv_2 = Av_2 + Bi_2 \tag{3.11a}$$

$$i_1 = \frac{1}{n} i_2 = Cv_2 + Di_2 \tag{3.11b}$$

The $ABCD$ matrix is

$$\begin{bmatrix} A & B \\ C & D \end{bmatrix} = \begin{bmatrix} n & 0 \\ 0 & \dfrac{1}{n} \end{bmatrix} \tag{3.11c}$$

E. Lossless Transmission Line

For a lossless transmission line as shown in Figure 3.8, we have

$$\begin{aligned} v_1 = v_2 e^{+j\beta l} &= v_2 \cos \beta l + jv_2 \sin \beta l \\ &= (\cos \beta l) v_2 + (jZ_0 \sin \beta l) i_2 \end{aligned} \tag{3.12a}$$

$$\begin{aligned} i_1 = i_2 e^{+j\beta l} &= i_2 \cos \beta l + ji_2 \sin \beta l \\ &= (jY_0 \sin \beta l) v_2 + (\cos \beta l) i_2 \end{aligned} \tag{3.12b}$$

The $ABCD$ matrix is thus given by

$$\begin{bmatrix} A & B \\ C & D \end{bmatrix} = \begin{bmatrix} \cos \beta l & jZ_0 \sin \beta l \\ jY_0 \sin \beta l & \cos \beta l \end{bmatrix} \tag{3.12c}$$

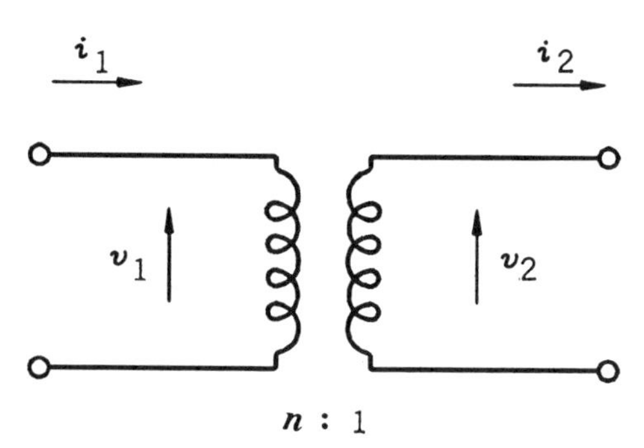

FIGURE 3.7 Transformer circuit.

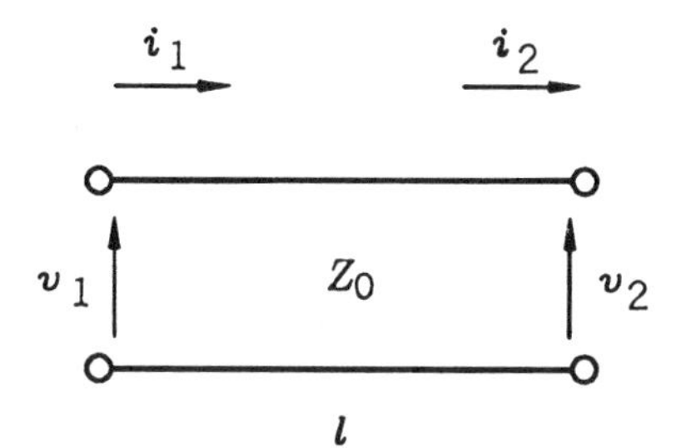

FIGURE 3.8 Lossless transmission line.

It should be noted that the derivation is arranged such that the *ABCD* matrix satisfies the requirements for a symmetric, reciprocal, and lossless circuit.

3.4 SCATTERING PARAMETERS FOR TWO-PORT NETWORKS

Although *ABCD* parameters are useful for circuit computation, they are difficult to measure at microwave frequencies (because the current and voltage are difficult to measure at microwave frequencies). It is thus necessary to define another set of parameters. Scattering parameters can be measured by reflectometers or network analyzers and can be related directly to *ABCD* parameters.

Figure 3.9 shows a two-port network and its S parameters. At the input port (port 1), the incident and reflected wave voltages are a_1 and b_1. At the output port (port 2), they are a_2 and b_2. The S parameters are defined by

$$b_1 = S_{11}a_1 + S_{12}a_2 \tag{3.13a}$$

$$b_2 = S_{21}a_1 + S_{22}a_2 \tag{3.13b}$$

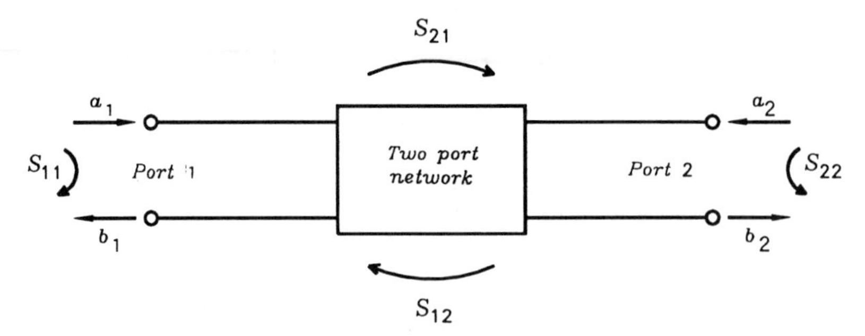

FIGURE 3.9 Two-port network and its S parameters which can be measured by using an HP-8510 network analyzer.

S parameters are complex variables relating the incident wave to the reflected wave. The S parameters are given by

$$S_{11} = \left.\frac{b_1}{a_1}\right|_{a_2=0} = \Gamma_1 = \text{reflection coefficient at port 1 with } a_2 = 0$$

$$S_{21} = \left.\frac{b_2}{a_1}\right|_{a_2=0} = T_{21} = \text{transmission coefficient from port 1 to port 2 with } a_2 = 0$$

$$S_{22} = \left.\frac{b_2}{a_2}\right|_{a_1=0} = \Gamma_2 = \text{reflection coefficient at port 2 with } a_1 = 0$$

$$S_{12} = \left.\frac{b_1}{a_2}\right|_{a_1=0} = T_{12} = \text{transmission coefficient from port 2 to port 1 with } a_1 = 0$$

The attenuation of the circuit is given by

$$\alpha = 20 \log \left|\frac{a_1}{b_2}\right| = 20 \log \left|\frac{1}{S_{21}}\right| \quad \text{dB} \tag{3.14}$$

and the phase shift is

$$\phi = \text{phase of } S_{21} \tag{3.15}$$

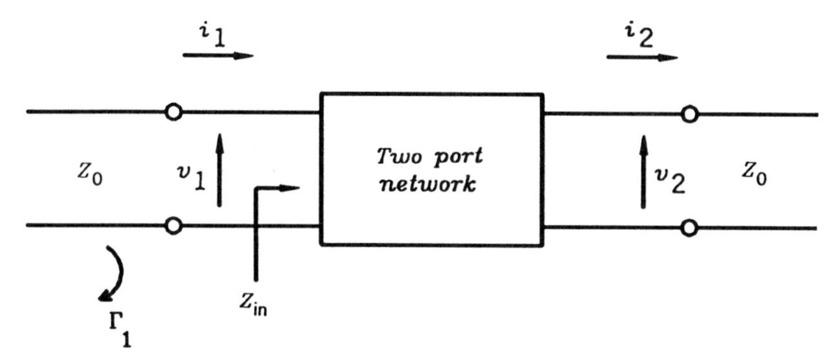

FIGURE 3.10 Two-port network.

3.5 DERIVATION OF [S] FROM [ABCD] AND VICE VERSA

S parameters can be derived from *ABCD* parameters, and vice versa. Considering the circuit shown in Figure 3.10, we have

$$v_1 = Av_2 + Bi_2 \tag{3.16a}$$

$$i_1 = Cv_2 + Di_2 \tag{3.16b}$$

The reflection coefficient Γ_1 is given by

$$\Gamma_1 = \frac{Z_L - Z_0}{Z_L + Z_0} \tag{3.17}$$

Now the load is Z_{in}, and (3.17) becomes

$$\Gamma_1 = \frac{Z_{\text{in}} - Z_0}{Z_{\text{in}} + Z_0} \tag{3.18}$$

From Figure 3.10 we have

$$\frac{v_2}{i_2} = Z_0 \quad \text{and} \quad \frac{v_1}{i_1} = Z_{\text{in}}$$

Therefore, Z_{in} can be written as

$$Z_{\text{in}} = \frac{v_1}{i_1} = \frac{Av_2 + Bi_2}{Cv_2 + Di_2} = \frac{Ai_2Z_0 + Bi_2}{Ci_2Z_0 + Di_2}$$

$$= \frac{AZ_0 + B}{CZ_0 + D} \tag{3.19}$$

TABLE 3.1 Conversions of Parameters

	S	z	y	h	ABCD
S	S_{11} S_{12} S_{21} S_{22}	$S_{11} = \frac{(z'_{11}-1)(z'_{22}+1) - z'_{12}z'_{21}}{\Delta_1}$ $S_{12} = \frac{2z'_{12}}{\Delta_1}$ $S_{21} = \frac{2z'_{21}}{\Delta_1}$ $S_{22} = \frac{(z'_{11}+1)(z'_{22}-1) - z'_{12}z'_{21}}{\Delta_1}$	$S_{11} = \frac{(1-y'_{11})(1+y'_{22}) + y'_{12}y'_{21}}{\Delta_2}$ $S_{12} = \frac{-2y'_{12}}{\Delta_2}$ $S_{21} = \frac{-2y'_{21}}{\Delta_2}$ $S_{22} = \frac{(1+y'_{11})(1-y'_{22}) + y'_{12}y'_{21}}{\Delta_2}$	$S_{11} = \frac{(h'_{11}-1)(h'_{22}+1) - h'_{12}h'_{21}}{\Delta_3}$ $S_{12} = \frac{2h'_{12}}{\Delta_3}$ $S_{21} = \frac{-2h'_{21}}{\Delta_3}$ $S_{22} = \frac{(1+h'_{11})(1-h'_{22}) + h'_{12}h'_{21}}{\Delta_3}$	$\frac{A'+B'-C'-D'}{\Delta_4}$ $\frac{2(A'D'-B'C')}{\Delta_4}$ $\frac{2}{\Delta_4}$ $\frac{-A'+B'-C'+D'}{\Delta_4}$
z	$z'_{11} = \frac{(1+S_{11})(1-S_{22}) + S_{12}S_{21}}{\Delta_5}$ $z'_{12} = \frac{2S_{12}}{\Delta_5}$ $z'_{21} = \frac{2S_{21}}{\Delta_5}$ $z'_{22} = \frac{(1-S_{11})(1+S_{22}) + S_{12}S_{21}}{\Delta_5}$	z_{11} z_{12} z_{21} z_{22}	$\frac{y_{22}}{\lvert y\rvert}$ $\frac{-y_{12}}{\lvert y\rvert}$ $\frac{-y_{21}}{\lvert y\rvert}$ $\frac{y_{11}}{\lvert y\rvert}$	$\frac{\lvert h\rvert}{h_{22}}$ $\frac{h_{12}}{h_{22}}$ $\frac{-h_{21}}{h_{22}}$ $\frac{1}{h_{22}}$	$\frac{A}{C}$ $\frac{\Delta_8}{C}$ $\frac{1}{C}$ $\frac{D}{C}$
y	$y'_{11} = \frac{(1-S_{11})(1+S_{22}) + S_{12}S_{21}}{\Delta_6}$ $y'_{12} = \frac{-2S_{12}}{\Delta_6}$ $y'_{21} = \frac{-2S_{21}}{\Delta_6}$ $y'_{22} = \frac{(1+S_{11})(1-S_{22}) + S_{12}S_{21}}{\Delta_6}$	$\frac{z_{22}}{\lvert z\rvert}$ $\frac{-z_{12}}{\lvert z\rvert}$ $\frac{-z_{21}}{\lvert z\rvert}$ $\frac{z_{11}}{\lvert z\rvert}$	y_{11} y_{12} y_{21} y_{22}	$\frac{1}{h_{11}}$ $\frac{-h_{12}}{h_{11}}$ $\frac{h_{21}}{h_{11}}$ $\frac{\lvert h\rvert}{h_{11}}$	$\frac{D}{B}$ $\frac{-\Delta_8}{B}$ $\frac{-1}{B}$ $\frac{A}{B}$

	S	z	y	h	ABCD
h	$h'_{11} = \dfrac{(1+S_{11})(1+S_{22}) - S_{12}S_{21}}{\Delta_7}$ $h'_{12} = \dfrac{2S_{12}}{\Delta_7}$ $h'_{21} = \dfrac{-2S_{21}}{\Delta_7}$ $h'_{22} = \dfrac{(1-S_{22})(1-S_{11}) - S_{12}S_{21}}{\Delta_7}$	$\dfrac{\|z\|}{z_{22}} \quad \dfrac{z_{12}}{z_{22}}$ $\dfrac{-z_{21}}{z_{22}} \quad \dfrac{1}{z_{22}}$	$\dfrac{1}{y_{11}} \quad \dfrac{-y_{12}}{y_{11}}$ $\dfrac{y_{21}}{y_{11}} \quad \dfrac{\|y\|}{y_{11}}$	$h_{11} \quad h_{12}$ $h_{21} \quad h_{22}$	$\dfrac{B}{D} \quad \dfrac{\Delta_8}{D}$ $\dfrac{-1}{D} \quad \dfrac{C}{D}$
$ABCD$	$A' = \dfrac{(1+S_{11})(1-S_{22}) + S_{12}S_{21}}{2S_{21}}$ $B' = \dfrac{(1+S_{11})(1+S_{22}) - S_{12}S_{21}}{2S_{21}}$ $C' = \dfrac{(1-S_{11})(1-S_{22}) - S_{12}S_{21}}{2S_{21}}$ $D' = \dfrac{(1-S_{11})(1+S_{22}) + S_{12}S_{21}}{2S_{21}}$	$\dfrac{z_{11}}{z_{21}} \quad \dfrac{\|z\|}{z_{21}}$ $\dfrac{1}{z_{21}} \quad \dfrac{z_{22}}{z_{21}}$	$\dfrac{-y_{22}}{y_{21}} \quad \dfrac{-1}{y_{21}}$ $\dfrac{-\|y\|}{y_{21}} \quad \dfrac{-y_{11}}{y_{21}}$	$\dfrac{-\|h\|}{h_{21}} \quad \dfrac{-h_{11}}{h_{21}}$ $\dfrac{-h_{22}}{h_{21}} \quad \dfrac{-1}{h_{21}}$	$A \quad B$ $C \quad D$

$\Delta_1 = (z'_{11} + 1)(z'_{22} + 1) - z'_{12}z'_{21}$ $\qquad z'_{11} = z_{11}/Z_0,\ z'_{12} = z_{12}/Z_0,\ z'_{21} = z_{21}/Z_0,\ z'_{22} = z_{22}/Z_0$

$\Delta_2 = (1 + y'_{11})(1 + y'_{22}) - y'_{12}y'_{21}$ $\qquad y'_{11} = y_{11}Z_0,\ y'_{12} = y_{12}Z_0,\ y'_{21} = y_{21}Z_0,\ y'_{22} = y_{22}Z_0$

$\Delta_3 = (h'_{11} + 1)(h'_{22} + 1) - h'_{12}h'_{21}$ $\qquad h'_{11} = h_{11}/Z_0,\ h'_{12} = h_{12},\ h'_{21} = h_{21},\ h'_{22} = h_{22}Z_0$

$\Delta_4 = A' + B' + C' + D'$ $\qquad A' = A,\ B' = B/Z_0,\ C' = CZ_0,\ D' = D$

$\Delta_5 = (1 - S_{11})(1 - S_{22}) - S_{12}S_{21}$ $\qquad |z| = z_{11}z_{22} - z_{12}z_{21}$

$\Delta_6 = (1 + S_{11})(1 + S_{22}) - S_{12}S_{21}$ $\qquad |y| = y_{11}y_{22} - y_{12}y_{21}$

$\Delta_7 = (1 - S_{11})(1 + S_{22}) + S_{12}S_{21}$ $\qquad |h| = h_{11}h_{22} - h_{12}h_{21}$

$\Delta_8 = AD - BC$

Source: Ref. 3. Permission from Prentice-Hall

Substituting (3.19) into (3.18) results in

$$\Gamma_1 = S_{11} = \frac{AZ_0 + B - Z_0(CZ_0 + D)}{AZ_0 + B + Z_0(CZ_0 + D)}$$

$$= \frac{A + BY_0 - CZ_0 - D}{A + BY_0 + CZ_0 + D} = \frac{A + BY_0 - CZ_0 - D}{\Delta} \tag{3.20}$$

where

$$\Delta = A + BY_0 + CZ_0 + D \tag{3.21}$$

Similar derivations apply to S_{22}, S_{12}, and S_{21}. The S matrix can be obtained from $ABCD$ parameters by the following equation:

$$[S] = \begin{bmatrix} S_{11} & S_{12} \\ S_{21} & S_{22} \end{bmatrix} = \begin{bmatrix} \dfrac{A + BY_0 - CZ_0 - D}{\Delta} & \dfrac{2(AD - BC)}{\Delta} \\ \dfrac{2}{\Delta} & \dfrac{-A + BY_0 - CZ_0 + D}{\Delta} \end{bmatrix} \tag{3.22}$$

To convert from S parameters to $ABCD$ parameters, we have

$$\begin{bmatrix} A & B \\ C & D \end{bmatrix} = \begin{bmatrix} \dfrac{(1 + S_{11})(1 - S_{22}) + S_{12}S_{21}}{2S_{21}} & Z_0 \dfrac{(1 + S_{11})(1 + S_{22}) - S_{12}S_{21}}{2S_{21}} \\ \dfrac{1}{Z_0} \dfrac{(1 - S_{11})(1 - S_{22}) - S_{12}S_{21}}{2S_{21}} & \dfrac{(1 - S_{11})(1 + S_{22}) + S_{12}S_{21}}{2S_{21}} \end{bmatrix} \tag{3.23}$$

The conversion relations between the S, z, y, h, and $ABCD$ parameters are given in Table 3.1.

3.6 EXAMPLES OF CIRCUIT REPRESENTATION USING S PARAMETERS

The S-parameter circuit representation of an element can be obtained from the conversion of $ABCD$ parameters. Some examples are given below.

A. Series Impedance

For the series impedance shown in Figure 3.11(a), the $ABCD$ matrix is given by Equation (3.8c) and repeated here.

$$\begin{bmatrix} A & B \\ C & D \end{bmatrix} = \begin{bmatrix} 1 & Z \\ 0 & 1 \end{bmatrix}$$

$$A = 1 \qquad B = Z \qquad C = 0 \qquad D = 1$$

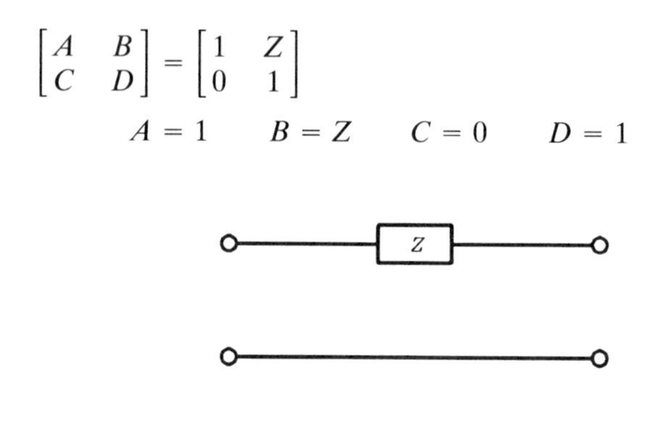

(a) Series impedance

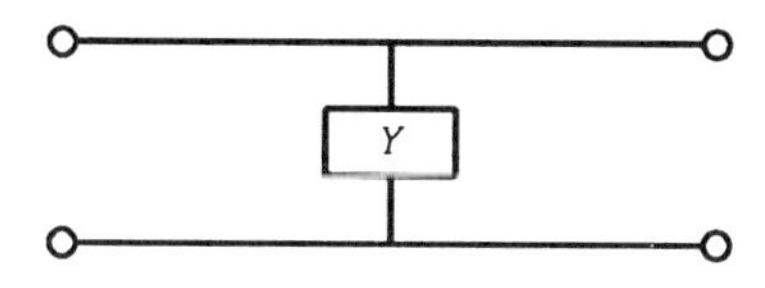

(b) Shunt admittance

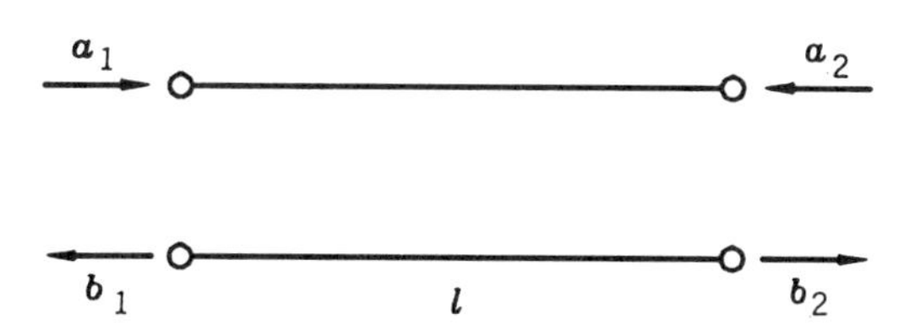

(c) Lossless transmission line

FIGURE 3.11 Some microwave circuit elements.

$$\Delta = A + BY_0 + CZ_0 + D = 2 + Y_0 Z$$

$$S_{11} = \frac{A + BY_0 - CZ_0 - D}{\Delta} = \frac{ZY_0}{2 + Y_0 Z} = \frac{Z}{Z + 2Z_0}$$

$$S_{12} = \frac{2(AD - BC)}{\Delta} = \frac{2}{2 + Y_0 Z} = \frac{2Z_0}{Z + 2Z_0}$$

$$S_{21} = \frac{2}{\Delta} = \frac{2}{2 + Y_0 Z} = \frac{2Z_0}{Z + 2Z_0}$$

$$S_{22} = \frac{-A + BY_0 - CZ_0 + D}{\Delta} = \frac{ZY_0}{2 + Y_0 Z} = \frac{Z}{Z + 2Z_0}$$

Therefore, the S matrix is

$$[S] = \begin{bmatrix} \dfrac{Z}{Z + 2Z_0} & \dfrac{2Z_0}{Z + 2Z_0} \\ \dfrac{2Z_0}{Z + 2Z_0} & \dfrac{Z}{Z + 2Z_0} \end{bmatrix} \tag{3.24}$$

B. Shunt Admittance

For the shunt admittance shown in Figure 3.11(b), the $ABCD$ matrix is given by [see Equation (3.9c)]

$$\begin{bmatrix} A & B \\ C & D \end{bmatrix} = \begin{bmatrix} 1 & 0 \\ Y & 1 \end{bmatrix}$$

$$A = 1 \qquad B = 0 \qquad C = Y \qquad D = 1$$

$$\Delta = A + BY_0 + CZ_0 + D = 2 + YZ_0$$

$$S_{11} = \frac{A + BY_0 - CZ_0 - D}{\Delta} = \frac{-YZ_0}{2 + YZ_0}$$

$$S_{12} = \frac{2(AD - BC)}{\Delta} = \frac{2}{2 + YZ_0}$$

$$S_{21} = \frac{2}{\Delta} = \frac{2}{2 + YZ_0}$$

$$S_{22} = \frac{-A + BY_0 - CZ_0 + D}{\Delta} = \frac{-YZ_0}{2 + YZ_0}$$

Therefore, the S matrix is

$$[S] = \begin{bmatrix} \dfrac{-YZ_0}{2 + YZ_0} & \dfrac{2}{2 + YZ_0} \\ \dfrac{2}{2 + YZ_0} & \dfrac{-YZ_0}{2 + YZ_0} \end{bmatrix} \tag{3.25}$$

C. Lossless Transmission-Line Section

For the lossless transmission line shown in Figure 3.11(c), the S matrix can be obtained from the $ABCD$ matrix using the conversion table. It can also be easily obtained by the following:

$$b_2 = a_1 e^{-j\beta l} \tag{3.26}$$

$$b_1 = a_2 e^{-j\beta l} \tag{3.27}$$

Thus we have

$$\begin{bmatrix} b_1 \\ b_2 \end{bmatrix} = \begin{bmatrix} 0 & e^{-j\beta l} \\ e^{-j\beta l} & 0 \end{bmatrix} \begin{bmatrix} a_1 \\ a_2 \end{bmatrix} \tag{3.28}$$

The S matrix is given by

$$[S] = \begin{bmatrix} 0 & e^{-j\beta l} \\ e^{-j\beta l} & 0 \end{bmatrix} \tag{3.29}$$

3.7 ATTENUATION AND PHASE SHIFT FOR CIRCUIT ELEMENTS

From S_{21} for a circuit element, one can obtain the phase shift and attenuation due to this element. The attenuation is derived from the magnitude of S_{21} and the phase shift from the phase of S_{21}. If $S_{21} = |S_{21}|e^{j\phi}$, we have

$$\text{attenuation (in dB)} = \alpha = 20 \log \left| \frac{1}{S_{21}} \right| \tag{3.30}$$

$$\text{phase shift (in degrees)} = \phi \tag{3.31}$$

A. Attenuation and Phase Shift due to a Series Impedance

The series impedance ($Z = R + jX$) shown in Figure 3.12(a) is connected to a transmission line with a characteristic impedance of Z_0. From Equation

(3.24), S_{21} is

$$S_{21} = \frac{2Z_0}{Z + 2Z_0} = \frac{2}{2 + ZY_0}$$

$$\frac{1}{S_{21}} = 1 + \frac{Z}{2Z_0}$$

The attenuation in dB is

$$\begin{aligned} \alpha &= 20 \log \left| \frac{1}{S_{21}} \right| \\ &= 20 \log \left| 1 + \frac{R + jX}{2Z_0} \right| \\ &= 20 \log \left[\left(1 + \frac{R}{2Z_0} \right)^2 + \left(\frac{X}{2Z_0} \right)^2 \right]^{1/2} \\ &= 10 \log \left[\left(1 + \frac{R}{2Z_0} \right)^2 + \left(\frac{X}{2Z_0} \right)^2 \right] \end{aligned} \tag{3.32}$$

To calculate the phase shift, we have

$$\begin{aligned} S_{21} &= |S_{21}| e^{j\phi} \\ &= \frac{2Z_0}{(R + 2Z_0) + jX} \end{aligned} \tag{3.33}$$

$$\phi = \tan^{-1} \frac{-X}{R + 2Z_0} \tag{3.34}$$

If Z is replaced by a diode, the impedance can be changed by varying the bias to the diode. One can obtain variable attenuation and phase shift by varying the bias to the diode. This arrangement can be used as a variable attenuator or phase shifter (see Chapter 8).

B. Attenuation and Phase Shift due to a Shunt Admittance

The attenuation and phase shift due to a shunt admittance can be obtained from S_{21} similarly. A shunt admittance with $Y = G + jB$ is connected to a transmission line with a characteristic admittance of Y_0 as shown in Figure

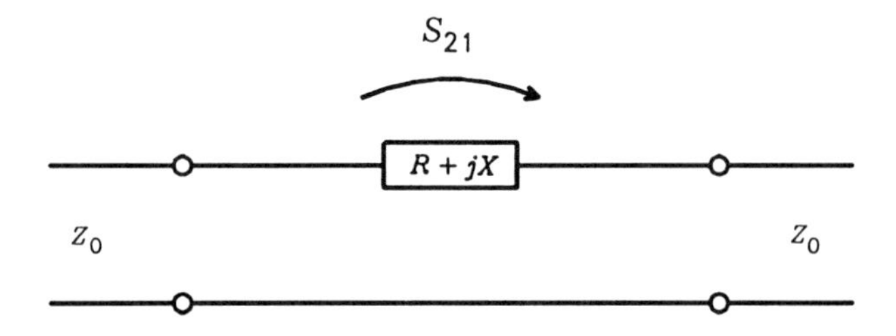

(*a*) Series impedance

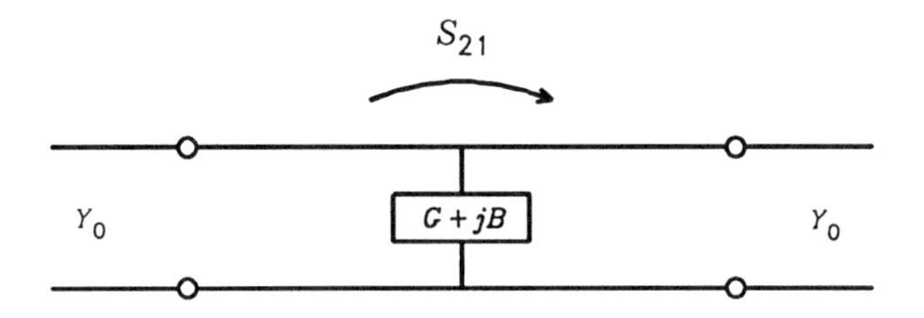

(*b*) Shunt admittance

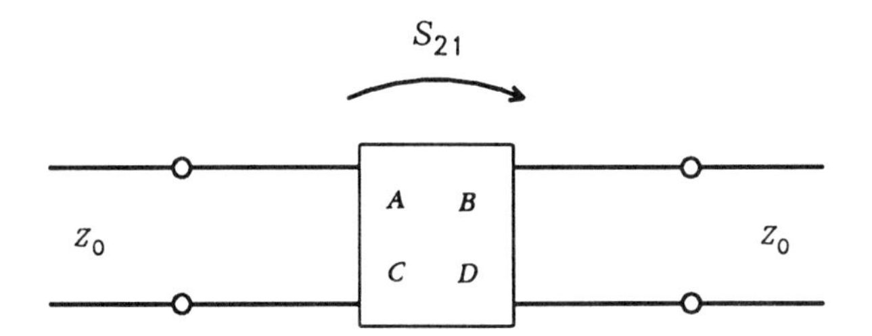

(*c*) General 2 – port network

FIGURE 3.12 Attenuation and phase shift due to circuit elements.

3.12(*b*). From Equation (3.25), S_{21} is given by

$$S_{21} = \frac{2}{2 + YZ_0} = \frac{2Y_0}{2Y_0 + Y}$$

The attenuation (in dB) is

$$\begin{aligned}
\alpha &= 20 \log \left| \frac{1}{S_{21}} \right| \\
&= 20 \log \left| \frac{2Y_0 + Y}{2Y_0} \right| \\
&= 20 \log \left| \left(1 + \frac{G}{2Y_0} \right) + j \frac{B}{2Y_0} \right| \\
&= 20 \log \left[\left(1 + \frac{G}{2Y_0} \right)^2 + \left(\frac{B}{2Y_0} \right)^2 \right]^{1/2} \\
&= 10 \log \left[\left(1 + \frac{G}{2Y_0} \right)^2 + \left(\frac{B}{2Y_0} \right)^2 \right] \qquad (3.35)
\end{aligned}$$

The phase shift is calculated from the phase of S_{21}.

$$S_{21} = |S_{21}| e^{j\phi} = \frac{2Y_0}{2Y_0 + G + jB}$$

$$\phi = \tan^{-1} \frac{-B}{G + 2Y_0} \qquad (3.36)$$

C. Attenuation and Phase Shift due to a General Element

For the general circuit element shown in Figure 3.12(*c*), the attenuation and phase shift can be expressed from the *ABCD* parameters of this element. From Equation (3.22), S_{21} can be expressed by *ABCD* parameters as

$$S_{21} = \frac{2}{A + BY_0 + CZ_0 + D} \qquad (3.37)$$

Note that A, B, C, and D could be complex. We could rearrange (3.37) such that

$$S_{21} = \frac{2}{\Sigma \text{ Re} + j\Sigma \text{ Im}} = \frac{2(\Sigma \text{ Re} - j\Sigma \text{ Im})}{(\Sigma \text{ Re})^2 + (\Sigma \text{ Im})^2} \tag{3.38}$$

Here Σ Re and Σ Im represent the summation of the real and imaginary parts of A, BY_0, CZ_0, and D, respectively. The attenuation and phase shift are given by

$$\alpha = \text{attenuation (in dB)} = 20 \log \left| \frac{1}{S_{21}} \right|$$

$$= 20 \log \frac{\left[(\Sigma \text{ Re})^2 + (\Sigma \text{ Im})^2 \right]^{1/2}}{2} \tag{3.39}$$

$$\text{phase shift } \phi = \tan^{-1} \left(- \frac{\Sigma \text{ Im}}{\Sigma \text{ Re}} \right) \tag{3.40}$$

3.8 S PARAMETERS FOR MULTIPLE-PORT CIRCUITS

The two-port S parameters can be extended to an N-port network as shown in Figure 3.13. In this case, the S matrix is given by the following equation:

$$\begin{bmatrix} b_1 \\ b_2 \\ b_3 \\ \vdots \\ b_N \end{bmatrix} = \begin{bmatrix} S_{11} & S_{12} & S_{13} & \cdots & S_{1N} \\ S_{21} & S_{22} & S_{23} & \cdots & S_{2N} \\ \vdots & \vdots & \vdots & & \vdots \\ S_{N1} & S_{N2} & S_{N3} & \cdots & S_{NN} \end{bmatrix} \begin{bmatrix} a_1 \\ a_2 \\ a_3 \\ \vdots \\ a_N \end{bmatrix} \tag{3.41}$$

or

$$[b] = [S][a] \tag{3.42}$$

The definition of each parameter is similar to that for a two-port network.

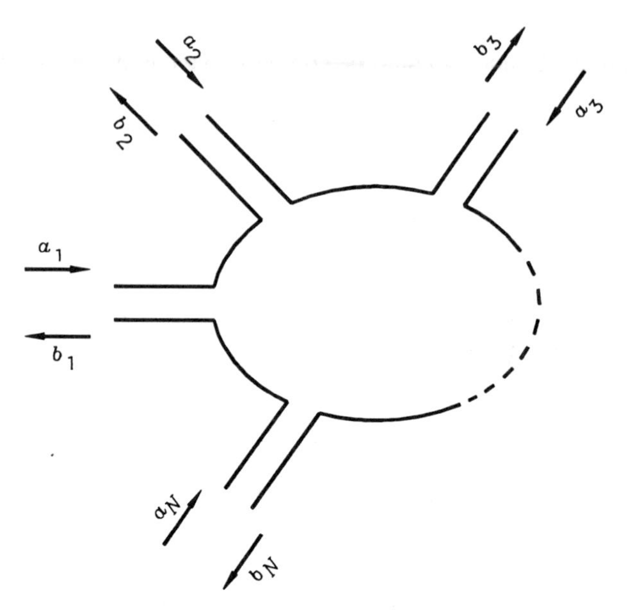

FIGURE 3.13 N-port network.

For example,

$$S_{11} = \left.\frac{b_1}{a_1}\right|_{a_2=0,\, a_3=0,\, \cdots} = \Gamma_{11} = \text{reflection coefficient at port 1 when all other ports are matched}$$

$$S_{12} = \left.\frac{b_1}{a_2}\right|_{a_1=0,\, a_3=0,\, \cdots} = T_{12} = \text{transmission coefficient from port 2 to port 1 when all other ports are matched}$$

The S parameters have some interesting properties:

1. For any matched port i, $S_{ii} = 0$.
2. For a reciprocal network, $S_{nm} = S_{mn}$.
3. For a passive circuit, $|S_{mn}| \leq 1$.
4. For a lossless and reciprocal network, one has for the ith port,

$$\sum_{n=1}^{N} |S_{ni}|^2 = \sum_{n=1}^{N} S_{ni} S_{ni}^* = 1 \tag{3.43}$$

or

$$|S_{1i}|^2 + |S_{2i}|^2 + |S_{3i}|^2 + \cdots + |S_{ii}|^2 + \cdots + |S_{Ni}|^2 = 1 \tag{3.44}$$

Equation (3.43) states that the inner product of any column of the S matrix with the conjugate of itself is equal to 1. Equation (3.43) is due to the power conservation of a lossless network. In Equation (3.43), the total power incident at the ith port is normalized and becomes 1, which equals the power reflected at the ith port plus the power transmitted into all other ports.

PROBLEMS

P3.1 Derive the S parameters for the transformer shown in Figure P3.1.

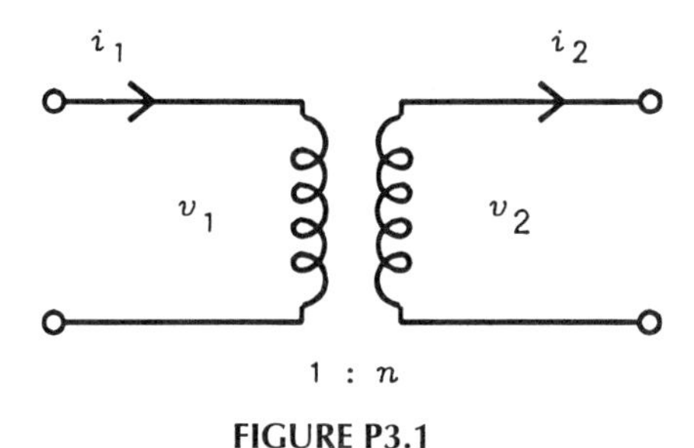

FIGURE P3.1

P3.2 Derive the $ABCD$ matrix and calculate the input VSWR for the circuit shown in Figure P3.2. The operating frequency is 1 GHz. The line is connected to a matched load.

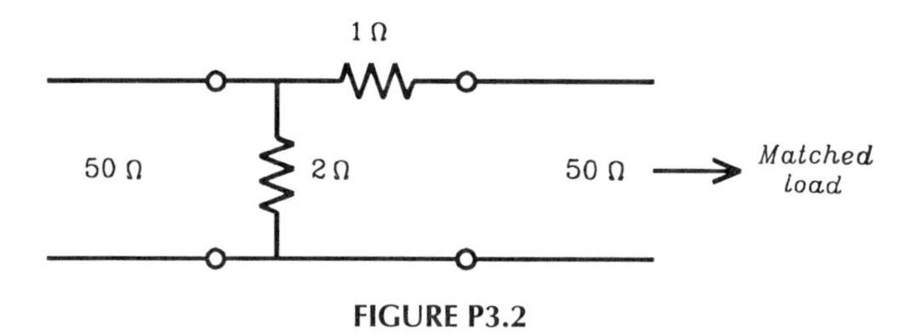

FIGURE P3.2

P3.3 A discontinuity with admittance of $G + jB$ is mounted in shunt with a 50-Ω transmission line (see Figure P3.3). From the S_{21} measurements, we have $S_{21} = 0.5e^{j180°}$ at 10 GHz. Find G and B. (Note that one can calculate G and B from the S_{21} measurements using this method.)

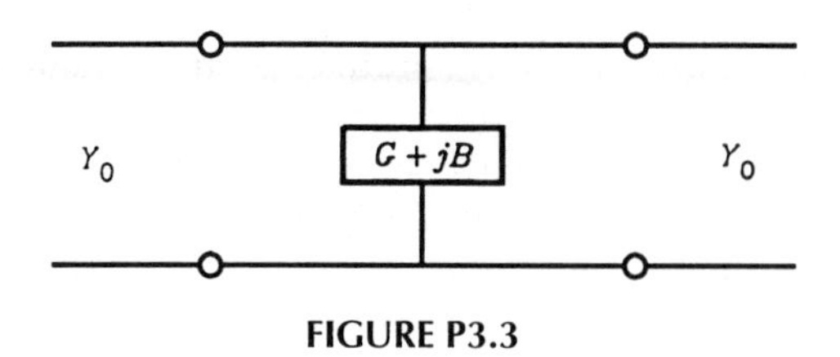

FIGURE P3.3

P3.4 Derive the S and $ABCD$ parameters for the circuit shown in Figure P3.4. $Z_L = 100\ \Omega$, $Z_0 = 50\ \Omega$, and $l = \frac{1}{4}\lambda_g$.

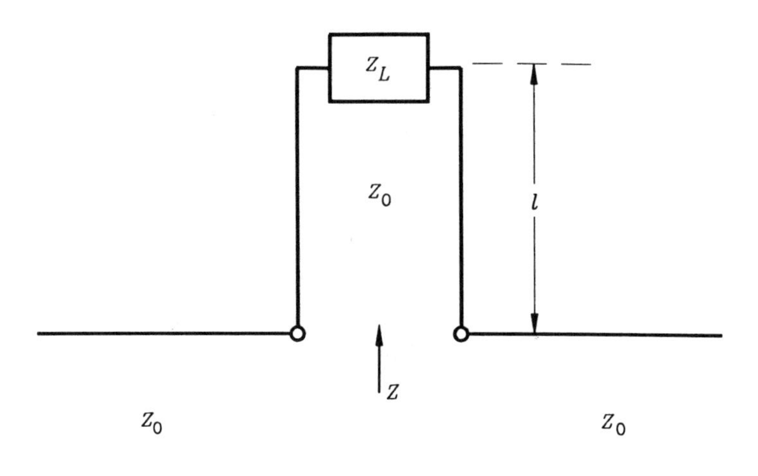

FIGURE P3.4

P3.5 Prove that the $ABCD$ parameters for the π network shown in Figure P3.5 are $A = 1 + Y_2/Y_3$, $B = 1/Y_3$, $C = Y_1 + Y_2 + Y_1Y_2/Y_3$, and $D = 1 + Y_1/Y_3$.

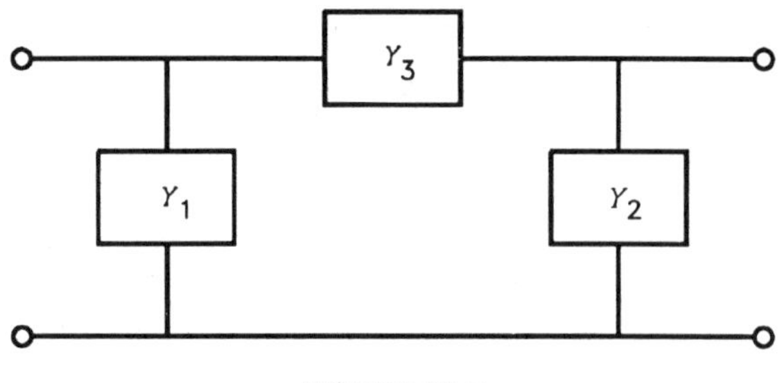

FIGURE P3.5

P3.6 Prove that the *ABCD* parameters for a T network shown in Figure P3.6 are $A = 1 + Z_1/Z_3$, $B = Z_1 + Z_2 + Z_1Z_2/Z_3$, and $D = 1 + Z_2/Z_3$.

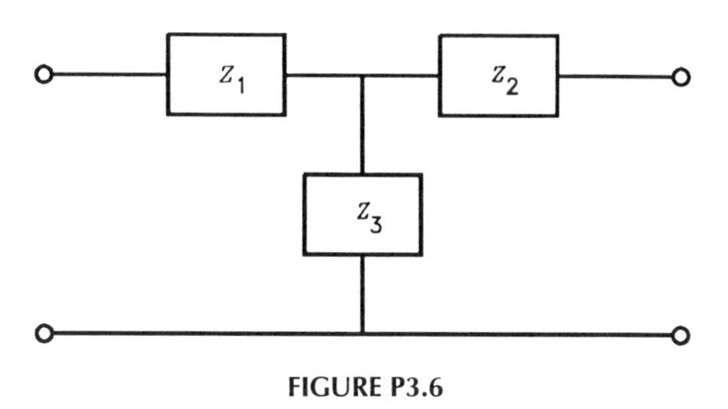

FIGURE P3.6

P3.7 For a lossless transmission line with length l and characteristic impedance Z_0, derive the *ABCD* matrix from the *S* matrix.

P3.8 Derive Equations (3.6) and (3.7).

P3.9 Find the *ABCD* matrix for the circuit shown in Figure P3.9 with $Z_0 = 50\ \Omega$. What are the VSWR and attenuation for this circuit?

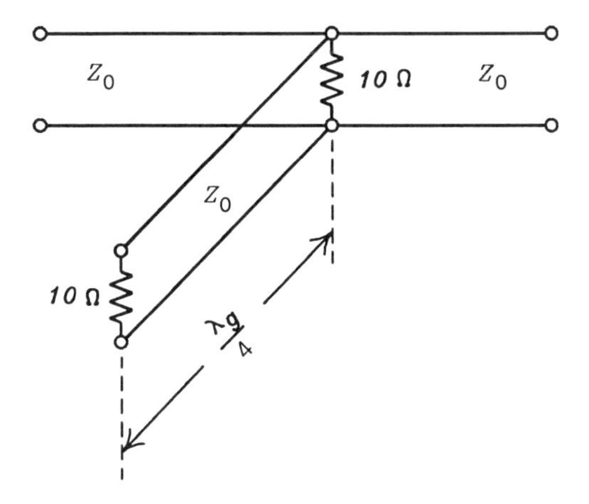

FIGURE P3.9

P3.10 Find the VSWR and attenuation in dB for the circuit shown in Figure P3.10. If $P_{in} = 10$ mW, what are the reflected power, transmitted power, and dissipated power?

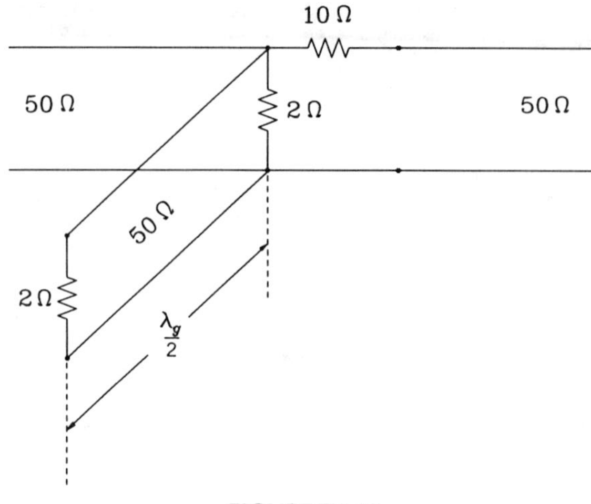

FIGURE P3.10

P3.11 Prove the conversion from Z parameters to $ABCD$ parameters as given in Table 3.1.

REFERENCES

1. R. E. Collin, *Foundations for Microwave Engineering*, 2nd ed., McGraw-Hill, New York, 1992, Chap. 4.
2. E. A. Wolff and R. Kaul, *Microwave Engineering and Systems Applications*, Wiley-Interscience, New York, 1988, Chap. 6.
3. G. Gonzalez, *Microwave Transistor Amplifiers: Analysis and Design*, Prentice Hall, Englewood Cliffs, N.J., 1984, Chap. 1.

CHAPTER 4

Review of Semiconductor Physics

4.1 INTRODUCTION

The properties of microwave solid-state devices depend on the physics of the semiconductor. Most microwave semiconductor devices are fabricated on a GaAs substrate because of its high mobility. A silicon substrate, on the other hand, has the advantages of low cost and high yield. For some special cases, InP substrates are used for high-frequency or high-efficiency applications due to its high peak-to-valley ratio and fast energy-transfer-time constant. Table 4.1 summarizes the various microwave solid-state devices and their applications. The basic configurations of these devices are given in Figure 4.1. In this chapter an overview of the properties and physics of semiconductors is given. These materials will serve as the basis for understanding the various devices.

4.2 INSULATORS, METALS, SEMICONDUCTORS, AND ENERGY BANDS

Most metals and semiconductors are crystalline structures. The energy–momentum (E–k) relationship of crystals can be obtained by solving the Schrödinger equation. The solution is periodic with the periodicity of the lattice. The potential is thus a periodic function in space, forming different energy bands. A band is formed because the energy states are closely spaced.

A material can be classified as an insulator, a metal, or a semiconductor, depending on its energy-band structure. An insulator is a very poor conductor of electricity. Figure 4.2(a) shows a typical energy-band structure of an insulator. The large forbidden band (e.g., 6 eV for diamond) separating the filled valence band and the empty conduction band prevents electrons from

TABLE 4.1 Devices and Their Applications

Device	Frequency Limitation	Geometry	Substrate Materials	Processing Techniques	Major Applications
IMPATT	< 300 GHz	Mesa	Si, GaAs, InP	Epitaxy Diffusion Ion-implantation	Transmitters Amplifiers
Gunn	< 140 GHz	Mesa	GaAs, InP	Epitaxy	Local oscillators Amplifiers Transmitters
FET, HEMT	< 100 GHz	Beam lead	GaAs, InP	Epitaxy Ion implantation	Oscillators Amplifiers Switches Mixers Phase shifters
Mixer	< 1000 GHz	Honeycomb/ beam lead	GaAs, Si	Epitaxy Proton Bombardment	Upconverters Downconverters
p-i-n	< 100 GHz	MESA/ beam lead	Si, GaAs	Epitaxy Diffusion Ion implantation	Switches Limiters Phase shifters Modulators
Varactor	< 300 GHz	Mesa/ beam lead	GaAs	Epitaxy Ion implantation	Multipliers Tuning Phase shifters Modulators

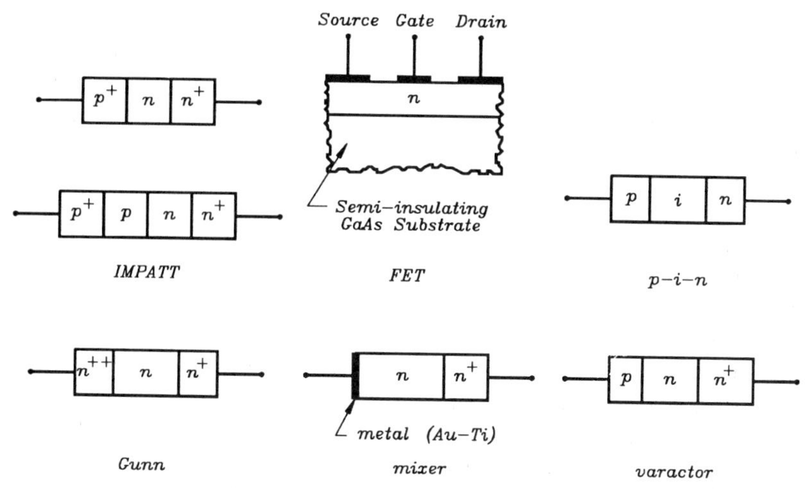

FIGURE 4.1 Active solid-state devices listed in Table 4.1 are shown here in schematic form. Note that in n-type semiconductors, electrons are the majority carriers, whereas in p-type material, holes are the majority carriers. These figures show only the structure of the devices; their use in different applications depends on the external circuits.

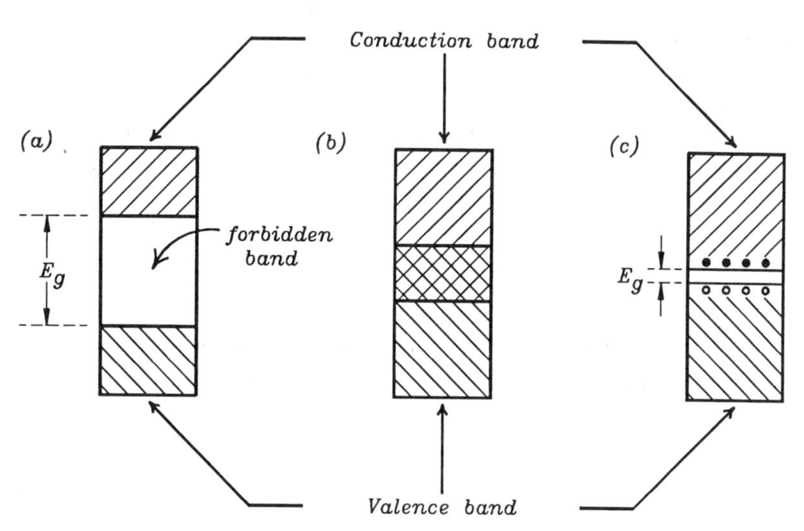

FIGURE 4.2 Energy-band structure for (a) an insulator, (b) a metal, and (c) a semiconductor.

moving from the valence band to the empty conduction band. Conduction is impossible by applying a field since electrons cannot acquire enough energy.

The band structure of a metal is shown in Figure 4.2(*b*), which does not contain a forbidden band. The valence band and conduction band overlap and electrons can move freely.

A semiconductor has a small forbidden band gap. For example, $E_g = 0.743$ eV for germanium (Ge), 1.17 eV for silicon (Si), and 1.519 eV for gallium arsenide (GaAs) at 0 K [1]. Energies of this magnitude are normally difficult to acquire; however, as the temperature is increased, some electrons in the valence band gain enough thermal energy (greater than E_g) and move to the conduction band. These electrons become free electrons that can move under a small applied field and are represented as solid dots in Figure 4.2(*c*). The absence of an electron in the valence band is a hole shown in Figure 4.2(*c*) as a small empty circle that can also be moved under an applied field. Due to these free electrons and holes, the insulator becomes slightly conducting and is thus called a *semiconductor*. A semiconductor without impurity atoms (or doping) is called an *intrinsic semiconductor*. Certain impurity atoms can be added to a semiconductor to allow energy states that lie in the forbidden energy gap. These impurities contribute to conduction, and a semiconductor with impurities is called an *extrinsic semiconductor* [2].

The band gap is a function of temperature, as shown in Figure 4.3. At room temperature and under a normal atmosphere, the values of the band gap are 0.66, 1.12, and 1.42 eV for Ge, Si, and GaAs, respectively.

4.3 MOBILITY, CONDUCTIVITY, AND FERMI LEVEL IN A CONDUCTOR

In a conductor, electrons are free to move under the influence of an applied field. An electron will be accelerated and the velocity will increase until it collides with the scatterers. At each collision with an ion, the electron loses energy and a steady-state condition is eventually reached where a finite drift speed v_d is attained. The drift speed v_d is proportional to the applied field E and can be written as*

$$v_d = \mu E \tag{4.1}$$

where μ is a proportionality constant called mobility with units of $\text{cm}^2/\text{v} \cdot \text{s}$.

The steady-state drift speed causes a directed flow of electrons that constitute a current. Thus

$$J = nqv_d = nq\mu E = \sigma E \tag{4.2}$$

*In equations (4.1), (4.2), (4.9), (4.35) and (4.37)–(4.41), E is the electric field. In the rest of this chapter, E is the energy level.

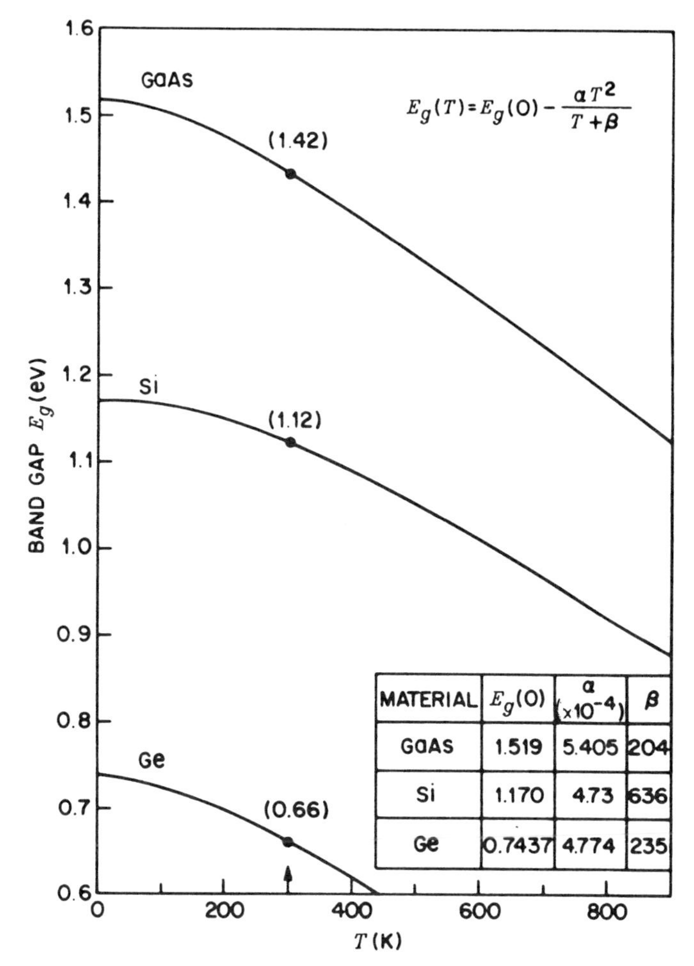

MATERIAL	$E_g(0)$	α (x10^{-4})	β
GaAs	1.519	5.405	204
Si	1.170	4.73	636
Ge	0.7437	4.774	235

FIGURE 4.3 Energy band gaps of Ge, Si, and GaAs as a function of temperature. (From Refs. 1 and 3 with permission.)

where q is the charge of an electron (1.6×10^{-19} C). σ is defined as the conductivity and is equal to $nq\mu$. The conductivity is thus proportional to the concentration of free electrons (n). For a good conductor, $n \approx 10^{28}$ m^{-3}. For an insulator, $n \approx 10^7$ m^{-3}. n lies between these two values for a semiconductor.

The number of free electrons per cubic meter whose energies lie in the energy interval dE can be given by

$$dn_E = \rho_E \, dE \tag{4.3}$$

where ρ_E is the density of electrons in the interval of dE. ρ_E can be

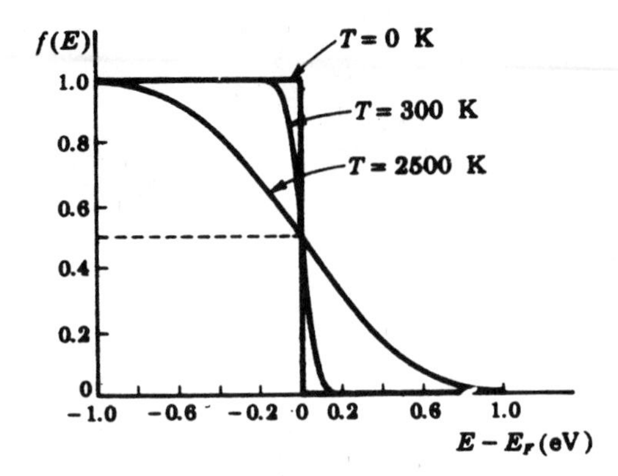

FIGURE 4.4 The Fermi–Dirac distribution function $f(E)$ gives the probability that a state of energy E is occupied.

represented by

$$\rho_E = f(E)N(E) \tag{4.4}$$

where $N(E)$ is the density of states (i.e., number of states per electron volt per cubic meter) in the conduction band and $f(E)$ is the probability that a state with energy E is occupied by an electron. The equation for $f(E)$, called the *Fermi–Dirac probability function*, has the following form obtained from quantum statistics [1]:

$$f(E) = \frac{1}{1 + e^{(E-E_F)/kT}} \tag{4.5}$$

where E_F is the Fermi level in eV, k is the Boltzmann constant (8.62×10^{-5} eV/K), and T is the temperature in K.

If $E = E_F$, $f(E) = \frac{1}{2}$ for any value of T. Therefore, the Fermi level represents the energy state with a 50% probability of being filled. Figure 4.4 shows $f(E)$ as a function of energy. The expression for $N(E)$ can be given by

$$N(E) = \gamma E^{1/2} \tag{4.6}$$

where γ is a constant equal to $(4\pi/h^3)(2m)^{3/2}(1.6 \times 10^{-19})^{3/2}$, m is the mass of an electron in kg, and h is Planck's constant in J · s ($h = 6.626 \times 10^{-34}$ J · s). From Equations (4.4) through (4.6), at absolute zero temperature, we have

$$\rho_E = \begin{cases} \gamma E^{1/2} & \text{for } E < E_F \\ 0 & \text{for } E > E_F \end{cases} \tag{4.7}$$

The total number of free electrons per cubic meter of metal is

$$n = \int_0^\infty \rho_E \, dE$$

$$= \int_0^{E_F} \gamma E^{1/2} \, dE = \tfrac{2}{3}\gamma E_F^{3/2} \tag{4.8}$$

Since E_F varies from metal to metal, n varies among metals.

4.4 MOBILITY, CONDUCTIVITY, AND FERMI LEVEL IN AN INTRINSIC SEMICONDUCTOR

In an intrinsic semiconductor, with each hole–electron created, both carriers contribute to the conduction and current flow. The current density is

$$J = (n\mu_n + p\mu_p)qE = \sigma E \tag{4.9}$$

where E is the applied electric field and μ_n and μ_p are the electron and hole mobility, respectively. For an intrinsic semiconductor,

$$n = p = n_i = \text{intrinsic concentration} \tag{4.10}$$

The intrinsic concentration depends on temperature (T) and the energy gap at absolute zero degree (E_{g0}), given by [2]

$$n_i^2 = A_0 T^3 e^{-E_{g0}/kT} \tag{4.11}$$

where A_0 is a constant.

The concentration of free electrons can be found from the following:

$$dn = \rho_E \, dE = f(E)N(E) \, dE \tag{4.12}$$

where

$$f(E) = \frac{1}{1 + e^{(E-E_F)/kT}} \tag{4.13}$$

$$N(E) = \gamma(E - E_c)^{1/2} \tag{4.14}$$

Equation (4.14) is a modification of Equation (4.6) since the lowest energy in the conduction band in semiconductor is E_c. The concentration of free electrons is

$$n = \int_{E_c}^\infty N(E)f(E) \, dE \tag{4.15}$$

For $E \geq E_c$ and $E - E_F \gg kT$, Equation (4.13) is simplified to

$$f(E) = e^{-(E-E_F)/kT} \tag{4.16}$$

The integral in Equation (4.15) is carried out, giving

$$n = N_c e^{-(E_c-E_F)/kT} \tag{4.17}$$

where

$$N_C = 2\left(\frac{2\pi m_n kT}{h^2}\right)^{3/2}(1.6 \times 10^{-19})^{3/2} \tag{4.18}$$

and m_n is the effective mass of electrons. Similarly,

$$p = N_v e^{-(E_F-E_v)/kT} \tag{4.19}$$

where N_v is given by Equation (4.18) with m_n replaced by m_p.

Since $n = n_i = p_i = p$ for an intrinsic semiconductor, from Equations (4.17) and (4.19) we have

$$E_{Fi} = \frac{E_c + E_v}{2} - \frac{kT}{2} \ln \frac{N_c}{N_v} \tag{4.20}$$

and

$$np = n_i^2 = N_c N_v e^{-E_G/kT} \tag{4.21}$$

where

$$E_G = E_c - E_v \tag{4.22}$$

Figure 4.5 shows n_i for Ge, Si, and GaAs as a function of temperature. If $m_n = m_p$, then $N_c = N_v$, and we have

$$E_{Fi} = \frac{E_c + E_v}{2}$$

The Fermi level lies in the middle of the band gap as shown in Figure 4.6.

4.5 EXTRINSIC SEMICONDUCTORS

Certain impurity atoms can be added into semiconductors to increase the conduction. A semiconductor with impurities is called an extrinsic semiconductor. Figure 4.7 shows three basic structures of a silicon semiconductor. Without impurities, each silicon atom shares its four valence electrons with

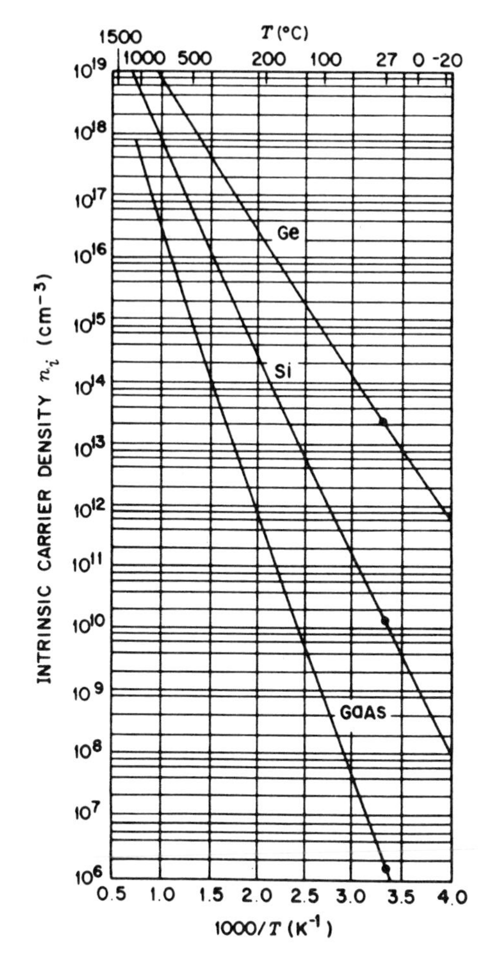

FIGURE 4.5 Intrinsic carrier densities of Ge, Si, and GaAs as a function of reciprocal temperature. (From Refs. 1 and 3 with permission.)

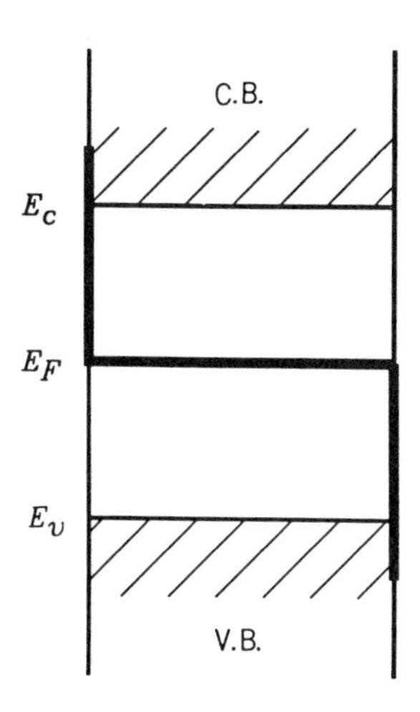

FIGURE 4.6 Fermi–Dirac distribution and energy-band diagram.

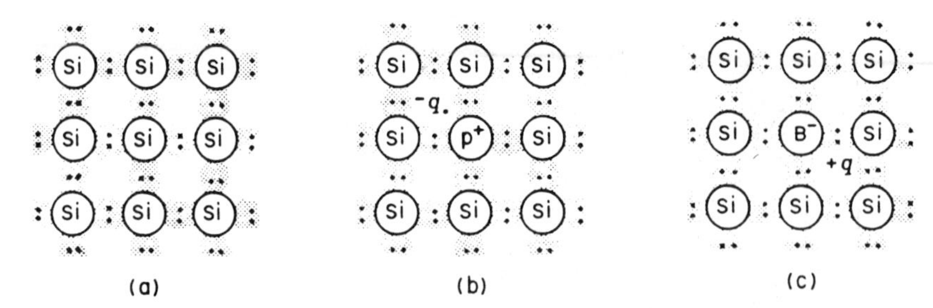

FIGURE 4.7 Three basic bond pictures of a semiconductor: (*a*) intrinsic Si with neglible impurities; (*b*) *n*-type Si with donor (phosphorus); (*c*) *p*-type Si with acceptor (boron). (From Ref. 1 with permission from Wiley.)

the four neighboring atoms forming four covalent bonds. A phosphorus atom with five valence electrons can replace a silicon atom. An extra electron is donated to the conduction band. The silicon is *n*-type with the addition of the electron and the phosphorus atom is called a *donor*. Similarly, a boron atom replacing a silicon atom needs to accept an additional electron to form the four covalent bonds. A positive charge carrier called a *hole* is created in the valence bond. The silicon is *p*-type and the boron atom is called an *acceptor*. Common impurities for *n*-type semiconductors are pentavalent atoms such as antimony (Sb), phosphorus (P), and arsenic (As). Impurities for *p*-type are trivalent atoms such as boron (B), gallium (Ga), and indium (In).

When donor or acceptor impurities are introduced into an intrinsic semiconductor, they create new allowed energy states inside the band gap. The donor impurities create a donor energy level near the conduction band, and the acceptor impurities create an acceptor energy level near the valence band as shown in Figure 4.8. If the separation between E_c and E_D is very small (0.05 eV in silicon, for example), almost all of the fifth electrons of the donor impurities are in the conduction band at room temperature. Therefore, an *n*-type semiconductor has many more electrons in its conduction band than holes in its valence band. Electrons are majority carriers in an *n*-type material, and holes are majority carriers in a *p*-type material. The doping of impurity atoms thus increases the conductivity and produces a semiconductor with predominantly electron or hole carriers.

For an extrinsic semiconductor, Equations (4.17) and (4.19) are still valid, and $np = n_i^2$. In an *n*-type semiconductor,

$$n = n_n \approx N_D \tag{4.23}$$

$$p = p_n \approx \frac{n_i^2}{N_D} \tag{4.24}$$

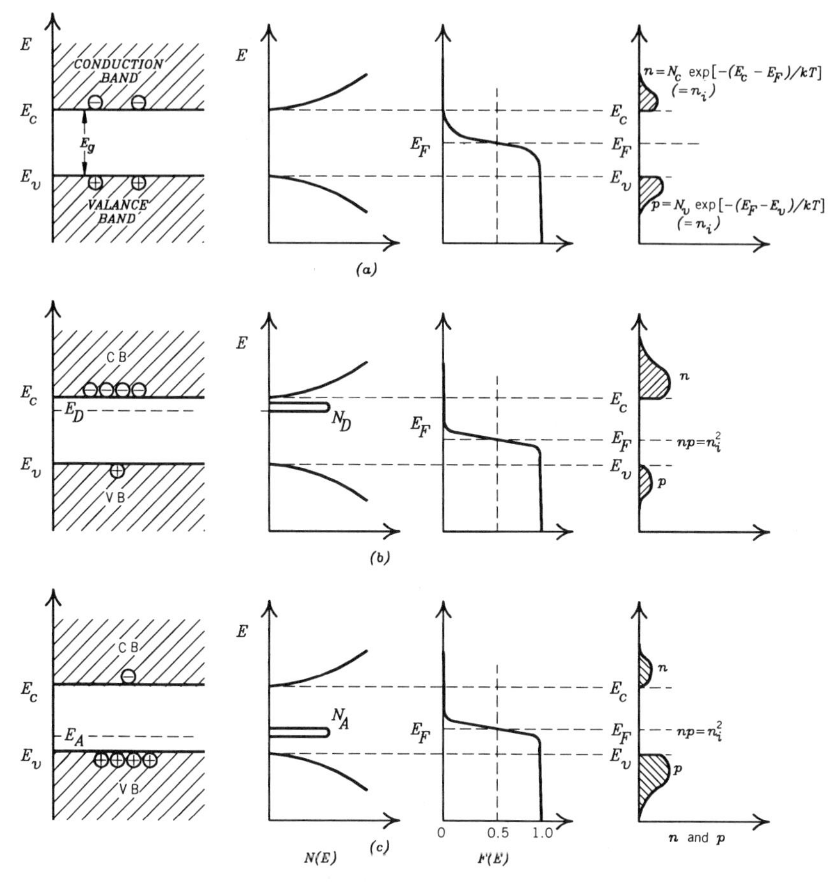

FIGURE 4.8 Schematic band diagram, density of states, Fermi–Dirac distribution, and the carrier concentrations for (*a*) intrinsic, (*b*) *n*-type, and (*c*) *p*-type semiconductors at thermal equilibrium. Note that $pn = n_i^2$ for all three cases. (From Ref. 1 with permission from Wiley.)

where N_D is the donor concentration. In a *p*-type semiconductor,

$$p = p_p \approx N_A \tag{4.25}$$

$$n = n_p \approx \frac{n_i^2}{N_A} \tag{4.26}$$

where N_A is the acceptor concentration.

Rewrite Equations (4.17) and (4.19) as

$$n = N_c e^{-(E_c - E_F)/kT} \tag{4.27}$$

$$p = N_v e^{-(E_F - E_v)/kT} \tag{4.28}$$

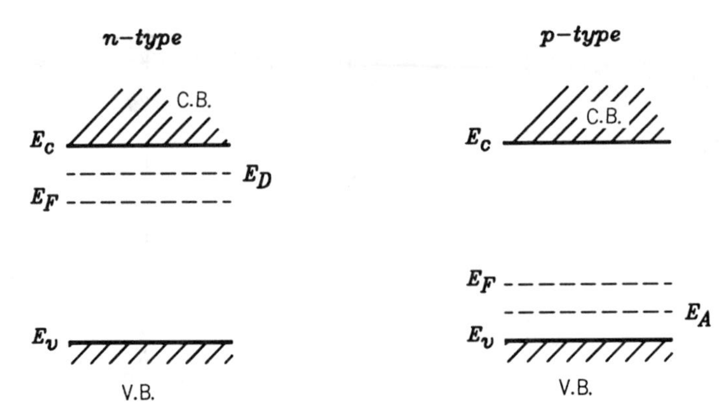

FIGURE 4.9 Positions of the Fermi level for different doping.

The only parameter that changes with impurity concentration in the equations above is the Fermi level E_F. In the n-type semiconductor, more electrons are in the conduction band and E_F moves up as seen by Equation (4.27). In the p-type semiconductor, more holes are in the valence band and E_F moves down.

To calculate E_F for an n-type semiconductor, using Equation (4.27) with $n \approx N_D$, we have

$$E_F = E_c - kT \ln \frac{N_c}{N_D} \tag{4.29}$$

Similarly, for a p-type semiconductor, we have

$$E_F = E_v + kT \ln \frac{N_v}{N_A} \tag{4.30}$$

Figure 4.9 shows the positions of the Fermi levels in n- and p-type semiconductors.

4.6 DIFFUSION AND CARRIER LIFETIME

In a semiconductor, the transport of charges due to a nonuniform concentration contributes to additional current flow. This mechanism, called *diffusion*, is not ordinarily encountered in metals. Diffusion is due to the nonuniform concentration of carriers inside a semiconductor. For example, the concentration p of holes varies with distance x inside the semiconductor. The holes will thus move from the high-density region toward the low-density region as shown in Figure 4.10. The diffusion hole current J_p in amperes per square

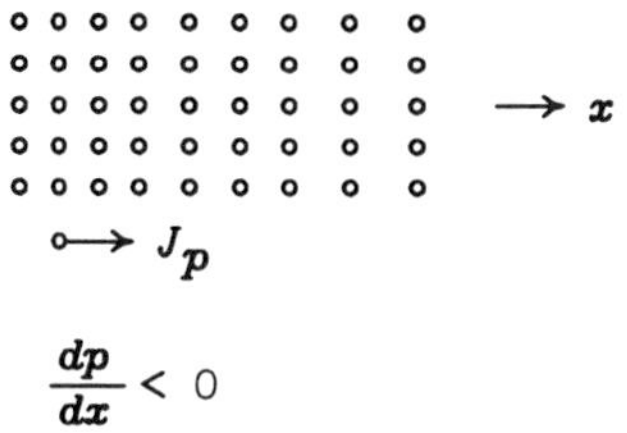

o o o o o o o o o
o o o o o o o o o →x
o o o o o o o o o
o o o o o o o o o
o o o o o o o o o

o→ J_p

$\frac{dp}{dx} < 0$

FIGURE 4.10 Diffusion mechanism in a semiconductor.

meter is proportional to the hole concentration gradient as

$$J_p = -qD_p \frac{dp}{dx} \tag{4.31}$$

where D_p is the diffusion constant for holes with units of m^2/s. A similar relationship applies to electrons with

$$J_n = qD_n \frac{dn}{dx} \tag{4.32}$$

The diffusion constants D_p and D_n are given by the Einstein equation:

$$\frac{D_p}{\mu_p} = \frac{D_n}{\mu_n} = kT = \frac{T}{11{,}600} \tag{4.33}$$

where T is the temperature in kelvin and kT in eV.

At any moment, thermal agitation continues to create new hole–electron pairs, while old hole–electron pairs may disappear due to recombination. On average, an electron will exist for τ_n seconds before recombination. τ_n is called the *mean lifetime* of an electron. Similarly, the mean lifetime for holes is τ_p. Carrier lifetimes vary from nanoseconds to hundreds of microseconds.

4.7 CONTINUITY EQUATION

The carrier concentration inside the semiconductor is a function of both time and distance. Consider the one-dimensional model shown in Figure 4.11 for simplicity. The infinitesimal element consists of holes with a concentration of p in holes/m^3. Due to the conservation of charge, the charge variation with time should equal the charge generation due to thermal agitation or other means, the charge recombination loss, and the current flow. This can be

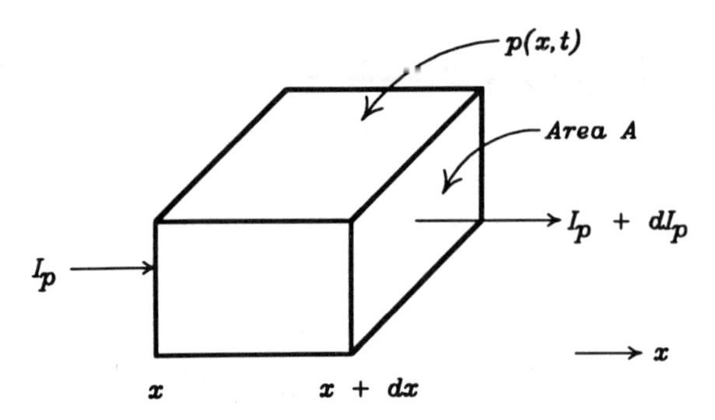

FIGURE 4.11 Infinitesimal length of semiconductor for the current derivation.

written in the following equation:

$$qA\,dx\,\frac{\partial p}{\partial t} = -qA\,dx\,\frac{p}{\tau_p} + qA\,dx\,g_p - dI_p \tag{4.34}$$

where g_p is the number of holes generated per unit volume. $A\,dx$ is the volume of the element shown in Figure 4.11. The one-dimensional hole current is the sum of the diffusion current and the drift current given by

$$J_p = -qD_p\frac{\partial p}{\partial x} + qp\mu_p E \tag{4.35}$$

Equation (4.34) becomes

$$\frac{\partial p}{\partial t} = -\frac{p}{\tau_p} + g_p - \frac{1}{qA}\frac{\partial I_p}{\partial x}$$

$$= -\frac{p}{\tau_p} + g_p - \frac{1}{q}\frac{\partial J_p}{\partial x} \tag{4.36}$$

$$= -\frac{p}{\tau_p} + g_p + D_p\frac{\partial^2 p}{\partial x^2} - \mu_p\frac{\partial(pE)}{\partial x} \tag{4.37}$$

Equations (4.36) and (4.37) are called *continuity equations.*

To generalize this to the three-dimensional case, Equation (4.36) is modified to

$$\frac{\partial p}{\partial t} = -\frac{p}{\tau_p} + g_p - \frac{1}{q}\boldsymbol{\nabla}\cdot\mathbf{J}_p \tag{4.38}$$

and

$$\mathbf{J}_p = -qD_p \nabla p + qp\mu_p \mathbf{E} \tag{4.39}$$

Equation (4.37) becomes

$$D_p \nabla^2 p - \mu_p \nabla \cdot (p\mathbf{E}) + g_p - \frac{p}{\tau_p} = \frac{\partial p}{\partial t} \tag{4.40}$$

Similarly, for electrons,

$$D_n \nabla^2 n + \mu_n \nabla \cdot (n\mathbf{E}) + g_n - \frac{n}{\tau_n} = \frac{\partial n}{\partial t} \tag{4.41}$$

Semiconductor devices are governed by Equations (4.40) and (4.41).

4.8 $p-n$ JUNCTION AND SCHOTTKY-BARRIER JUNCTION

When a p-type semiconductor is connected to an n-type semiconductor, a p–n junction (or a diode) is formed. The details are discussed in Chapter 5. The derivation of current–voltage characteristics for an ideal p–n junction can be found in many books [1, 2]. The current is given by

$$I = I(V) = I_s(e^{qV/\eta kT} - 1) \tag{4.42}$$

where

I_s = reverse saturation current

q = charge of an electron

V = voltage across junction

η = a constant between 1 and 2

k = Boltzmann constant

T = absolute temperature

Equation (4.42) is a well-known nonlinear equation governing diode behavior. A sketch of Equation (4.42) is given in Figure 4.12. I_s is normally very small (in μA) and the range of forward currents (in mA) is many orders of magnitude larger than the reverse saturation current. The dynamic resistance, which is defined as dV/dI, is small in forward bias and very large under reverse bias. At a junction voltage V_B, a large reverse current occurs and the diode is in the breakdown region, where Equation (4.42) is no longer valid.

Another useful property of a p–n junction is its nonlinear C–V characteristics. When a p–n junction is formed, a space-charge region is introduced.

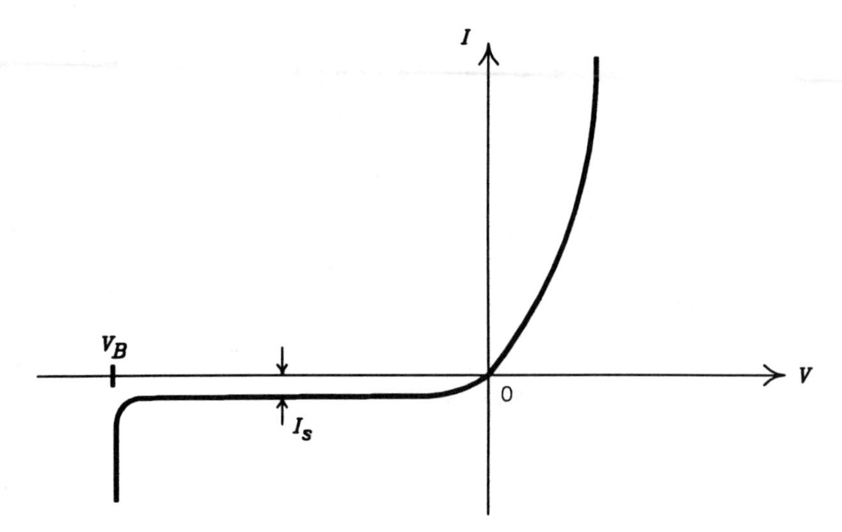

FIGURE 4.12 Nonlinear I–V characteristics of a diode.

The width of this space-charge region is a function of the applied voltage. The junction capacitance, which is inversely proportional to the space-charge width, is controlled by the applied voltage. Figure 4.13 shows a typical nonlinear C–V characteristics for a diode. These are derived and discussed in detail in Chapter 5.

The Schottky barrier consists of a metal–semiconductor contact. The metal–semiconductor contact could be a Schottky barrier or an ohmic

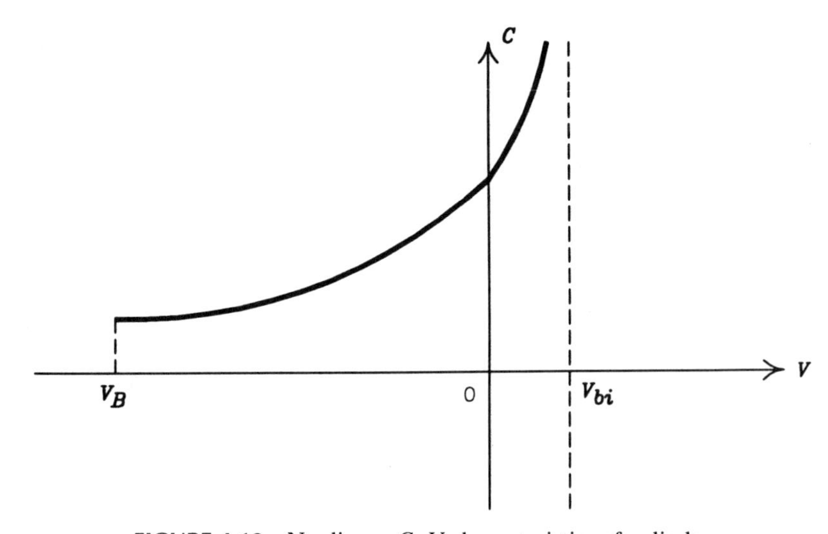

FIGURE 4.13 Nonlinear C–V characteristics of a diode.

contact, depending on the contact materials. The Schottky barrier has a rectifying property similar to that of a p–n junction. As a matter of fact, the point-contact rectifier had been used for many decades in various forms before the p–n junction was invented. The semiconductor used could be either silicon or GaAs. The Schottky barrier has I–V and C–V characteristics similar to those of a p–n junction [1]. Since the Schottky-barrier device is a unipolar (one-carrier) device with only majority-carrier flow and no minority-carrier storage, it is a much faster diode. It is generally preferred for microwave detectors or mixer devices. The Schottky barrier is also used as the gate electrode of a metal–semiconductor field-effect transistor (MESFET) or electrodes in IMPATT devices.

4.9 MICROWAVE APPLICATIONS OF $p-n$ AND SCHOTTKY-BARRIER JUNCTION

Many devices have been developed using the nonlinear I–V and C–V characteristics of the p–n or Schottky-barrier junction. Various applications are summarized below.

Nonlinear I–V Characteristics

- Detection
- Frequency mixing
- Frequency conversion
- Harmonic generation
- Switching
- Modulation
- Limiting

Nonlinear C–V Characteristics

- Frequency multiplication
- Voltage-tuned oscillator
- Voltage-tuned filter
- Frequency conversion
- Harmonic generation
- Parametric amplification

Avalanche Breakdown

- Power generation
- Power amplification
- Noise generation

Devices using nonlinear I–V characteristics are detectors (Chapter 6), mixers (Chapter 6), and *p-i-n* switches (Chapter 8). Devices using nonlinear C–V characteristics are tuning or multiplier varactors (Chapter 5). The device using avalanche breakdown is the IMPATT diode (Chapter 11). The details of these devices and their applications are covered in the following chapters.

REFERENCES

1. S. M. Sze, *Physics of Semiconductor Devices*, 2nd ed., Wiley, New York, 1981.
2. J. Milliman, *Microelectronics*, McGraw-Hill, New York, 1979.
3. C. D. Thurmond, "The Standard Thermodynamic Function of the Formation of Electrons and Holes in Ge, Si, GaAs and GaP," *Journal of the Electrochemical Society*, Vol. 122, No. 8, August 1975, pp. 1133–1141.

CHAPTER 5

Varactor Devices and Circuits

5.1 INTRODUCTION

The varactor diode is one of the old microwave solid-state devices. It is also called a parametric diode [1]. The varactor diode is a nonlinear device and provides a voltage-dependent variable capacitance. Varactors are generally semiconductor *p–n* junctions, Schottky-barrier junctions, or point-contact diodes made from gallium arsenide or silicon. Most varactors are fabricated on *n*-type semiconductors with *p*-type diffusion to form junctions.

Varactors are useful for many applications: frequency tuning for active and passive circuits, frequency multiplication, frequency conversion, harmonic generation, and parametric amplification. In this chapter we give an overview of a varactor's principle of operation and its applications.

5.2 SPACE-CHARGE CAPACITANCE AND PRINCIPLE OF OPERATION

Consider the *p–n* junction shown in Figure 5.1. A *p–n* junction is formed if donors are introduced into one side (the *n* side) and acceptors into the other side (the *p* side) of a semiconductor. Because a density gradient forms across the junction when the two layers are brought together, holes will diffuse to the right and electrons to the left across the junction. The electrons and holes near the junction are thus combined and an unneutralized ion region is created. This region is called the *space-charge* or *depletion region*. As a result of the displacement of charges near the junction, an electric field and a potential are built up across the junction. This process continues until the potential is large enough to prevent further movement. This potential is

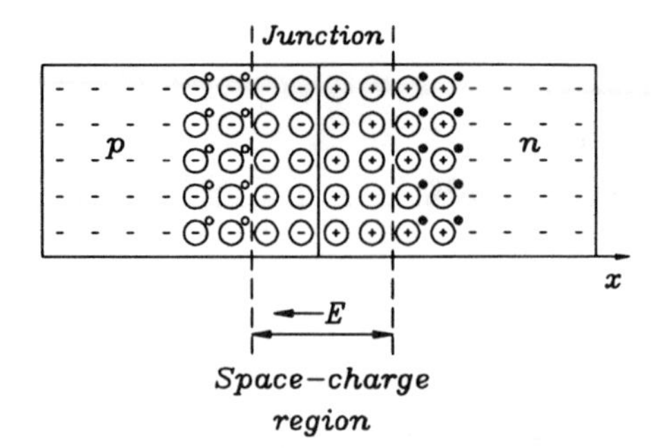

FIGURE 5.1 A p–n junction and space-charge region.

called the *built-in potential* [2]:

$$\text{built-in potential} = V_{\text{bi}} = -\int E\,dx \tag{5.1}$$

For GaAs, this potential is about 1.3 V.

If we apply an external reverse bias as shown in Figure 5.2(a), the external bias voltage will break the balance and cause more electrons to move to the right and holes to the left. The space-charge region is increased. The increase in space charge with the applied voltage is a capacitive effect given by

$$\text{junction capacitance } C_j = \left|\frac{dQ}{dV}\right| \tag{5.2}$$

where Q is the space charge and V is the applied voltage.

Considering the p–n junction shown in Figure 5.3(a), we assume that the donor concentration (N_D) is higher than that of the acceptors (N_A). The space-charge concentration is shown in Figure 5.3(b), and the E-field and

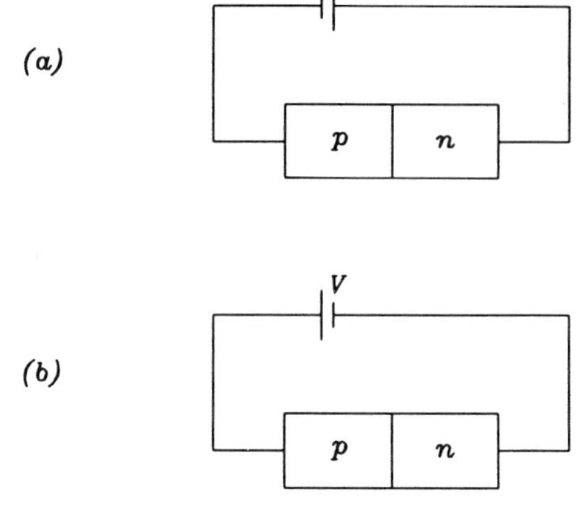

FIGURE 5.2 External bias: (a) reverse bias: the space-charge region increases; (b) forward bias: the space-charge region decreases.

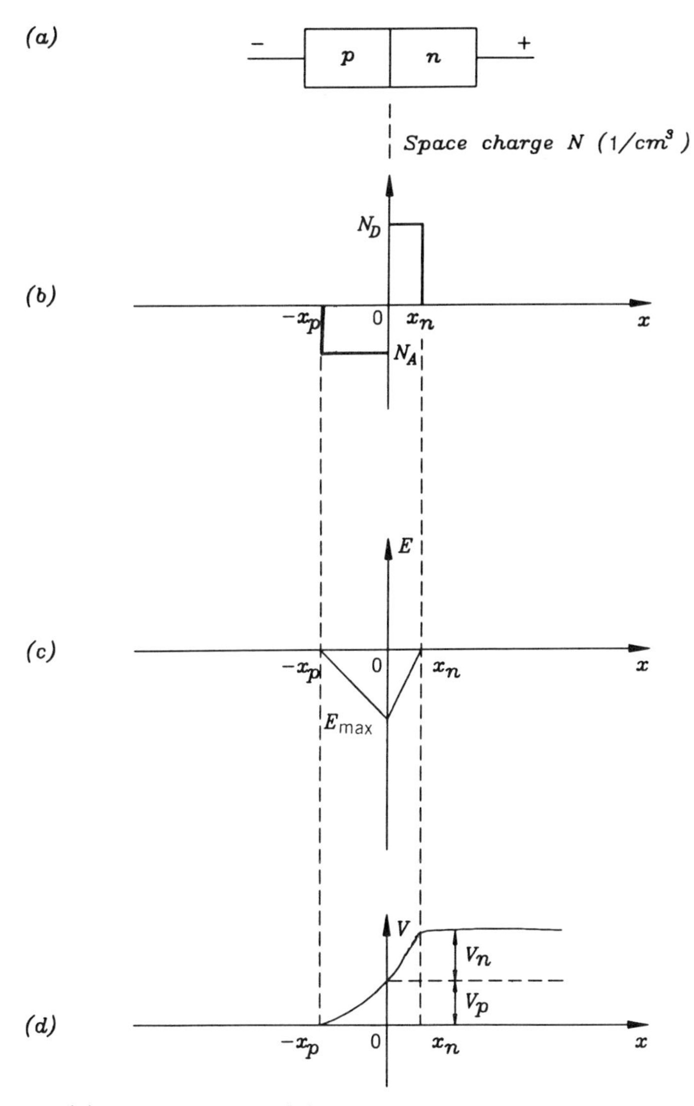

FIGURE 5.3 (a) A p–n junction: (b) space-charge concentration; (c) electric field; (d) potential across the junction.

potential, in Figure 5.3(c) and (d). The total space-charge region width is equal to $x_p + x_n$.

Since the charges on the left and right sides are equal, we have

$$Q_A = Q_D \tag{5.3}$$

Therefore,

$$AqN_A x_p = AqN_D x_n \tag{5.4}$$

where q is the charge of an electron (or hole) and A is the cross-sectional area. From (5.4) we have

$$N_A x_p = N_D x_n \tag{5.5}$$

Considering Poisson's equation for the one-dimensional case, we have

$$\frac{d^2V}{dx^2} = -\frac{dE}{dx} = -\frac{\rho}{\varepsilon_s} = \begin{cases} -\dfrac{qN_D}{\varepsilon_s} & \text{for } 0 < x < x_n \\ \dfrac{qN_A}{\varepsilon_s} & \text{for } -x_p \le x < 0 \end{cases} \tag{5.6}$$

where ε_s is the permittivity of the semiconductor and ρ is the charge density.
Equation (5.6) can be written as

$$E(x) = \begin{cases} \dfrac{q}{\varepsilon_s} N_D \displaystyle\int_{x_n}^{x} dx = \dfrac{qN_D}{\varepsilon_s}(x - x_n) & \text{for } 0 < x < x_n \quad (5.7) \\ -\dfrac{q}{\varepsilon_s} N_A \displaystyle\int_{-x_p}^{x} dx = -\dfrac{qN_A}{\varepsilon_s}(x + x_p) & \text{for } -x_p \le x < 0 \quad (5.8) \end{cases}$$

The maximum field occurs at $x = 0$:

$$E_m = E(x)|_{x=0} = -\frac{qN_D x_n}{\varepsilon_s} = -\frac{qN_A x_p}{\varepsilon_s} \tag{5.9}$$

The potential due to the n side is

$$V_n = -\frac{q}{\varepsilon_s} N_D \int_0^{x_n} (x - x_n)\, dx = \frac{qN_D}{2\varepsilon_s} x_n^2 \tag{5.10}$$

Similarly for the p side,

$$V_p = \frac{q}{\varepsilon_s} N_A \int_{-x_p}^{0} (x + x_p)\, dx = \frac{qN_A}{2\varepsilon_s} x_p^2 \tag{5.11}$$

The summation of the bias and built-in voltages should equal $V_n + V_p$. Therefore,

$$
\begin{aligned}
V_{\mathrm{bi}} + |V| = V_{\mathrm{bi}} - V &= V_n + V_p \\
&= \frac{qN_D}{2\varepsilon_s} x_n^2 + \frac{qN_A}{2\varepsilon_s} x_p^2 \\
&= x_n^2 \left[\frac{q}{2\varepsilon_s} \frac{N_D(N_A + N_D)}{N_A} \right]
\end{aligned} \tag{5.12}
$$

where from Equation (5.5), $x_p = (N_D/N_A)x_n$. V is negative for reverse bias. A relationship between space-charge width and voltage can be obtained from Equation (5.12). On the n side the width is

$$
x_n = \left[\frac{2\varepsilon_s (V_{\mathrm{bi}} - V) N_A}{qN_D(N_A + N_D)} \right]^{1/2} \tag{5.13}
$$

Similarly, the width on the p side is

$$
x_p = \left[\frac{2\varepsilon_s (V_{\mathrm{bi}} - V) N_D}{qN_A(N_A + N_D)} \right]^{1/2} \tag{5.14}
$$

The total space-charge width is

$$
W = x_n + x_p \tag{5.15}
$$

Substituting (5.13) and (5.14) into (5.15) gives

$$
W = \left[\frac{2\varepsilon_s (V_{\mathrm{bi}} - V)}{q} \frac{N_A + N_D}{N_A N_D} \right]^{1/2} \tag{5.16}
$$

Note that for reverse bias, V is negative.

Considering three special cases with the doping concentration shown in Figure 5.4, we can simplify Equation (5.16).

1. If $N_A = N_D = N$ with a symmetrical double-sided junction,

$$
W = \left[\frac{4\varepsilon_s (V_{\mathrm{bi}} - V)}{qN} \right]^{1/2} \tag{5.17}
$$

2. If $N_D \gg N_A$ with a single-sided n^+–p junction,

$$W \approx x_p = \left[\frac{2\varepsilon_s (V_{\text{bi}} - V)}{qN_A} \right]^{1/2} \qquad x_p \gg x_n \tag{5.18}$$

3. If $N_A \gg N_D$ with a single-sided p^+–n junction (or a Schottky-barrier junction),

$$W \approx x_n = \left[\frac{2\varepsilon_s (V_{\text{bi}} - V)}{qN_D} \right]^{1/2} \qquad x_n \gg x_n \tag{5.19}$$

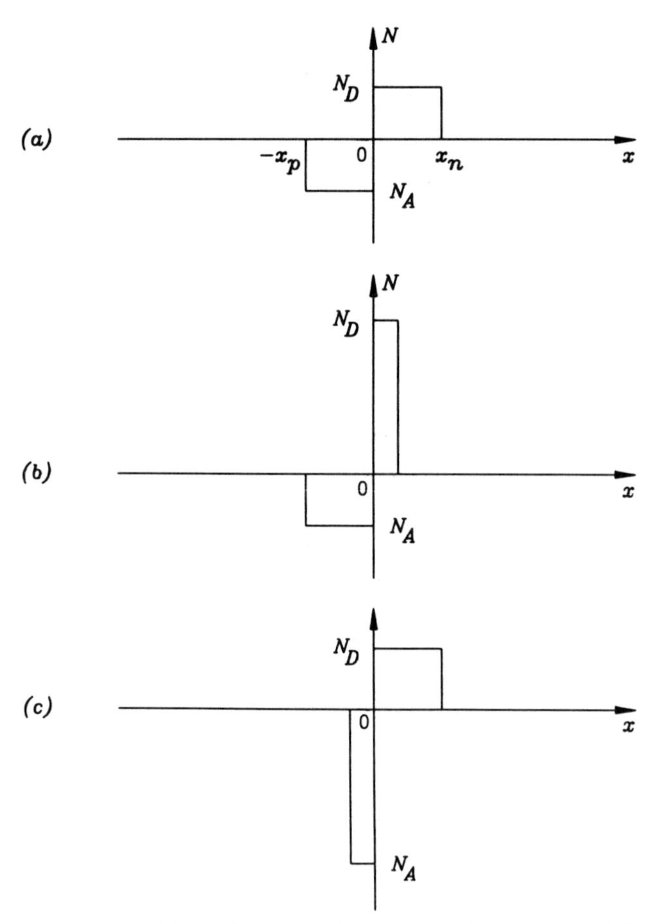

FIGURE 5.4 Doping profiles: (a) symmetrical double-side junction; (b) single-sided n^+–p junction; (c) single-sided p^+–n junction.

The junction capacitance is defined as

$$C_j(V) = \left| \frac{dQ}{dV} \right| \tag{5.20}$$

Since the total charge in one side is

$$\begin{aligned} Q &= qAN_A x_p \\ &= qAN_A \left[\frac{2\varepsilon_s (V_{\text{bi}} - V) N_D}{qN_A (N_A + N_D)} \right]^{1/2} \end{aligned} \tag{5.21}$$

the junction capacitance can be derived from Equations (5.20) and (5.21):

$$\begin{aligned} C_j(V) &= \left| \frac{dQ}{dV} \right| \\ &= \frac{qAN_A}{2} \left[\frac{2\varepsilon_s (V_{\text{bi}} - V) N_D}{qN_A (N_A + N_D)} \right]^{-1/2} \frac{2\varepsilon_s N_D}{qN_A} \frac{1}{N_A + N_D} \\ &= \frac{A}{2} \left(\frac{2q\varepsilon_s N_A N_D}{N_A + N_D} \right)^{1/2} (V_{\text{bi}} - V)^{-1/2} \end{aligned} \tag{5.22}$$

From Equation (5.16) we find that

$$C_j(V) = \frac{\varepsilon_s A}{W(V)} \tag{5.23}$$

Let us define C_{j0} as C_j at $V = 0$, or

$$C_{j0} = C_j(0) = \frac{A}{2} \left[\frac{2q\varepsilon_s N_A N_D}{(N_A + N_D) V_{\text{bi}}} \right]^{1/2} \tag{5.24}$$

Equation (5.22) can be rewritten as

$$C_j(V) = C_{j0} \left(1 - \frac{V}{V_{\text{bi}}} \right)^{-1/2} = C'_{j0} (V_{\text{bi}} - V)^{-1/2} \tag{5.25}$$

Here $C'_{j0} = C_{j0} \sqrt{V_{\text{bi}}}$.

Equation (5.25) is an important equation governing varactor operation. It gives the junction capacitance as a function of bias voltage for an abrupt p–n junction. Equation (5.25) is plotted in Figure 5.5. Varactor diodes are usually operated under reverse bias conditions.

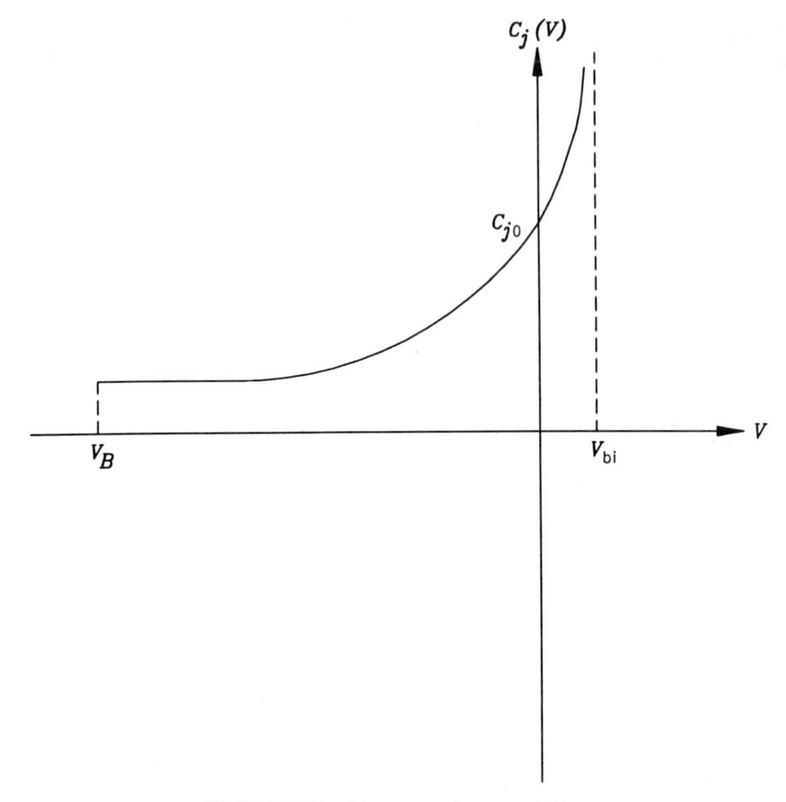

FIGURE 5.5 Varactor characteristics.

So far all derivations have been for an abrupt p–n junction. The derivation can be modified for an arbitrary junction. In general,

$$C_j(V) = C_{j0}\left(1 - \frac{V}{V_{\text{bi}}}\right)^{-\gamma} = C'_{j0}(V_{\text{bi}} - V)^{-\gamma} \tag{5.26}$$

where γ is a parameter depending on the doping profile and $C'_{j0} = C_{j0}V^{r}_{\text{bi}}$. For a general p^+–n junction, the doping in the n side is

$$N_D = Bx^m \tag{5.27}$$

When $m = 0$, we have an abrupt junction with $N_D = B =$ constant. It can be proved that (see Problem P5.1)

$$\gamma = \frac{1}{m + 2} \tag{5.28}$$

Therefore, we have

If $m = 0$, then $\gamma = \frac{1}{2}$ for an abrupt junction.
If $m = 1$, then $\gamma = \frac{1}{3}$ for a linear junction.
If $m = -1$, then $\gamma = 1$ for a hyper-abrupt junction.
If $m = -\frac{3}{2}$, then $\gamma = 2$ for a hyper-abrupt junction.

It can be seen from Equation (5.26) that the hyper-abrupt junction provides a wider variation in capacitance than does the abrupt junction.

5.3 PACKAGING CONSIDERATIONS AND EQUIVALENT CIRCUITS

Varactors can be packaged in various configurations. The selection of package depends on the operating frequency, transmission line used, cost, and applications. In this section, several commonly used packages for microwave solid-state devices are presented.

A. Pill Package

The pill package is mechanically sturdy and easy to handle. A typical configuration is shown in Figure 5.6(*a*). The two metal disks are used for making contact with external bias and the ceramic or quartz ring is used as an insulator. The bonding wire makes contact with the semiconductor device and metal cap. The equivalent circuit of this device and package is shown in Figure 5.6(*b*). C_p represents the package capacitance due to the ceramic or quartz ring. L_s is the inductance due to the bonding wire. R_s is the series resistance associated with the semiconductor, $C_j(V)$ is the variable capacitance due to the *p–n* junction, and $R_j(V)$ is the junction resistance.

Some typical values for microwave operation are: $C_p = 1$ pF, $L_s = 1$ nH, $R_s = 1\ \Omega$, and $C_j(V)$ from 0.5 to 2.5 pF. $R_j(V) > 10$ M Ω for reverse bias and can be neglected.

Pill packaged devices are preferred in waveguide and coaxial-line circuits. They can also be used in a microstrip circuit in the shunt arrangement; however, a hole needs to be drilled for mounting the device in a microstrip circuit. For ease of mounting, pill devices with a post or screw connected to the metal disks can be used.

B. Planar Package

For integrated circuit application, a planar flat package is preferred. Figure 5.7 shows a planar package with two leads for device mounting. The equivalent circuit is the same as in Figure 5.6(*b*) except that the package parasitics are of different values.

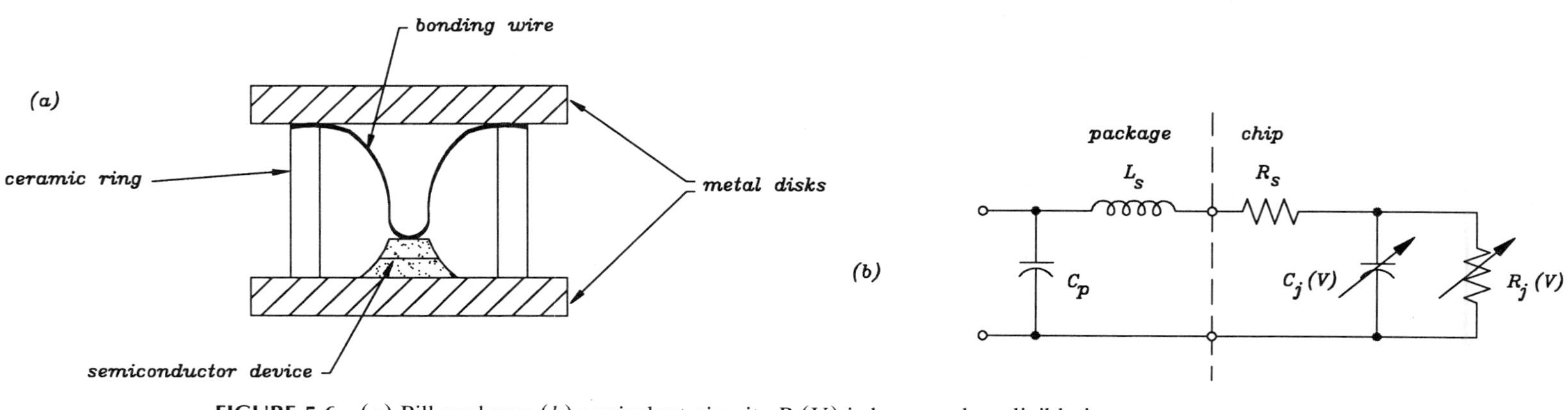

FIGURE 5.6 (*a*) Pill package; (*b*) equivalent circuit. $R_j(V)$ is large and negligible in reverse bias.

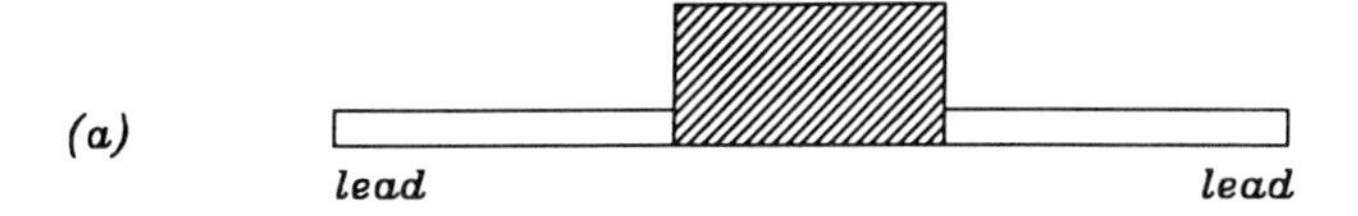

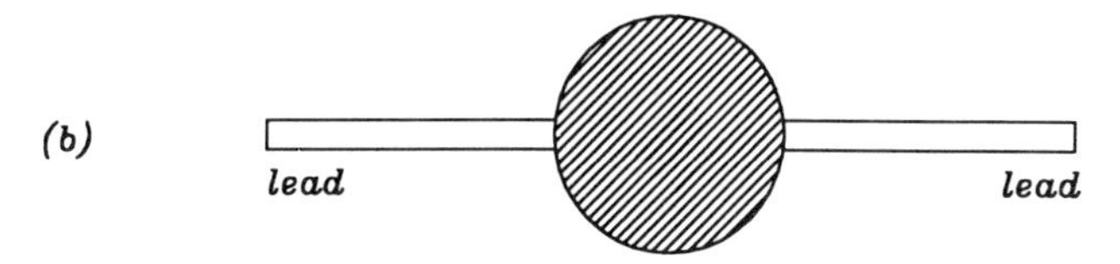

FIGURE 5.7 Flat package: (*a*) side view; (*b*) top view.

C. Beam-Lead Package

For high-frequency applications using integrated circuits, a beam-lead package is more attractive for its low package parasitics. The beam-lead package is a planar, surface-oriented, open package Figure 5.8 shows the planar configuration and possible dimensions. Passivation (a dielectric cover layer) can be used to seal and protect the device.

Shown in the same figure is the equivalent circuit. L_s is due to lead inductance and C_p is attributed to the stray capacitance. Their values are very small (e.g., $C_p = 0.01$ pF and $L_s = 0.2$ nH).

D. Chip

A varactor chip without a package can be used to eliminate parasitics. In this case, a bonding wire is used to connect the chip to the circuit.

5.4 PRACTICAL CONSIDERATIONS

In practice, one would like the varactor to have the following features:

1. High Q factor (figure of merit) or cutoff frequency
2. High breakdown voltage (V_B)
3. High sensitivity (S)

High Q means low loss or a small R_s. Q is defined as

$$Q(V) = \frac{f_{\text{co}}}{f_0} = \frac{1}{2\pi R_s C_j(V) f_0} \tag{5.29}$$

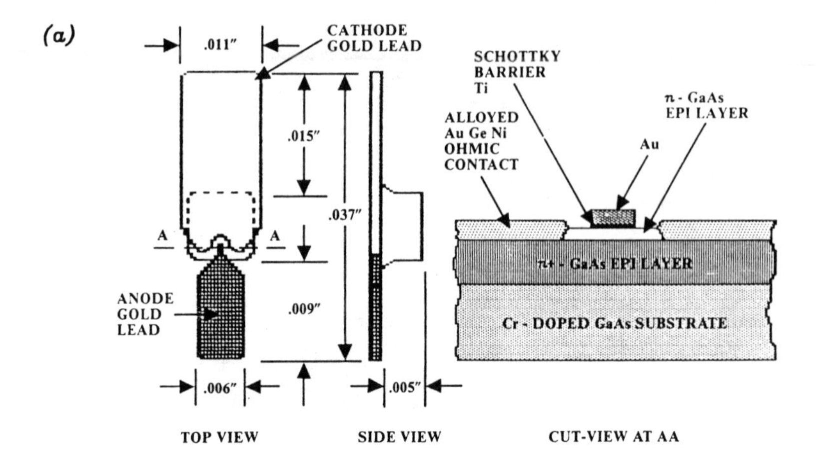

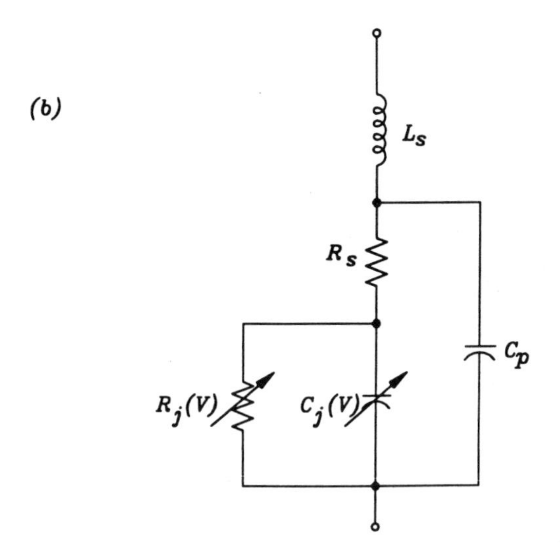

FIGURE 5.8 Beam-lead package: (*a*) configuration; (*b*) equivalent circuit. $R_j(V)$ is large and negligible in reverse bias.

where f_0 is the operating frequency and f_{co} is the static cutoff frequency at $V = 0$. Since C_j varies, it is more meaningful to define a dynamic cutoff frequency given by

$$f_c = \frac{1}{2\pi R_s}\left[\frac{1}{C_j(\text{min})} - \frac{1}{C_j(\text{max})}\right] \tag{5.30}$$

The sensitivity S is defined as

$$S = -\frac{\text{ratio change of } C_j}{\text{ratio change of } V} = -\frac{dC_j/C_j}{dV/V} = -\frac{dC_j}{C_j}\frac{V}{dV} \tag{5.31}$$

Since the capacitance decreases as the negative voltage increases, a negative sign is introduced here to keep S positive.

Substituting $C_j = C'_{j0}(V_{\text{bi}} - V)^{-\gamma}$ into Equation (5.31), we have

$$\begin{aligned} S &= -\frac{dC_j}{C_j}\frac{V}{dV} = -C'_{j0}\gamma(V_{\text{bi}} - V)^{-(\gamma+1)}\frac{V}{C'_{j0}(V_{\text{bi}} - V)^{-\gamma}} \\ &= -\gamma\frac{V}{V_{\text{bi}} - V} \end{aligned} \tag{5.32}$$

If $V \gg V_{\text{bi}}$, Equation (5.32) becomes

$$S \approx -\gamma\frac{V}{-V} = \gamma \tag{5.33}$$

The larger S is, the wider is the capacitance variation.

Example 5.1 A GaAs-abrupt junction beam-lead varactor has $C_{j0} = 0.075$ pF, a breakdown voltage of -50 V, and $R_s = 1\ \Omega$. The varactor operates between $V = -2$ and -30 V. What is the capacitance variation ratio?

SOLUTION:

$$C_j(V) = C_{j0}\left(1 - \frac{V}{V_{\text{bi}}}\right)^{-1/2}$$

$$V_{\text{bi}} = 1.3 \text{ V} \qquad C_{j0} = 0.075 \text{ pF}$$

$$C_j(-2) = 0.075\left(1 + \frac{2}{1.3}\right)^{-1/2} = 0.047 \text{ pF}$$

$$C_j(-30) = 0.075\left(1 + \frac{30}{1.3}\right)^{-1/2} = 0.015 \text{ pF}$$

$$\text{ratio} = \frac{C_j(-2)}{C_j(-30)} = 3.13$$

5.5 VARACTOR-TUNED OSCILLATOR AND FILTER CIRCUITS

A microwave oscillator can be frequency tuned electronically by varying the bias applied to the active device or by incorporating a varactor in the circuit. The bias tuning approach is simple, but the tuning is nonlinear and the power variation over the tuning range is large. Varactor tuning is more complicated, but the output power is fairly constant. An electronic tuning oscillator is called a VCO (voltage-controlled oscillator) or VTO (voltage-tuned oscillator).

VCOs have applications in frequency-modulation (FM) systems and frequency-agile systems commonly used in radar and communications. They are also widely used in instrumentation, electronic warfare (EW), and electronic counter measurement (ECM) systems.

Figure 5.9 shows a possible oscillator circuit as an example to illustrate the varactor tuning. A microwave active device with a negative resistance of $-R_D$ and a capacitance of C_D is connected to an external circuit with an inductance of L_L and a load of R_L. Without the varactor [Figure 5.9(*a*)], the oscillating frequency, which equals the resonant frequency, is

$$f_0 = f_r = \frac{1}{2\pi\sqrt{L_L C_D}} \tag{5.34}$$

where C_D depends on the bias voltage of the active device. By varying the bias, one can change C_D and the oscillating frequency. With the varactor included [Figure 5.9(*b*)], the oscillating frequency is

$$f_0 = f_r = \frac{1}{2\pi\sqrt{L_L C_T}} \tag{5.35}$$

where

$$C_T = \frac{C_j(V)C_D}{C_j(V) + C_D} \tag{5.36}$$

For simplicity, the varactor is represented only by its variable junction capacitance.

From Equations (5.35) and (5.36), it is clear that the oscillation frequency can be tuned by varying $C_j(V)$. Of course, the actual microwave varactor tuned network is more complicated than that shown in Figure 5.9. The effectiveness of tuning normally depends on the varactor, coupling circuit, load, and active device. More examples are given later in chapters dealing with active devices.

Varactors can be used to tune the passband of a bandpass filter [3, 4] or the resonant frequency of a resonator [5, 6]. In this case, one has an

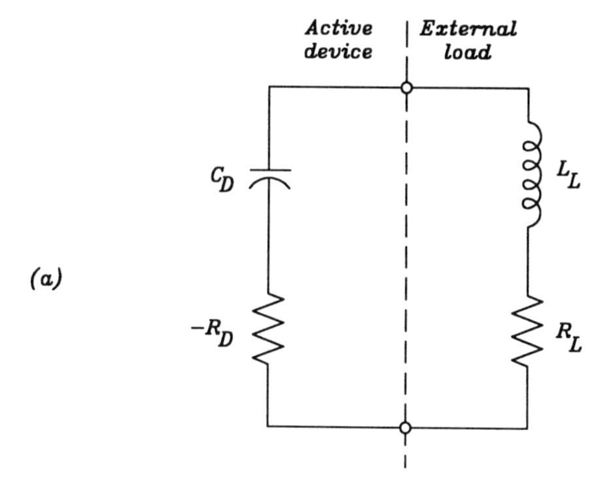

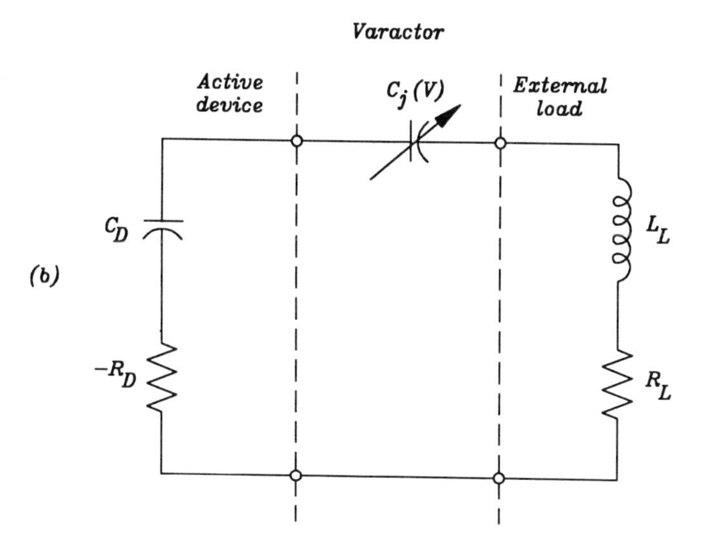

FIGURE 5.9 Varactor-tuned oscillator: (*a*) without a varactor; (*b*) with a varactor.

electronically tunable bandpass filter. Figure 5.10 shows how a varactor tunes a resonator. The resonant frequency is

$$f_0 = \frac{1}{2\pi\sqrt{L_0 C_T}} \tag{5.37}$$

$$C_T = C_j(V) + C_0 \tag{5.38}$$

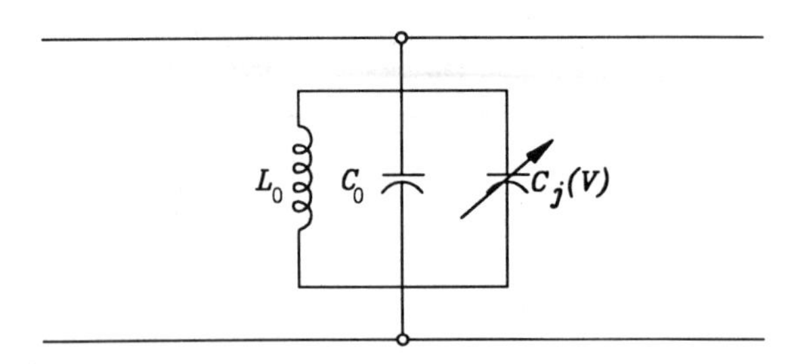

FIGURE 5.10 Varactor-tuned resonator.

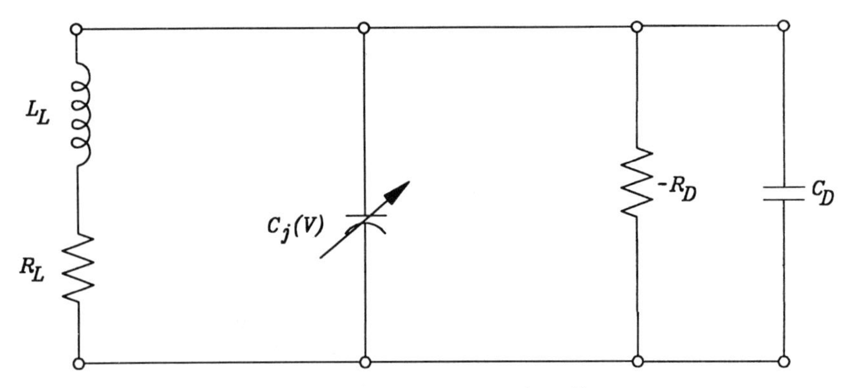

FIGURE 5.11 Varactor-tuned oscillator.

Example 5.2 A GaAs linear *p–n* junction varactor is used to tune the oscillator circuit shown in Figure 5.11. At zero bias voltage, the varactor has a junction capacitance of 3 pF. $V_{\mathrm{bi}} = 1.3$ V for GaAs. The active device has $C_D = 1$ pF and $R_D = 8\ \Omega$. The load has $L_L = 10$ nH and $R_L = 5\ \Omega$. What is the oscillation frequency at (a) -20 V and (b) 0 V?

SOLUTION:

$$C_j = C_{j0}\left(1 - \frac{V}{V_{\mathrm{bi}}}\right)^{-1/3}$$

$$= 3\left(1 + \frac{20}{1.3}\right)^{-1/3}$$

$$= 1.18 \text{ pF} \qquad \text{at} \quad V = -20 \text{ V}$$

$$C_j = 3 \text{ pF} \qquad \text{at} \quad V = 0 \text{ V}$$

(a) $V = -20$ V.

$$f_0 = f_r = \frac{1}{2\pi\sqrt{L_L C_T}}$$

$$C_T = C_j(-20) + C_D$$

$$= 1.18 \text{ pF} + 1 \text{ pF} = 2.18 \text{ pF}$$

$$f_0 = f_r = \frac{1}{2\pi\sqrt{10 \times 10^{-9} \times 2.18 \times 10^{-12}}} = 1.078 \text{ GHz}$$

(b) $V = 0$ V.

$$f_0 = f_r = \frac{1}{2\pi\sqrt{L_L C_T}}$$

$$C_T = C_j(0) + C_D = 3 + 1 = 4 \text{ pF}$$

$$f_0 = f_r = \frac{1}{2\pi\sqrt{10 \times 10^{-9} \times 4 \times 10^{-12}}} = 0.796 \text{ GHz}$$

5.6 MULTIPLIER AND HARMONIC GENERATOR CIRCUITS

A varactor can be used for frequency conversion or multiplication, due to its nonlinear characteristics. The use of an efficient multiplier together with a low-frequency oscillator or VCO provides a feasible alternative for generating a high-frequency signal. This approach is especially useful to generate signals at high millimeter-wave frequencies where fundamental sources are difficult to obtain.

Figure 5.12 shows circuit symbols representing frequency multipliers. The $\times 2$ multiplier is normally called a doubler, and the $\times 3$, a tripler. The multiplier consists basically of a low-pass filter, a bandpass (or high-pass) filter, a varactor circuit, and input- and output-matching networks. Figure 5.13 shows a block diagram. The low-pass filter, located in the input side, passes the fundamental and rejects all higher harmonics. The varactor is a nonlinear device that provides harmonic generation. The bandpass or high-pass filter at the output side passes the only desired harmonic and rejects all other signals. The operating considerations of a multiplier are:

1. A nonlinear device is necessary to provide frequency conversion.
2. For high conversion efficiency, the varactor and circuits should have low loss.

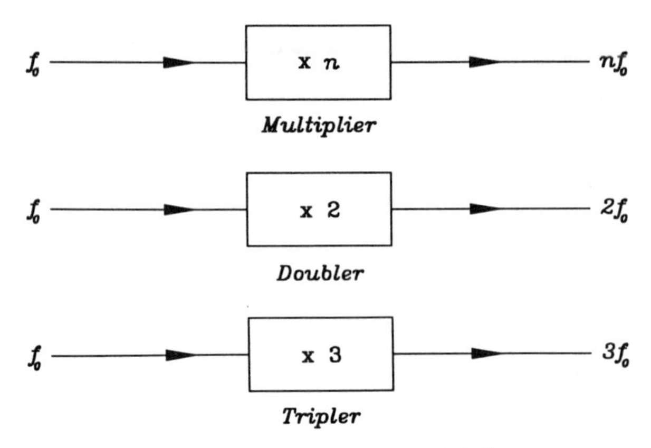

FIGURE 5.12 Frequency multipliers.

3. The input and output frequencies should appear at separate terminals. The input and output circuits should be conjugately matched to the impedance of the varactor at their respective harmonic frequencies.
4. The multiplier should be open circuited at all harmonic frequencies except the input and output frequencies.

The basic model of the varactor shown in Figure 5.13 is represented by a resistance R_s in series with a variable capacitance whose value is a function of bias voltage. For accurate modeling, the package parasitics of the varactor should be included.

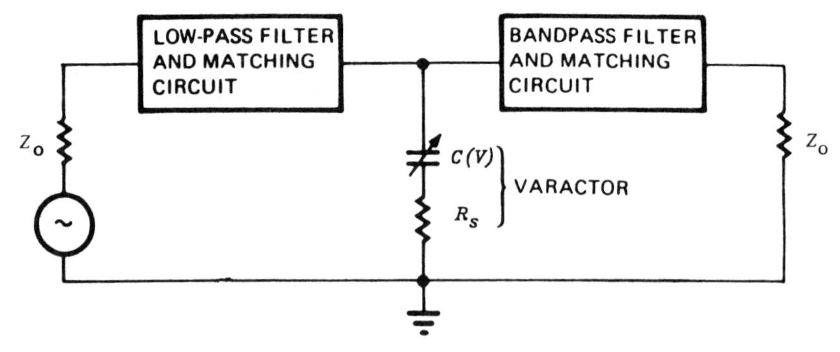

FIGURE 5.13 Multiplier circuit schematic. Z_0 is the load impedance or characteristic impedance of a transmission line.

The conversion efficiency (η) and conversion loss (L_c) are defined as

$$\eta = \frac{P_{\text{out}}}{P_{\text{in}}} \tag{5.39a}$$

$$L_c(\text{in dB}) = 10 \log \frac{P_{\text{in}}}{P_{\text{out}}} \tag{5.39b}$$

P_{in} is the input power of the fundamental frequency, and P_{out} is the output power of the desired harmonic. For a given multiplier, the conversion efficiency is maximized for a unique output-circuit load impedance. This impedance and the maximum value of the efficiency are functions of the input power level and the varactor characteristics. The conversion efficiency can be written as [7, 8]

$$\eta = \exp\left(-\alpha \frac{f_{\text{out}}}{f_c}\right) \tag{5.40}$$

where f_{out} is the output frequency and α is a parameter related to the harmonic order of the output and to the input drive level. Typically, for multipliers driven so that the RF voltage swings between 0 and V_B, α is about 10 for a doubler and about 16.5 for a tripler [7].

Curves are available for estimating the conversion efficiency. Figure 5.14 shows the maximum efficiency versus frequency ratio $f_{\text{in}}/f_{\text{co}}$ for a doubler as an example [9]. The actual efficiency is lower, due to other circuit losses.

Many practical multipliers have been reported. Some examples are given here. Figure 5.15 shows a multiplier circuit consisting of input and output waveguide ports. The varactor diode is located at the output waveguide with a sliding short tuned for optimum output matching. The output waveguide also serves as a high-pass filter rejecting the input signal. A low-pass filter and input-matching circuit are fabricated on the circuit board. The low-pass filter is necessary to prevent the desired output frequency from leaking back into the input port. Using this circuit, a conversion loss of less than 6.5 dB has been achieved for a 40- to 80-GHz doubler with a 1.5-GHz 3-dB down bandwidth [10].

Doublers with output frequency between 100 and 260 GHz have also been reported [7, 11]. Similar circuits were employed by Takada and Hirayama [12] for doublers and triplers with 300- and 450-GHz output frequencies. The doubler has a conversion loss of 10.7 dB and the tripler a loss of 19.4 dB. This work was later extended to 600 GHz with a quadrupler with a 20 dB conversion loss [13]. The 3-dB down bandwidth was also improved to more than 10 GHz [14]. Most of these multiplier are similar to the configuration shown in Figure 5.16. A whisker-contact varactor chip is used to reduce the parasitics.

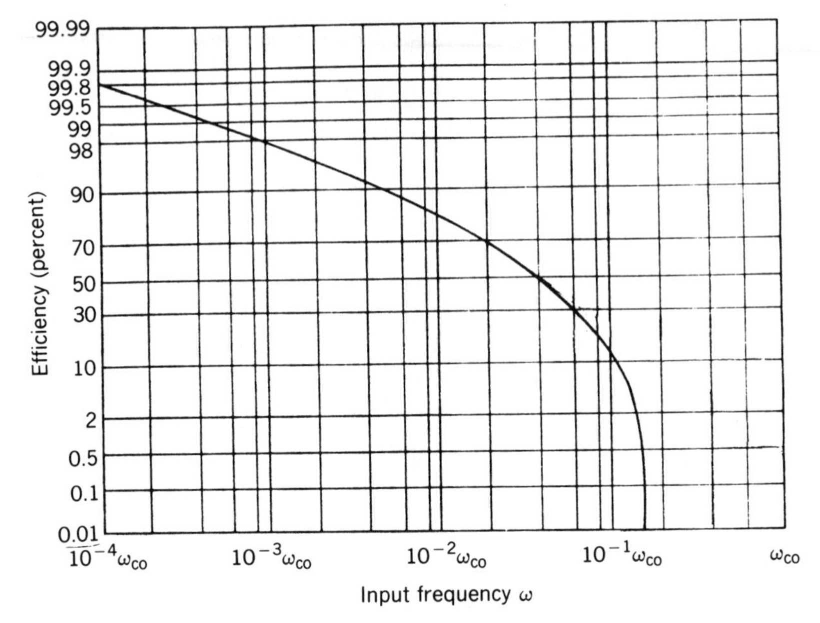

FIGURE 5.14 Efficiency of a doubler. (From Ref. 9 with permission from IEEE.)

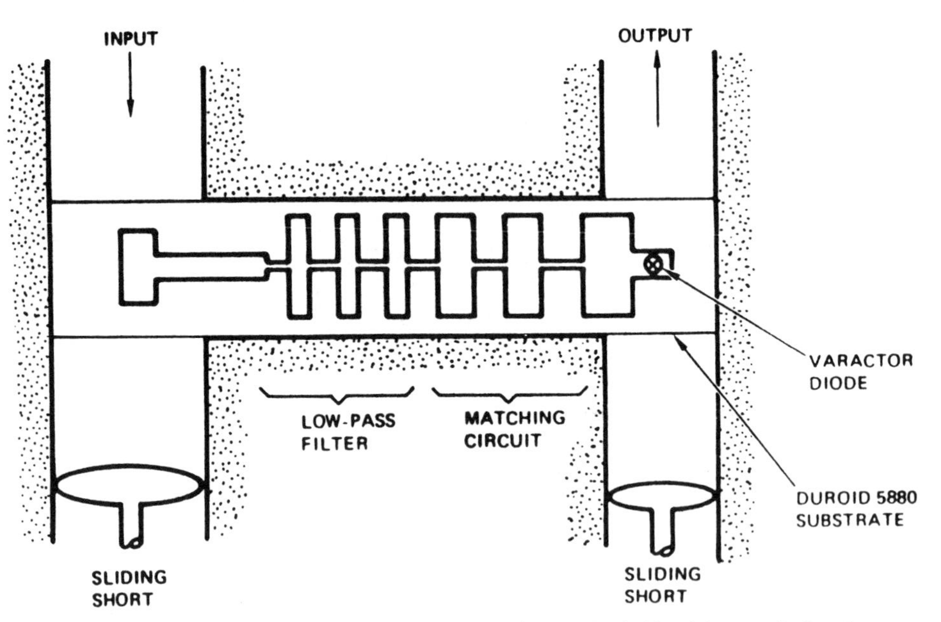

FIGURE 5.15 Multiplier circuit configuration. (From Ref. 10 with permission from IEEE.)

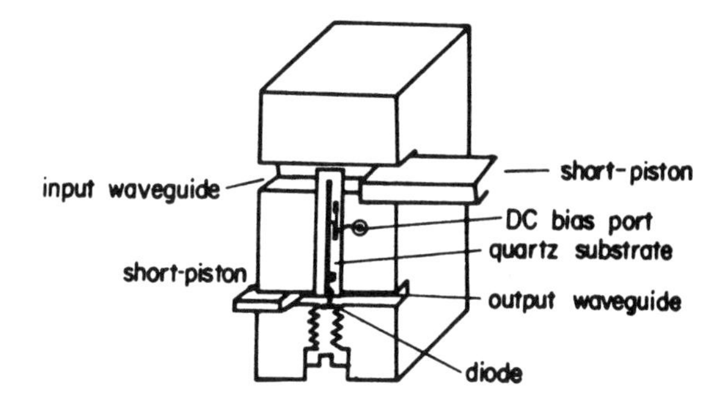

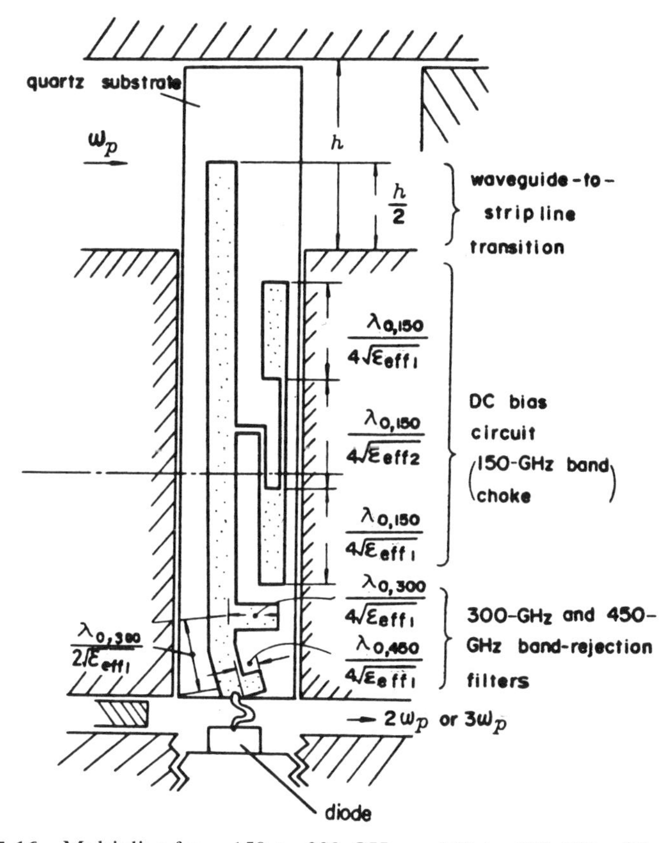

FIGURE 5.16 Multiplier from 150 to 300 GHz or 150 to 450 GHz. (From Ref. 12 with permission from IEEE.)

5.7 PARAMETRIC AMPLIFIER

The parametric amplifier uses the time variation of a reactive parameter to produce amplification. Van der Ziel first analyzed the nonlinear capacitances in 1948, and the first realization of a microwave parametric amplifier was by Weiss. In the parametric amplifier, normally three frequencies are present: the pump frequency f_p, the signal frequency f_s, and an idler frequency $f_i = f_p \pm f_s$. The output frequency is the same as the input. Figure 5.17 shows a schematic configuration of a parametric amplifier. The varactor is connected as a common element between the signal, idler, and pump ports. Ideally, current should not flow at any other frequencies which are terminated in open circuits. If this condition is satisfied, the only noise sources internal to the amplifier are the varactor series resistance and any resistance in the idler circuit. As a consequence of current flowing at the idler frequency, a negative resistance is presented to the external circuit at the signal frequency [15]. This negative resistance contributes to the signal amplification.

Parametric amplifiers provide very low noise amplification and were useful devices before field-effect-transistor (FET) amplifiers were developed in 1970s. FETs have replaced parametric amplifiers in many applications, due to the advantages of small size, low cost, and good performance. FET amplifiers are discussed in Chapter 14.

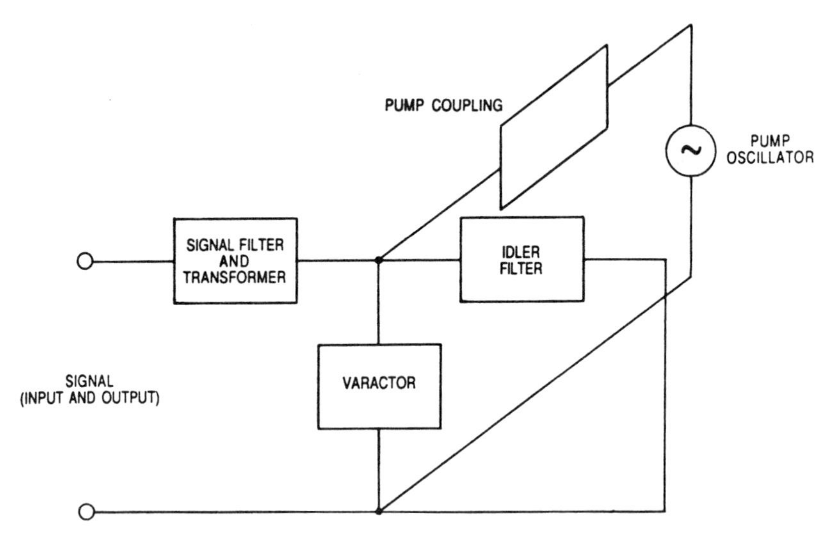

FIGURE 5.17 Schematic diagram of a parametric amplifier. (From Ref. 15 with permission from Wiley.)

PROBLEMS

P5.1 For a device with the structure shown in Figure P5.1, if $N_D = Bx^m$, show that:

(a) $W \simeq \left[\dfrac{\varepsilon_s(m+2)(V_{bi}-V)}{qB}\right]^{1/(m+2)}$

(b) $C_j = A\left[\dfrac{qB\varepsilon_s^{m+1}}{(m+2)(V_{bi}-V)}\right]^{1/(m+2)} = C'_{j0}(V_{bi}-V)^{-\gamma}$

where $\gamma = 1/(m+2)$.

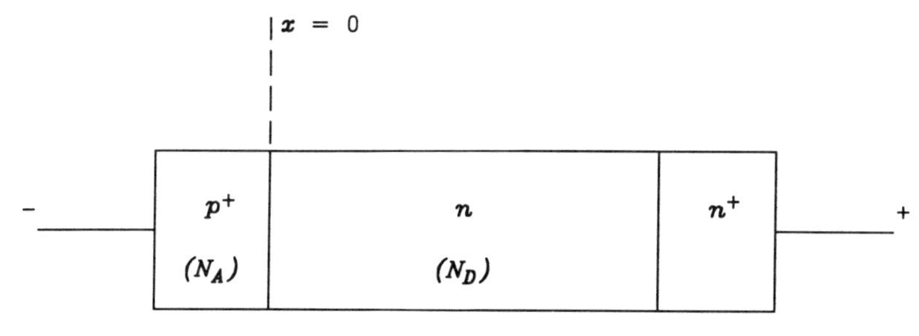

FIGURE P5.1

P5.2 Show that for any concentration of donors in the n layer of a p^+–n junction, the donor concentration as a function of W (the space-charge width) is given by

$$N_D(W) = \frac{-C_j^3(V)}{q\varepsilon_s A^2(dC_j/dV)}$$

This can be used as a method to measure $N_D(W)$.

P5.3 A linear junction GaAs varactor is used to tune an active oscillator as shown in Figure P5.3. At the bias voltage of −20 V, the varactor has a junction capacitance of 2 pF. $V_{bi} = 1.3$ V for a GaAs varactor. The active device and load have a combined impedance of

$$Z = -3 + j3\omega \times 10^{-10}\ \Omega$$

where ω is the angular frequency. The bonding wire connecting the varactor to Z has an inductance of 0.5 nH. Determine the oscillating frequency when the varactor is at **(a)** −20 V and **(b)** −10 V.

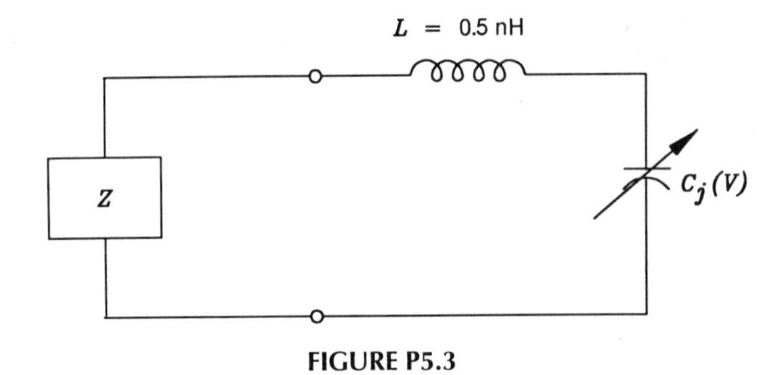

FIGURE P5.3

P5.4 A varactor has $R_s = 0.5\ \Omega$ and $C_j(V) = 2$ pF at 0 V. If this varactor is used as a doubler to convert a 15-GHz signal to 30 GHz, estimate the maximum efficiency using Figure 5.14.

P5.5 A GaAs abrupt junction varactor has a junction capacitance of 0.1 pF at a reverse bias of $V = -10$ V. The varactor is used to tune an oscillator with an impedance of $Z_D = -6 + j\omega L$ (ohms) where $L =$ 1 nH. The varactor is connected in series with the oscillator. Determine the operating frequency at $V = -20$ V.

P5.6 Explain how the circuit shown in Figure 5.16 works.

P5.7 For a GaAs varactor diode with $V_{\text{bi}} = 1.3$ V, plot $C_j(V)$ as a function of voltage and $N_D(x)$ as a function of x for the following two cases: **(a)** a linear junction with $m = 1$; **(b)** a hyperabrupt junction with $m = -1$. Assume that $C_{j0} = 4$ pF for both cases.

P5.8 Consider a linearly graded junction with the impurity distribution shown in Figure P5.8. The impurity density in the space-charge region can be represented by

$$\rho(x) = qax \qquad -\frac{w}{2} \leq x \leq \frac{w}{2}$$

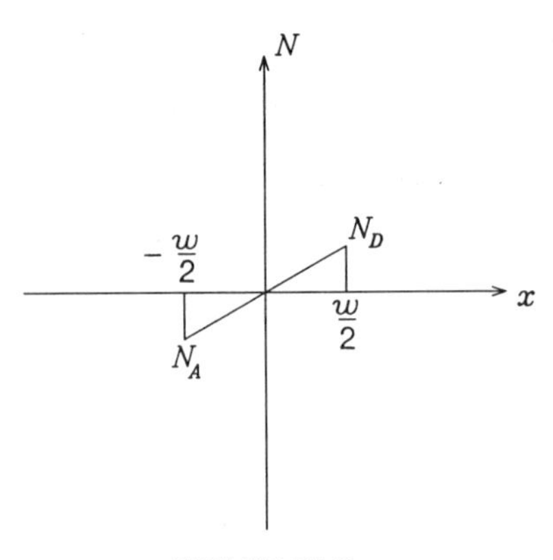

FIGURE P5.8

where q is the charge and a is a constant. Find $E(x)$ and E_{max}, and plot $E(x)$ as a function of x.

P5.9 A GaAs abrupt junction varactor is used to tune a resonator (filter) as shown in Figure P5.9. The varactor has a junction capacitance of 1 pF at zero bias. What are the resonant frequencies for the bias voltages of **(a)** 0 V, **(b)** −10 V, and **(c)** −20 V?

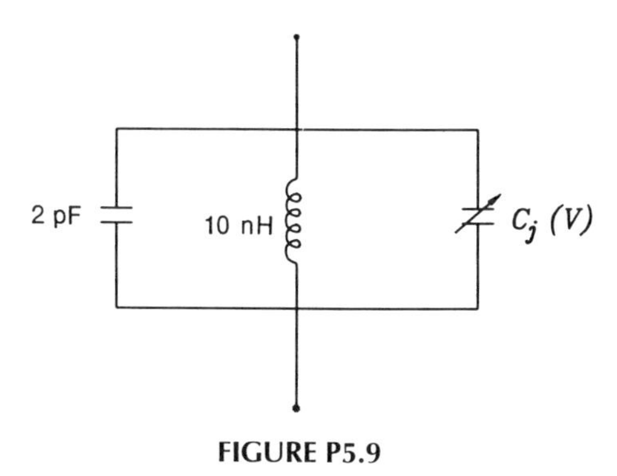

FIGURE P5.9

REFERENCES

1. K. K. N. Chang, *Parametric and Tunnel Diodes*, Prentice Hall, Englewood Cliffs, N.J., 1964.
2. S. M. Sze, *Physics of Semiconductor Devices*, 2nd ed., Wiley-Interscience, New York, 1981, Chap. 2.
3. J. Uher and W. J. R. Hoefer, "Tunable Microwave and Millimeter-Wave Band-Pass Filters," *IEEE Transactions on Microwave Theory and Techniques*, Vol. MTT-39, April 1991, pp. 643–653.
4. Y. Shu, J. A. Navarro, and K. Chang, "Electronically Switchable and Tunable Coplanar Waveguide-Slotline Band-Pass Filters," *IEEE Transactions on Microwave Theory and Techniques*, Vol. MTT-39, March 1991, pp. 548–554.
5. K. Chang, T. S. Martin, F. Wang, and J. L. Klein, "On the Study of Microstrip Ring and Varactor-Tuned Ring Circuits," *IEEE Transactions on Microwave Theory and Techniques*, Vol. MTT-35, December 1987, pp. 1288–1295.
6. T. S. Martin, F. Wang, and K. Chang, "Theoretical and Experimental Investigation of Novel Varactor-Tuned Switchable Microstrip Ring Resonator Circuits," *IEEE Transactions on Microwave Theory and Techniques*, Vol. MTT-36, December 1988, pp. 1733–1739.
7. J. W. Archer, "Millimeter Wavelength Frequency Multipliers," *IEEE Transactions on Microwave Theory and Techniques*, Vol. MTT-29, June 1981, pp. 552–557.

8. J. W. Gewartowski, "Varactor Applications," in *Microwave Semiconductor Devices and Their Applications*, H. A. Watson, Editor, McGraw-Hill, New York, 1969, Chap. 8.
9. C. H. Tang, "An Exact Analysis of Varactor Frequency Multipliers," *IEEE Transactions on Microwave Theory and Techniques*, Vol. MTT-14, April 1966, pp. 210–212.
10. R. S. Tahim, G. M. Hayashibara, and K. Chang, "High Frequency Q-W Band MIC Frequency Doubler," *Electronics Letters*, Vol. 19, No. 6, March 17, 1983, pp. 219–220.
11. J. W. Archer, "A High Performance Frequency Doubler for 80 to 120 GHz," *IEEE Transactions on Microwave Theory and Techniques*, Vol. MTT-30, May 1982, pp. 824–825.
12. T. Takada and M. Hirayama, "Hybrid Integrated Frequency Multipliers at 300 and 450 GHz," *IEEE Transactions on Microwave Theory and Techniques*, Vol. MTT-26, October 1978, pp. 733–737.
13. T. Takada and M. Ohmori, "Frequency Triplers and Quadruplers with GaAs Schottky-Barrier Diodes at 450 and 600 GHz," *IEEE Transactions on Microwave Theory and Techniques*, Vol. MTT-27, May 1979, pp. 519–523.
14. T. Takada, T. Makimura, and M. Ohmori, "Hybrid Integrated Frequency Doublers and Triplers to 300 and 450 GHz," *IEEE Transactions on Microwave Theory and Techniques*, Vol. MTT-28, September 1980, pp. 966–973.
15. J. W. Archer and R. A. Batchelor, "Multipliers and Parametric Devices," in *Handbook of Microwave and Optical Components*, K. Chang, Editor, Wiley, New York, 1990, Vol. 2, Chap. 3.

FURTHER READING

1. S. M. Sze, *Physics of Semiconductor Devices*, 2nd ed., Wiley, New York, 1981, Chap. 2.
2. J. W. Archer and R. A. Batchelor, "Multipliers and Parametric Devices," in *Handbook of Microwave and Optical Components*, K. Chang, Editor, Wiley, New York, 1990, Vol. 2, Chap. 3.
3. S. Yrgvesson, *Microwave Semiconductor Devices*, Kluwer, Norwell, Mass., 1991, Chap. 9.
4. I. Bahl and P. Bhartia, *Microwave Solid-State Circuit Design*, Wiley, New York, 1988, Chap. 13.
5. P. Penfield and R. P. Rafuse, *Varactor Applications*, MIT Press, Cambridge, Mass., 1962.
6. C. B. Burckhardt, "Analysis of Varactor Frequency Multipliers for Arbitrary Capacitance Variation and Drive Level," *Bell System Technical Journal*, Vol. 44, April 1965, pp. 675–692.
7. R. E. Collin, *Foundations for Microwave Engineering*, 2nd ed., McGraw-Hill, New York, 1992, Chap. 11.

CHAPTER 6

Detector and Mixer Devices and Circuits

6.1 INTRODUCTION

A detector is a device that converts a microwave signal into a dc voltage or demodulates a modulated microwave signal to recover a modulating low-frequency information-bearing signal. The detector normally suffers from $1/f$ noise (flicker noise) [1]. The sensitivity of a microwave receiver can be greatly increased by using the heterodyne principle to avoid the $1/f$ noise problem. In heterodyne microwave systems there is a need to convert frequencies up from an initial baseband frequency to a transmitted microwave frequency and then to reverse the process at the receiver. These conversions are done by mixers (downconverters or upconverters). Figure 6.1 shows the functions of these devices. In the downconverter, the microwave signal (RF) received is mixed with a local oscillator (LO) signal in a nonlinear resistance, and a difference signal [IF (intermediate-frequency) signal] is generated. An IF signal is easily amplified by an IF amplifier. The amplified IF signal will be further processed through mixing and detection to recover the information. The upconverter is used to generate a high-frequency RF signal for transmission from a low-frequency IF signal. A downconverter is normally used in a receiver and an upconverter in a transmitter.

Detector and mixer diodes use the nonlinear voltage–current (I–V) characteristics of a p–n junction or Schottky-barrier junction (i.e., metal–semiconductor junction). Low minority-carrier injection, a low junction capacitance, and a low series resistance are required for good performance. The requirement of low minority-carrier injection makes the Schottky barrier a better candidate for many applications. GaAs diodes are preferred for high-frequency uses.

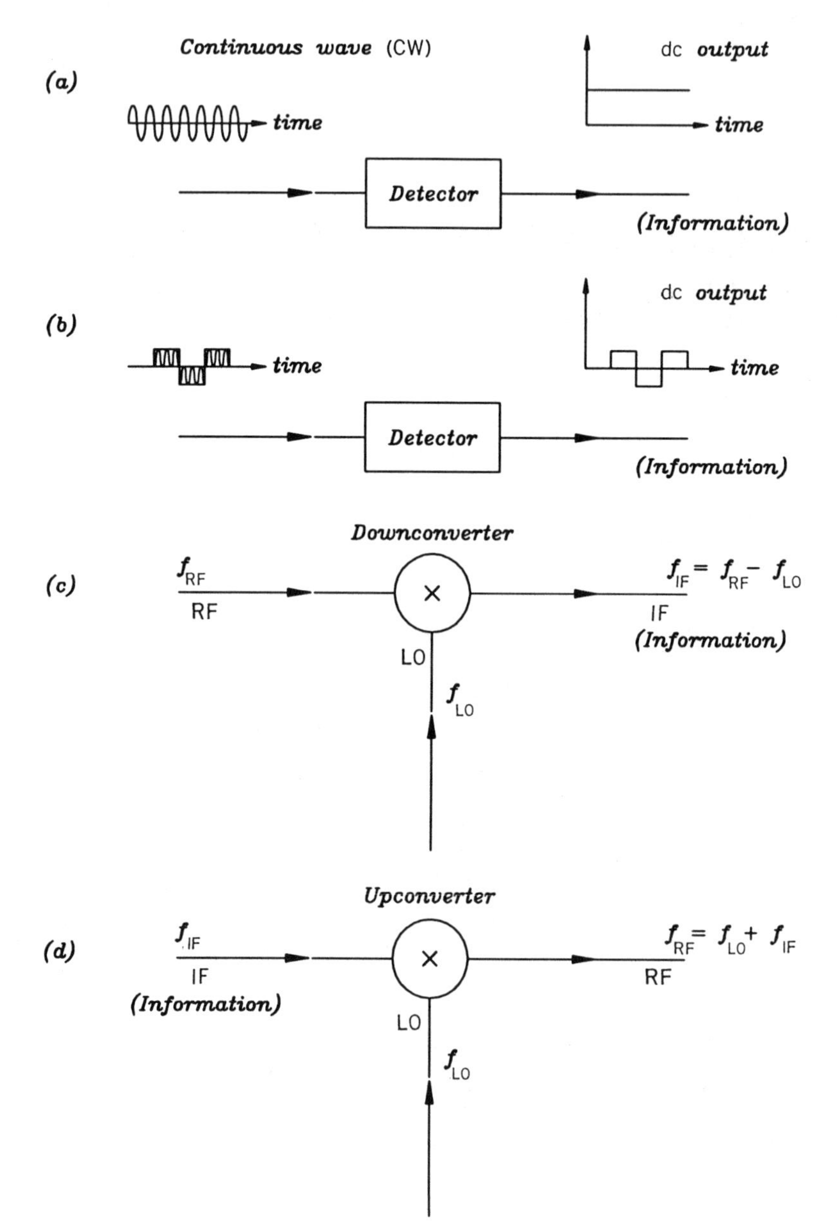

FIGURE 6.1 Functions of detectors and mixers.

Detector and mixer diodes can be operated in self-biased or forward-biased arrangements. In the self-biased circuit, the rectified dc current creates a forward bias, and no external bias is needed. The self-biased circuit has the advantages of simplicity and convenience. Detectors may be biased at an operating point that provides the best sensitivity. In a mixer, external bias can be used to improve performance if the LO power is too low.

Many detector and mixer circuits have been developed operating from low microwave frequencies to high millimeter-wave or submillimeter-wave frequencies. In this chapter we discuss the operating theory, circuit design, and various applications.

6.2 DEVICE PACKAGING CONSIDERATIONS

The choice of device packaging depends on the operating frequency and applications. Similar packages are used for the varactor devices given in Section 5.3. At high frequencies, beam-lead or whisker-contact devices are used for their small package parasitics. At lower frequencies, pill or flat enclosed packages are used. The pill and flat packages are generally more rugged and can be hermetically sealed. Pill or flat packages can be used for frequencies below 20 GHz, beam-lead packages for frequencies up to 100 GHz, and whisker-contact or open packages for frequencies up to 1000 GHz.

A. Pill, Flat, and Beam-Lead Packages

These package configurations are similar to those described in Chapter 5 for varactor diodes. The equivalent circuits are similar to those of the varactor except that $R_j(v)$ cannot be neglected since the diodes are forward biased. Figure 6.2(*a*) shows a beam-lead GaAs Schottky-barrier mixer diode.

B. Whisker-Contact Device

To reduce the package parasitics for high-frequency applications, a whisker-contact device is used. Figure 6.2(*b*) shows a whisker-contact mixer built in a rectangular waveguide and its equivalent circuit. The semiconductor chip consists of many dots of metallization for contact. The dot is about 1 to 2 μm in diameter. The device is thus called a honeycomb mixer diode. The whisker is a tiny metal wire that makes contact with one of the circular dots. Since this is an open, packageless diode, there is no C_p in the equivalent circuit [Figure 6.2(*b*)]. C_j and R_j are the junction capacitance and resistance of the device. R_s is the series resistance and L_s is the inductance due to the bonding whisker.

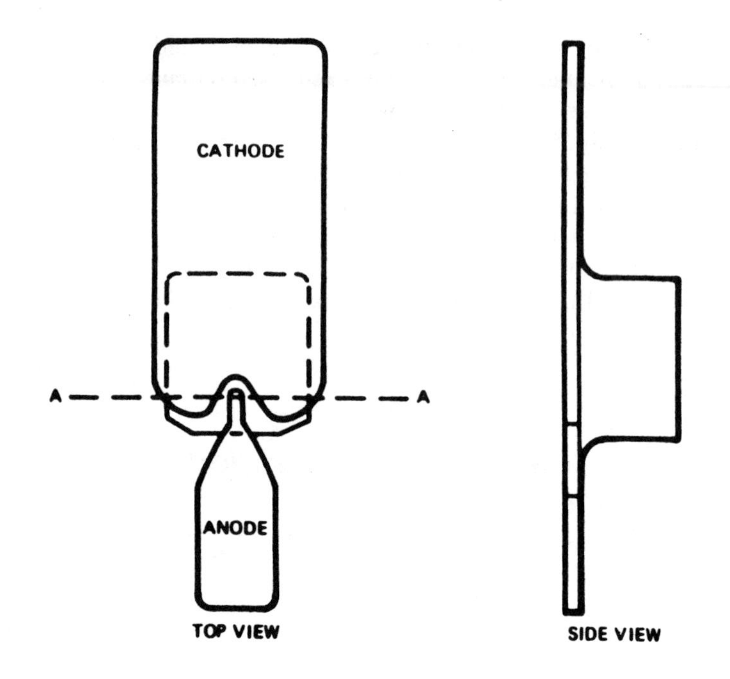

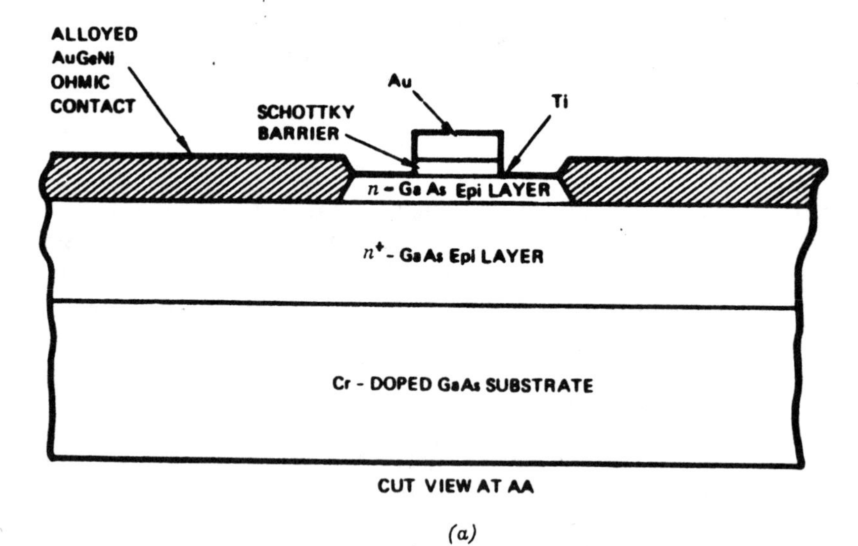

(a)

FIGURE 6.2 (*a*) Beam-lead GaAs Schottky-barrier mixer diode; (*b*) whisker-contact mixer device and its equivalent circuit.

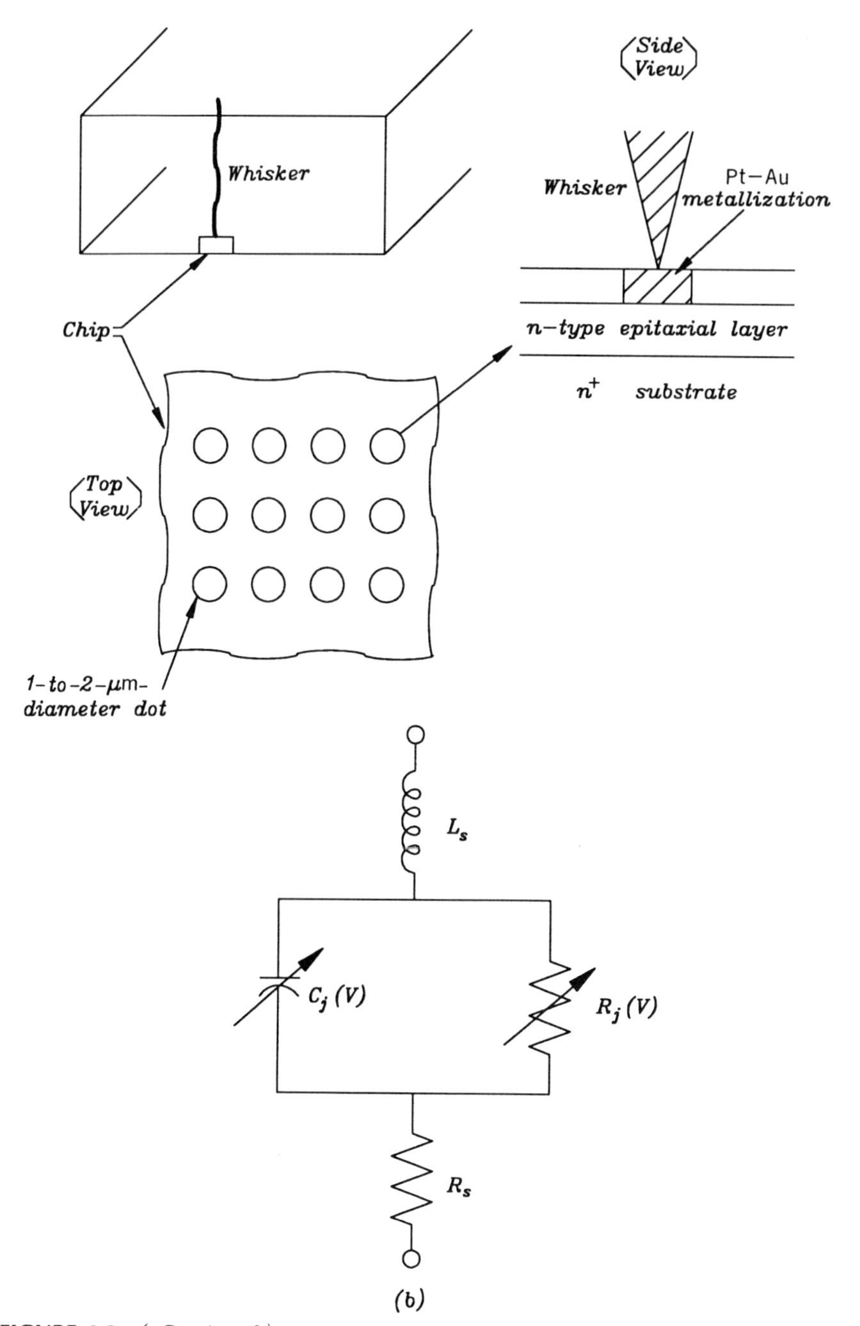

FIGURE 6.2 (*Continued.)*

6.3 DESIRABLE DEVICE PROPERTIES

For good performance, the detector or mixer diode should have the following features:

1. *Variable Nonlinear Resistance* (i.e., a varistor).
2. *High Cutoff Frequency*. The cutoff frequency is defined as $f_c = 1/2\pi R_s C_{j0}$. A device with a low series resistance and junction capacitance will have a high cutoff frequency. Normally, one would like to operate the mixer at an RF frequency of less than $\frac{1}{10}$ of f_c.
3. *Low Barrier Height*. A device with a low barrier height will require less LO pump power.
4. *High Breakdown Voltage*. A device with a high breakdown voltage can handle high RF power and has a high burned-out level.
5. *Quick Recovery After Saturation Caused by a Large Signal*. Using a unipolar device (majority carrier) with no minority-carrier storage is desirable. For this reason, the metal–semiconductor junction is more desirable than the p–n junction.

6.4 DETECTOR OPERATING THEORY

A detector diode has a nonlinear I–V characteristic, as shown in Figure 6.3. The characteristic is given by

$$I = I(V) = I_s(e^{qV/\eta kT} - 1) \tag{6.1}$$

where

I_s = diode saturation current

q = charge of an electron

V = voltage across junction

η = a constant between 1 and 2

k = Boltzmann's constant

T = absolute temperature

The voltage applied to the junction consists of a dc bias voltage V_o superimposed on a small ac voltage δV.

$$V = V_o + \delta V \tag{6.2}$$

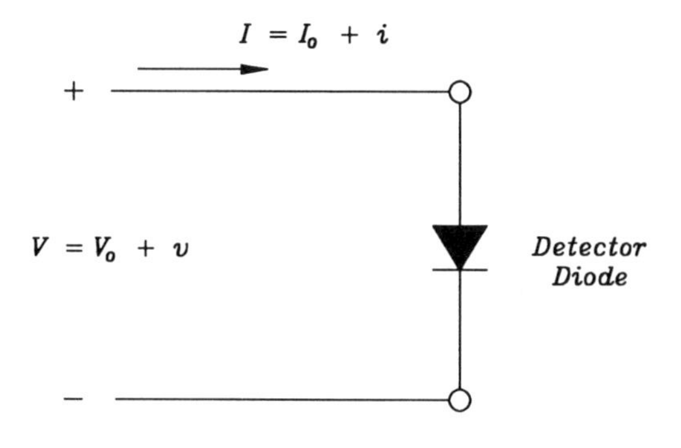

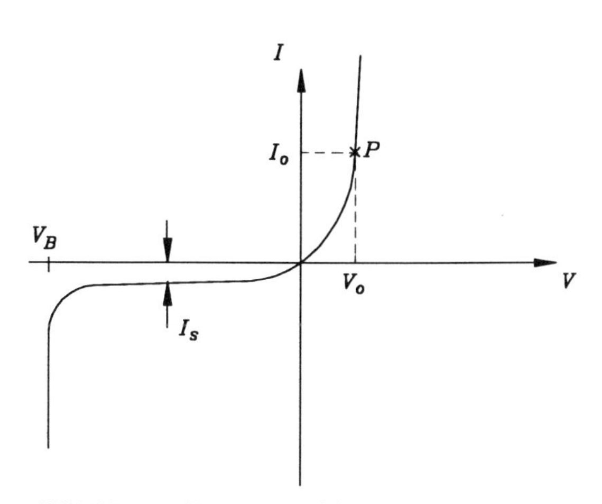

FIGURE 6.3 Detector and its I–V characteristic.

Equation (6.1) can be expanded in a Taylor series around V_o by substituting (6.2) into (6.1):

$$\begin{aligned} I &= I(V) = I(V_o + \delta V) \\ &= I(V_o) + \left.\frac{dI}{dV}\right|_{V_o} \delta V + \frac{1}{2} \left.\frac{d^2 I}{dV^2}\right|_{V_o} (\delta V)^2 + \cdots \\ &= I_o + \delta I \end{aligned} \tag{6.3}$$

where I_o is a dc bias current and δI is a small ac current.

Let $\delta V = v$, $\delta I = i$, $I_o = I(V_o)$, $dI/dV|_{V_o} = a_1$, $\frac{1}{2}d^2I/dV^2|_{V_o} = a_2$, and so on. Equation (6.3) becomes

$$i = a_1 v + a_2 v^2 + a_3 v^3 + \cdots \tag{6.4}$$

What follows are descriptions of how Equation (6.4) can be applied to unmodulated and amplitude modulated signals.

A. Unmodulated Signal

The detector can be used to convert an unmodulated microwave signal into a dc voltage or current. In the case of an unmodulated ac signal v applied to the terminal of a detector diode, the voltage can be written as

$$v = A \cos \omega_{\mathrm{RF}} t \quad \text{or} \quad A \sin \omega_{\mathrm{RF}} t \tag{6.5}$$

If v is very small, the current is

$$\begin{aligned} i &= a_1 v + a_2 v^2 + a_3 v^3 + \cdots \\ &\approx a_1 v = a_1 A \cos \omega_{\mathrm{RF}} t \end{aligned} \tag{6.6}$$

No dc or a very small dc output is detected.

If v is increased such that we only need to keep the first two terms in Equation (6.4), the current is

$$\begin{aligned} i &= a_1 v + a_2 v^2 + a_3 v^3 + \cdots \\ &\approx a_1 v + a_2 v^2 \end{aligned} \tag{6.7}$$

Substituting v into (6.7) gives

$$\begin{aligned} i &\approx a_1 A \cos \omega_{\mathrm{RF}} t + a_2 A^2 \cos^2 \omega_{\mathrm{RF}} t \\ &= a_1 A \cos \omega_{\mathrm{RF}} t + \frac{a_2 A^2}{2} + \frac{a_2 A^2}{2} \cos 2\omega_{\mathrm{RF}} t \end{aligned} \tag{6.8}$$

A dc current (i_{dc}) proportional to A^2 exists. This current is given by

$$i_{\mathrm{dc}} = \frac{a_2 A^2}{2} \propto A^2 \tag{6.9}$$

The detector is said to operate in the square-law region. Most detectors are designed to operate in this region. If a dc current meter or an oscilloscope is connected with the detector diode, one can read this current level.

B. Amplitude-Modulated Signal

A detector can be used to demodulate an amplitude-modulated (AM) signal to recover the modulating low-frequency information-bearing signal as shown in Figure 6.4. In this case, the ac signal applied to the terminal is [2]

$$\begin{aligned} v &= \text{AM signal} \\ &= A(1 + m \sin \omega_m t) \sin \omega_{RF} t \qquad \omega_{RF} \gg \omega_m \end{aligned} \tag{6.10}$$

where A is the amplitude of the carrier, m the modulation index, ω_m the angular frequency of the modulating signal, and ω_{RF} the angular frequency of the RF carrier. The condition $m \leq 1$ sets an upper limit on how heavily the carrier can be modulated. With $m = 1$, the modulation is 100%. Note that $\omega_{RF} \gg \omega_m$. f_{RF} is in the microwave frequency range (GHz) and f_m is in the audio or video frequency range (kHz or MHz).

Equation (6.10) can be rewritten as

$$v = A\left\{\sin \omega_{RF} t + \frac{m}{2}\left[\cos(\omega_{RF} - \omega_m)t - \cos(\omega_{RF} + \omega_m)t\right]\right\} \tag{6.11}$$

Assuming that the detector is operating in the square-law region, substituting (6.11) into (6.7) gives (see Problem P6.4)

$$\begin{aligned} i \approx a_1 v + a_2 v^2 = Aa_1 &\left[\sin \omega_{RF} t + \frac{m}{2} \cos(\omega_{RF} - \omega_m)t \right. \\ &\left. \quad - \frac{m}{2} \cos(\omega_{RF} + \omega_m)t\right] \\ + \frac{1}{2} a_2 A^2 &\left[\left(1 + \frac{m^2}{2}\right) + 2m \sin \omega_m t - \frac{m^2}{2} \cos 2\omega_m t \right. \\ &+ \frac{m^2}{4} \cos 2(\omega_{RF} - \omega_m)t + m \sin(2\omega_{RF} - \omega_m)t \\ &- \left(1 + \frac{m^2}{2}\right) \cos 2\omega_{RF} t - m \sin(2\omega_{RF} + \omega_m)t \\ &\left. + \frac{m^2}{4} \cos 2(\omega_{RF} + \omega_m)t\right] \end{aligned} \tag{6.12}$$

The frequency spectrum of this output is illustrated in Figure 6.5. The desired output signal $\sin \omega_m t$ is obtained by using a low-pass filter. The output signal is

$$i_{out} = a_2 A^2 m \sin \omega_m t \propto A^2 \tag{6.13}$$

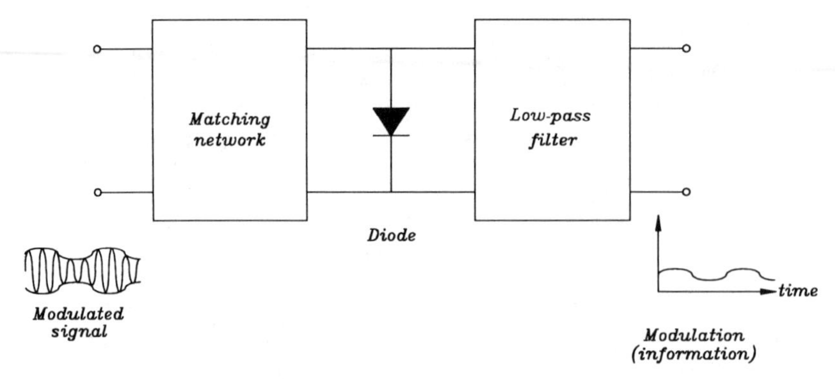

FIGURE 6.4 Detector circuit.

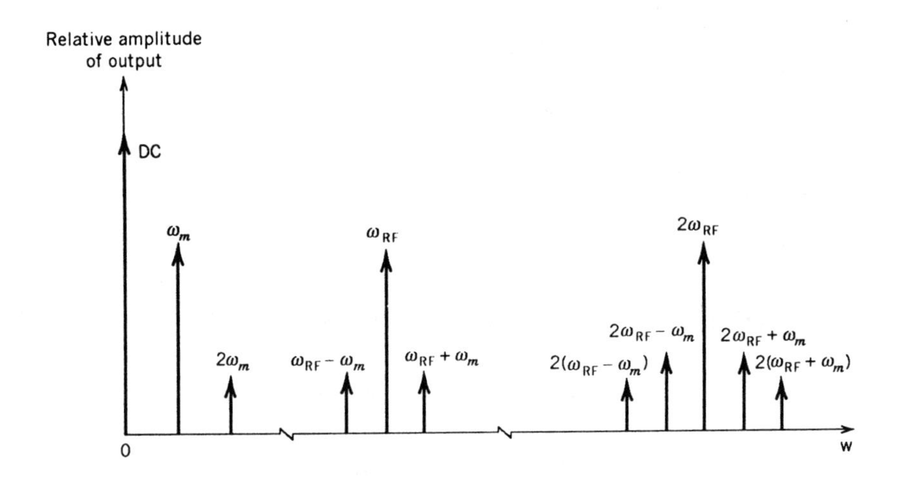

Angular Frequency	Relative Amplitude (a_1 A)	Angular Frequency	Relative Amplitude ($\frac{1}{2}a_2$ A^2)
ω_{RF}	1	dc	$1+\frac{m^2}{2}$
$\omega_{RF} - \omega m$	$\frac{m}{2}$	ω_m	$2m$
$\omega_{RF} + \omega m$	$\frac{-m}{2}$	$2\omega_m$	$\frac{-m^2}{2}$
		$2(\omega_{RF} - \omega_m)$	$\frac{m^2}{4}$
		$2\omega_{RF} - \omega_m$	m
		$2\omega_{RF}$	$-(1+\frac{m^2}{2})$
		$2\omega_{RF} + \omega_m$	$-m$
		$2(\omega_{RF} + \omega_m)$	$\frac{m^2}{4}$

FIGURE 6.5 Spectrum and relative amplitude of a detector diode operating at square-law region.

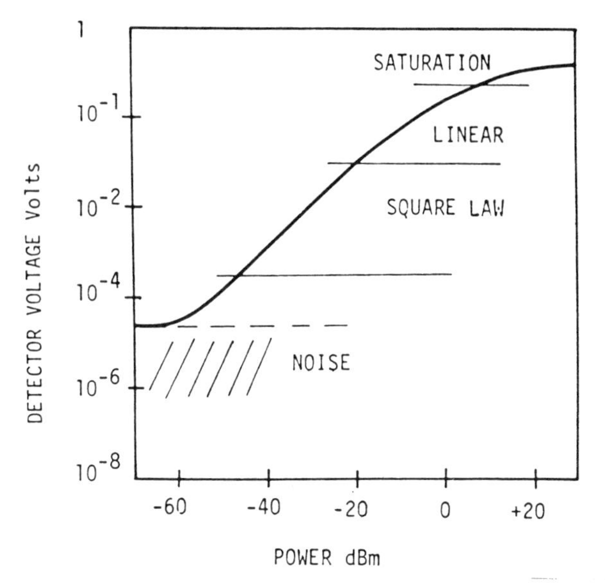

FIGURE 6.6 Typical diode detector output characteristic. (From Ref. 3.)

The output signal has the waveform of the modulating signal and the amplitude is proportional to the power of the input RF signal. The diode is operating in the square-law region. This region is valid only for a restricted range of input power levels, as shown in Figure 6.6. At higher signal levels, the detector will become linear, and at still higher levels the detector saturates.

6.5 DETECTOR SENSITIVITY

A detector is normally used for low-level detection (square-law region) in the microwave receivers to detect signal levels in the range from -60 to -30 dBm. A typical circuit representation is shown in Figure 6.7. The input-matching network is used to match the input line impedance to the diode and mounting circuit impedance. The diode is usually forward biased in the region 10 to 100 μA for optimum performance.

The current sensitivity of a detector is defined as

$$\beta_i = \frac{i_{\text{dc}}}{P_a} \tag{6.14}$$

where P_a is the input power absorbed by the diode and i_{dc} is the dc output

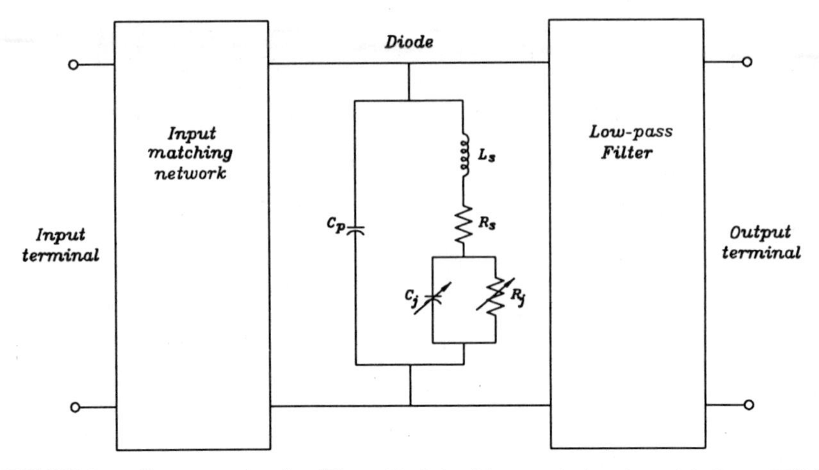

FIGURE 6.7 Detector circuits. (From Ref. 1 with permission from McGraw-Hill.)

current. From Figure 6.7 one can prove that

$$\beta_i = \frac{q}{2\eta kT} \frac{1}{(1 + R_s/R_j)^2} \frac{1}{1 + \omega^2 C_j^2 R_s R_j^2/(R_s + R_j)} \qquad (6.15)$$

The loss in input signal power delivered to R_j is given by

$$L \text{ (in dB)} = 10 \log\left(1 + \frac{R_s}{R_j} + \omega^2 C_j^2 R_s R_j\right) \qquad (6.16)$$

Example 6.1 Prove Equations (6.15) and (6.16), assuming that the package parasitics C_p and L_s are negligible, as shown in Figure 6.8.

SOLUTION: The rectified dc current can be found from Equation (6.9) as

$$i_{\text{dc}} = \frac{A^2}{2} a_2 = \frac{A^2}{2}\left(\frac{1}{2}\frac{d^2 I}{dV^2}\right)\bigg|_{V_o} = \frac{A^2}{4}\frac{d^2 I}{dV^2}\bigg|_{V_o}$$

where $\delta V = v = A \cos \omega_{\text{RF}} t$ and $I = I_s(e^{qV/\eta kT} - 1)$. The derivative of I is

$$\frac{dI}{dV}\bigg|_{V_o} = \frac{I_s q}{\eta kT} e^{qV_o/\eta kT}$$

$$= \frac{q}{\eta kT}(I_o + I_s) = \frac{1}{R_j}$$

Here we use the relation of

$$I_o = I_s(e^{qV_o/\eta kT} - 1)$$

The second derivative of I is

$$\left.\frac{d^2I}{dV^2}\right|_{V_o} = \frac{I_s q}{\eta kT}\frac{q}{\eta kT}e^{qV_o/\eta kT}$$

$$= \left(\frac{q}{\eta kT}\right)^2 (I_o + I_s) = \frac{q}{\eta kT}\frac{1}{R_j}$$

The dc current is thus given by

$$i_{\mathrm{dc}} = \frac{A^2}{4}\left.\frac{d^2I}{dV^2}\right|_{V_o} = \frac{A^2}{4}\frac{q}{\eta kT}\frac{1}{R_j}$$

To account for the series resistance effects on a nonideal junction, we need to replace R_j by $R_j + R_s$ [4] in the equation above. Therefore,

$$i_{\mathrm{dc}} = \frac{A^2}{4}\frac{q}{\eta kT}\frac{1}{R_s + R_j}$$

The power absorbed by the diode is given by

$$P_a = \frac{|V_d|^2}{2}\mathrm{Re}[Y_d]$$

Y_d is the diode admittance shown in Figure 6.8, given by

$$Y_d = \frac{1/R_j + j\omega C_j}{j\omega C_j R_s + (1 + R_s/R_j)}$$

and

$$\mathrm{Re}[Y_d] = \frac{(1/R_j)(1 + R_s/R_j) + (\omega C_j)^2 R_s}{(1 + R_s/R_j)^2 + (\omega C_j R_s)^2}$$

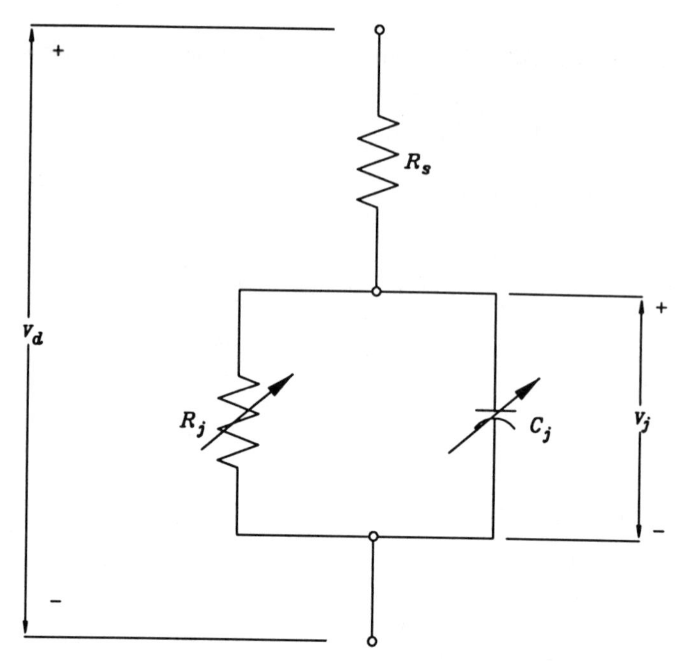

FIGURE 6.8 Diode equivalent circuit.

The power absorbed by the diode is thus written as

$$P_a = \frac{|V_d|^2}{2} \frac{(1/R_j)(1 + R_s/R_j) + (\omega C_j)^2 R_s}{(1 + R_s/R_j)^2 + (\omega C_j R_s)^2}$$

The junction voltage V_j can be related to V_d by the following equation:

$$V_j = \frac{V_d}{(1 + R_s/R_j) + j\omega C_j R_s}$$

The magnitude of V_j is

$$|V_j| = A = \frac{|V_d|}{\left[(1 + R_s/R_j)^2 + (\omega C_j R_s)^2\right]^{1/2}}$$

Substituting this into P_a, we have the power absorbed by the diode:

$$P_a = \frac{A^2}{2}\left[\frac{1}{R_j}\left(1 + \frac{R_s}{R_j}\right) + (\omega C_j)^2 R_s\right]$$

The total power available is

$$P_t = \frac{A^2}{2}\frac{1}{R_j}$$

The loss is calculated by

$$L\ (\text{in dB}) = 10\log\frac{P_a}{P_t}$$
$$= 10\log\left(1 + \frac{R_s}{R_j} + \omega^2 C_j^2 R_s R_j\right)$$

which is Equation (6.16). The detector current sensitivity is given by

$$\beta_i = \frac{i_{\text{dc}}}{P_a} = \left(\frac{A^2}{4}\frac{q}{\eta kT}\frac{1}{R_s + R_j}\right)\left[\frac{2}{A^2}\frac{1}{(1/R_j)(1 + R_s/R_j) + (\omega C_j)^2 R_s}\right]$$
$$= \frac{q}{2\eta kT}\frac{1}{(1 + R_s/R_j)^2}\frac{1}{1 + \omega^2 C_j^2 R_s R_j^2/(R_s + R_j)}$$

which is Equation (6.15).

β_i is generally expressed in μA/μW. The highest values are usually obtained when the dc bias is between 10 and 100 μA. An open-circuit voltage sensitivity can be defined as the voltage drop across the junction resistance when the diode is open circuited. We have

$$\beta'_v = \beta_i R_j$$

To correct for noninfinite load resistors, the sensitivity must be multiplied by $R_L/(R_v + R_L)$, where R_L is the load resistance and R_v is the diode video resistance, which is equal to $R_s + R_j$ [5]. The detector voltage sensitivity is expressed by

$$\beta_v = \beta'_v \frac{R_L}{R_v + R_L} \tag{6.17}$$

Typical values for β_v are 1000 mV/mW at 10 GHz and 500 mV/mW at 95 GHz.

Detector sensitivity can be expressed in different ways. One common term is *tangential sensitivity*. The measurement of tangential sensitivity requires some degree of judgment and is normally done using an oscilloscope. As shown in Figure 6.9, the observer sets the input power level at the value

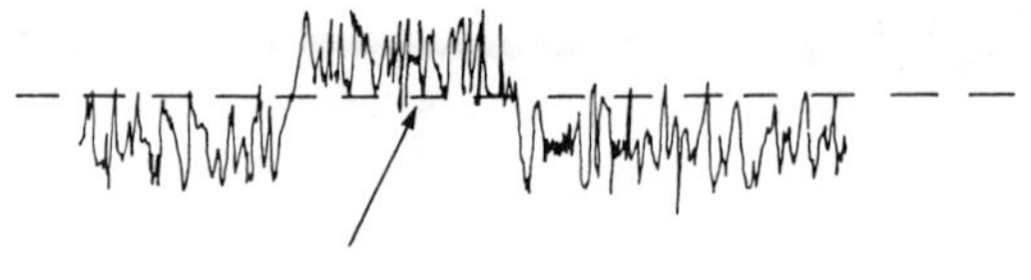

FIGURE 6.9 Measurement of the tangential sensitivity. (From Ref. 3.)

where, in his or her opinion, the highest noise peaks in the absence of a signal are at the same level as the lowest noise peaks in the presence of the signal. The input signal level (in dBm) is then the tangential sensitivity measured.

6.6 DETECTOR CIRCUITS

Many detectors in coaxial, waveguide, or microstrip circuits are commercially available. Figure 6.10 shows an example of the circuit arrangement. The bias circuit is designed to set the current through the diode equal to I_o for zero input power. In this example, the output low-frequency video signal is amplified by a low-noise video amplifier.

Figure 6.11 illustrates a waveguide detector circuit. The detector diode is mounted at the end of a metal post in a reduced-height waveguide. The reduced-height waveguide has a lower guide impedance than the regular height waveguide and thus provides better impedance matching to the diode. A sliding short is used for impedance matching and a coaxial-line low-pass filter is incorporated at the detector output circuit. Figure 6.12 shows a picture of this type of detector. Detectors can also be built in coaxial-line circuits (Figure 6.13).

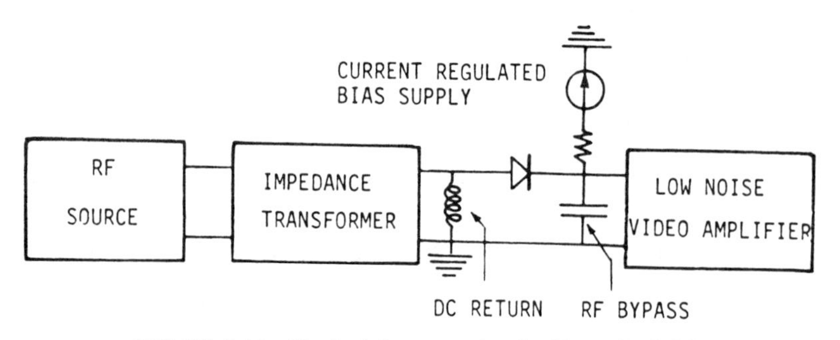

FIGURE 6.10 Typical detector circuit. (From Ref. 3.)

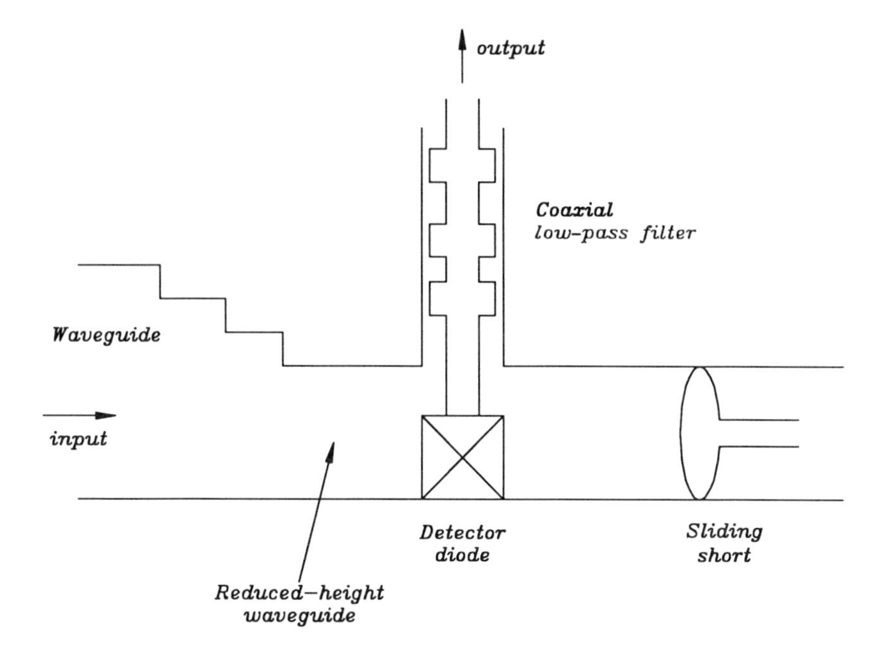

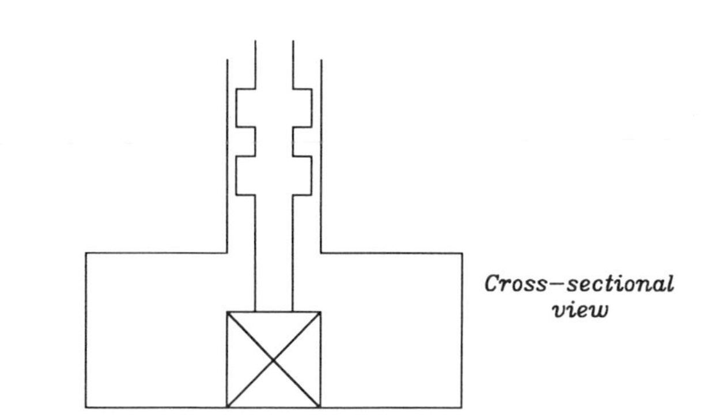

FIGURE 6.11 Waveguide detector circuit.

6.7 RECTENNAS

Rectenna stands for rectifying antenna. It converts RF microwave power into dc power in a microwave power transmission system. Figure 6.14 shows a free-space microwave power transmission system that converts dc power to RF power, transmits the RF power through free space, and then converts the RF power back to dc power. The microwave power transmission technique provides a feasible alternative to delivering power to a remote location [6].

FIGURE 6.12 Waveguide detectors. (Courtesy of Hughes Aircrafts Co.)

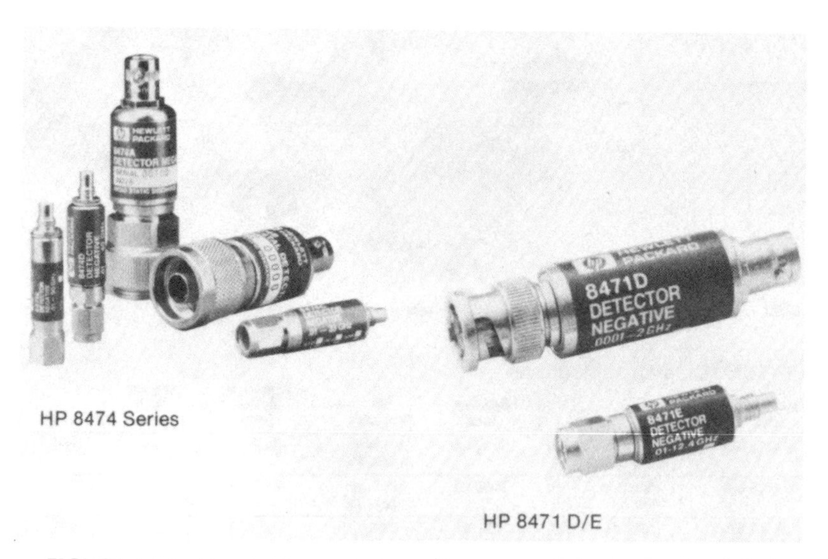

FIGURE 6.13 Coaxial-line detectors. (Courtesy of Hewlett-Packard Co.)

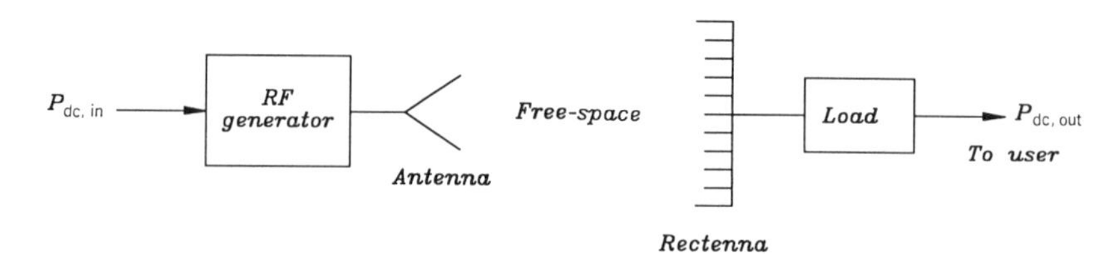

FIGURE 6.14 Microwave power transmission system block diagram.

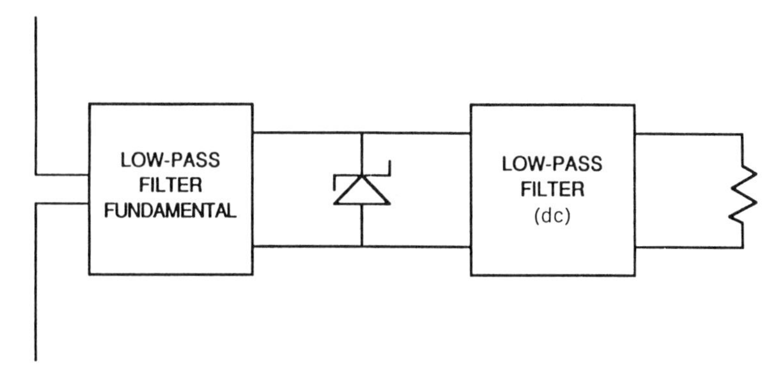

FIGURE 6.15 Simplified block diagram for a rectenna element.

A rectenna consists of an array of rectenna elements. Figure 6.15 shows a simple model for a single element. The main components of the rectenna element are a half-wavelength dipole and a rectifying diode. Between the dipole and the diode is a low-pass filter that passes the fundamental frequency of the incident radiation on the dipole but blocks harmonics generated by the diode from radiating from the dipole. Between the diode and the load is another low-pass filter that passes dc current from the diode to the load. To achieve high current or high voltage, several rectenna elements can be connected in parallel or series to a common load.

For rectenna applications, the diode is normally operated in a large-signal region with a high level of input signal. Conversion efficiency is used to measure the performance of a rectenna. The conversion efficiency is defined as

$$\eta = \frac{P_{\mathrm{dc}}}{P_{\mathrm{RF}}} \tag{6.18}$$

where P_{RF} is the RF power input to the rectenna element and P_{dc} is the output dc power. At 2.45 GHz, over 80% conversion efficiency has been achieved using integrated circuit technology [6]. At 35 GHz, a rectenna was recently developed with about 40% conversion efficiency [7]. The 35-GHz rectenna is shown in Figure 6.16.

6.8 MIXER (DOWNCONVERTER) OPERATING THEORY

The sensitivity of a detector is limited by the noise generated in the diode and the amplifier circuit that follows the detector. As shown in Figure 6.17, at low modulation frequencies, this noise is inversely proportional to the modulation frequency (called $1/f$ *noise*). As the modulation frequencies

FIGURE 6.16 A 35-GHz rectenna element. (From Ref. 7 with permission from IEEE.)

become sufficiently high, the "white" noise becomes the major noise source. The sensitivity of a microwave receiver can be improved significantly by using the heterodyne scheme, which alleviates the $1/f$ noise problem [1].

In the heterodyne scheme shown in Figure 6.18, the microwave signal received (RF) is first mixed with a local oscillator signal (LO) in a nonlinear circuit to generate an intermediate difference signal (IF). The IF signal can

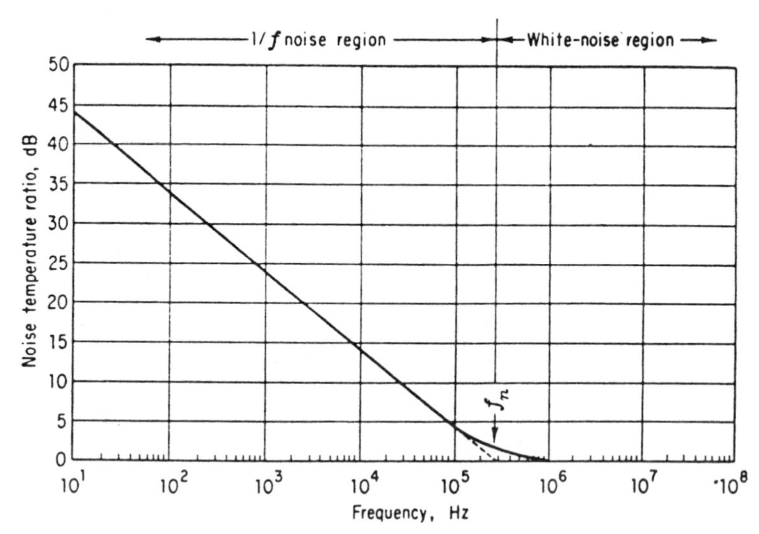

FIGURE 6.17 Noise spectrum for a gallium arsenide detector diode. Noise-temperature ratio is defined as the ratio of output noise of the device to output noise of an equivalent resistor. (From Ref. 1 with permission from McGraw-Hill.)

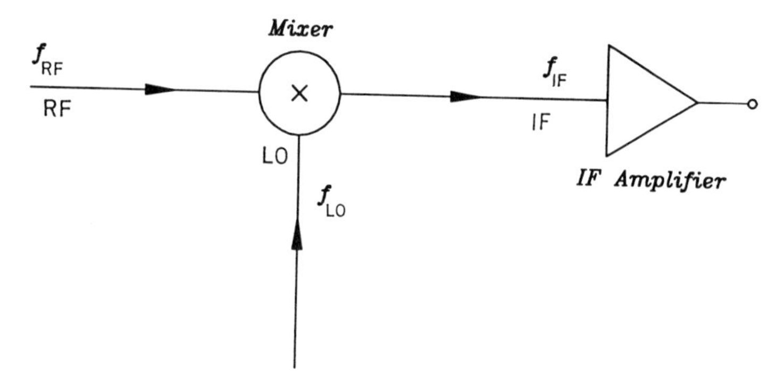

FIGURE 6.18 Heterodyne receiver.

easily be amplified by a low-noise IF amplifier before it is processed. The IF signal is selected at a frequency much higher than 1 MHz to avoid the $1/f$ noise problem. The mixer thus provides much better sensitivity than a detector. The price paid is that the mixer circuit is more complicated and a LO source is needed.

The mixer diode is the same as a detector diode, which is a p–n junction or Schottky-barrier device. At the forward bias (point P in Figure 6.3), the device is represented by a nonlinear resistance. The forward bias is normally achieved by the LO pump power, which is rectified by the diode. External bias is needed only when the LO pump power is too small for optimum bias. The ac current and voltage are related by Equation (6.4), which is rewritten here:

$$i = a_1 v + a_2 v^2 + a_3 v^3 + \cdots$$

Now the voltage can be expressed as

$$v = v_{\mathrm{RF}} \sin \omega_{\mathrm{RF}} t + v_{\mathrm{LO}} \sin \omega_{\mathrm{LO}} t$$

where v_{RF} and v_{LO} are the amplitudes of the signals. Therefore, the current becomes

$$\begin{aligned} i &= a_1 v + a_2 v^2 + \cdots \\ &= a_1(v_{\mathrm{RF}} \sin \omega_{\mathrm{RF}} t + v_{\mathrm{LO}} \sin \omega_{\mathrm{LO}} t) \\ &\quad + a_2(v_{\mathrm{RF}} \sin \omega_{\mathrm{RF}} t + v_{\mathrm{LO}} \sin \omega_{\mathrm{LO}} t)^2 \cdots \\ &= a_1(v_{\mathrm{RF}} \sin \omega_{\mathrm{RF}} t + v_{\mathrm{LO}} \sin \omega_{\mathrm{LO}} t) \\ &\quad + a_2\left\{\tfrac{1}{2} v_{\mathrm{RF}}^2 (1 - \cos 2\omega_{\mathrm{RF}} t) + v_{\mathrm{LO}} v_{\mathrm{RF}} \left[\cos(\omega_{\mathrm{RF}} - \omega_{\mathrm{LO}}) t \right.\right. \\ &\qquad \left.\left. - \cos(\omega_{\mathrm{RF}} + \omega_{\mathrm{LO}}) t\right] + \tfrac{1}{2} v_{\mathrm{LO}}^2 (1 - \cos 2\omega_{\mathrm{LO}} t)\right\} + \cdots \end{aligned} \tag{6.19}$$

Since ω_{RF} is usually near ω_{LO}, the difference frequency ($\omega_{RF} - \omega_{LO}$) can be extracted from the mixer by using a low-pass filter. All other frequencies are trapped and remixed with the LO to generate the desired IF signal or dissipated in the device and circuit resistors. The difference or intermediate frequency is

$$\omega_{IF} = \omega_{RF} - \omega_{LO} \quad \text{or} \quad \omega_{LO} - \omega_{RF} \tag{6.20}$$

A mixer circuit consists of matching networks for the RF, LO, and IF ports and a low-pass filter in the IF port. Sometimes, an RF bandpass filter or LO bandpass filter is incorporated to improve the isolation. Figure 6.19 shows a single-ended mixer circuit.

The performance of a mixer is judged by its conversion loss. The conversion loss is defined as

$$L_c \text{ (in dB)} = 10 \log \frac{P_{RF}}{P_{IF}} \tag{6.21}$$

This conversion loss is the single-sideband (SSB) conversion loss since we consider only the conversion to the IF frequency, which is either $f_{RF} - f_{LO}$ or $f_{RF} + f_{LO}$. The double-sideband (DSB) conversion loss is 3 dB lower than

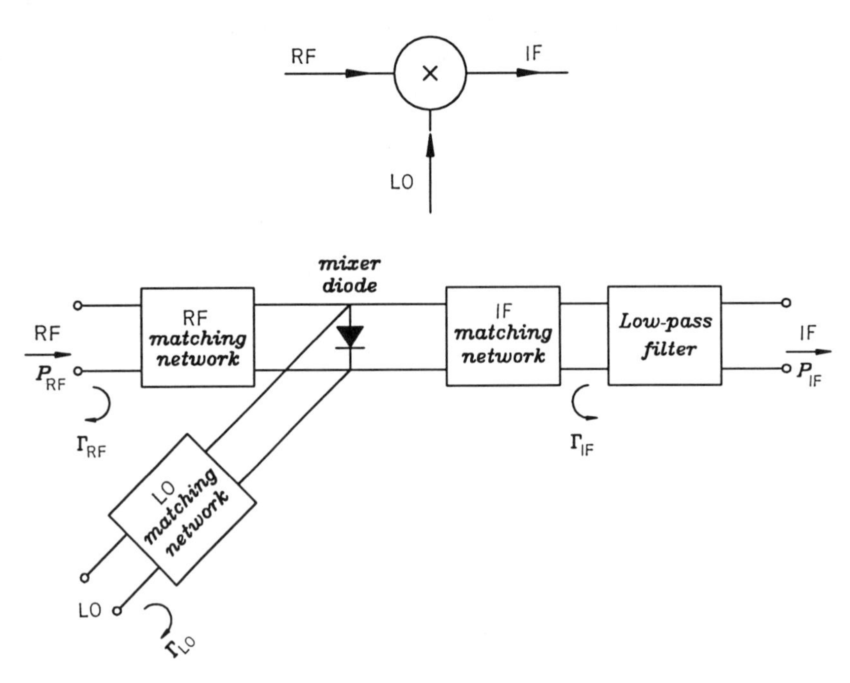

FIGURE 6.19 Single-ended mixer circuit block diagram.

the SSB conversion loss. Unless otherwise noted, the conversion loss throughout this book is the SSB conversion loss. For example, if $P_{RF} = 1$ mW and $P_{IF} = 0.1$ mW, the conversion loss is 10 dB. The conversion loss is attributed to three major losses L_1, L_2, and L_3, which are given below.

$$L_c = L_1 + L_2 + L_3 \qquad \text{dB} \tag{6.22}$$

where

$$L_1 = \text{mismatch loss} = 10 \log\left[\left(1 - |\Gamma_{RF}|^2\right)\left(1 - |\Gamma_{IF}|^2\right)\right]^{-1} \tag{6.23}$$

$$L_2 = \text{parasitic loss} = 10 \log\left[1 + \frac{R_s}{R_j} + \left(\omega C_j\right)^2 R_s R_j\right] \tag{6.24}$$

L_3 = intrinsic loss due to the mixing action. The mixer generates many components, but only $\omega_{RF} - \omega_{LO}$ is useful.

Note that Equation (6.24) is the same as Equation (6.16). Equation (6.23) is derived from the fact that the transmitted power is related to the input power at a mismatched port by

$$P_t = P_i\left(1 - |\Gamma|^2\right) \qquad \text{transmitted power}$$

$$P_r = P_i|\Gamma|^2 \qquad \text{reflected power}$$

where P_i is the input power and Γ is the reflection coefficient at this particular port. The power loss for this port is thus

$$\frac{P_i}{P_t} = \frac{1}{1 - |\Gamma|^2} \tag{6.25}$$

With the RF and IF ports in cascade, the total mismatch loss is equal to

$$L_1 = 10 \log\left[\left(\frac{P_i}{P_t}\right)_{RF}\left(\frac{P_i}{P_t}\right)_{IF}\right] = 10 \log\left[\frac{1}{\left(1 - |\Gamma_{RF}|^2\right)\left(1 - |\Gamma_{IF}|^2\right)}\right]$$

Other important parameters in designing a mixer are RF-to-LO isolation and RF-to-IF isolation. RF-to-LO isolation indicates that the amount of RF power is leaking into the LO port, or vice versa. The leakage will increase the conversion loss and cause the receiver to radiate at the LO frequency. Both results are not desirable. Normally, one would like to have this isolation greater than 20 dB, meaning less than 1% of leakage. The same argument applies to the RF-to-IF isolation.

For a receiver or mixer design, other parameters, such as noise figure, 1-dB compression point, third-order intermodulation, and dynamic range, are of interest. These parameters are discussed in Chapter 7.

The typical performance for a mixer is 4 to 5 dB conversion loss at 10 GHz, 5 to 6 dB at 45 GHz, and 7 to 9 dB at 94 GHz. The conversion loss depends on the operating bandwidth: the wider the bandwidth, the higher the conversion loss.

Example 6.2 Give specifications of mixers used at 20 GHz for communication applications (medium bandwidth) and at 94 GHz for radar applications (narrow bandwidth).

SOLUTION: At 20 GHz for communication applications:

RF bandwidth	19.0 to 21.0 GHz
Conversion loss	5 dB maximum
LO/RF isolation	20 dB minimum
LO pump power	10 mW maximum
External bias	No
RF match (VSWR)	1.5 maximum (i.e., 14 dB return loss)
IF match (VSWR)	1.5 maximum

At 94 GHz for radar applications:

RF bandwidth	94 to 94.5 GHz
Conversion loss	8 dB maximum
LO/RF isolation	20 dB minimum
LO pump power	3 mW maximum
External bias	may be required
RF match (VSWR)	2.0 maximum
IF match (VSWR)	2.0 maximum

Example 6.3 A mixer has the following parameters: $R_s = 2\ \Omega$, VSWR at RF port = 2.0, $C_j = 0.2$ pF, VSWR at IF port = 3.0, $R_j = 100\ \Omega$.

(a) Calculate the conversion loss at 10 GHz assuming that the package parasitics of the device are negligible and that the mixer has an intrinsic loss of 3 dB.

(b) If a RF signal of 1 mW is incident on the RF port, what is the output IF signal in mW?

SOLUTION: (a) At the RF port, VSWR = 2.0.

$$|\Gamma_{RF}| = \frac{\text{VSWR} - 1}{\text{VSWR} + 1} = \frac{1}{3}$$

At the IF port, VSWR = 3.0.

$$|\Gamma_{IF}| = \frac{\text{VSWR} - 1}{\text{VSWR} + 1} = \frac{1}{2}$$

$$L_1 = 10 \log\left[\left(1 - |\Gamma_{RF}|^2\right)\left(1 - |\Gamma_{IF}|^2\right)\right]^{-1} = 1.76 \text{ dB}$$

$$L_2 = 10 \log\left[1 + \frac{R_s}{R_j} + \left(\omega C_j\right)^2 R_s R_j\right]$$

$$= 10 \log\left[1 + \frac{2}{100} + (6.28 \times 10^{10} \times 0.2 \times 10^{-12})^2 \times 2 \times 100\right]$$

$$= 10 \log(1.052) = 0.22 \text{ dB}$$

$$L_3 = 3 \text{ dB}$$

$$L_c = L_1 + L_2 + L_3 = 1.7 + 0.2 + 3 = 4.9 \text{ dB}$$

(b) $P_{RF} = 1 \text{ mW} = 0 \text{ dBm}$

$$P_{IF} = P_{RF} - L_c = -4.9 \text{ dBm} = 0.324 \text{ mW}$$

6.9 MIXER CIRCUITS

Three types of mixer circuits shown in Figure 6.20—single-ended, single-balanced, and double-balanced—are commonly used in microwave application. Single-ended mixers are the simplest type and require very low LO power since they use only one diode. The disadvantages are small dynamic range, poor intermodulation (IM) suppression, and low LO/RF isolation. The definitions of dynamic range, 1-dB compression point, and intermodulation are discussed in Chapter 7.

The single-balanced configuration has two diodes arranged so that the LO pump is 180° out of phase and the RF pump is in phase at the diodes, or vice versa. This balanced operation results in LO noise suppression. Although the LO power required is twice that for the single-ended mixer, the single-balanced mixer provides a larger dynamic range and good IM suppression. The noise is reduced dramatically compared with the single-ended mixer because the AM noise from the local oscillator at the signal frequency is canceled at the IF output, provided that the diodes are well matched.

The double-balanced mixer is composed of two single-balanced mixers. Double-balanced mixers theoretically generate only one-fourth of the possible IM products. Because four diodes are required, the LO power required for double-balanced mixers is twice that of single-balanced mixers. Hence the 1-dB compression point of a double-balanced mixer is higher than that of a

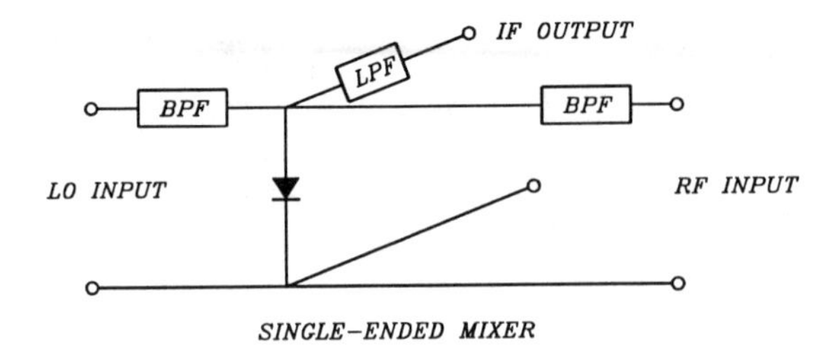

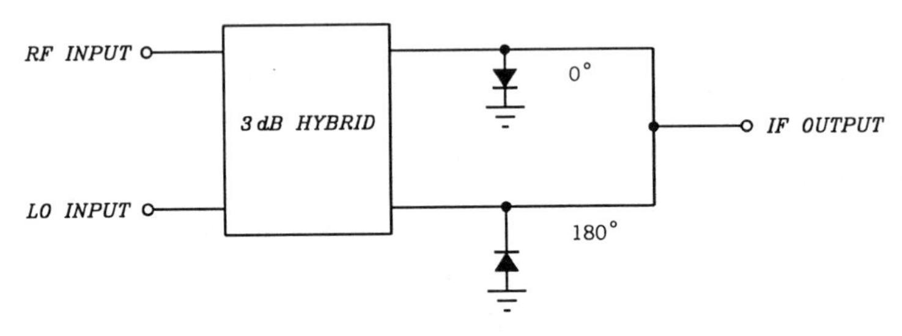

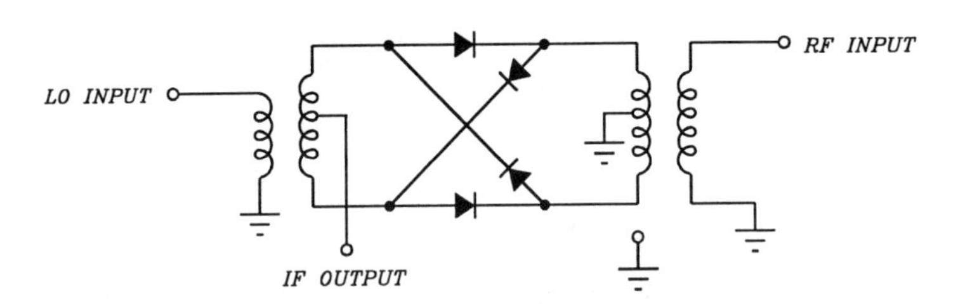

FIGURE 6.20 Mixer configurations.

single-balanced mixer, causing greater dynamic range and IM suppression. Some examples of mixer circuits are given next.

A. Microstrip Branch-Line Balanced Mixers

The branch-line mixer uses a 3-dB branch-line coupler to divide the LO power and RF signal into two ports, each coupled to a mixer diode. The IF output is recombined and coupled out through an IF filter. Figure 6.21 shows a circuit layout of a branch-line mixer. The branch line coupler has the property that power at the input port is split into two components with equal amplitudes but 90° out of phase. The mixer has been developed successfully at 20 GHz [8] with a 6-dB conversion loss over a 1-GHz bandwidth. At higher

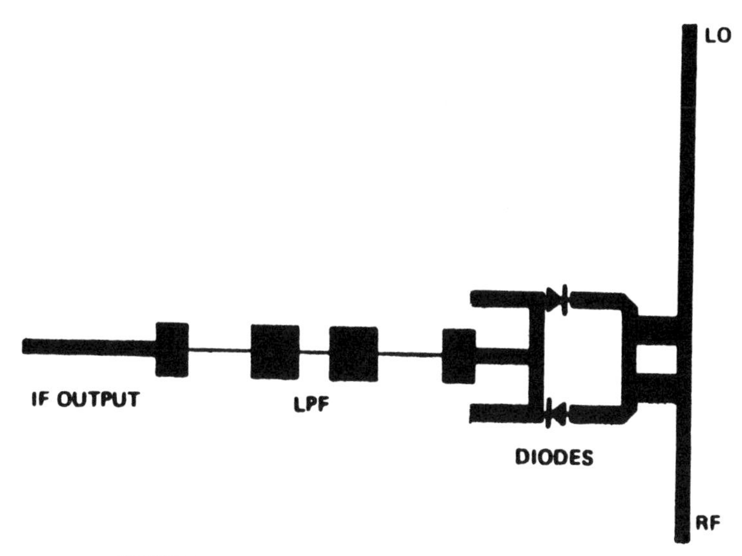

FIGURE 6.21 Circuit layout of a branch-line mixer.

frequencies, the design and fabrication of a good branch-line coupler become increasingly difficult. A conversion loss of 7 to 8 dB has been achieved at 60 GHz over a 1-GHz bandwidth.

B. Microstrip Rat-Race Hybrid Ring Mixers

The rat-race hybrid ring mixer shown in Figure 6.22 consists of a ring circuit or power splitter, two dc blocks, two mixer diodes, two RF chokes, and a low-pass filter. The RF input is split equally into two mixer diodes. The LO input is also split equally but is 180° out of phase at the mixer diodes. Both the LO and RF are mixed in these diodes, which generate signals that are then combined through the ring and taken out through a low-pass filter. The RF chokes (e.g., open stub circuits) provide a tuning mechanism and prevent the RF signal from feeding into ground. The IF output can be taken out of the ring (Figure 6.22) or from the two diodes through a combiner circuit.

The rat-race ring shown in Figure 6.23 consists of a ring 1.5 wavelengths in circumference, with four arms separated by 60° of angular rotation. Two input and output arms are spaced $\lambda_g/4$ from one another. At the center frequency, the input power from arm A splits equally into arms B and D. Because of the length of the electric path, the signal in arm B is 180° out of phase with that in arm D. For the power input from arm C, the output splits equally in phase to arms B and D. Ports A and C are isolated to each other (see Problem P6.9). The design of the ring requires that the impedance of the main ring be equal to $\sqrt{2}$ times the characteristic impedance of each arm for good impedance matching. For a 50-Ω system, this main ring impedance is 70.7 Ω.

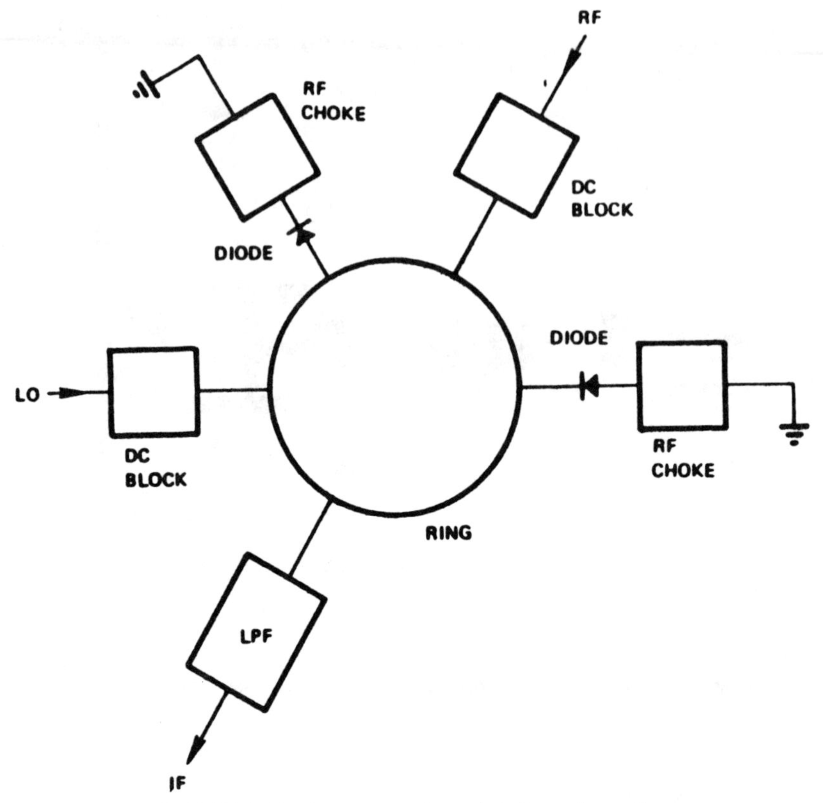

FIGURE 6.22 Rat-race mixer circuit configuration.

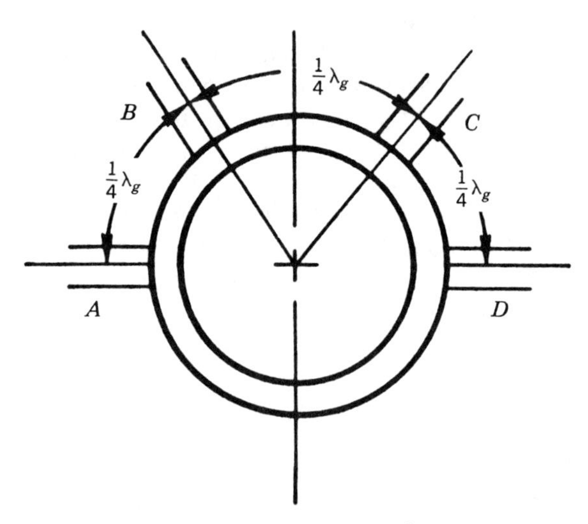

FIGURE 6.23 Rat-race ring.

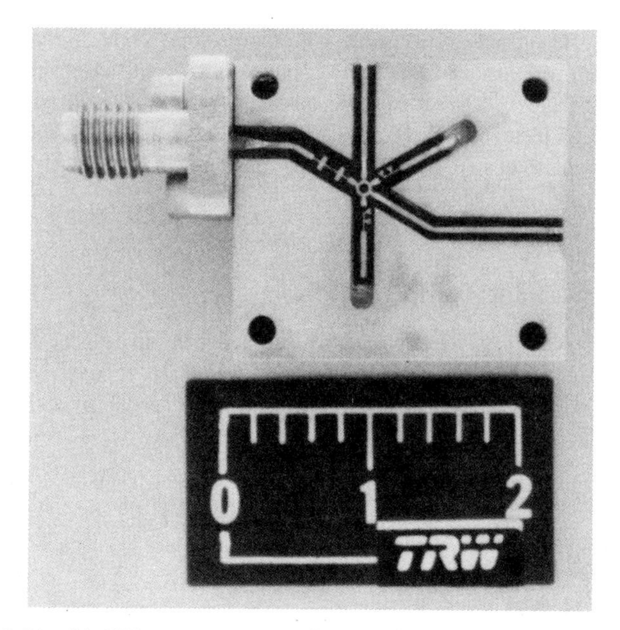

FIGURE 6.24 94-GHz rat-race mixer. (From Ref. 9 with permission from IEEE.)

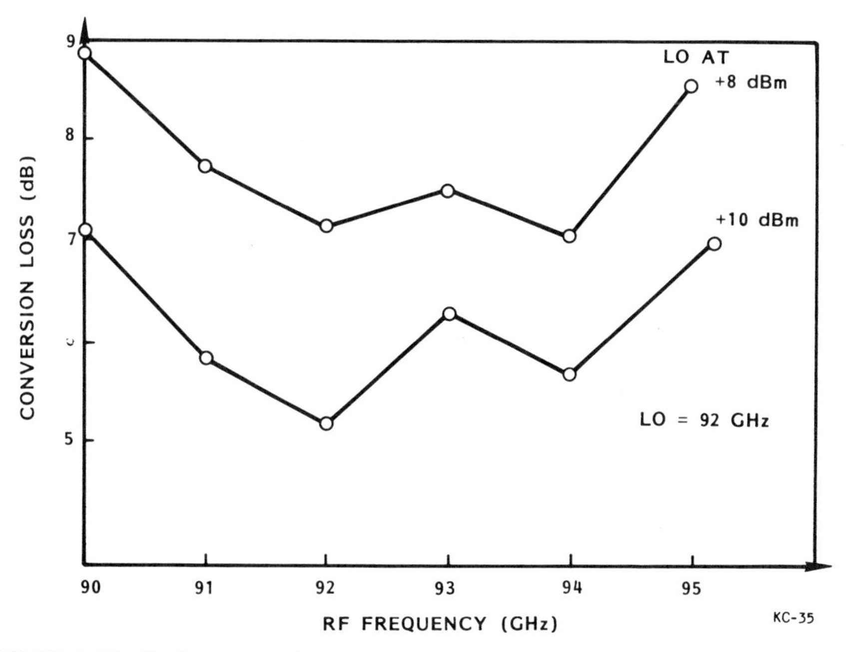

FIGURE 6.25 Performance of a 94-GHz rat-race mixer. (From Ref. 9 with permission from IEEE.)

Rat-race mixers have been demonstrated up to 94 GHz. Because the ring is bandwidth limited, only a 10 to 20% bandwidth has been achieved using rat-race mixers. Figure 6.24 shows the circuit of a 94-GHz rat-race mixer. A conversion loss of less than 8 dB was achieved over a 3-GHz RF bandwidth using a LO pump power of +8 dBm, and less than 6.5 dB with a LO pump power of +10 dBm [9] (Figure 6.25).

C. Crossbar Waveguide / Strip-Line Mixers

Most millimeter-wave suspended strip-line mixers are in a crossbar-type structure (Figure 6.26). The RF signal is applied to mixer diodes from a waveguide perpendicular to the circuit board. The crossbar configuration is

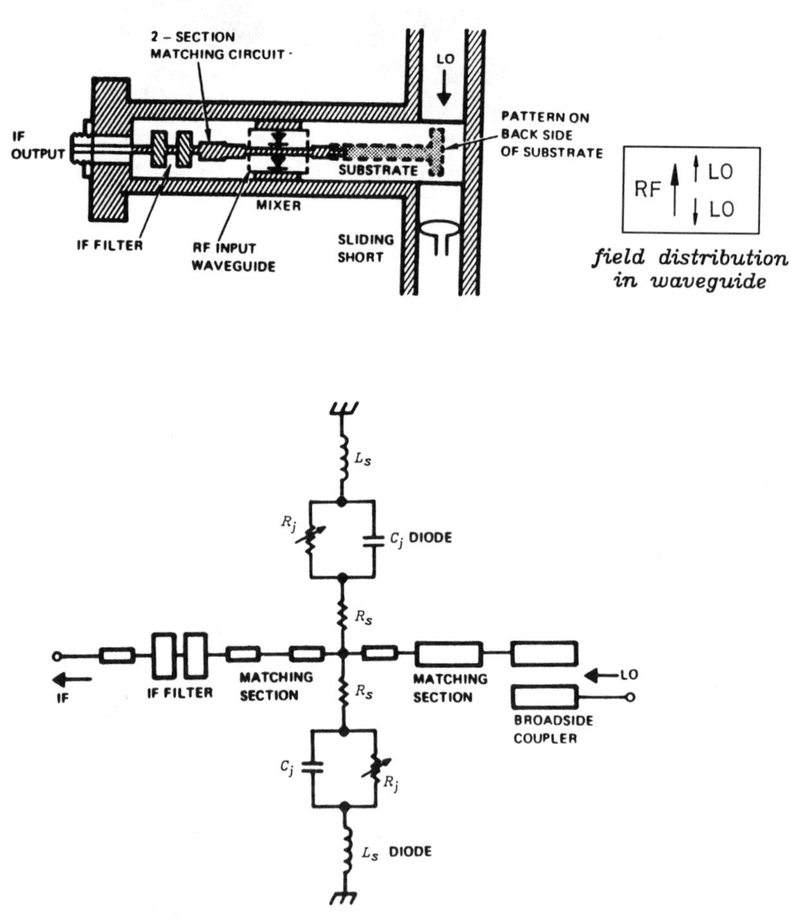

FIGURE 6.26 Circuit configuration and equivalent circuit of crossbar stripline mixer. (From Ref. 10 with permission from IEEE.)

formed by two mixer diodes with opposite polarities connected in series across the broadwall of the waveguide. The mixer diodes are thus in series with respect to the RF signal and in parallel with respect to the IF circuit. The IF signal is extracted via a low-pass filter and the LO signal is injected from the other side through a broadside coupler and an electric probe type of transition. RF matching is achieved by a reduced-height waveguide transformer and a backshort behind the diodes. The RF waveguide port and LO port are orthogonal to each other, thus presenting inherent RF-to-LO isolation.

Neglecting the case capacitance, the equivalent circuit of the crossbar mixer can be represented as in Figure 6.26. Conversion loss is minimized by optimizing the RF and IF circuit matching. The junction resistance R_j of the mixer diode is varied with the LO voltage, and its value can be as low as 100 to 150 Ω under fully turned-on conditions. The waveguide impedance is in the range 400 to 600 Ω and can be matched to the diode impedance by a reduced-height taper transformer. A sliding short on the opposite side of the RF port tunes out the reactive part of the diode. Another sliding short is used to match the LO port.

IF and LO matching designs are aided by computer analysis of the equivalent circuit shown in Figure 6.26. An IF filter passes the IF frequency band and rejects the LO and RF signals. The connecting transmission line between the IF and LO ports can be optimized to provide matched conditions at the LO and IF ports.

A broadside coupler is designed to present an open circuit to the IF frequencies to prevent dissipation of IF power in the LO port. The coupler

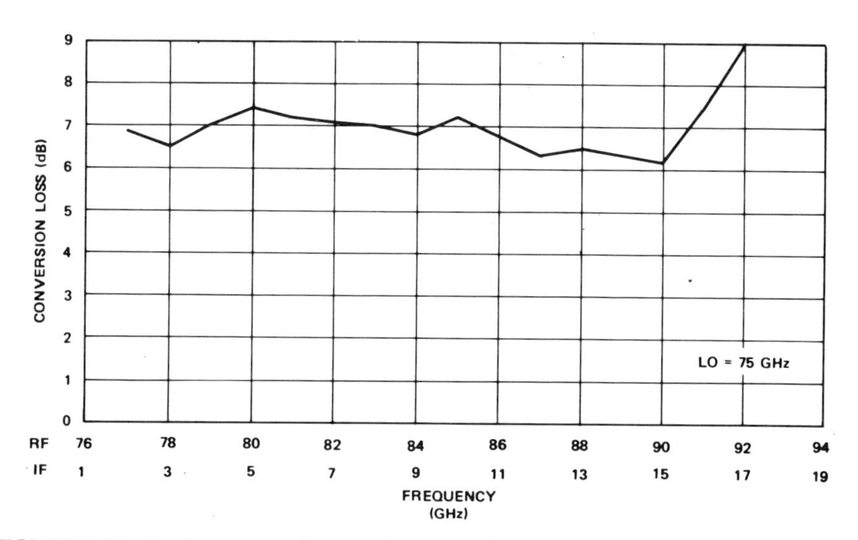

FIGURE 6.27 Performance of a crossbar mixer. (From Ref. 10 with permission from IEEE.)

also serves as a dc block if the mixer is integrated with an MIC local oscillator. A two-section matching circuit is located between the diodes and the IF filter to achieve the wideband IF matching.

Crossbar mixers have been built up to 140 GHz. Very broadband performance can be achieved with this circuit. Figure 6.27 shows that a conversion loss of less than 7.5 dB for a 15-GHz IF bandwidth was achieved with an LO at 75 GHz [10]. The RF signal was varied from 76 to 91 GHz.

D. Image-Enhanced Mixers

The image frequency ($f_{\mathrm{im}} = 2f_{\mathrm{RF}} - f_{\mathrm{LO}}$ or $2f_{\mathrm{LO}} - f_{\mathrm{RF}}$) can be troublesome since its frequency is near the RF and LO frequencies. The conversion loss in a mixer can be minimized if the image frequency is terminated properly. Image-rejection or image-enhanced mixers can be built using hybrid couplers or filters [5].

6.10 HARMONIC AND SUBHARMONIC MIXERS

Harmonic and subharmonic mixers offer a simple way of downconverting a high-frequency RF signal using a low-frequency local oscillator. Downconversion is accomplished by mixing the high-frequency signal with the appropriate harmonic of the LO generated in the mixer itself. The IF frequency is determined by

$$\omega_{\mathrm{IF}} = n\omega_{\mathrm{LO}} - \omega_{\mathrm{RF}} \tag{6.26}$$

The conversion loss is higher in the harmonic mixer than that in the fundamental frequency mixer.

For a fundamental frequency mixer, $n = 1$. The conversion loss is increased by approximately 3 dB when n is increased by 1. Figure 6.28 gives an example of a $\times 2$ harmonic mixer. Applications of harmonic mixers are the

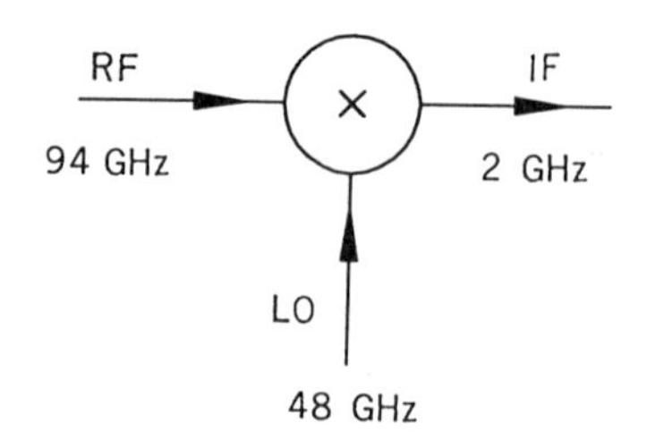

FIGURE 6.28 Harmonic ($\times 2$) mixer.

mixer used in a spectrum analyzer for millimeter-wave measurements or the mixer used in phase-locked-loop operation.

For applications in which sensitivity is important, a subharmonically pumped mixer using an antiparallel diode pair gives much better performance. Figure 6.29 shows a schematic of a subharmonically pumped mixer circuit. Since the bias voltage is zero volts, the LO voltage swings over the I–V characteristic so as to produce a modulated small-signal conductance $g(t)$ with a modulation rate which is twice that of the LO frequency. Hence frequency conversion occurs only for frequencies close to twice the LO frequency, and no fundamental mixing occurs near the LO frequency [3]. The advantages of the subharmonically pumped mixer with antiparallel diodes are: (1) low conversion loss; (2) low noise through LO noise suppression; (3) low LO frequency, which is particularly important for millimeter-wave applications; and (4) inherent self-protection against large peak inverse voltage burnout.

Figure 6.30 shows a subharmonic mixer operating at W-band (75 to 110 GHz). The RF signal is incident in a WR-10 waveguide and the LO in a WR-19 waveguide. A conversion loss of less than 6 dB at 100 GHz was achieved.

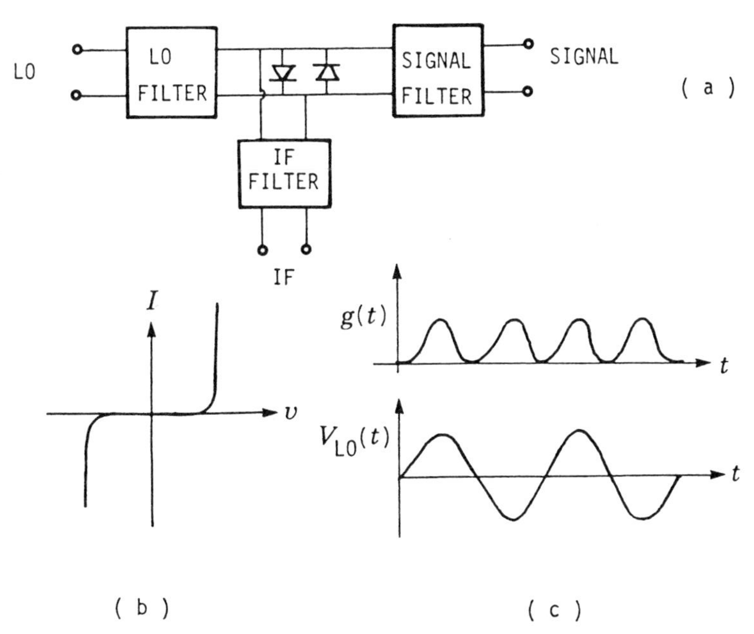

FIGURE 6.29 Subharmonically pumped mixer using an antiparallel diode pair: (*a*) mixer circuit; (*b*) I–V characteristic; (*c*) time dependence of the local oscillator voltage and the differential conductance. (From Ref. 3.)

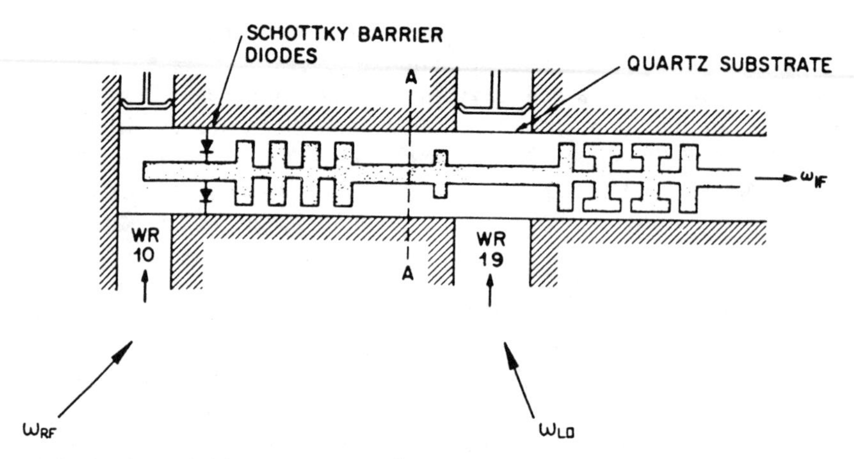

FIGURE 6.30 Subharmonic mixer. (From Ref. 11 with permission from IEEE.)

6.11 UPCONVERTERS

The upconverter is basically a downconverter operating in reverse order. As shown in Figure 6.31, the output RF signal has a frequency given by

$$\omega_{\mathrm{RF}} = \omega_{\mathrm{LO}} + \omega_{\mathrm{IF}} \tag{6.27}$$

For example, if the IF frequency is 2 GHz and the LO frequency is 92 GHz, the RF frequency is 94 GHz. The upconverter is used in a transmitter to generate an information-bearing RF carrier. The operating principle of the upconverter is similar to that of the downconverter.

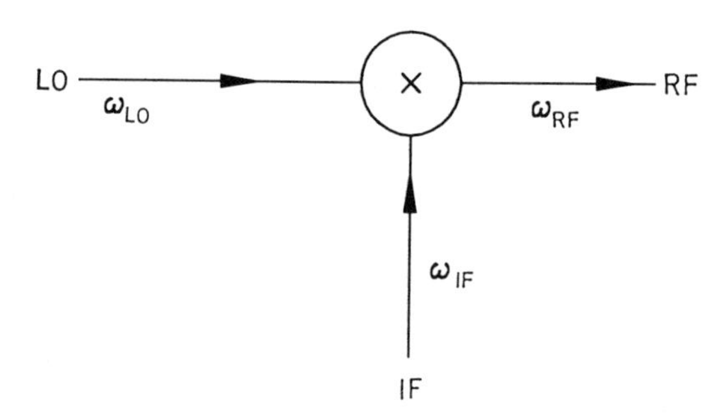

FIGURE 6.31 Upconverter configuration.

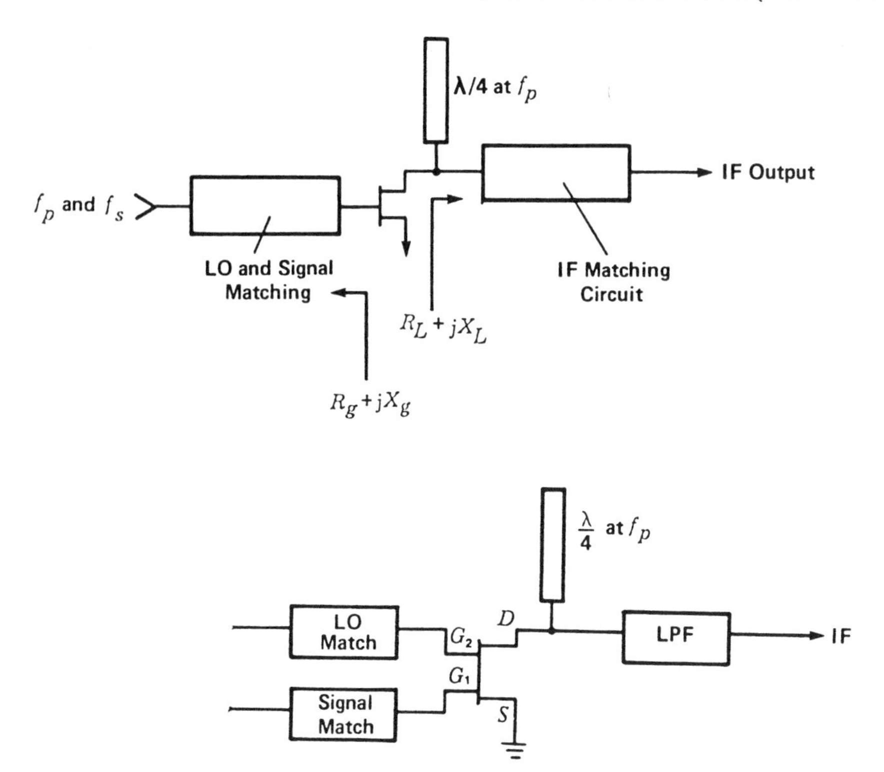

FIGURE 6.32 Single- and dual-gate FET mixers. (From Ref. 5.)

6.12 FET MIXERS

FETs can be used as mixers due to their nonlinear I–V characteristics. The major advantage of a FET mixer is that a conversion gain can be provided instead of a conversion loss. The FET mixer is thus called an *active mixer*. Both single- and dual-gate FETs can be used as mixers (Figure 6.32), but dual-gate FET mixers can provide higher conversion gain than single-gate mixers by utilizing the additional nonlinearities associated with the second gate. Table 6.1 gives a comparison between active and passive mixers. More details are given in Chapter 16.

6.13 NONLINEAR ANALYSIS TECHNIQUES

In a mixer, the junction conductance and capacitance are time-varying components. They can be expressed using Fourier series as follows. Assuming

TABLE 6.1 Comparison between Active and Passive Mixers

Parameter	Single-Gate FET Active Mixer	Dual-Gate FET Active Mixer	Schottky-Diode Passive Mixer
Conversion factor	Low gain	Higher gain	Loss
Noise figure (typical)	6–10 dB	8–12 dB	4–6 dB
RF/LO-to-IF isolation	Requires filter	Requires filter	Requires filter
LO-to-RF isolation	Requires coupler	20 dB	Requires coupler
Bandwidth	Octave	Multioctave	Octave
Bias	Yes	Yes	No

Source: Ref. 5.

that the LO signal is much greater than the RF signal, we have

$$V = V_o + V_{\mathrm{LO}} \cos \omega_{\mathrm{LO}} t \tag{6.28}$$

V_o is the dc bias voltage. Equation (6.1) can be rewritten as

$$I = V = I_s(e^{\alpha V} - 1) \tag{6.29}$$

where $\alpha = q/\eta kT = 40\ \mathrm{V}^{-1}$ at room temperature for an ideal diode. The transconductance $g(t)$ is given by

$$g(t) = \frac{dI}{dV} = \alpha I_s e^{\alpha V} \tag{6.30}$$

Substituting Equation (6.28) into (6.30) and using $e^x = 1 + x + x^2/2! + x^3/3! + \cdots$, we have

$$g(t) = g_0 + \sum_{n=1}^{\infty} 2g_n \cos(n\omega_{\mathrm{LO}} t) \qquad g_n = g_{-n} \tag{6.31}$$

Similarly, the junction capacitance can be expressed as

$$C_j(t) = C_0 + \sum_{n=1}^{\infty} 2C_n \cos(n\omega_{\mathrm{LO}} t) \qquad C_n = C_{-n} \tag{6.32}$$

Since $\cos(n\omega_{\mathrm{LO}} t) = \frac{1}{2}[\exp(jn\omega_{\mathrm{LO}} t) + \exp(-jn\omega_{\mathrm{LO}} t)]$, Equation (6.31) can be written as

$$g(t) = \sum_{n=-\infty}^{\infty} g_n \exp(jn\omega_{\mathrm{LO}} t) \tag{6.33}$$

Several methods have been developed to solve the waveforms $g(t)$ and $C_j(t)$. The most comprehensive analysis was presented by Held and Kerr [12].

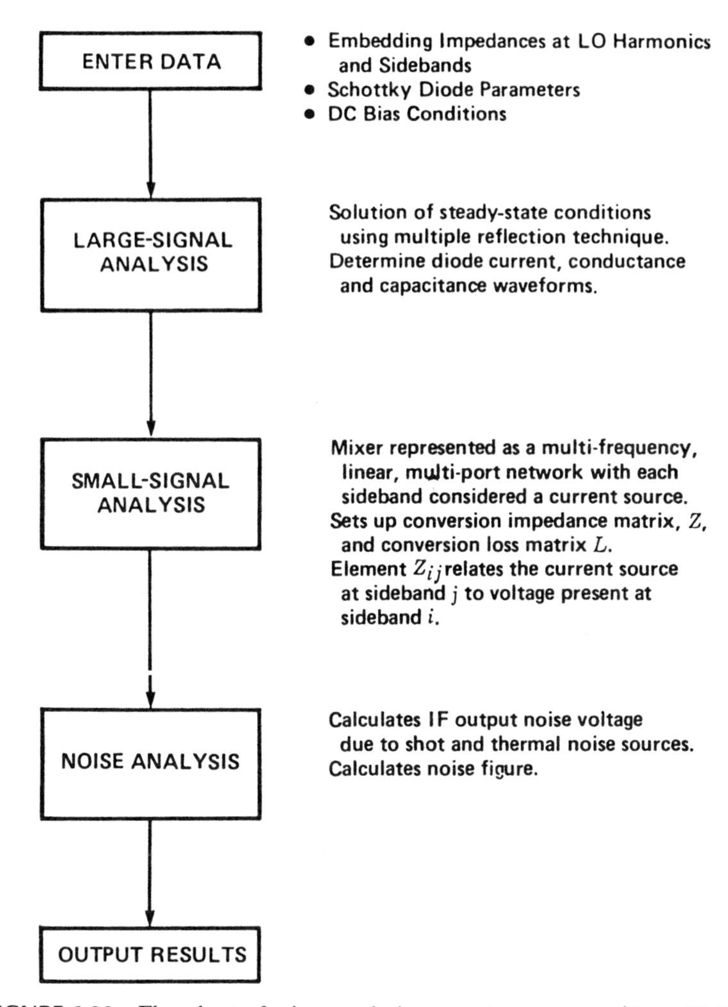

FIGURE 6.33 Flowchart of mixer analysis computer program. (From Ref. 5.)

The analysis is divided into small-signal, large-signal, and noise analysis sections. The diode current and voltage are determined by the LO signal since the LO voltage is normally much higher than the RF signal voltage. Arbitrary embedding impedances are allowed at the harmonics of the LO and at all the sideband frequencies. Computer programs have been written and are available for these analyses. Figure 6.33 shows a flowchart of such a program.

PROBLEMS

P6.1 A mixer (shown in Figure P6.1) has a parasitic loss of 3 dB and an intrinsic loss of 3 dB. The RF port has a VSWR of 2 and the IF port

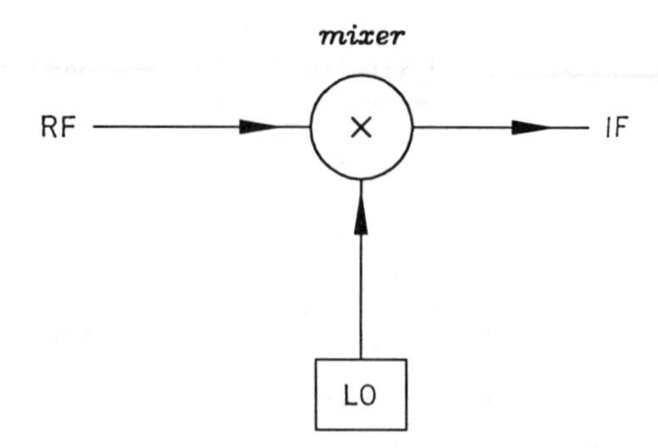

FIGURE P6.1

has a VSWR of 1.5. If an RF signal of 1 mW is incident into the RF port, what is the IF power at the IF port?

P6.2 A mixer has a VSWR equal to 2 at its RF and IF ports. Assuming that the mixer has an intrinsic loss of 3 dB with the parameters $R_s = 1\ \Omega$, $C_j = 1$ pF, and $R_j = 100\ \Omega$, what is the conversion loss at 10 GHz?

P6.3 Considering two RF signals f_1 and f_2 incident into a mixer, derive the third-order intermodulation products from the a_3v^3 terms. Prove the amplitudes for frequencies $2f_1 - f_2$ and $2f_2 - f_1$ given in Figure P6.3. v_1 and v_2 are amplitudes of incident frequencies f_1 and f_2. The intermodulation products $2f_1 - f_2$ and $2f_2 - f_1$ are spurious re-

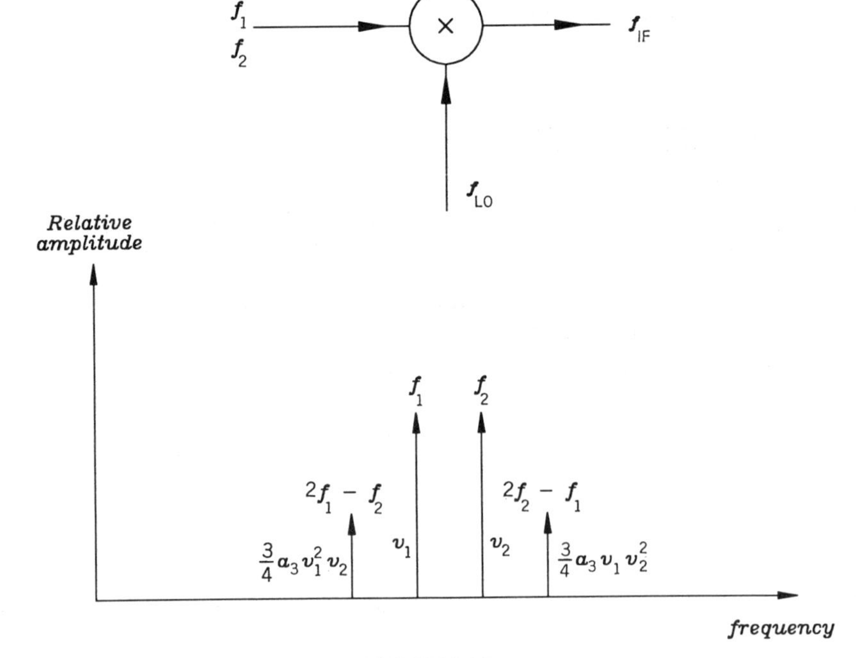

FIGURE P6.3

sponses which are difficult to remove by filtering since f_1 and f_2 are close to each other. (*Hint:* Use $4\sin^3 x = 3\sin x - \sin 3x$ and assume that $f_{LO} > f_2 > f_1$.)

P6.4 Prove Equation (6.12).

P6.5 What is the conversion loss in dB for the mixer shown in Figure P6.5? The VSWR at the RF port is 2. The reflected power at the RF port is measured to be 0.1 mW. The power detected at the IF port is 0.1 mW. The IF and LO ports are perfectly matched.

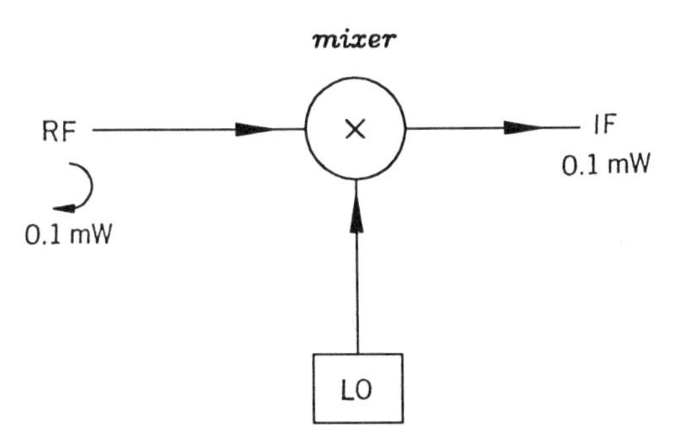

FIGURE P6.5

P6.6 A nonlinear mixer diode is used as an upconverter. Assume that the output current of the diode is $i = a_0 + a_1 v + a_2 v^2$. What are the frequencies of all signals at port A as shown in Figure P6.6?

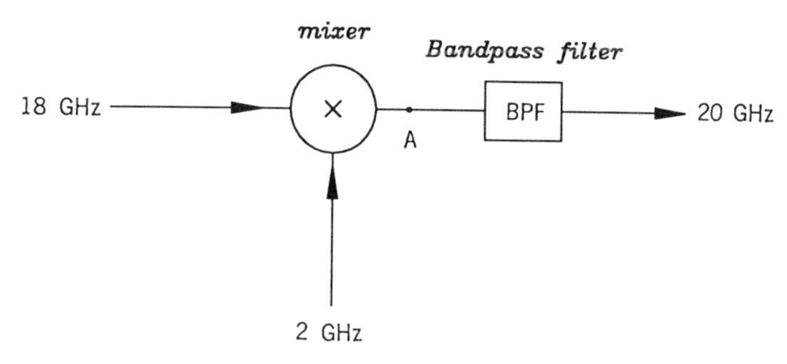

FIGURE P6.6

P6.7 A mixer has a conversion loss of 10 dB in a circuit with a VSWR at the RF port = 1.5 and a VSWR at the IF port = 1.5. What is the conversion loss if the mixer is inserted into a circuit with a VSWR at the RF port = 3.0 and a VSWR at the IF port = 3.0?

P6.8 Explain why the RF and LO ports are isolated in a crossbar waveguide/strip-line mixer.

P6.9 Explain why ports A and C of a rat-race ring (Figure 6.23) are isolated to each other.

REFERENCES

1. E. E. Elder and V. J. Glinski, "Detector and Mixer Diodes and Circuits," in *Microwave Semiconductor Devices and Their Applications*, H. A. Watson, Editor, McGraw-Hill, New York, 1969, Chap. 12.
2. A. B. Carlson, *Communication Systems*, McGraw-Hill, New York, 1968, Chap. 5.
3. E. L. Kollberg, "Mixers and Detectors," in *Handbook of Microwave and Optical Components*, K. Chang, Editor, Wiley, New York, 1990, Vol. 2, Chap. 2.
4. H. C. Torrey and C. A. Whitner, *Crystal Rectifiers*, McGraw-Hill, New York, 1948.
5. I. Bahl and P. Bhartia, *Microwave Solid State Circuit Design*, Wiley, New York, 1988, Chap. 11.
6. W. C. Brown, "The History of Power Transmission by Radio Waves," *IEEE Transactions on Microwave Theory and Techniques*, Vol. MTT-32, September 1984, pp. 1230–1242.
7. T. Yoo and K. Chang, "Theoretical and Experimental Development of 10 and 35 GHz Rectennas," *IEEE Transactions on Microwave Theory and Techniques*, Vol. MTT-40, June 1992, pp. 1259–1266.
8. T. Araki and M. Hirayama, "A 20-GHz Integrated Balanced Mixer," *IEEE Transactions on Microwave Theory and Techniques*, Vol. MTT-19, July 1971, pp. 638–643.
9. K. Chang, et al., "W-Band (75–110 GHz) Microstrip Components," *IEEE Transactions on Microwave Theory and Techniques*, Vol. MTT-33, December 1985, pp. 1375–1382.
10. R. S. Tahim, G. M. Hayashibara, and K. Chang, "Design and Performance of W-Band Broad-Band Integrated Circuit Mixers," *IEEE Transactions on Microwave Theory and Techniques*, Vol. MTT-31, March 1983, pp. 277–283.
11. E. R. Carlson, M. V. Schneider, and T. F. McMaster, "Subharmonically Pumped Millimeter-Wave Mixers," *IEEE Transactions on Microwave Theory and Techniques*, Vol. MTT-26, October 1978, pp. 706–715.
12. D. N. Held and A. R. Kerr, "Conversion Loss and Noise of Microwave and Millimeter-Wave Mixers, Part 1, Theory," *IEEE Transactions on Microwave Theory and Techniques*, Vol. MTT-26, February 1978, pp. 49–55.

FURTHER READING

1. S. A. Maas, *Microwave Mixers*, Artech House, Norwood, Mass., 1986.
2. S. Yngvesson, *Microwave Semiconductor Devices*, Kluwer, Norwell, Mass., 1991.
3. S. A. Maas, *Nonlinear Microwave Circuits*, Artech House, Norwood, Mass., 1988.

CHAPTER 7

Receiver Noise Figure and Dynamic Range

7.1 INTRODUCTION

The noise that occurs in a receiver acts to mask weak signals and to limit the ultimate sensitivity of the receiver. Noise can be generated in a receiver itself or incident from external sources. To detect a signal, the desired signal should have a strength much greater than the noise floor of the system. The noise sources in thermionic and solid-state devices may be divided into three major types:

1. Thermal, Johnson, or Nyquist Noise. This noise is caused by the random fluctuations produced by thermal agitation of the bound charges. The rms value of the thermal resistance noise voltage of V_n over a frequency range B is given by

$$V_n^2 = 4kTBR \tag{7.1}$$

where

k = Boltzmann constant = 1.38×10^{-23} J/K
T = resistor absolute temperature, K
B = bandwidth, Hz
R = resistance, Ω

From Equation (7.1), the noise power exists in a given bandwidth regardless of the center frequency. The distribution of noise per unit bandwidth anywhere is called *white noise*.

2. *Shot Noise*. The fluctuations in the number of electrons emitted from a source constitute shot noise. Shot noise occurs in tubes or solid-state devices.
3. *Flicker or* $1/f$ *Noise*. A large number of physical phenomena, such as mobility fluctuations, electromagnetic radiation, and quantum noise [1], exhibit a noise power that varies inversely with frequency ($1/f$). The $1/f$ noise is important from 1 Hz to 1 MHz. Beyond 1 MHz, the thermal noise is more noticeable.

7.2 NOISE FIGURE

The noise figure is a figure of merit specifying quantitatively how noisy a component or system is. The noise figure of a system depends on a number of factors, such as losses in the circuit, the solid-state devices, bias applied, and amplification. The noise figure of a two-port network is defined as

$$F = \frac{\text{signal-to-noise ratio at input}}{\text{signal-to-noise ratio at output}} = \frac{S_i/N_i}{S_o/N_o} \tag{7.2}$$

Figure 7.1 shows a two-port network with a gain (or loss) G. We have

$$S_o = GS_i \tag{7.3}$$

Note that $N_o \neq GN_i$; instead, the output noise $N_o = GN_i$ + noise generated by the network. The noise added by the network is

$$N_n = N_o - GN_i \qquad \text{W} \tag{7.4}$$

Substituting (7.3) into (7.2), we have

$$F = \frac{S_i/N_i}{GS_i/N_o} = \frac{N_o}{GN_i} \tag{7.5}$$

Therefore,

$$N_o = FGN_i \qquad \text{W} \tag{7.6}$$

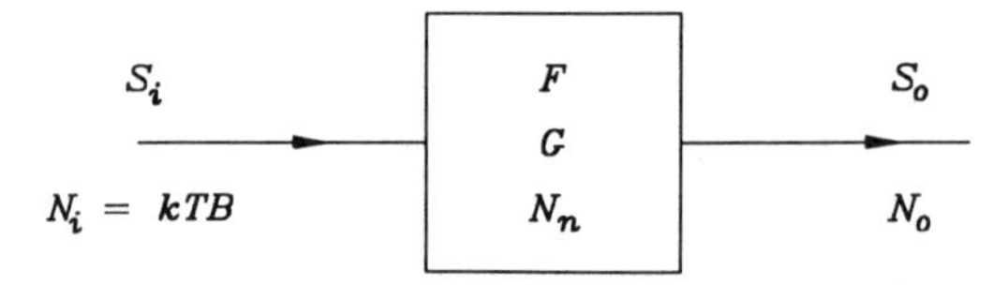

FIGURE 7.1 Two-port network with gain G and added noise power N_n.

Equation (7.6) implies that the input noise N_i (in dB) is raised by the noise figure F (in dB) and the gain (in dB).

Since the noise figure of a component should be independent of the input noise, define F based on a standard input noise source N_i at room temperature in a bandwidth B. N_i is given by

$$N_i = kTB \qquad \text{W} \tag{7.7}$$

where k is the Boltzmann constant, T is 290 K (room temperature), and B is the bandwidth. Then Equations (7.4) and (7.5) become

$$N_n = N_o - GkTB = (F - 1)GkTB \tag{7.8}$$

$$F = \frac{N_o}{GkTB} \tag{7.9}$$

7.3 NOISE FIGURE IN CASCADED CIRCUITS

Now consider two networks with gain values G_1 and G_2 and noise figures of F_1 and F_2, respectively (Figure 7.2). The combined circuit has a noise figure F_{12} and gain G_{12}. From Equation (7.9), we have

$$N_o = F_{12}G_{12}kTB \tag{7.10}$$

where $G_{12} = G_1G_2$, but $F_{12} \neq F_1F_2$.

For the first network, from the definition of a noise figure, one has, from Equations (7.4), (7.7), and (7.9),

$$\begin{aligned} N_1 &= F_1G_1kTB = N_iG_1 + N_{n1} \\ &= G_1kTB + N_{n1} \end{aligned} \tag{7.11}$$

From Equation (7.8), the noise at the output due to the second network is

$$N_{n2} = (F_2 - 1)G_2kTB \tag{7.12}$$

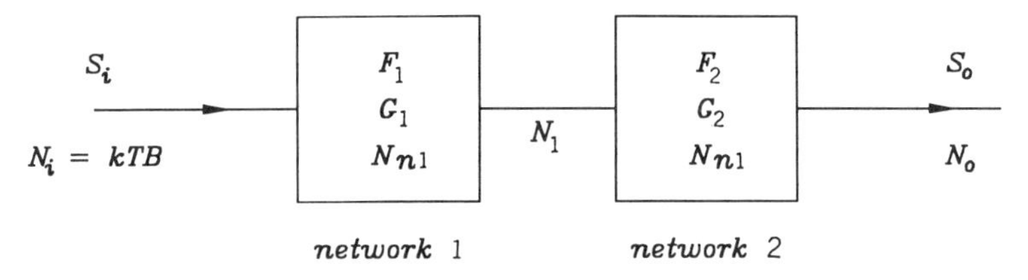

FIGURE 7.2 Cascaded circuit with two networks.

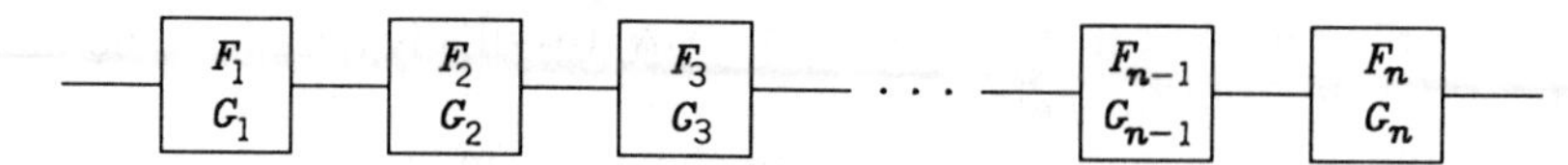

FIGURE 7.3 Cascaded circuit with n networks.

The total noise power at the output is

$$\begin{aligned} N_o &= N_1 G_2 + N_{n2} \\ &= F_1 G_1 G_2 kTB + (F_2 - 1) G_2 kTB \end{aligned}$$

From Equation (7.10), the noise figure of the cascaded circuit is

$$\begin{aligned} F_{12} &= \frac{N_o}{G_{12} kTB} = \frac{N_o}{G_1 G_2 kTB} \\ &= F_1 + \frac{F_2 - 1}{G_1} \end{aligned} \tag{7.13}$$

The analysis for two networks can be extended to a general cascaded circuit with N networks as shown in Figure 7.3. The overall noise figure is

$$F = F_1 + \frac{F_2 - 1}{G_1} + \frac{F_3 - 1}{G_1 G_2} + \cdots + \frac{F_n - 1}{G_1 G_2 \cdots G_{n-1}} \tag{7.14}$$

Equation (7.14) allows for calculation of the noise figure of a general cascaded system. From Equation (7.14) it is clear that the gain and noise figure in the first stage are critical in achieving a low overall noise figure. It is very desirable to have a low noise figure and high gain in the first stage. To use Equation (7.14), all F's and G's are expressed as ratios, not in dB.

Example 7.1 Calculate the overall gain and noise figure for the system shown in Figure 7.4.

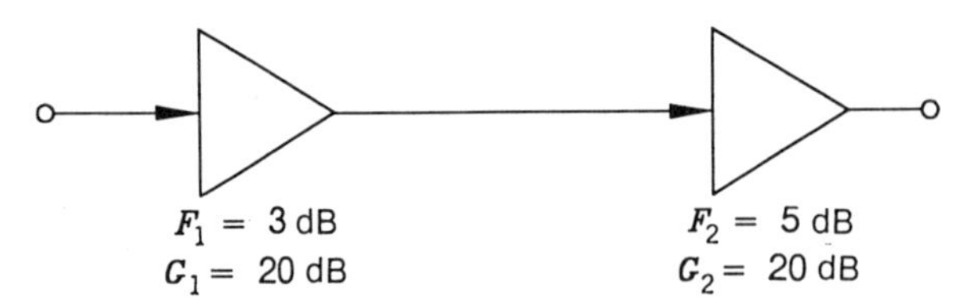

FIGURE 7.4 Cascaded amplifiers.

SOLUTION:

$$F_1 = 3 \text{ dB} = 2 \qquad F_2 = 5 \text{ dB} = 3.162$$
$$G_1 = 20 \text{ dB} = 100 \qquad G_2 = 20 \text{ dB} = 100$$
$$G = G_1 G_2 = 10{,}000 = 40 \text{ dB}$$
$$F = F_1 + \frac{F_2 - 1}{G_1} = 2 + \frac{3.162 - 1}{100}$$
$$= 2 + 0.0216 = 2.0216 = 3.06 \text{ dB}$$

Note that $F \approx F_1$, due to the high gain in the first stage.

7.4 NOISE FIGURE FOR A MIXER CIRCUIT

Figure 7.5 shows a mixer assembly. The mixer is generally followed by an IF amplifier, which amplifies the IF output from the mixer. The input RF matching network has a loss L_{RF}.

For a lossy circuit with a loss L, the gain $G = 1/L$ and the noise figure $F \approx L$. Therefore, we have

$$G_1 = \frac{1}{L_{\text{RF}}} \qquad F_1 = L_{\text{RF}}$$
$$G_2 = \frac{1}{L_c} \qquad F_2 = L_c$$
$$G_3 = G_{\text{IF}} \qquad F_3 = F_{\text{IF}}$$

Assuming that the loss in the matching network is small (i.e., well matched) and that no RF filter is used, we have $L_{\text{RF}} \approx 1$ and $G_1 = 1/F_1 = 1$. The

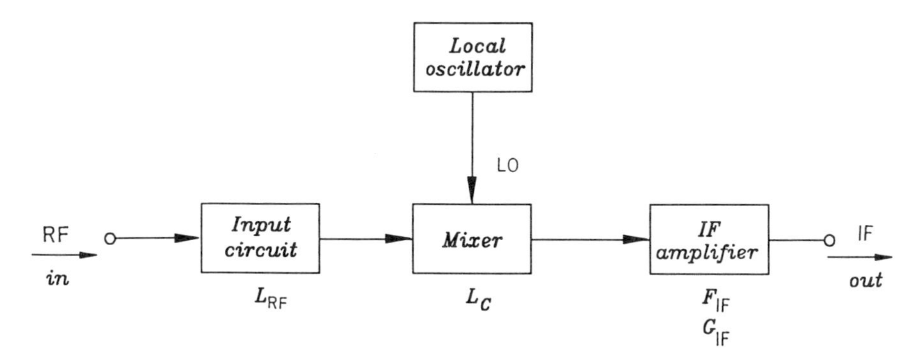

FIGURE 7.5 Mixer/IF amplifier assembly.

overall gain is

$$G = G_2 G_3 = \frac{G_{\text{IF}}}{L_c} \tag{7.15}$$

or

$$G = G_{\text{IF}} - L_c \qquad \text{dB} \tag{7.16}$$

The noise figure is

$$\begin{aligned} F &= F_1 + \frac{F_2 - 1}{G_1} + \frac{F_3 - 1}{G_1 G_2} \\ &= L_{\text{RF}} + L_{\text{RF}}(L_c - 1) + L_{\text{RF}} L_c (F_{\text{IF}} - 1) \\ &= L_{\text{RF}} L_c F_{\text{IF}} \approx L_c F_{\text{IF}} \end{aligned} \tag{7.17}$$

or

$$F = L_c + F_{\text{IF}} \qquad \text{dB} \tag{7.18}$$

Equation (7.18) says that the overall noise figure for a mixer/IF amplifier assembly is equal to the summation of the conversion loss in the mixer and the noise figure of the IF amplifier, all expressed in dB.

Example 7.2 Calculate the overall noise figure and gain for the system shown in Figure 7.6.

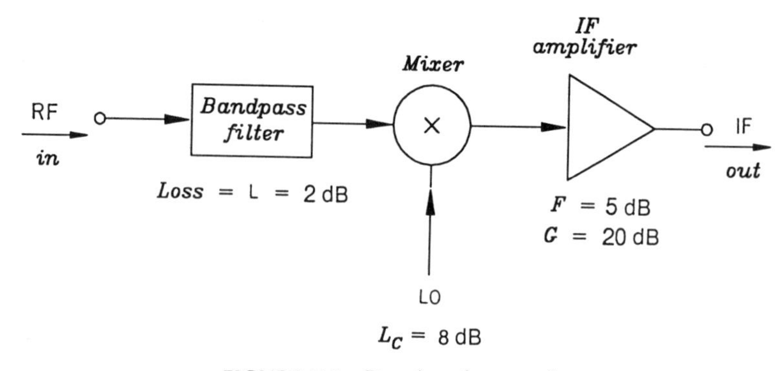

FIGURE 7.6 Receiver front end.

SOLUTION:

$$G_1 = \frac{1}{L} = -2 \text{ dB} = 0.63 \qquad F_1 = 2 \text{ dB} = 1.58$$

$$G_2 = \frac{1}{L_c} = -8 \text{ dB} = 0.158 \qquad F_2 = L_c = 6.31$$

$$G_3 = G_{\text{IF}} = 20 \text{ dB} = 100 \qquad F_3 = 5 \text{ dB} = 3.162$$

$$F = F_1 + \frac{F_2 - 1}{G_1} + \frac{F_3 - 1}{G_1 G_2}$$

$$= 1.58 + \frac{6.31 - 1}{0.63} + \frac{3.16 - 1}{0.63 \times 0.158} = 31.71$$

$$F \text{ (in dB)} = 10 \log 31.71 = 15 \text{ dB}$$

or

$$F = L_{\text{RF}} + L_c + F_{\text{IF}}$$

$$= 2 + 8 + 5 = 15 \text{ dB}$$

$$G = 20 - 8 - 2 = 10 \text{ dB}$$

7.5 DYNAMIC RANGE, 1-dB COMPRESSION POINT, AND MINIMUM-DETECTABLE SIGNAL

In a mixer, an amplifier, or a receiver, operation is normally in a region where the output power is linearly proportional to the input power. The proportionality constant is the conversion loss or the gain. This region, called the *dynamic range*, is shown in Figure 7.7. If the input power is above this range, the output starts to saturate. If the input power is below this range, the noise dominates. The dynamic range is defined as the range between the 1-dB compression point and the minimum detectable signal (MDS). The range could be specified in terms of input power (as shown in Figure 7.7) or output power. For a mixer, amplifier, or receiver system, we would like to have a high dynamic range so that the system can operate over a wide range of input power levels.

The 1-dB compression point is shown in Figure 7.7. Consider an example of a mixer. Beginning at the low end of the dynamic range, just enough RF power is fed into the mixer to cause the IF signal to be barely discernible above the noise. Increasing the RF input power causes the IF output power to increase 1 dB for 1 dB of input power; this continues until the RF input power reaches a level at which the IF output power begins to roll off, causing an increase in conversion loss. The input power level at which the conversion loss increases by 1 dB, called the 1-dB compression point, is generally taken

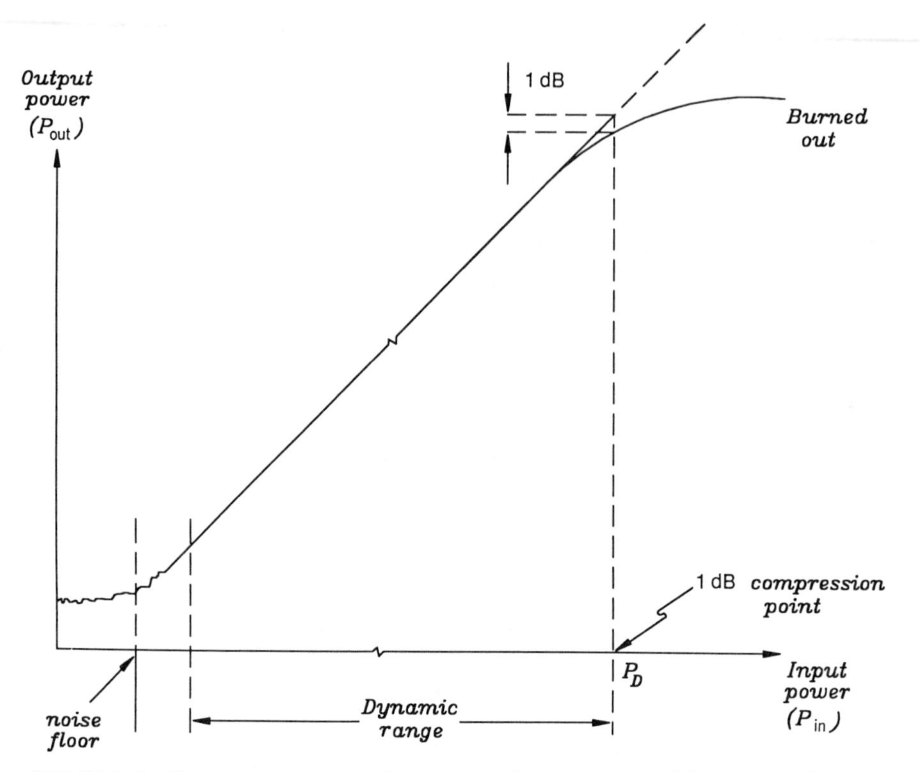

FIGURE 7.7 Output power versus input power for mixers, amplifiers, or receivers.

to be the top limit of the dynamic range. Beyond this range, the conversion loss is higher, and the input RF power not converted into the desired IF output power is converted into heat and higher-order intermodulation products.

From the 1-dB compression point, gain, bandwidth, and noise figure, the dynamic range (DR) of the mixer can be calculated. The DR can be defined as the difference between the input signal level that causes 1 dB of gain compression and the minimum input signal level that can be detected above the noise level. The minimum-detectable signal is defined as 3 dB above the noise level.*

The noise floor due to a matched resistive load is

$$N_i = kTB \tag{7.19}$$

where k is the Boltzmann constant ($k = 1.38 \times 10^{-23}$ J/K). At room tem-

* In a radar system, 10 to 16 dB above noise level is used to increase the probability of detection and to lower the false alarm rate.

perature (290 K) and 1 MHz bandwidth, we have

$$\begin{aligned} kT &= 1.38 \times 10^{-23} \text{ J/K} \times 290 \text{ K} \\ &= 4 \times 10^{-21} \text{ W/Hz} \\ N_i &= kTB = 4 \times 10^{-21} \text{ W/Hz} \times 10^6 \text{ Hz} \\ &= 4 \times 10^{-12} \text{ mW} \\ N_i &= 10 \log kTB = 10 \log 4 \times 10^{-12} \text{ mW} \\ &= -114 \text{ dBm} \end{aligned} \tag{7.20}$$

The minimum-detectable signal (MDS) is defined as 3 dB above the noise floor and is given by

$$\begin{aligned} \text{MDS} &= -114 \text{ dBm} + 3 \text{ dB} \\ &= -111 \text{ dBm} \end{aligned} \tag{7.21}$$

Therefore, MDS is -111 dBm (or 7.94×10^{-12} mW) in a megahertz bandwidth at room temperature.

For a mixer (or a receiver) with an operating bandwidth BW instead of 1 MHz at room temperature, the noise floor and MDS will be modified by 10 log BW, where BW is in MHz.

$$\text{MDS} = -111 \text{ dBm} + 10 \log \text{BW} \tag{7.22}$$

Note that

$$\begin{aligned} &\text{If BW} = 1 \text{ MHz}; && \text{MDS} = -111 \text{ dBm} \\ &\text{If BW} = 1 \text{ GHz}; && \text{MDS} = -81 \text{ dBm} \\ &\text{If BW} = 1 \text{ Hz}; && \text{MDS} = -171 \text{ dBm} \end{aligned}$$

In a system (mixer, amplifier, or receiver) with a noise figure of F, extra noise will be contributed from the system. The noise floor and MDS need to be modified by F in dB. [See Equation (7.6) for the derivation.] The MDS due to a system with a noise figure of F at room temperature is

$$\text{MDS} = -111 \text{ dBm} + 10 \log \text{BW} + F \tag{7.23}$$

with BW in MHz, F in dB, and MDS in dBm. Note that if BW $= 1$ MHz and $F = 0$, MDS $= -111$ dBm, which is 3 dB above the level of -114 dBm.

Example 7.3 Calculate the minimum detectable signal at room temperature (290 K) for a system with (a) BW $= 1$ MHz, $F = 10$ dB; (b) BW $= 1$ GHz, $F = 10$ dB; (c) BW $= 10$ GHz, $F = 10$ dB; and (d) BW $= 1$ kHz, $F = 10$ dB.

SOLUTION: $\mathrm{MDS} = -111\ \mathrm{dBm} + 10\log \mathrm{BW} + F$.

(a) $\mathrm{BW} = 1\ \mathrm{MHz} \qquad F = 10\ \mathrm{dB}$

$$\mathrm{MDS} = -111 + 10\log 1 + 10 = -101\ \mathrm{dBm} = 7.94 \times 10^{-11}\ \mathrm{mW}$$

(b) $\mathrm{BW} = 1\ \mathrm{GHz} \qquad F = 10\ \mathrm{dB}$

$$\mathrm{MDS} = -111 + 10\log 1000 + 10 = -71\ \mathrm{dBm} = 7.94 \times 10^{-8}\ \mathrm{mW}$$

or

$$\begin{aligned}\mathrm{MDS} &= 10\log kTBF + 3 \\ &= 2kTBF \\ &= 2 \times 1.38 \times 10^{-23}\ \mathrm{J/K} \times 290\ \mathrm{K} \times 10^{9}\ \mathrm{Hz} \times 10 \\ &= 8 \times 10^{-11}\ \mathrm{W} = 8 \times 10^{-8}\ \mathrm{mW}\end{aligned}$$

(c) $\mathrm{BW} = 10\ \mathrm{GHz} \qquad F = 10\ \mathrm{dB}$

$$\begin{aligned}\mathrm{MDS} &= -111 + 10\log 10{,}000 + 10 \\ &= -111 + 40 + 10 = -61\ \mathrm{dBm} = 7.94 \times 10^{-7}\ \mathrm{mW}\end{aligned}$$

(d) $\mathrm{BW} = 1\ \mathrm{kHz} \qquad F = 10\ \mathrm{dB}$

$$\begin{aligned}\mathrm{MDS} &= -111 + 10\log 0.001 + 10 \\ &= -111 - 30 + 10 = -131\ \mathrm{dBm} = 7.94 \times 10^{-14}\ \mathrm{mW}\end{aligned}$$

The input signal power or drive power (P_D) in dBm that produces 1-dB gain compression is shown in Figure 7.7 and is given by

$$P_D = P_{\mathrm{out}} - \mathrm{gain} + 1\ \mathrm{dB} \tag{7.24}$$

for an amplifier or a receiver with gain. For a mixer with conversion loss, P_D is given by

$$P_D = P_{\mathrm{out}} + L_c + 1\ \mathrm{dB} \tag{7.25}$$

or one can use Equation (7.24) with a negative gain. Note that P_D and P_{out} are in dBm, and gain and L_c are in dB. P_{out} is the output power at the 1-dB compression point, and P_D is the input power at the 1-dB compression point.

The dynamic range is the difference between P_D and MDS:

$$\underset{\text{(in dB)}}{\mathrm{DR}} = \underset{\text{(in dBm)}}{P_D} - \underset{\text{(in dBm)}}{\mathrm{MDS}} \tag{7.26}$$

Therefore,

$$\mathrm{DR} = (P_{\mathrm{out}} - \text{gain} + 1) - (-111\ \mathrm{dBm} + 10 \log \mathrm{BW} + F) \tag{7.27}$$

where P_{out} is in dBm, gain and F in dB, BW in MHz, and DR in dB.

Another definition of dynamic range is the "spurious-free" region that characterizes the receiver with more than one signal applied to the input. For the case of input signals at equal level, the spurious-free dynamic range $\mathrm{DR}_{\mathrm{sf}}$ is given by

$$\mathrm{DR}_{\mathrm{sf}} = \tfrac{2}{3}(\mathrm{IP} - \text{gain} - \mathrm{MDS}) \tag{7.28}$$

where IP is the output power at the third-order, two-tone intercept point in dBm. The definition of third-order two-tone intercept point is given in the next section.

7.6 INTERMODULATION AND INTERCEPT POINT

Two-tone, third-order intermodulation products (IM) result from two signals applied simultaneously to the mixer RF input. These signals generate harmonics that mix with one another and then beat with the mixer local oscillator according to the expressions (see Problem P6.3)

$$(2\mathrm{RF}_1 - \mathrm{RF}_2) - \mathrm{LO} = \mathrm{IM}_1 \tag{7.29}$$

$$(2\mathrm{RF}_2 - \mathrm{RF}_1) - \mathrm{LO} = \mathrm{IM}_2 \tag{7.30}$$

The IM_1 and IM_2 are shown in Figure 7.8 together with fundamental products IF_1 and IF_2 generated by the mixer.

$$\mathrm{RF}_1 - \mathrm{LO} = \mathrm{IF}_1 \tag{7.31}$$

$$\mathrm{RF}_2 - \mathrm{LO} = \mathrm{IF}_2 \tag{7.32}$$

Note that the frequency separation is

$$\mathrm{RF}_1 - \mathrm{RF}_2 = \mathrm{IM}_1 - \mathrm{IF}_1 = \mathrm{IF}_1 - \mathrm{IF}_2 = \mathrm{IF}_2 - \mathrm{IM}_2 \tag{7.33}$$

These intermodulation products are usually of primary interest because of their relatively large magnitudes and because they are difficult to filter from the desired mixer outputs (IF_1 and IF_2).

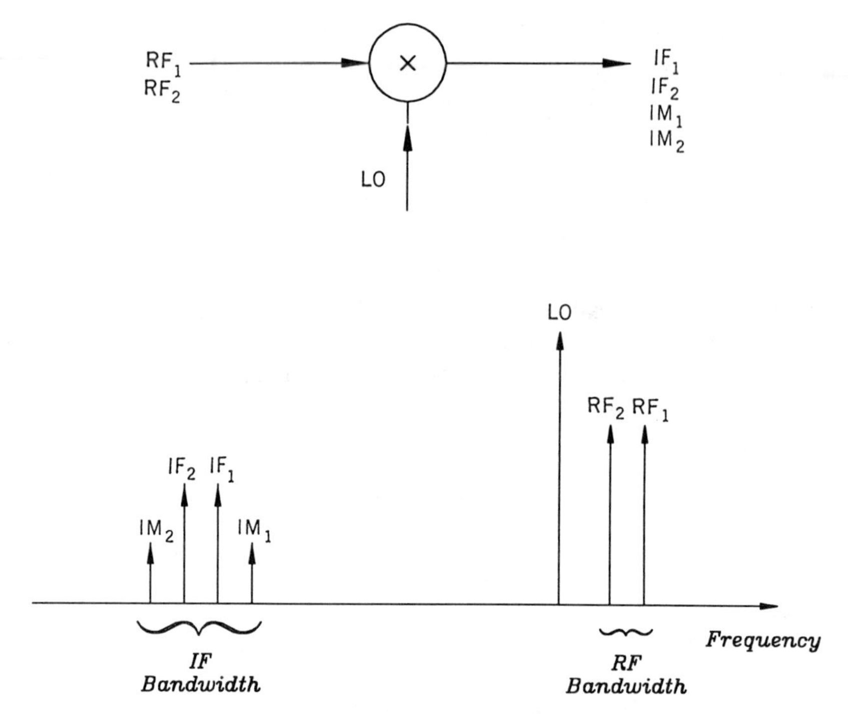

FIGURE 7.8 Intermodulation products in a mixer.

The intercept point, measured in dBm, is a figure of merit for intermodulation product suppression. A high intercept point indicates a high suppression of undesired intermodulation products. A convenient method for determining the two-tone third-order performance of a mixer is the third-order intercept-point measurement.

The measurement test setup for the third-order intercept point is shown in Figure 7.9. Two signals, RF_1 and RF_2, are adjusted to have the same power levels at the mixer input. Third-order intermodulation products are described by $(2RF_2 - RF_1) - LO$ and $(2RF_1 - RF_2) - LO$. These signals are measured using a spectrum analyzer. The third-order intercept point can be determined based on the third-order intermodulation products. This measurement can be correlated to a plot of RF input power versus IF output power. The power saturation caused by an increased RF input level can also be used to determine the third-order intercept point. A typical measurement for a V-band (50 to 75 GHz) crossbar strip-line mixer [2] is shown in Figure 7.10. The 1-dB compression point is also shown in Figure 7.10. It can be seen that the 1-dB compression point occurs at the input power of +8 dBm. The third-order intercept point occurs at the input power of +16 dBm and the mixer will suppress third-order products over 55 dB with both signals at −10 dBm. With both input signals at 0 dBm, the third-order products are suppressed over 35 dB. The mixer operates with the LO at 57 GHz and the RF swept from 60 to 63 GHz. The conversion loss is less than 6.5 dB.

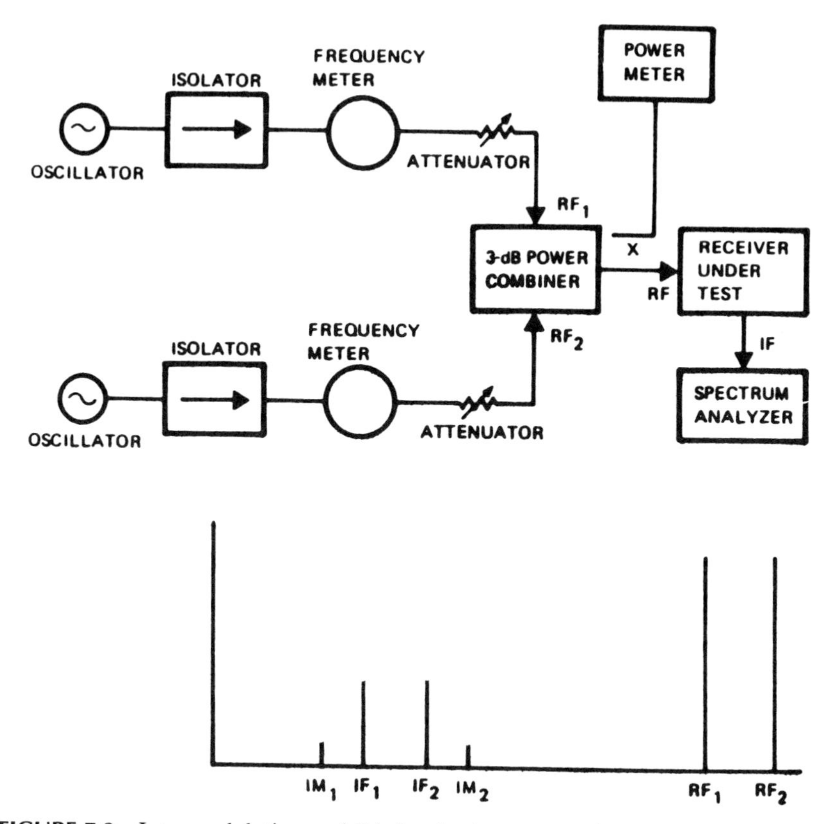

FIGURE 7.9 Intermodulation and third-order intercept-point measurement setup.

Example 7.4 A W-band receiver and its third-order intermodulation are shown in Figure 7.11. The receiver operates at a bandwidth of 10 GHz with noise figure of 15 dB. Find P_D, MDS, DR, and DR_{sf} at room temperature.

SOLUTION: From Figure 7.11, the 1-dB compression point is at the IF output of +13 dBm with a corresponding RF input power of −2 dBm. The average gain is 15 dB. The input signal that produces 1 dB of gain compression is

$$P_D = P_{\text{out}} - \text{gain} + 1 = 13 - 15 + 1 = -1\ \text{dBm}$$

$$\begin{aligned}\text{MDS} &= -111\ \text{dBm} + 10\log \text{BW} + F \\ &= -111\ \text{dBm} + 10\log 10^4 + 15 \\ &= -56\ \text{dBm}\end{aligned}$$

$$\text{DR} = P_D - \text{MDS} = -1 - (-56) = 55\ \text{dB}$$

$$\text{IP} = +27\ \text{dBm} \qquad \text{output power}$$

$$\begin{aligned}\text{DR}_{\text{sf}} &= \tfrac{2}{3}(\text{IP} - \text{gain} - \text{MDS}) \\ &= \tfrac{2}{3}(27 - 15 + 56) = 45\ \text{dB}\end{aligned}$$

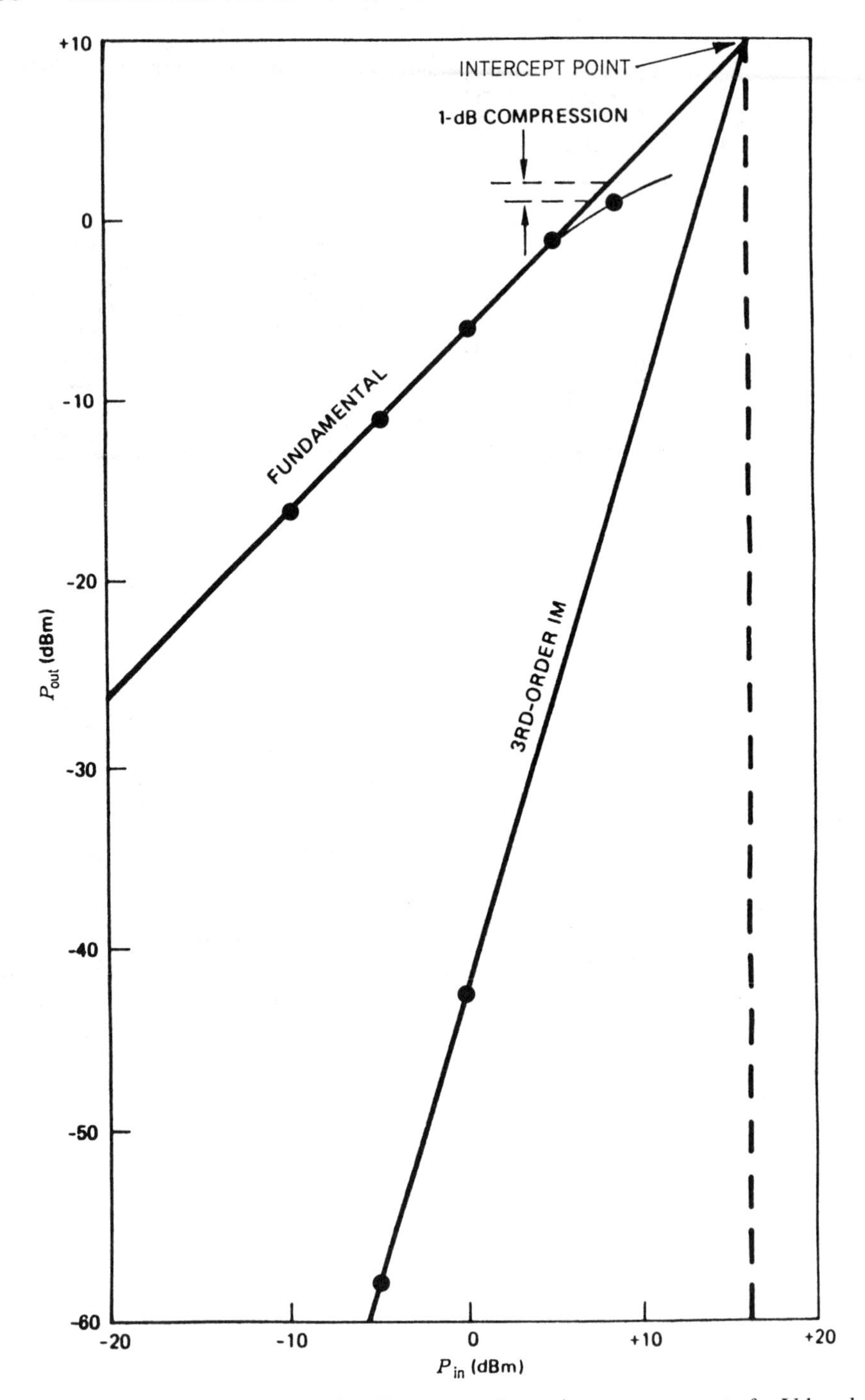

FIGURE 7.10 Intercept point and 1-dB compression point measurement of a V-band crossbar strip-line mixer. (From Ref. 2 with permission from IEEE.)

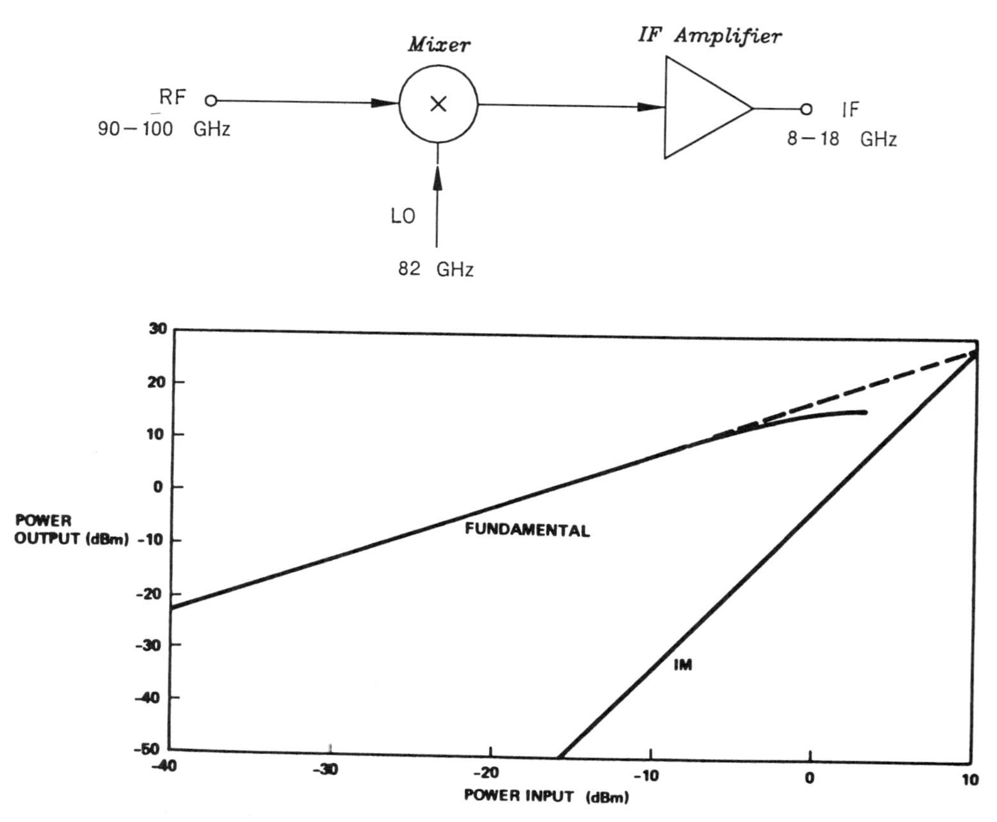

FIGURE 7.11 W-band receiver and its third-order intermodulation measurements.

PROBLEMS

P7.1 For a two-stage cascaded network (Figure P7.1) with gain values of G_1 and G_2, noise figures of F_1 and F_2 as shown, the input noise

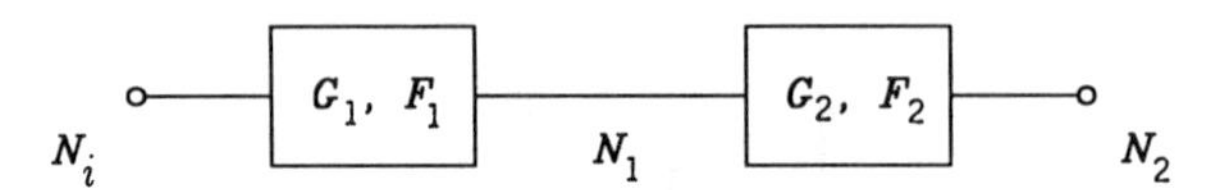

FIGURE P7.1 Two-stage cascaded network.

power is $N_i = kTB$. The output noise power is N_1 and N_2 at the output of the first and second stages. Are the following expressions correct? If not, give a simple explanation.

(a) $F_1 = \dfrac{N_1}{G_1 N_i}$ (b) $F_2 = \dfrac{N_2}{G_2 N_1}$

(c) $F_{12} = \dfrac{N_2}{G_1 G_2 N_i}$ (d) $F_2 = \dfrac{N_2}{G_1 G_2 F_1 N_i}$

P7.2 A receiver has the block diagram shown in Figure P7.2. Calculate (a) the overall noise figure in dB, (b) the total gain or loss in dB, and (c) the minimum detectable signal power level at both the input and output ports in dBm.

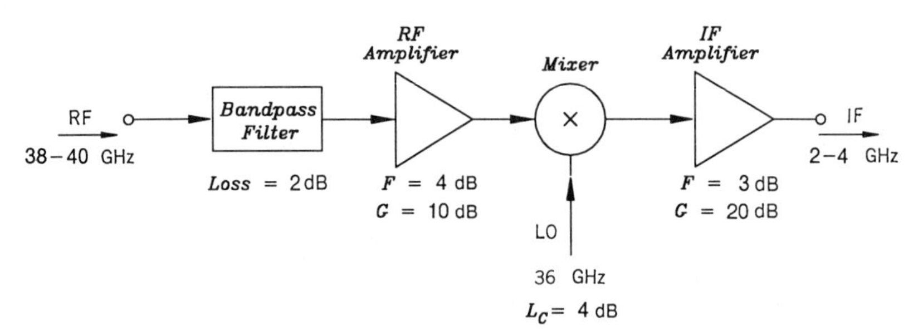

FIGURE P7.2 Block diagram of a receiver.

P7.3 Two satellite receiver systems are shown in Figure P7.3 for a design trade-off study. The components have the following specifications:

RF amplifier	F = 5 dB, gain = 10 dB
Mixer	L_c = 5 dB,
IF amplifier	F = 2 dB, gain = 15 dB

System A

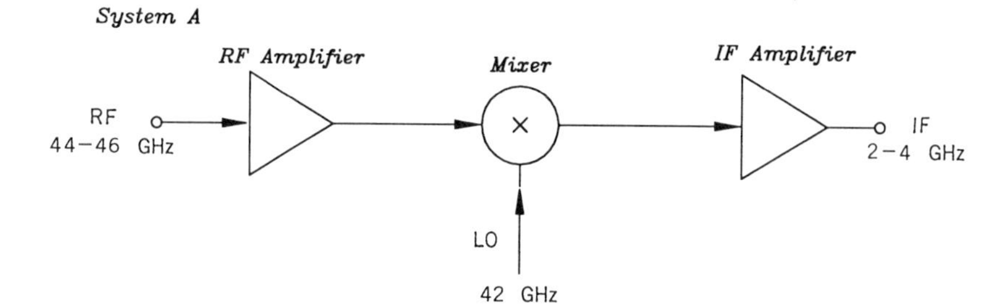

System B

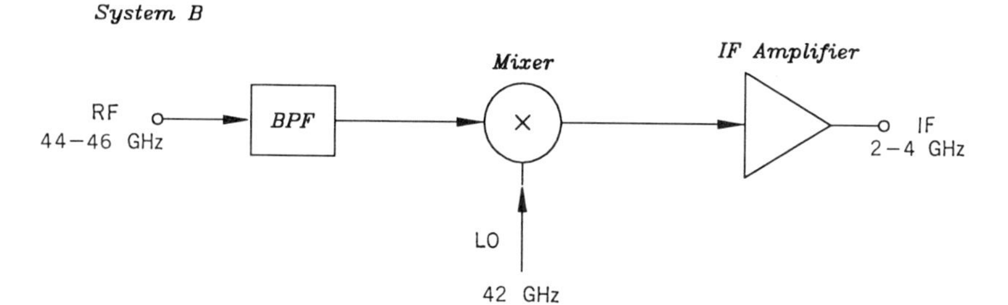

FIGURE P7.3 Two receiver systems for comparison.

Bandpass filter (BPF) Insertion loss = 2 dB

(a) Calculate the noise figure in dB for the two systems.

(b) Calculate the overall gain in dB for the two systems.

(c) Calculate the input minimum-detectable signal for the two systems.

(d) If the output power at the 1-dB compression point is 10 mW for both systems, calculate the dynamic range for the two systems.

P7.4 The following receiver system (Figure P7.4) is used for communications. The mixer has $R_s = 5\ \Omega$, $C_j = 0.1$ pF, $R_j = 100\ \Omega$, VSWR = 2.0 at RF and IF ports, and an intrinsic loss of 3 dB. The RF amplifier has a noise figure of 4 dB and gain of 5 dB. The IF amplifier has a noise figure of 2 dB and a gain of 15 dB. Calculate the noise figure and gain of the overall system in dB.

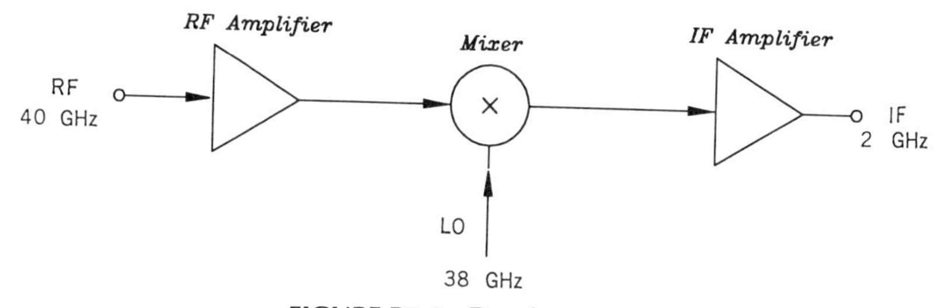

FIGURE P7.4 Receiver system.

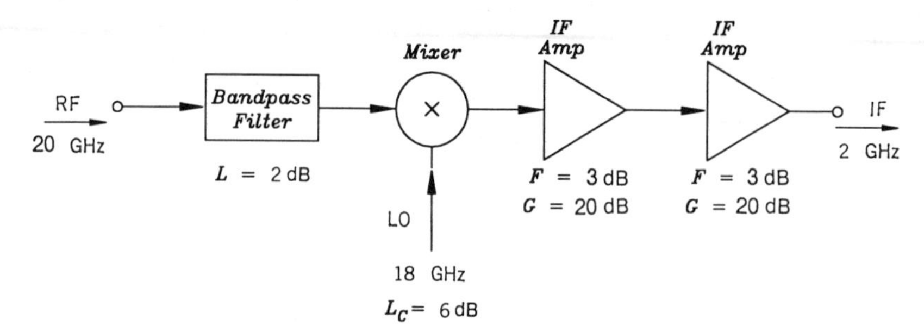

FIGURE P7.5 Receiver system.

P7.5 Calculate the overall noise figure and gain for the receiver system shown in Figure P7.5.

P7.6 Calculate the input minimum-detectable signal in watts at 100 K for a receiver with a bandwidth of 10 GHz and a noise figure of 10 dB.

P7.7 Calculate the input minimum-detectable signal in mW at room temperature for a receiver with **(a)** BW = 1 GHz, $F = 10$ dB; and **(b)** BW = 100 MHz, $F = 5$ dB.

P7.8 Explain why $F = L$ for a lossy circuit.

P7.9 The receiver system shown in Figure P7.9 is operating with a RF input signal of 10 to 11 GHz. Calculate **(a)** the overall gain or loss of the system in dB, **(b)** the overall noise figure in dB, and **(c)** the input and output minimum-detectable signal levels in mW for the receiver at room temperature (290 K).

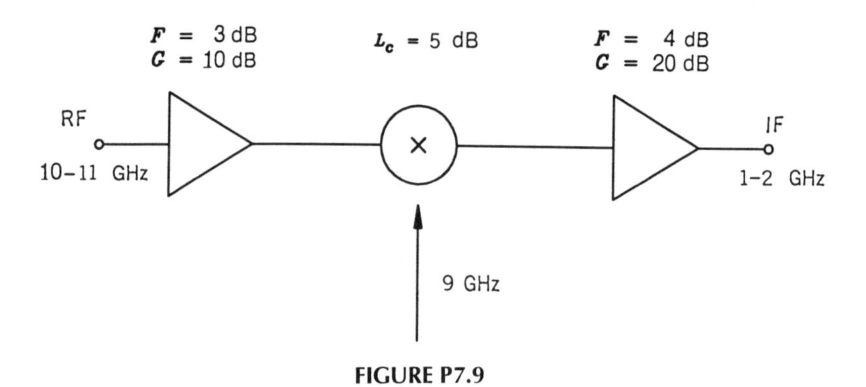

FIGURE P7.9

P7.10 For the receiver system shown in Figure P7.10, calculate **(a)** the overall system gain in dB, **(b)** the overall noise figure in dB, **(c)** the minimum-detectable signal in mW, and **(d)** the output power (P_{out}).

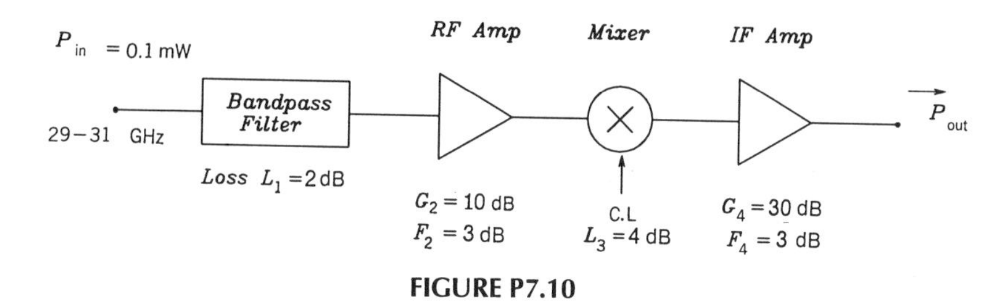

FIGURE P7.10

REFERENCES

1. S. Yugvesson, *Microwave Semiconductor Devices*, Kluwer, Norwell, Mass., 1991, Chap. 8.
2. K. Chang et al., "V-Band Low-Noise Integrated Circuit Receiver," *IEEE Transactions on Microwave Theory and Techniques*, Vol. MTT-31, February 1983, pp. 146–154.

CHAPTER 8

p-i-n Diodes and Control Devices

8.1 INTRODUCTION

The two principal means of providing electronic control of the phase and amplitude of microwave/millimeter-wave signals are the diode and ferrite devices. Phase shifting and switching with ferrites are usually accomplished by changing the magnetic permeability, which occurs with the application of a magnetic biasing field. The change in relative permeability for the ferrite comes about because of the difference in propagation constants of right- and left-circularly-polarized RF magnetic fields.

In contrast with ferrite devices, *p-i-n* diodes or FETs are small compared with the operating wavelength. The *p-i-n* (or PIN) diode has its name because of its semiconductor doping profile which consists of a lightly doped "intrinsic" region sandwiched between two doped *p* and *n* regions. Phase shifting and switching result both from reactive switching of the diode capacitance and the resultant rerouting of microwave currents through circuits containing the diodes. To maintain the lowest loss, the variable resistance of the *p-i-n* diode obtainable by varying the bias is avoided in phase shifters, and discrete switching between forward and reverse bias is used. Although ferrite switches and phase shifters have the advantages of higher power-handling capability and lower insertion loss and VSWR, diode switches and phase shifters generally have the advantages of small size, low cost, high switching speed, simple drive, and less sensitivity to temperature changes. Table 8.1 summarizes the comparison between the ferrite and *p-i-n* control devices. The FET switches and phase shifters can have the additional advantage of signal gain instead of insertion loss.

TABLE 8.1 Comparison between Ferrite and *p-i-n* Diode Control Devices

Parameter	Ferrite	*p-i-n*
Speed	Low (ms)	High (μs)
Loss	Low (0.2 dB)	High (0.5 dB/diode)
Cost	High	Low
Weight	Heavy	Light
Driver	Complicated	Simple
Size	Large	Small
Power handling	High	Low

In this chapter, the operation principles of *p-i-n* diodes and their related circuit applications are discussed. FET control devices are described in Chapter 16.

8.2 *p-i-n* DIODES

The *p-i-n* diode is similar to the *pn* diode but with a smaller junction capacitance. Figure 8.1 shows *p-i-n* diodes and their profiles. Since the width of the depletion zone is inversely proportional to the resistivity (or doping concentration) of the *p* or *n* region [whichever has the lower impurity doping concentration (see Chapter 5)], the depletion region in a *p-i-n* diode is wider than that in a *pn* diode. The wider depletion region corresponds to a smaller junction capacitance. This effect is very useful for a diode used as a microwave switch because the lower the capacitance, the higher the impedance of the diode under reverse bias and the more effective the device is as an "open circuit" [1].

The *i* ("intrinsic") region normally consists of impurities because, in practice, no material is without impurities. If the *i* region consists of *n* impurities, the diode is called *ν-type p-i-n* (or $p\nu n$). For the *i* region of *p* impurities, it is called *π-type* or $p\pi n$. As shown in Figure 8.1(b), the *i* region is of sufficiently high resistivity so that the few impurities in the region are ionized and the depletion region extends throughout the *i* region and includes a small penetration into both the p^+ and n^+ regions. Because of the heavy doping of the p^+ and n^+ regions, the depletion does not extend very far into them, and the depletion width is essentially equal to the *i* region width. The junction capacitance in the reverse bias is determined by this width.

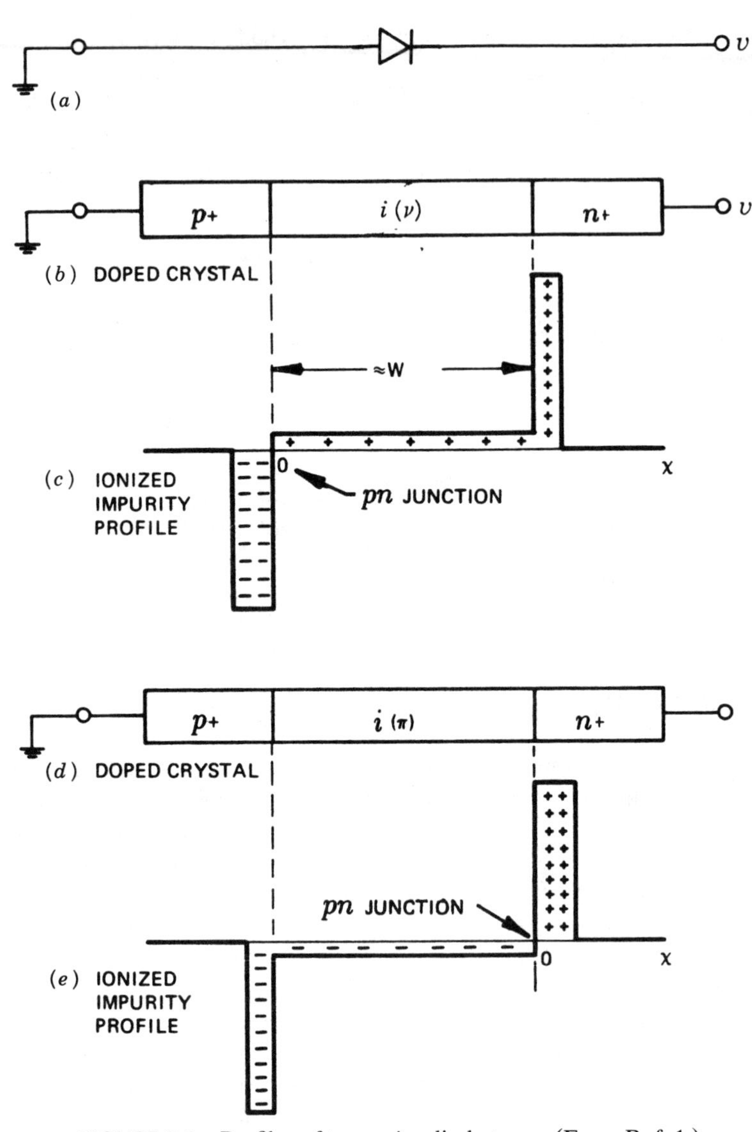

FIGURE 8.1 Profiles of two *p-i-n* diode types. (From Ref. 1.)

P-i-n diodes can be packaged in a pill-type enclosure or in a beam-lead structure similar to those discussed in Chapters 5 and 6 for varactor and mixer diodes. The equivalent circuit can be represented as shown in Figure 8.2. The arrow is connected to R_j in the forward bias and C_j in the reverse bias. For example, a *p-i-n* diode could have parasitics of $R_s = 0.3\ \Omega$, $L_s = 0.1$ nH, and $C_p = 0.3$ pF. $C_j(V)$ and $R_j(V)$ will depend on the applied

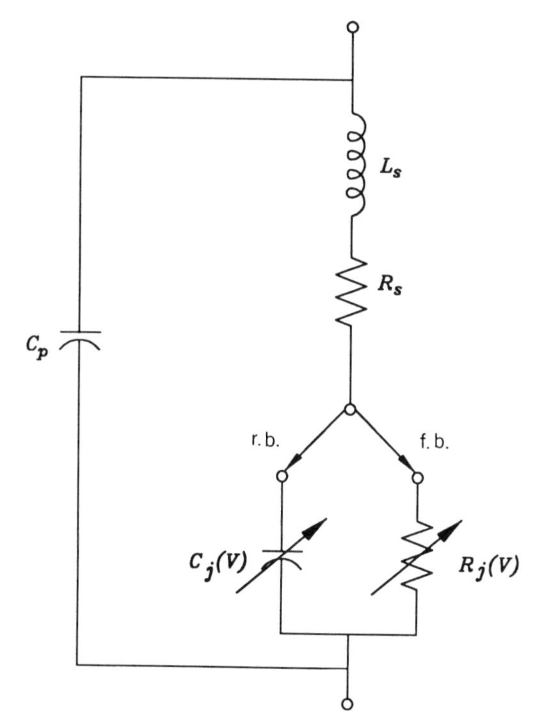

FIGURE 8.2 Equivalent circuit of a *p-i-n* diode.

bias as shown in the I–V curve of Figure 8.3. Let us consider two cases:

1. At forward bias (point A in Figure 8.3, for example):

$$C_j(V) \approx 1 \text{ pF}$$

$$R_j(V) = \frac{dV}{dI} = 0.5\ \Omega$$

$$Z_c = -j\frac{1}{\omega C_j} = -j160\ \Omega \quad \text{and} \quad |Z_c| \gg R_j \text{ at 1 GHz}$$

Neglecting the package effects of L_s and C_p, the equivalent circuit is shown in Figure 8.4. This circuit is almost a short circuit since $R_j + R_s = 0.8\ \Omega$.

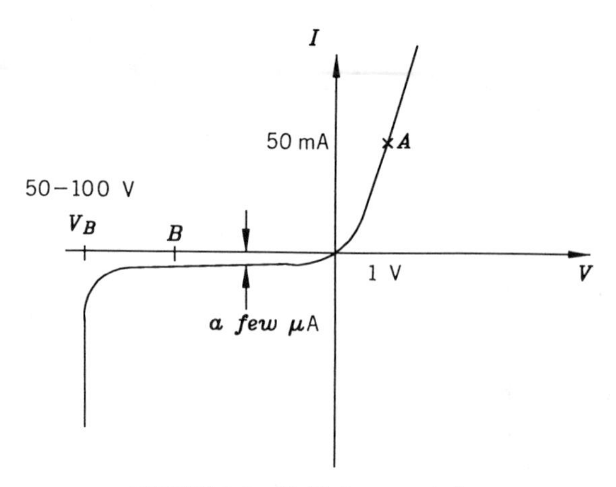

FIGURE 8.3 I–V characteristics.

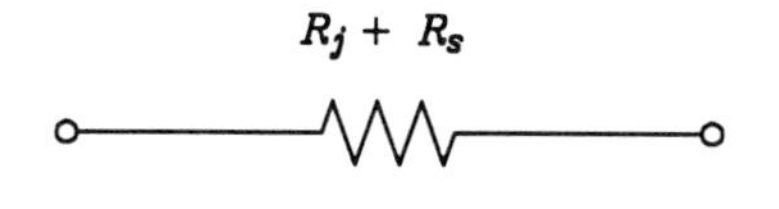

FIGURE 8.4 *p-i-n* diode equivalent circuit in a forward bias.

2. At reverse bias (point B in Figure 8.3, for example):

$$C_j(V) \approx 0.2 \text{ pF}$$

$$R_j(V) \approx 20 \text{ k}\Omega$$

$$Z_c = -j\frac{1}{\omega C_j} = -j796\ \Omega \text{ and } |Z_c| \ll 20 \text{ k}\Omega \qquad \text{at 1 GHz}$$

Neglecting the effects of L_s and C_p, the equivalent circuit is shown in Figure 8.5. Since Z_c is much higher than the 50-Ω transmission-line impedance, the circuit acts as an open circuit. Although this example uses simplified models, it gives some insight into *p-i-n* diode operation, which basically provides nearly open or short circuits by switching the bias.

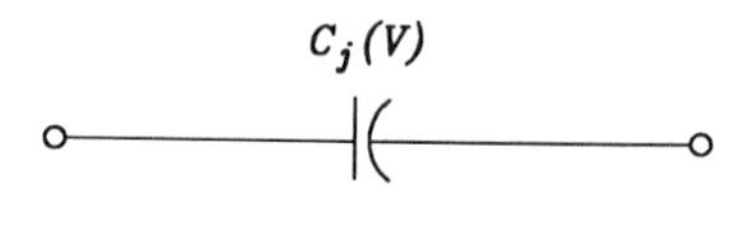

FIGURE 8.5 *p-i-n* diode equivalent circuit in a reverse bias.

8.3 REVIEW OF *ABCD* MATRICES FOR CIRCUIT BUILDING BLOCKS

In Chapter 3 the *ABCD* matrix for a microwave network was defined and shown for several basic circuits. The S_{21} parameters for these circuits were also calculated.

For the convenience of the following discussions, the *ABCD* matrix and the attenuation for a series impedance Z ($= R + jX$) in Figure 8.6(a) are rewritten here as

$$\begin{bmatrix} A & B \\ C & D \end{bmatrix} = \begin{bmatrix} 1 & Z \\ 0 & 1 \end{bmatrix} \tag{8.1}$$

$$\alpha = 20 \log \left| \frac{1}{S_{21}} \right| = 10 \log \left[\left(1 + \frac{R}{2Z_0} \right)^2 + \left(\frac{X}{2Z_0} \right)^2 \right] \quad \text{dB} \tag{8.2}$$

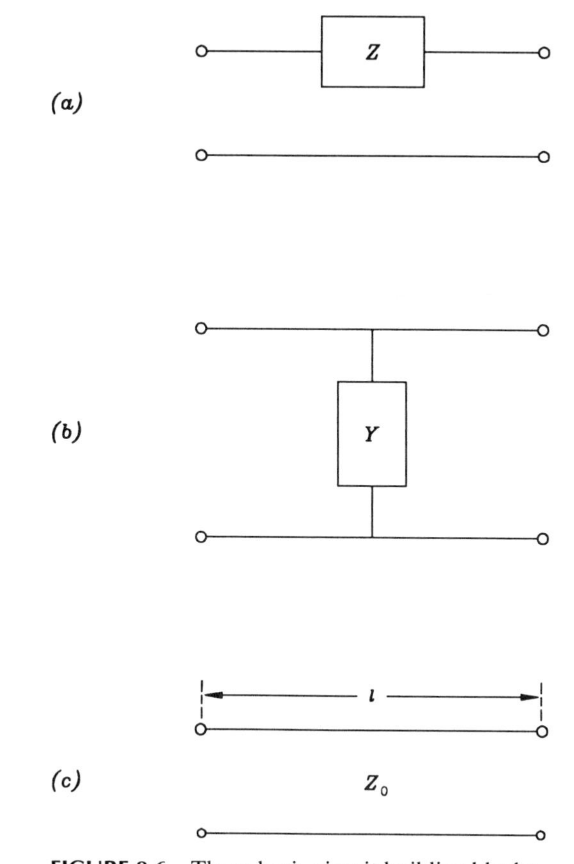

FIGURE 8.6 Three basic circuit building blocks.

For a shunt admittance Y $(= G + jB)$ as shown in Figure 8.6(b), the $ABCD$ matrix and the attenuation are

$$\begin{bmatrix} A & B \\ C & D \end{bmatrix} = \begin{bmatrix} 1 & 0 \\ Y & 1 \end{bmatrix} \tag{8.3}$$

$$\alpha = 20 \log \left| \frac{1}{S_{21}} \right| = 10 \log \left[\left(\frac{G}{2Y_0} + 1 \right)^2 + \left(\frac{B}{2Y_0} \right)^2 \right] \quad \text{dB} \tag{8.4}$$

The $ABCD$ matrix for a lossless transmission line with a characteristic impedance Z_0 and length l is

$$\begin{bmatrix} A & B \\ C & D \end{bmatrix} = \begin{bmatrix} \cos \beta l & jZ_0 \sin \beta l \\ jY_0 \sin \beta l & \cos \beta l \end{bmatrix} \tag{8.5}$$

The derivation of Equations (8.1) through (8.5) can be found in Chapter 3.

8.4 SWITCHES

The use of a *p-i-n* diode as a switch is based on the impedance difference between the diode's reverse- and forward-biased characteristics. The diode appears as a very small impedance under forward-biased conditions and as a very large impedance under zero- or reverse-biased conditions. Switches have many applications in microwave systems. Some examples are:

1. To protect the receiver from the transmitter in radar applications. Figure 8.7(a) shows a transmitter and a receiver sharing the same antenna. In the transmit mode, the switch is thrown to port 1. In the receive mode, the switch is at port 2.
2. To use in a digital modulator in communication systems as shown in Figure 8.7(b). In this case, the switch serves as a gate to pass or stop the signal flow.
3. To serve as a switch in a wideband system, as shown in Figure 8.7(c). For instance, narrowband sources can be connected to a switch to create a wideband sweeper.
4. To use in a wideband receiver for channel selection. Figure 8.7(d) shows that two matrix switches are used for channel selection in a wideband receiver.
5. To use for signal path control in measurement systems. For example, a switch can be used for monitoring a signal using a power meter and a spectrum analyzer, as shown in Figure 8.7(e).

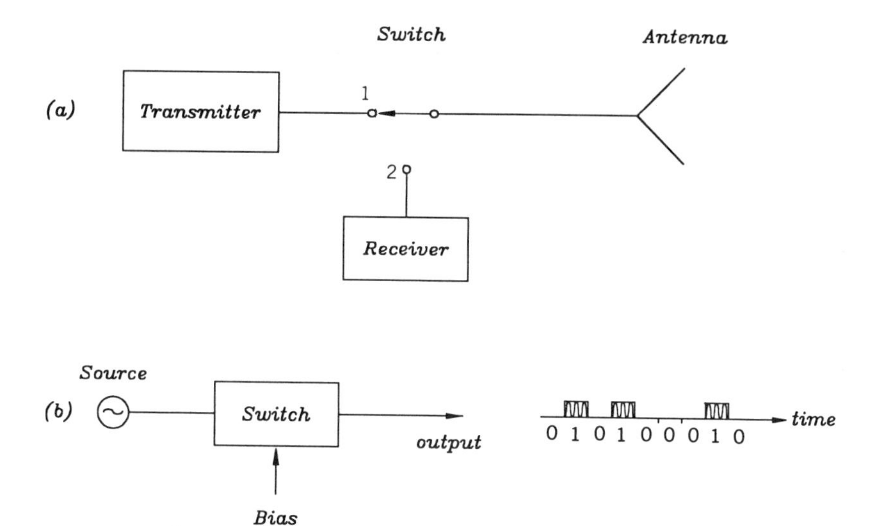

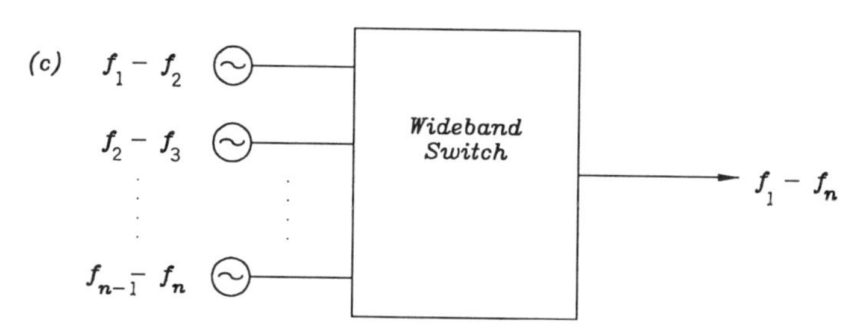

FIGURE 8.7 Switches and their applications.

6. To use in a radar system for built-in test equipment as shown in Figure 8.7(f). A simulated return can be used to test and calibrate the radar.

A switch can be classified as a single-pole single-throw (SPST), a single-pole double-throw (SPDT), a single-pole triple-throw (SP3T), and so on, as shown in Figure 8.8. Figure 8.7(b) uses a SPST switch. Figure 8.7(a), (e), and (f) use SPDT switches. Figure 8.7(c) and (d) use single-pole multiple-throw (SPMT) switches.

Switches can be designed with the diodes series mounted or shunt mounted. Figure 8.9 shows the series-mounted circuits using microstrip line. In SPST case, the signal passes through if the bias is positive. In the SPDT case, the

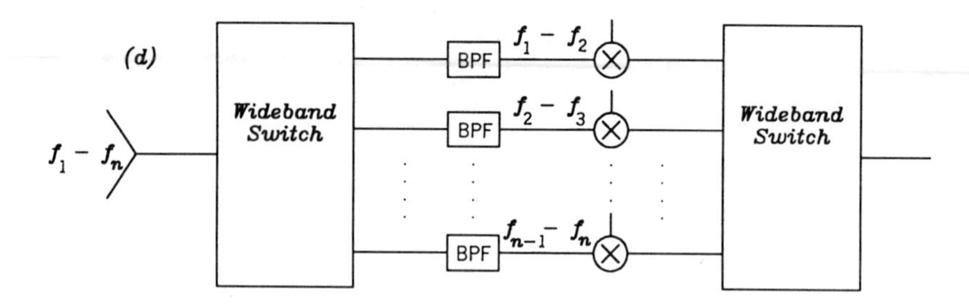

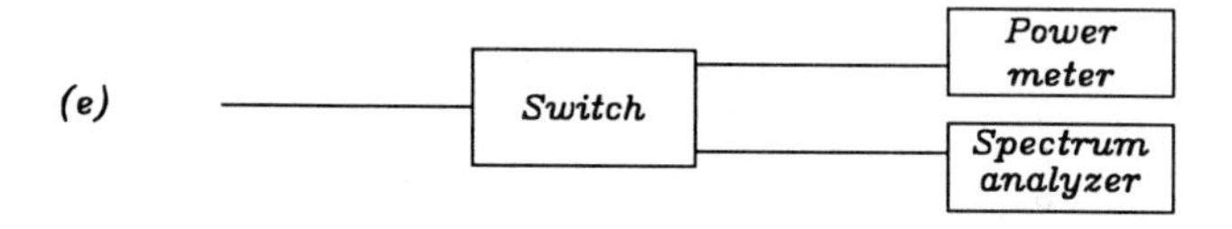

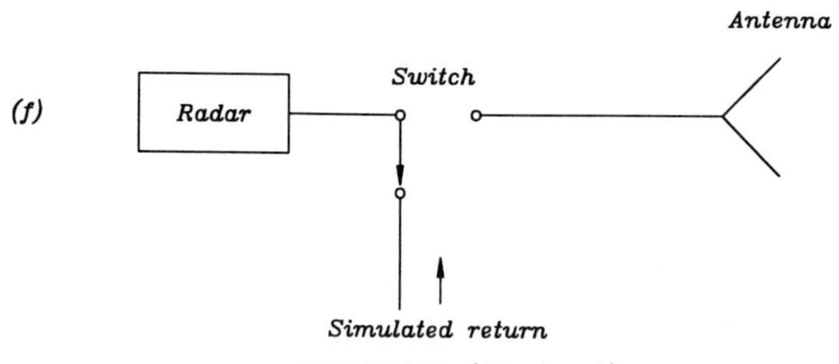

FIGURE 8.7 (*Continued.*).

signal passes through from port 1 to port 2 if the bias is positive and from port 1 to port 3 if the bias is negative.

The insertion loss and isolation for a series-mounted circuit can be derived by using the *ABCD* matrix for a series impedance Z_D $(= R + jX)$. Z_D is the impedance of the *p-i-n* diode equivalent circuit. From Equation (8.2) we have the attenuation

$$\alpha = 10 \log\left[\left(1 + \frac{R}{2Z_0}\right)^2 + \left(\frac{X}{2Z_0}\right)^2\right] \tag{8.6}$$

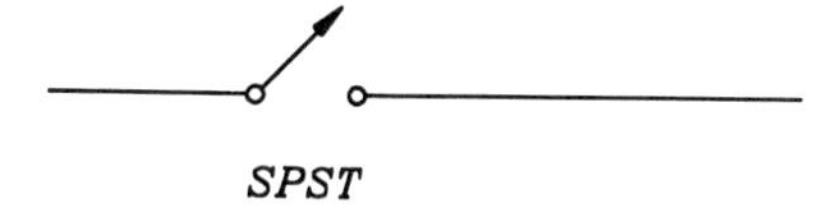

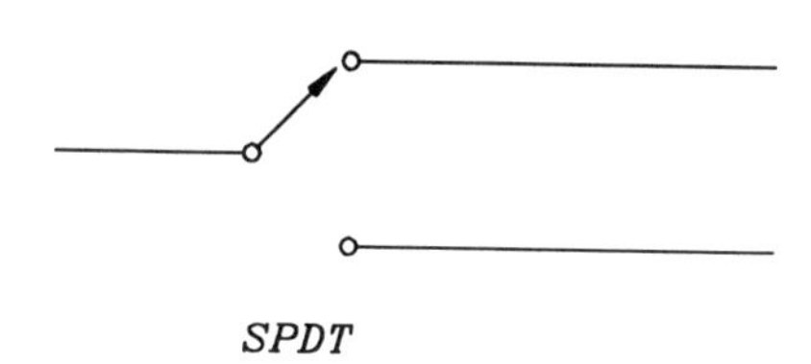

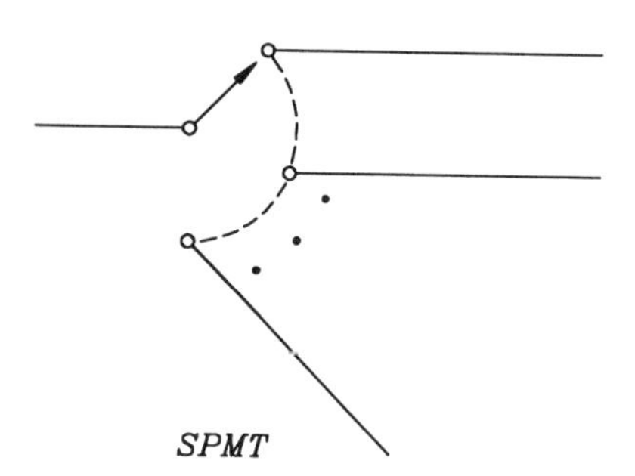

FIGURE 8.8 Switches and their output ports.

R and X can be calculated from Figure 8.10(b). For simplicity, let us assume that the packaged parasitics are negligible. From Figure 8.10(c), for forward bias, we have

$$Z_D \approx R_s + R_j = R + jX \tag{8.7}$$

$$X = 0 \quad \text{and} \quad R = R_s + R_j \tag{8.8}$$

$$\alpha = \alpha_L = \text{insertion loss (in dB)} = 20 \log\left(1 + \frac{R_s + R_j}{2Z_0}\right) \tag{8.9}$$

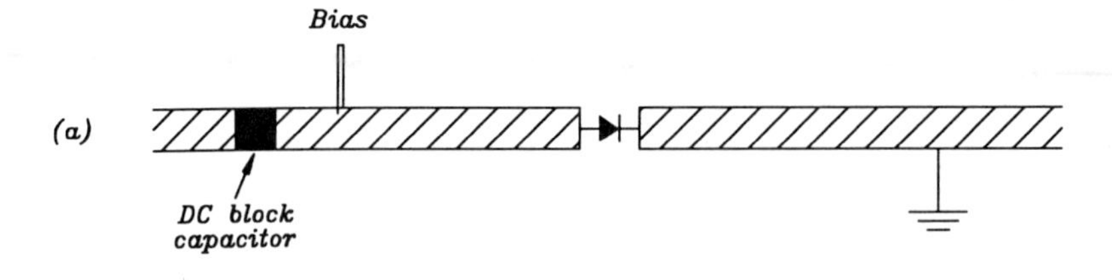

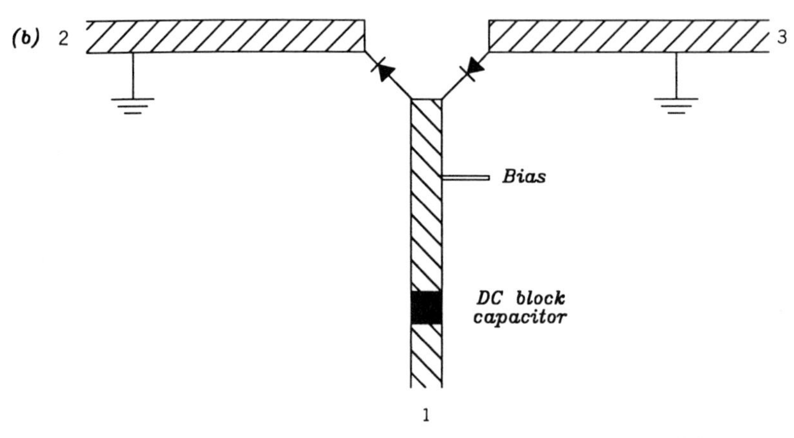

FIGURE 8.9 Series-mounted *p-i-n* switches in microstrip circuits: (*a*) SPST; (*b*) SPDT.

For reverse bias [Figure 8.10(*d*)], we have

$$Z_D \approx -j\frac{1}{\omega C_j} = R + jX \tag{8.10}$$

$$\frac{R}{Z_0} = 0 \quad \text{and} \quad X = -\frac{1}{\omega C_j} \tag{8.11}$$

$$\alpha = \alpha_I = \text{isolation (in dB)}$$

$$= 10\log\left[1 + \left(\frac{1}{4\pi f C_j Z_0}\right)^2\right] \tag{8.12}$$

For forward bias, the diode acts like a nearly shorted circuit. The power passes through with an insertion loss of α_L and the switch is ON. For reverse bias, the diode acts like a nearly open circuit. The power is reflected and the switch is OFF. A small portion leaks through with an isolation of α_I. For an

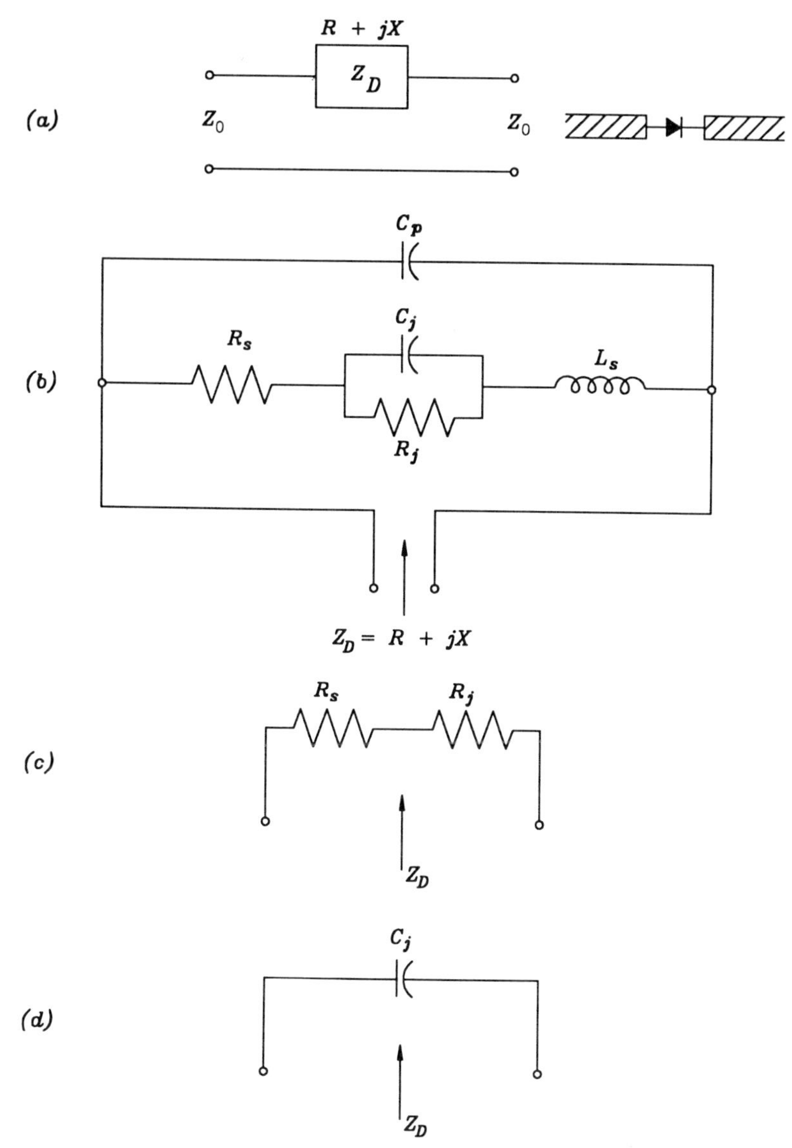

FIGURE 8.10 Series-mounted *p-i-n* switch: (*a*) SPST circuit; (*b*) diode impedance Z_D; (*c*) diode equivalent circuit in forward bias (neglecting the package parasitics); (*d*) diode equivalent circuit in reverse bias (neglecting the package parasitics).

ideal switch, one would like to have $\alpha_L = 0$ and $\alpha_I = \infty$, which mean that $R_s + R_j = 0$ and $C_j = 0$.

Example 8.1 Derive α_L and α_I in dB for a shunt-mounted SPST switch assuming that the package parasitics are negligible.

SOLUTION: As shown in Figure 8.11(a), the *p-i-n* diode is in shunt with a microstrip line. The equivalent circuit of the diode in Figure 8.11(b) is simplified to become Figure 8.11(c) and (d), respectively, if the package parasitics are negligible. From Equation (8.4), the insertion loss (in dB) is

$$\alpha = 10\log\left[\left(\frac{G}{2Y_0} + 1\right)^2 + \left(\frac{B}{2Y_0}\right)^2\right]$$

At reverse bias,

$$Y_D = G + jB = j\omega C_j,$$

$$B = \omega C_j \quad \text{and} \quad G \approx 0$$

$$\alpha = \alpha_L = \text{insertion loss} = 10\log\left[1 + (\pi f C_j Z_0)^2\right] \quad \text{dB} \qquad (8.13)$$

At forward bias,

$$Y_D = G + jB = \frac{1}{R_s + R_j}$$

$$G = \frac{1}{R_s + R_j} \quad \text{and} \quad B = 0$$

$$\alpha = \alpha_I = \text{isolation} = 20\log\left[\frac{Z_0}{2(R_s + R_j)} + 1\right] \quad \text{dB} \qquad (8.14)$$

At reverse bias, the diode acts as a nearly open circuit, the signal passes through the microstrip line with an insertion loss α_L which is due to the finite value of C_j. The switch is in the ON state. At forward bias, the diode acts as a nearly short circuit which causes the microwave power to be reflected totally. The isolation α_I is due to the finite value of $R_s + R_j$ when the switch is in the OFF state. Note that $\alpha_I = \infty$ and $\alpha_L = 0$ for an ideal diode with $C_j = 0$ and $R_s = R_j = 0$.

Example 8.2 A *p-i-n* diode has the following parameters: $L_s = 5$ nH, $R_s = 5\ \Omega$, $C_p = 0.05$ pF, $R_j = 0$ (in forward bias), $C_j = 0.2$ pF (in reverse

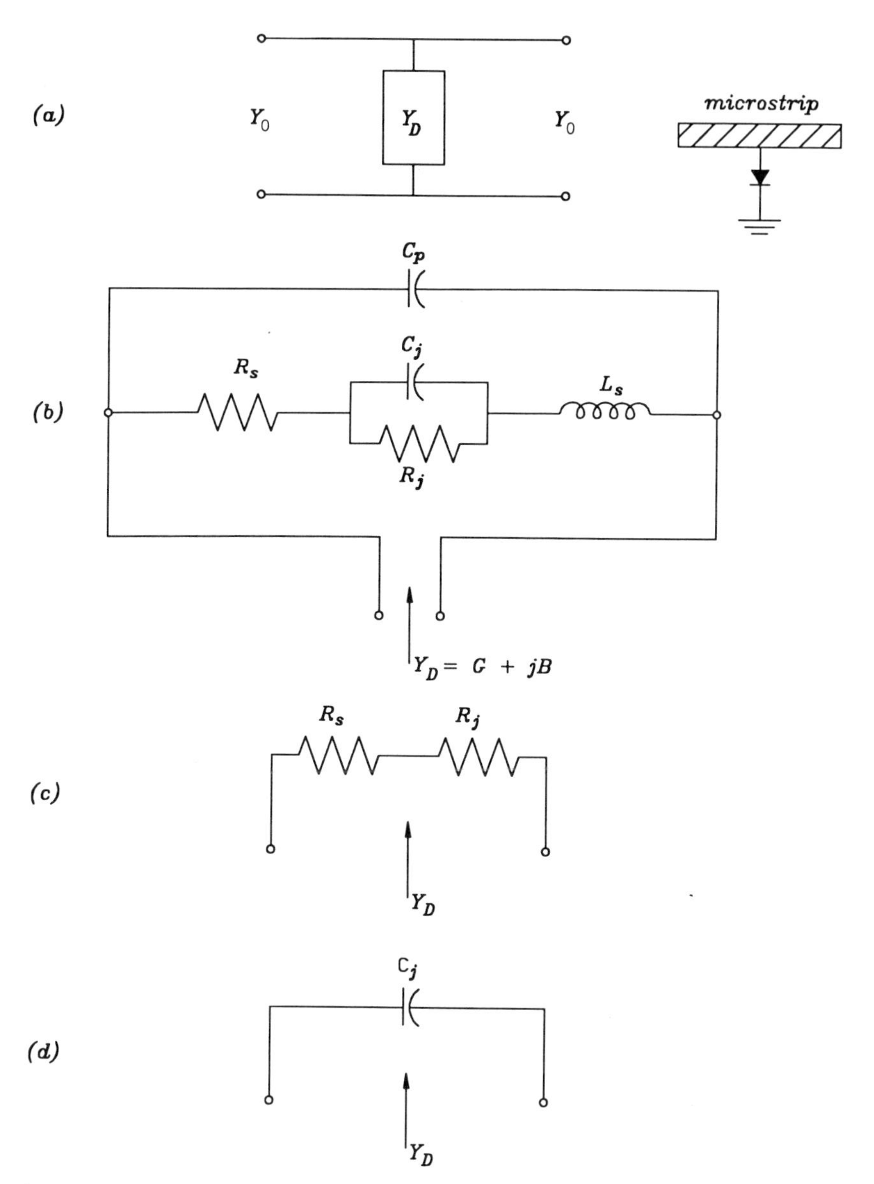

FIGURE 8.11 Shunt-mounted *p-i-n* switch: (*a*) SPST circuit; (*b*) diode admittance Y_D; (*c*) diode equivalent circuit in forward bias; (*d*) diode equivalent circuit in reverse bias.

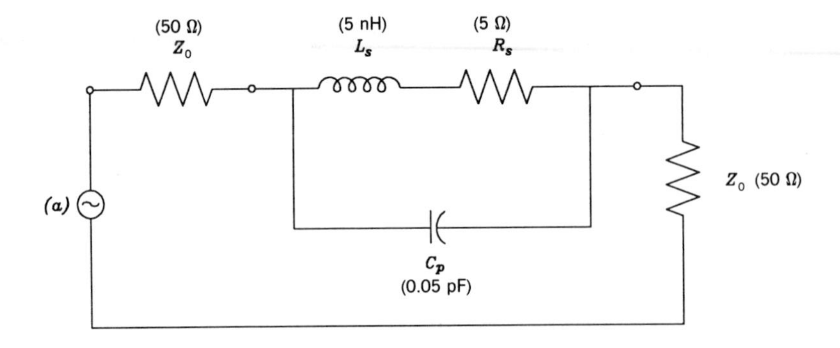

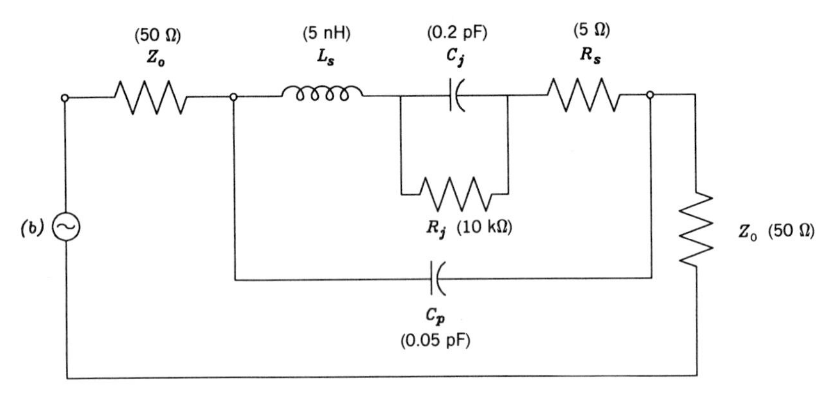

FIGURE 8.12 Equivalent circuits at (*a*) forward bias and (*b*) reverse bias.

bias), and $R_j = 10$ kΩ (in reverse bias). If the *p-i-n* is mounted in series with a 50-Ω microstrip line, calculate the isolation and insertion loss in dB at 1 GHz.

SOLUTION: At forward bias, the equivalent circuit is shown in Figure 8.12(*a*). The switch is ON and the insertion loss is

$$\alpha = \alpha_L = 10\log\left[\left(1 + \frac{R}{2Z_0}\right)^2 + \left(\frac{X}{2Z_0}\right)^2\right]$$

$$Y_D = j\omega C_p + \frac{1}{5 + j\omega L_s} = 0.0049 - j0.0307$$

$$Z_D = \frac{1}{Y_D} = R + jX = 5.070 + j31.76$$

$$\text{insertion loss} = \alpha_L = 10\log\left[\left(1 + \frac{5.070}{100}\right)^2 + \left(\frac{31.76}{100}\right)^2\right] = 0.809 \text{ dB}$$

At reverse bias, the equivalent circuit is shown in Figure 8.12(*b*). The switch is OFF and the isolation is

$$\alpha = \alpha_I = 10 \log\left[\left(1 + \frac{R}{2Z_0}\right)^2 + \left(\frac{X}{2Z_0}\right)^2\right]$$

Now

$$R_s + j\omega L_s + \frac{1}{(1/R_j) + j\omega C_j}$$

$$= 5 + 62.93 + j(31.4 - 791)$$

$$= 67.93 - j759.6$$

$$Y_D = j\omega C_p + \frac{1}{R_s + j\omega L_s + 1/\left[(1/R_j) + j\omega C_j\right]}$$

$$= j0.314 \times 10^{-3} + 116.8 \times 10^{-6} + j1.306 \times 10^{-3}$$

$$= 116.8 \times 10^{-6} + j1.62 \times 10^{-3}$$

$$Z_D = \frac{1}{Y_D} = R + jX = 44.27 - j614$$

$$\text{isolation} = \alpha_I = 10 \log\left[\left(\frac{44.27}{100} + 1\right)^2 + \left(\frac{614}{100}\right)^2\right]$$

$$= 16 \text{ dB}$$

Example 8.3 In Example 8.2, if a source generates 10 mW, calculate the power level reaching the load for (a) the ON and (b) the OFF states.

SOLUTION: (a) For the ON state, the *p-i-n* diode is forward biased. From Example 8.2, the insertion loss is

$$\alpha_L = 0.809 \text{ dB}$$

$$\text{input power } P_i = 10 \text{ mW} = 10 \text{ dBm}$$

$$\text{output power } P_o = P_i - \alpha_L$$

$$= 10 \text{ dBm} - 0.809 \text{ dB} = 9.191 \text{ dBm} = 8.3 \text{ mW}$$

(b) For the OFF state, the *p-i-n* diode is reversed biased. The isolation is

$$\alpha_I = 16 \text{ dB}$$

$$\text{output power} = P_i - \alpha_I$$

$$= 10 \text{ dBm} - 16 \text{ dB} = -6 \text{ dBm} = 0.25 \text{ mW}$$

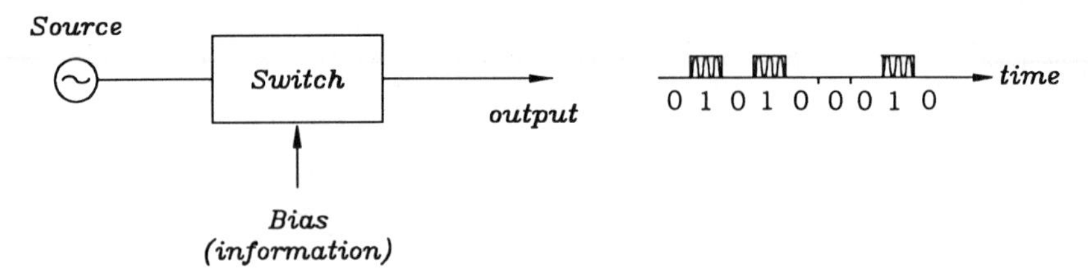

FIGURE 8.13 Digital amplitude modulators.

8.5 MODULATORS AND ATTENUATORS

A switch can be used as a digital modulator that generates a "1" or "0" output as a function of time according to the transmit information. As shown in Figure 8.13 when the switch is ON, the power will pass through the switch. When the switch is OFF, the power is reflected and no power will appear at the output port. By controlling the bias to the switch, information can be transmitted using amplitude modulation.

Instead of operating in the two states ON and OFF, the switch can be used as an analog modulator. By varying the bias continuously, different levels of output power can be achieved. In this case, the circuit becomes an analog modulator or a variable attenuator.

8.6 BIASING TECHNIQUES

The ideal biasing arrangement should not affect the RF performance of a microwave circuit. In other words, one would like to design the biasing network such that the RF signal will not leak into the biasing circuit and the biasing circuit will not disturb the microwave transmission.

Figure 8.14(*a*) shows a possible biasing arrangement for a series-mounted *p-i-n* switch [2]. The chip capacitors are used as dc bias blocks. Ideally, the capacitor should present a short circuit at the operating microwave frequency. The capacitance value should be selected such that $2\pi fC$ is very large. Therefore, $Z_c = -j(1/\omega C)$ is very small and the signal will pass through the capacitor with little loss or reflection. For example, at 10 GHz, a 100-pF capacitor will give $Z_c = -j0.16 \approx 0$, which is almost a short circuit.

At high microwave frequencies, a gap in a microstrip line could serve as a dc-block capacitor. The gap is usually designed in a finger shape, as shown in Figure 8.14(*b*), to increase the capacitance. Similar configurations could be realized in coaxial line center conductors.

In Figure 8.14(*a*), the bias is connected to the 50-Ω microstrip line through a low-impedance pad and a $\lambda_g/4$ high-impedance line. The low–high line serves as a low-pass filter which prevents the microwave signal from leaking into the bias. The $\lambda_g/4$ line transforms the low impedance at the bias

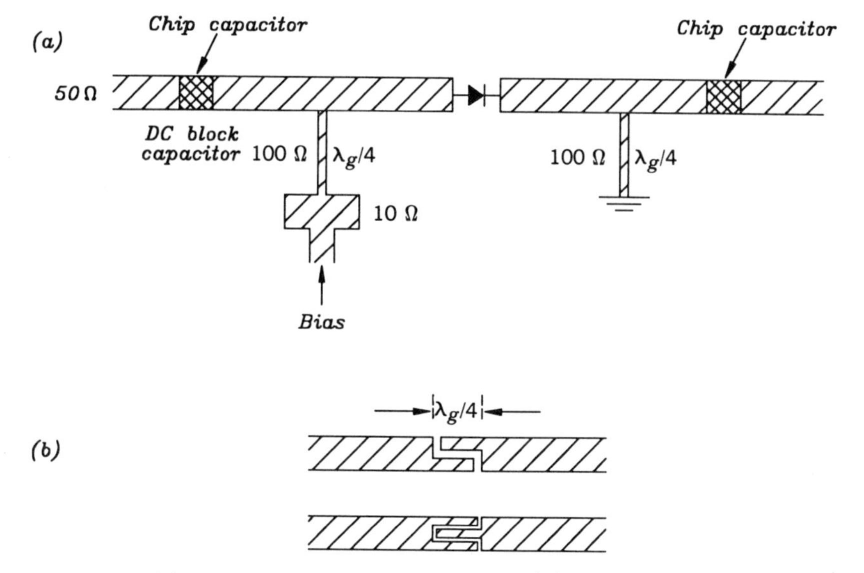

FIGURE 8.14 (a) Example of biasing arrangement; (b) capacitance formed by a gap in the microstrip line.

pad to a high impedance (like an open circuit) at the microstrip line. Since the bias line presents an RF open circuit to the microstrip line, the bias line does not disturb the microwave circuit. Similarly, a very high impedance quarter-wavelength line (100 Ω) is used to connect the ground to the microstrip line. At the connection, the microstrip line sees an RF open circuit.

8.7 PHASE SHIFTERS

One major application of phase shifters is in phased-array antennas. In a phased-array antenna, each antenna element is connected to a phase shifter. By varying the phase of each element, one can steer the array radiation beam to the desired direction electronically.

Several forms of phase-shifter circuits have been used in the past. The three most common methods are the hybrid-coupler phase-shifter circuit, the direct transmission loaded-line phase-shifter circuit, and the switched-line phase-shifter circuit. Their circuit configurations are shown in Figure 8.15.

The advantage of the hybrid-coupler phase shifter is that it uses the fewest diodes (two per bit), and any phase-shift increment can be obtained with proper design of the terminating circuit. A hybrid coupler provides a 3-dB power split for the two output arms with a 90° phase difference. This coupler can be a branch-line hybrid coupler, a backward-wave, or a proximity-

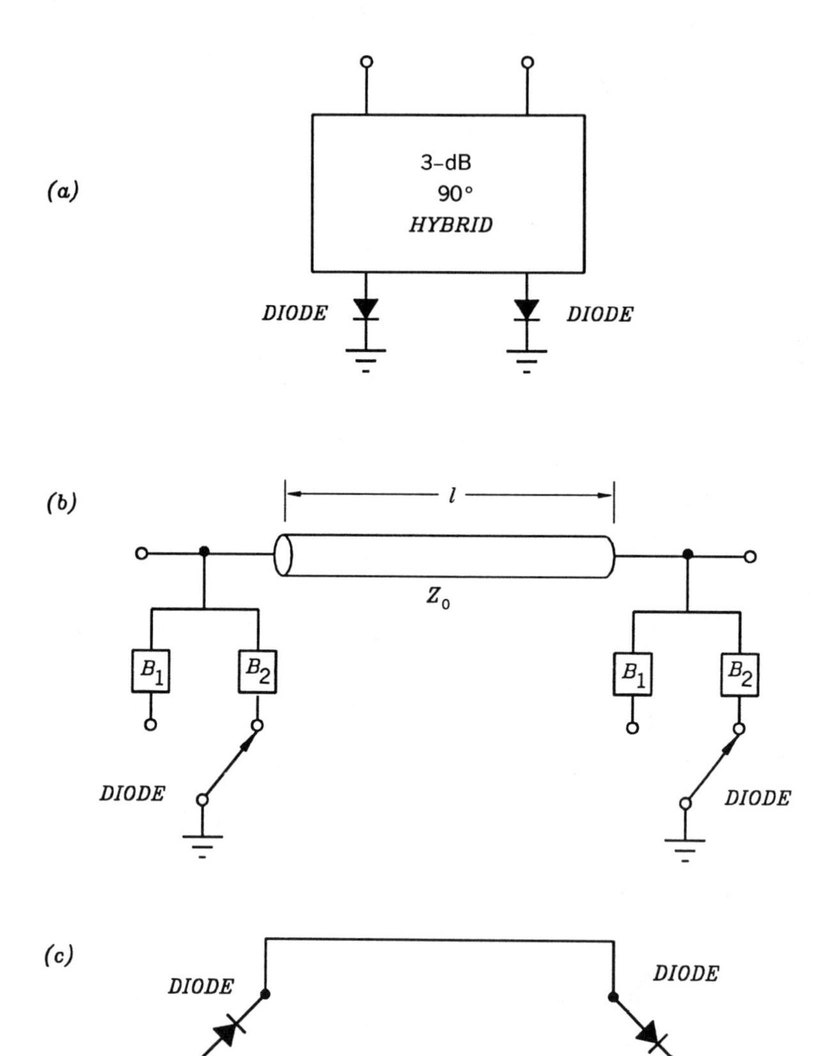

FIGURE 8.15 Three different types of phase-shifter circuits (shown are one-bit phase shifters): (*a*) hybrid-coupler phase shifter; (*b*) loaded-line phase shifter; (*c*) switched-line phase shifter.

coupled hybrid. At frequencies below 40 GHz, such couplers can easily be fabricated in a microstrip line with 1- to 2-dB insertion loss per bit. For frequencies above 40 GHz, it is difficult to achieve low loss in hybrid couplers.

The loaded-line phase shifter is more attractive for high-frequency applications. As shown in Figure 8.15(*b*), the phase shifter is realized by placing diode-controlled, switched reactances a quarter-wavelength apart on a transmission line. The phase-shifter design is based on two factors. First, any symmetric pair of quarter-wavelength-spaced shunt susceptances will have mutually canceling reflections if their normalized susceptances are small compared with unity. The second factor is that shunt capacitance elements electrically lengthen a transmission line, whereas inductive elements shorten it. Thus switching from inductive to capacitive elements produces an increase in electrical length with a corresponding phase shift. The loading susceptances of these elements are controlled with switching diodes to electrically shorten or lengthen the transmission line.

The third type of phase shifter uses switching action to obtain insertion phase by providing alternative transmission paths, the difference in electrical lengths being the desired phase shift. This approach offers the opportunity for true time delay rather than steady-state phase control. The disadvantage of this circuit is its high insertion loss at millimeter-wave frequencies. Another disadvantage is that four diodes are needed per bit.

The switched-line phase shifter is easily understood. As shown in Figure 8.16, if the bias is positive, the signal flows through the upper line with a path length of l_1. If the bias is negative, the signal flows through the lower line with a path length of l_2. The phase difference between the two bias states is

$$\Delta\phi = \frac{2\pi}{\lambda_g}(l_1 - l_2) \tag{8.15}$$

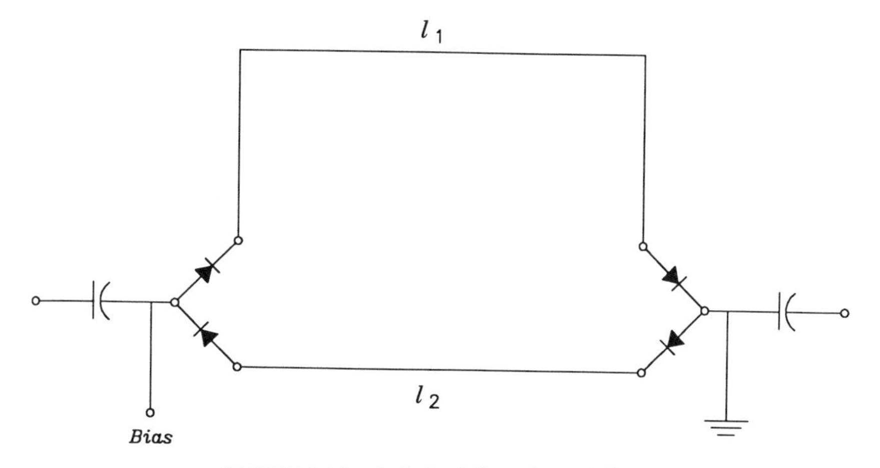

FIGURE 8.16 Switched-line phase shifter.

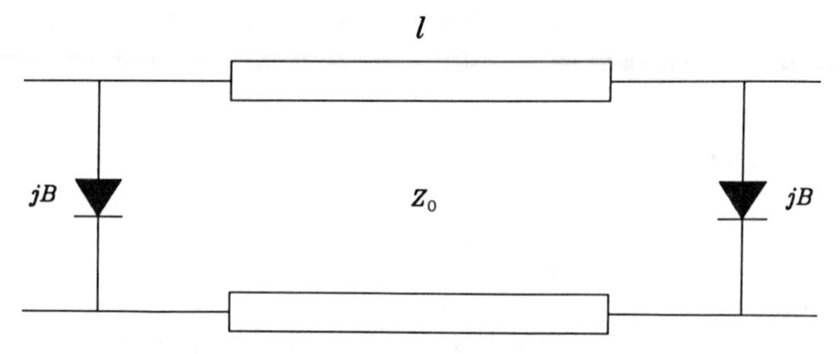

FIGURE 8.17 Loaded-line phase shifter.

The loaded-line phase shifter can be analyzed using the *ABCD* matrices as discussed in Chapter 3 and Section 8.3. For simplicity, assume that the package parasitics are negligible and that $R_j + R_s = 0$. The diode acts as a short circuit at forward bias and is equivalent to a capacitor C_j at reverse bias. The loaded-line phase-shifter circuit is shown in Figure 8.17 with

$$jB = \begin{cases} jB_1 = 0 & \text{at forward bias} \\ jB_2 = j\omega C_j & \text{at reverse bias} \end{cases}$$

The overall *ABCD* matrix is

$$\begin{bmatrix} A_t & B_t \\ C_t & D_t \end{bmatrix} = \begin{bmatrix} 1 & 0 \\ jB & 1 \end{bmatrix} \begin{bmatrix} \cos\theta & jZ_0 \sin\theta \\ jY_0 \sin\theta & \cos\theta \end{bmatrix} \begin{bmatrix} 1 & 0 \\ jB & 1 \end{bmatrix}$$
$$= \begin{bmatrix} \cos\theta - BZ_0 \sin\theta & jZ_0 \sin\theta \\ j\left[2B\cos\theta + \left(Y_0 - B^2 Z_0\right)\sin\theta\right] & \cos\theta - BZ_0\sin\theta \end{bmatrix} \quad (8.16)$$

where $\theta = \beta l$. The S_{21} can be calculated from the *ABCD* matrix by using the following equation:

$$S_{21} = \frac{2}{A_t + B_t Y_0 + C_t Z_0 + D_t} = \text{Re}(S_{21}) + j\,\text{Im}(S_{21})$$
$$= |S_{21}| e^{j\phi} = \frac{2}{E + jF} \quad (8.17)$$
$$E = 2(\cos\theta - BZ_0 \sin\theta)$$
$$F = 2\left(\sin\theta + Z_0 B\cos\theta - \frac{B^2 Z_0^2}{2}\sin\theta\right)$$

The phase shift is

$$\phi = \tan^{-1}\left(-\frac{F}{E}\right) = \tan^{-1}\left[-\frac{\sin\theta + Z_0 B\cos\theta - \left(B^2 Z_0^2/2\right)\sin\theta}{\cos\theta - BZ_0\sin\theta}\right]$$

$$= \tan^{-1}\left\{\frac{-(B/Y_0) - \left[1 - \frac{1}{2}\left(B^2/Y_0^2\right)\right]\tan\theta}{1 - (B/Y_0)\tan\theta}\right\} \tag{8.18}$$

If $l = \frac{1}{4}\lambda_g$, $\theta = \pi/2$ and $\tan\theta \to \infty$, we have

$$\phi = \tan^{-1}\left(\frac{2Y_0^2 - B^2}{2Y_0 B}\right) \tag{8.19}$$

The differential phase shift between the two states (forward and reverse biased) is

$$\Delta\phi = \tan^{-1}\frac{2Y_0^2 - B_2^2}{2Y_0 B_2} - \tan^{-1}\frac{2Y_0^2 - B_1^2}{2Y_0 B_1} \tag{8.20}$$

The operating principle of a 3 dB branch-line hybrid-coupler, reflection-type phase shifter is explained in the following. As shown in Figure 8.18, the signal exhibits an additional 90° phase shift when it crosses the hybrid from port 1 to 3, 3 to 1, 2 to 4, or 4 to 2 (because of $\lambda_g/4$-longer path) as compared to the signal moving from port 1 to 2 or 3 to 4. If ports 2 and 3 are connected to a ground (i.e., a short circuit), the input power will split equally into ports 2 and 3. The reflected signals from ports 2 and 3 will recombine in phase at port 4 but 180° out of phase at port 1. Therefore, the output appears at port 4. Now, let us consider that *p-i-n* diodes are connected to ports 2 and 3 as shown in Figure 8.18. When the bias to the diodes is positive, the diodes are ON. We have the phase shift from port 1 to port 4:

$$\text{phase of signal at output} - \text{phase of signal at input} = \phi_1$$

When the bias is negative, the diodes are OFF. The split input signals will each travel an additional length of $2l$. We have

$$\text{phase of signal at output} - \text{phase of signal at input} = \phi_1 + \frac{2\pi}{\lambda_g}\cdot 2l$$

The differential phase shift is thus given by

$$\Delta\phi = \frac{2\pi}{\lambda_g}\cdot 2l = \frac{4\pi l}{\lambda_g} \tag{8.21}$$

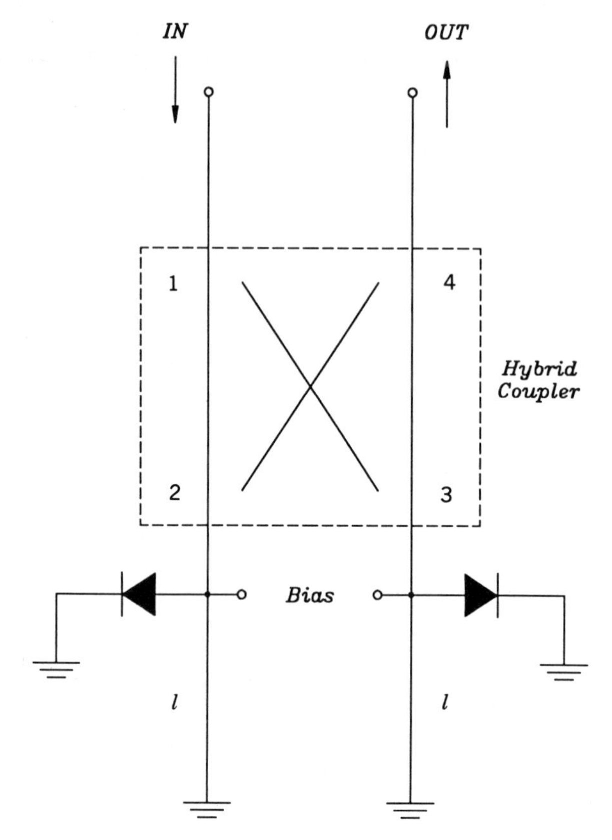

FIGURE 8.18 Hybrid coupler reflection type of phase shifter.

Example 8.4 A circulator can be used with a *p-i-n* switch to form a reflection-type phase shifter as shown in Figure 8.19. Explain how it works.

SOLUTION: As shown in Figure 8.19, when the bias is positive, the diode is forward biased. A nearly shorted circuit is presented at point A and the path length is $2l_1$. When the bias is negative, the diode is reverse biased. The signal travels to point B before it is reflected back to the circulator. The path length is $2(l_1 + l_2)$. The differential phase shift between these two states is

$$\Delta\phi = \beta\,\Delta l = \frac{2\pi}{\lambda_g}[2(l_1 + l_2) - 2l_1]$$

$$= \frac{4\pi l_2}{\lambda_g}$$

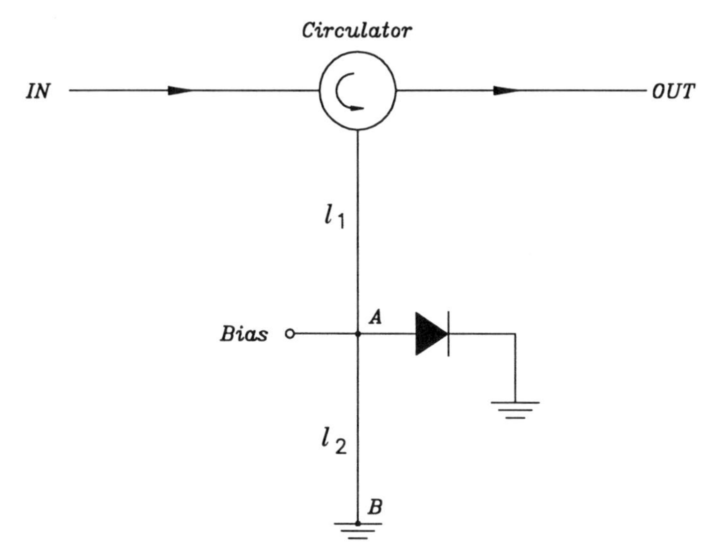

FIGURE 8.19 Circulator-coupled phase shifter.

8.8 EXAMPLES OF PRACTICAL CIRCUITS

Switches, modulators, and phase shifters have been built in coaxial line, waveguide, microstrip, strip-line, and fin-line circuits. In this section we look at a few practical circuits developed up to millimeter-wave applications.

In a typical waveguide circuit, the *p-i-n* diode is mounted at the end of a bias post. Figure 8.20 shows a microstrip *p-i-n* attenuator or SPST switch circuit configuration. Two diodes are mounted across the microstrip line in a shunt configuration. Bias voltages are applied through two bias pads. This circuit has an insertion loss of less than 1 dB and maximum attenuation of 25 dB. The attenuator can be used for an SPST switch if only two bias states are employed rather than continuous variation of the bias current. A similar circuit has been used by Tokumitsu et al. [4] for an amplitude shift keying (ASK) modulator.

A biphase switch using a combination of microstrip and slot line was demonstrated by Ogawa et al. [5] at 27 GHz and by Grote and Chang at 60 GHz [6]. Figure 8.21 shows the schematic circuit of the biphase switch. The 60-GHz signal at the microstrip input is transferred to the slot line via the microstrip-to-slot line transition. The bias states of the Schottky-barrier diodes then determine which path the signal takes as the data alternately switch the Schottky diodes on and off. The signal takes path 1 or path 2, producing a biphase output signal since the direction of the electric field at the output junction is 180° out of phase. A second slot line-to-microstrip transition is used to transfer the 60-GHz modulated signal back to microstrip.

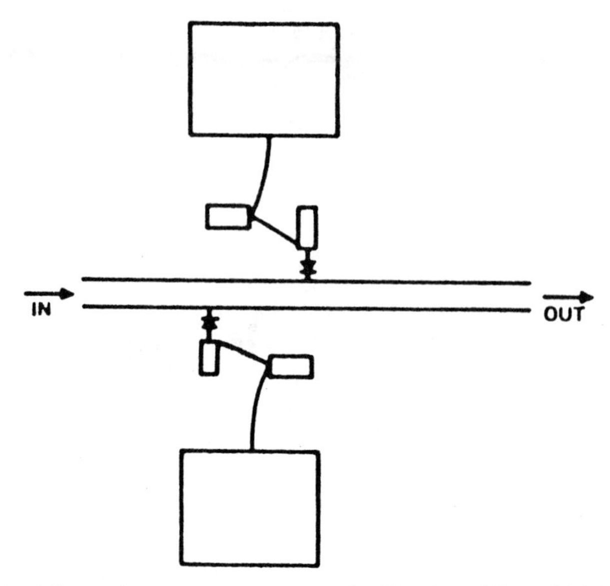

FIGURE 8.20 Microstrip *p-i-n* attenuator or SPST switch. (From Ref. 3 with permission from SPIE.)

This switch was designed for very high speed switching with data rates up to multigigabit.

A microstrip *p-i-n* switch has been demonstrated at W-band. Sisson et al. [7] reported greater than 20 dB isolation and 1 dB insertion loss over the frequency range 80 to 90 GHz.

P-i-n switches have been integrated with resonators and filters to form a switchable bandpass filter. Integration with a microstrip ring resonator and a coplanar waveguide bandpass filter has recently been reported [8,9].

Fin line has been considered for *p-i-n* switch applications because of its low-loss and broadband characteristics. A series-mounted *p-i-n* switch has been devised to tune out the reactance of the diode, thus achieving a pole in the attenuation. For simplicity and good performance, most circuits are in a shunt-mounted configuration. Figure 8.22 shows a two-diode switch. Two diodes are mounted across the gap in a shunt configuration. The lower side of the fin line is grounded. Bias voltage is applied through a resistor. The equivalent circuit of this switch is shown in Figure 8.23. With this equivalent circuit, the mismatch loss and isolation can be predicted. Fin-line attenuators and switches have been demonstrated successfully by Meier [10] and Meinel and Rembold [11] for Ka-band and by Chang and Sun [3] for W-band. Fin-line SPDT and SP3T switches can be formed by mounting diodes in three- or four-port junctions.

Significant strides have been made in the development and application of FETs to microwaves and millimeter waves since 1971. As a switching device,

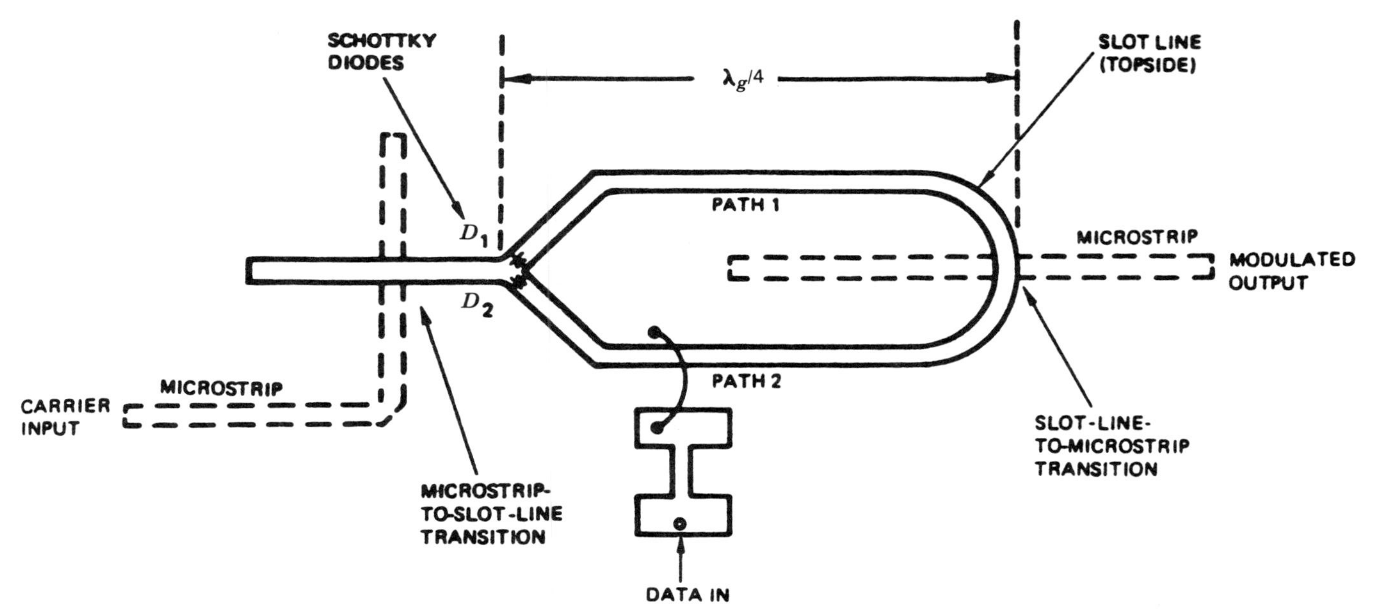

FIGURE 8.21 Schematic of a biphase switch. (From Ref. 6 with permission from IEEE.)

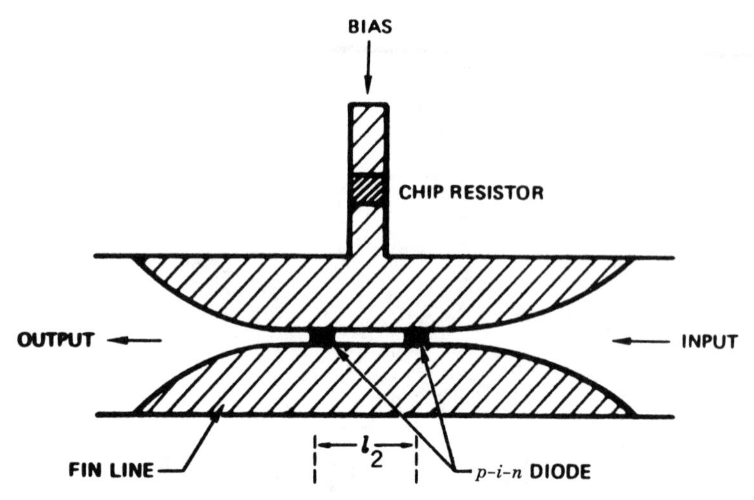

FIGURE 8.22 Two-diode attenuator or switch. (From Ref. 3 with permission from SPIE.)

the FET is an attractive alternative to the *p-i-n* diode because it offers the possibility of high-speed switching with gain. Other advantages of the FET are good on-to-off isolation and little or no impedance variation from the input or output ports in either the ON or OFF states. The FET switch or phase shifter has another advantage that it can eventually be translated to monolithic circuits. FET switches and phase shifters are discussed later.

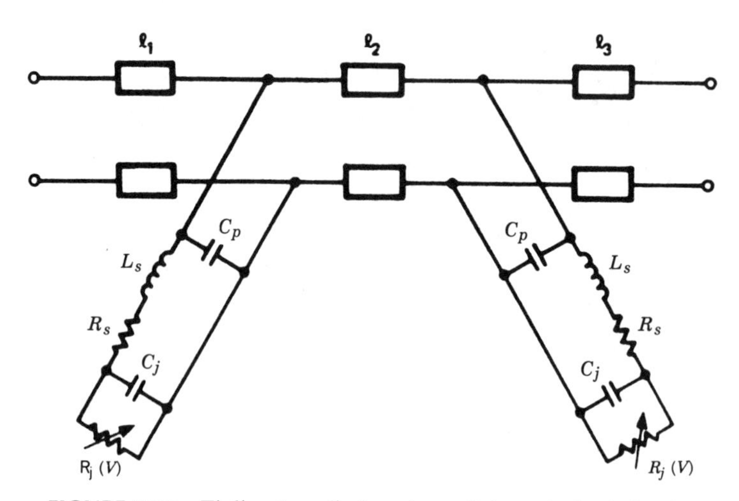

FIGURE 8.23 Finline two-diode *p-i-n* switch equivalent circuit.

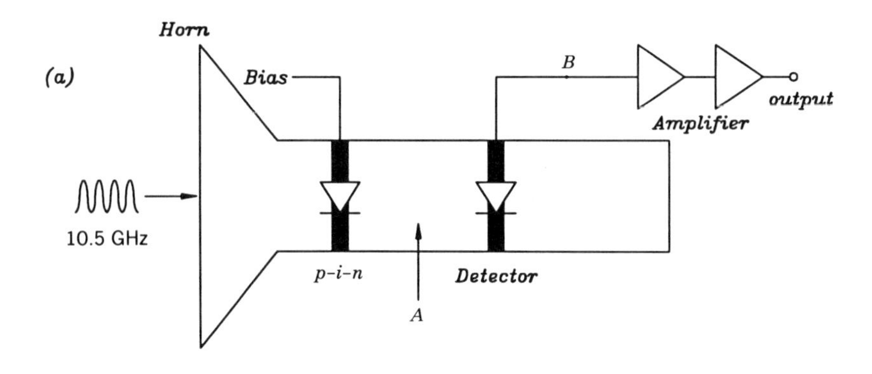

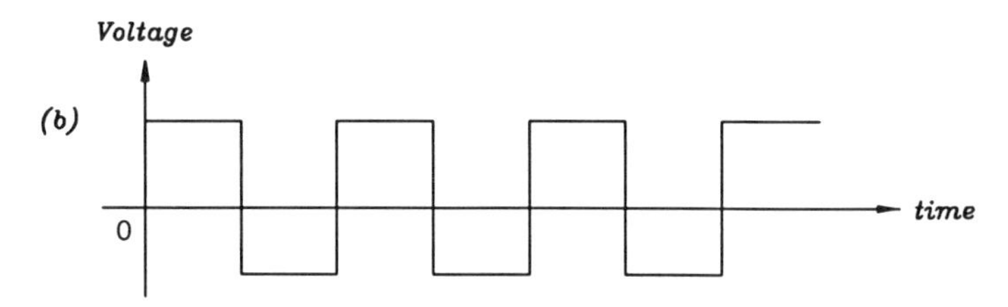

FIGURE 8.24 Radar detector: (a) receiver configuration; (b) bias to the *p-i-n* diode.

Example 8.5 A 10.5-GHz police radar detector consists of a *p-i-n* diode followed by a detector diode. The circuit can be built in a waveguide using a horn antenna as shown in Figure 8.24(a). The police radar transmits a 10.5-GHz CW signal. Draw the waveforms after the *p-i-n* diode and the detector and explain how it works. The bias to the *p-i-n* is shown in Figure 8.24(b). The pulse repetition rate is in kHz.

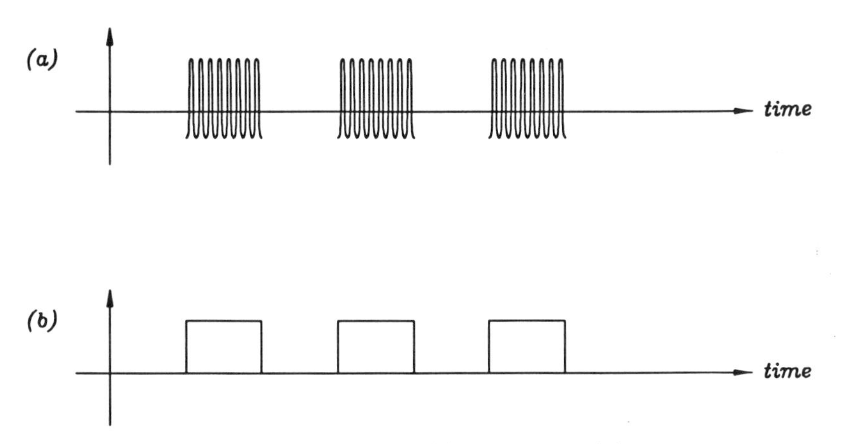

FIGURE 8.25 Waveforms at (a) point A and (b) point B.

SOLUTION: At forward bias, the *p-i-n* diode acts as a short circuit in shunt with the waveguide. All power is reflected. At reverse bias, the *p-i-n* diode acts as an open circuit and all power passes through. The waveform after the *p-i-n* diode (at point A) is shown in Figure 8.25(a). The detector acts as a demodulator. A dc output appears at the detector output port when there is a CW signal incident into the detector. The waveform after the detector (at point B) is shown in Figure 8.25(b).

When a police radar signal is present, there is an output pulse train at the detector output. The output is amplified and integrated. If the output is greater than the threshold, an alarm is activated.

PROBLEMS

P8.1 In Example 8.2, calculate the isolation and insertion loss in dB at 1 GHz using the approximate formulas neglecting L_s, C_p, and R_j.

P8.2 A *p-i-n* diode is mounted in series with a 50-Ω microstrip line. The *p-i-n* diode has $R_s = 1.3\ \Omega$, $R_j = 0.2\ \Omega$ at forward bias and 100 kΩ at reverse bias, and $C_j = 0.1$ pF. L_s and C_p are negligible. At 1 GHz, calculate (**a**) the attenuation in dB when the diode is forward biased, (**b**) the attenuation in dB when the diode is reverse biased, and (**c**) the output power in mW when the diode is forward and reverse biased. The input power is 10 mW, and there is no mismatch loss.

P8.3 For an ideal switch (i.e., insertion loss $\alpha_L = 0$ and isolation $\alpha_I = \infty$), we would like to see a *p-i-n* diode with $R_s =$ _____ ohms, $R_j =$ _____ ohms, $C_j =$ _____ pF, $C_p =$ _____ pF, and $L_s =$ _____ nH.

P8.4 Derive the phase shift ϕ and attenuation α for the circuit shown in Figure P8.4.

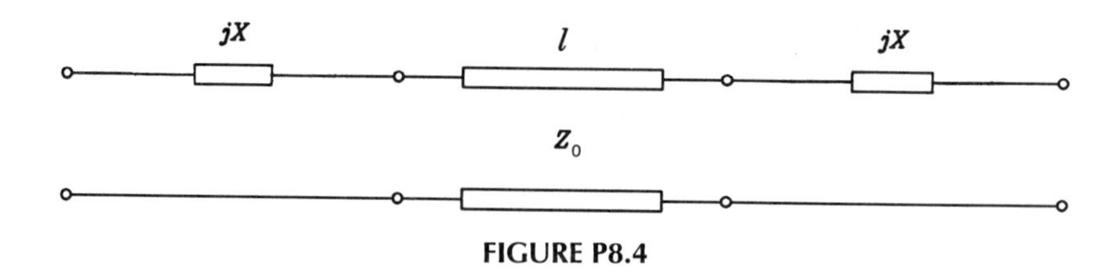

FIGURE P8.4

P8.5 For the circuit shown in Figure P8.5, fill in the signal paths for different bias conditions. How many different states of differential phase shift ($\Delta\phi$) can one achieve if $l_1 = \lambda_g/8$, $l_2 = \lambda_g/8$, and $l_3 = \lambda_g/4$?

Bias 1	*Bias 2*	*Signal path*
+	+	$2l_1$
+	−	?
−	+	?
−	−	?

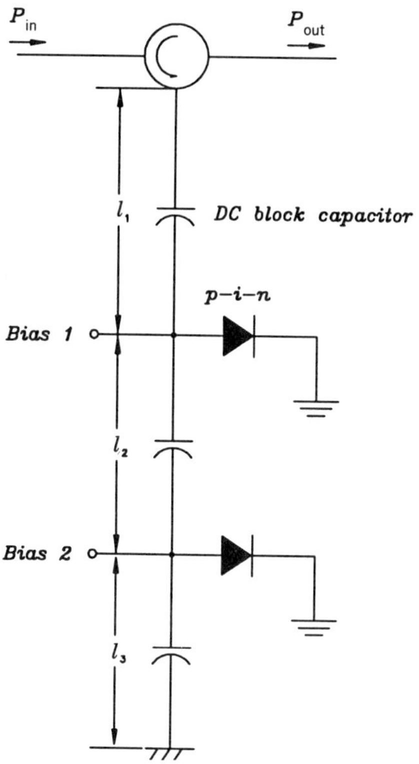

FIGURE P8.5

P8.6 At 1 GHz, a variable phase shifter is built with a varactor mounted in series with a 50-Ω transmission line as shown in Figure P8.6. An abrupt junction varactor is used with a junction capacitance of 20 pF at 0 V, a built-in voltage of 1 V, and a series resistance of 2 Ω. What is the phase shift when the bias is (**a**) 0 V and (**b**) −10 V?

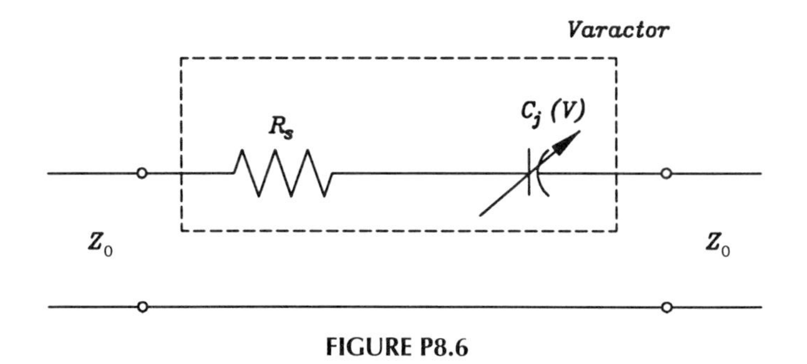

FIGURE P8.6

P8.7 Derive the S parameters for the circuit shown in Figure P8.7. What are the attenuation values in dB and the phase shift in degrees for this circuit?

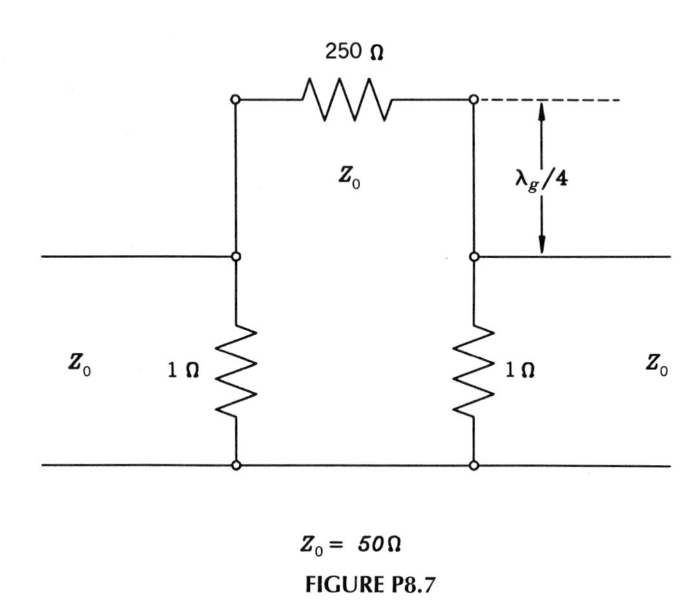

FIGURE P8.7

P8.8 A varactor is mounted in shunt with a transmission line of characteristic impedance of 50 Ω as shown in Figure P8.8. Calculate the phase shift at 10 GHz if $G = 0.2$ S and $C_j(V)$ is (**a**) 10 pF and (**b**) 20 pF.

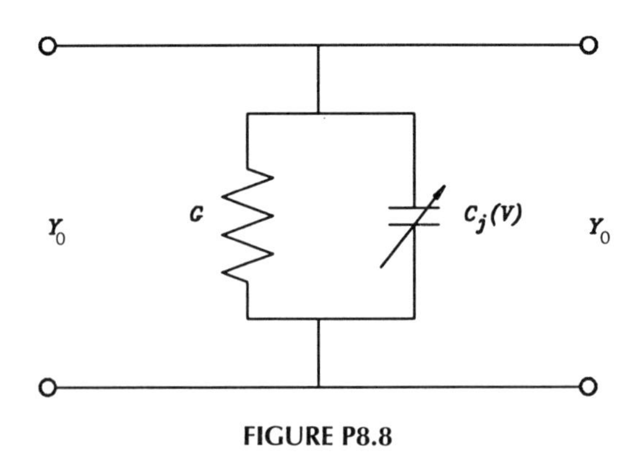

FIGURE P8.8

P8.9 A *p-i-n* diode has $R_s = 1\ \Omega$, $C_j = 0.1$ pF, and $R_j = 1\ \Omega$ at forward bias. $R_j = 100$ kΩ at reverse bias. The diode is mounted in series with a 50-Ω microstrip line to form a switch as shown in Figure P8.9. Neglect the package parasitics.

(a) Calculate the insertion loss and isolation in dB at 5 GHz when the diode is forward and reverse biased.

(b) What is the reflection coefficient Γ when the diode is forward biased?

(c) What is the insertion loss and the isolation if two of these diodes are connected in series?

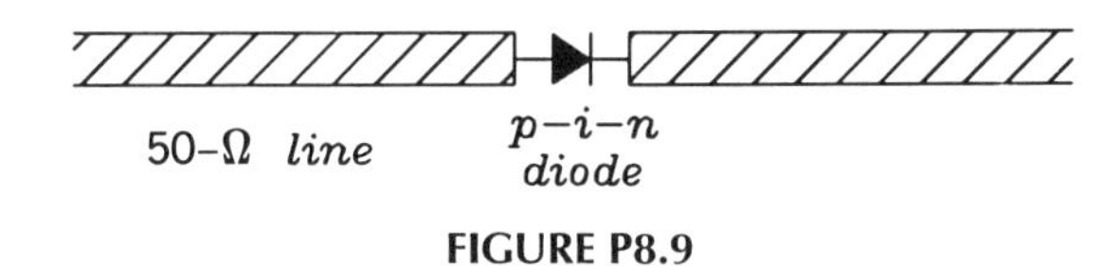

FIGURE P8.9

P8.10 A dc block capacitor of 100 pF is used in a bias circuit for a solid-state device as shown in Figure P8.10. The microstrip line has a characteristic impedance of 50 Ω. At 10 GHz, what are the reflection coefficient and VSWR values due to this capacitor? Is this a good dc block? (The capacitor is in series with the microstrip line.)

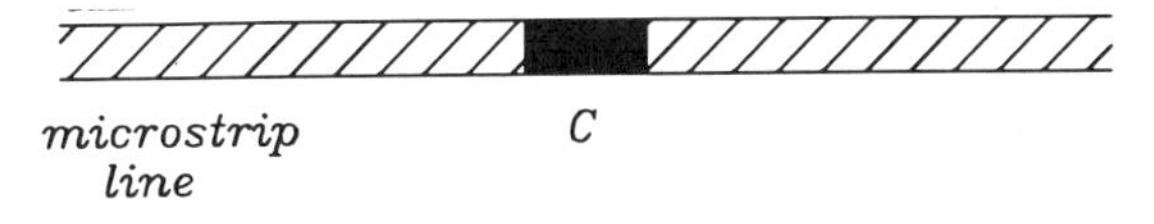

FIGURE P8.10

REFERENCES

1. J. F. White, "Semiconductor Control Devices: PIN Diodes," in *Handbook of Microwave and Optical Components*, K Chang, Editor, Wiley, New York, 1990, Vol. 2, Chap. 4.
2. E. A. Wolff and R. Kaul, *Microwave Engineering and Systems Applications*, Wiley, New York, 1988, Chap. 12.
3. K. Chang and C. Sun, "Millimeter-Wave Integrated Circuit Components," *Millimeter-Wave Technology II*, Vol. 423, *Proceedings of SPIE (International Society of Optical Engineering)*, August 1983, San Diego, Calif., pp. 50–57.

4. Y. Tokumitsu, M. Ishizaki, M. Iwakuni, and T. Saito, "50 GHz IC Components Using Alumina Substrates," *IEEE Transactions on Microwave Theory and Techniques*, Vol. MTT-31, February 1983, pp. 121–128.
5. H. Ogawa, M. Aikawa, and M. Akaike, "Integrated Balanced BPSK and QPSK Modulators for the Ka-Band," *IEEE Transactions on Microwave Theory and Techniques*, Vol. MTT-30, March 1982, pp. 227–234.
6. A. Grote and K. Chang, "60 GHz Integrated Circuit High Data Rate Quadriphase Shift Keying Exciter and Modulator," *IEEE Transactions on Microwave Theory and Techniques*, Vol. MTT-32, December 1984, pp. 1663–1667.
7. M. J. Sisson et al., "Microstrip Devices for Millimeter-Wave Frequencies," in *IEEE International Microwave Symposium Technical Digest*, 1982, pp. 212–214.
8. T. S. Martin, F. Wang, and K. Chang, "Theoretical and Experimental Investigation of Novel Varactor Tuned Switchable Microstrip Ring Resonator Circuits," *IEEE Transactions on Microwave Theory and Techniques*, Vol. 36, December 1988, pp. 1733–1739.
9. Y. Shu, J. A. Navarro, and K. Chang, "Electronically Switchable and Tunable Coplanar Waveguide-Slotline Bandpass Filters," *IEEE Transactions on Microwave Theory and Techniques*, Vol. 39, March 1991, pp. 548–554.
10. P. J. Meier, "Integrated Fin-Line Millimeter Components," *IEEE Transactions on Microwave Theory and Techniques*, Vol. MTT-22, December 1974, pp. 1209–1216.
11. M. Meinel and B. Rembold, "New Millimeter-Wave Fin-Line Attenuators and Switches," *IEEE International Microwave Symposium Technical Digest*, 1979, pp. 249–252.

FURTHER READING

1. J. F. White, *Microwave Semiconductor Engineering*, Van Nostrand Reinhold, New York, 1982.
2. W. A. Davis, *Microwave Semiconductor Circuit Design*, Van Nostrand Reinhold, New York, 1984, Chap. 13.
3. P. Bhartia and I. J. Bahl, *Millimeter-Wave Engineering and Applications*, Wiley, New York, 1984, Chap. 8.
4. R. Stockton and A. I. Screenivas, "Semiconductor Control Devices: Phase Shifters and Switches," in *Handbook of Microwave and Optical Components*, K. Chang, Editor, Wiley, New York, 1990, Vol. 2, Chap. 5.

CHAPTER 9

Oscillator and Amplifier Circuits Using Two-Terminal Devices

9.1 INTRODUCTION

Before introducing the IMPATT and Gunn devices in the next two chapters, the general theory of oscillator and amplifier circuits using two-terminal devices should be presented. Both IMPATT and Gunn devices are two-terminal devices and their major applications are to build oscillators and amplifiers.

9.2 GENERAL THEORY OF OSCILLATORS

Two-terminal active devices can be represented in terms of impedances or admittances at the frequency range of operation. The device impedance or admittance has a negative real part in the frequency range of interest. A detailed oscillator theory is described by Kurokawa [1].

The impedance of the device at the fundamental frequency can be expressed as

$$Z_D = -R_D + jX_D \qquad \text{with } R_D > 0 \tag{9.1}$$

Assuming only a fundamental component of RF current present in addition to the dc current, one can write

$$I = I_0 + I_{\mathrm{RF}} \sin \omega t \tag{9.2}$$

FIGURE 9.1 General oscillator circuit.

where I_{RF} is the amplitude of the fundamental component of the RF current.

The device impedance is a function of frequency (f), dc current (I_0), RF current (I_{RF}), and temperature (T). Thus,

$$Z_D = Z_D(f, I_0, I_{RF}, T) \tag{9.3}$$

A similar expression can be written for the device admittance.

A general oscillator circuit is shown in Figure 9.1, where Z_D is the device impedance and Z_C is the circuit impedance looking at the device terminals (driving point). The transformer network includes the diode package and embedding circuits. Z_C can be expressed as

$$Z_C(f) = R_C(f) + jX_C(f) \tag{9.4}$$

The oscillating frequency is determined by the resonant frequency of the overall circuit. At resonance, the total reactance (or admittance) equals zero. Another condition of oscillation requires that the net resistance (or conductance) of the circuit be negative. Negative resistance is a common characteristic of all two-terminal solid-state active devices. The voltage drop across a positive resistance is positive and a power of I^2R is dissipated in the resistance. The voltage drop across a negative resistance is negative and a power of $-I^2R$ is generated in the resistance. This generated power is converted from the dc power supplies. The two oscillation conditions are written as

$$\text{Im}(Z_D) = -\text{Im}(Z_C) \tag{9.5}$$

$$|\text{Re}(Z_D)| \geq \text{Re}(Z_C) \tag{9.6}$$

Im and Re mean imaginary and real parts, respectively. Thus at the oscillating frequency f_0,

$$R_C(f_0) \leq R_D(f_0, I_0, I_{RF}, T) \tag{9.7}$$

$$X_C(f_0) = -X_D(f_0, I_0, I_{RF}, T) \tag{9.8}$$

Equation (9.8) can be used to calculate the oscillating frequency. The condition of Equation (9.7) controls the output power [2].

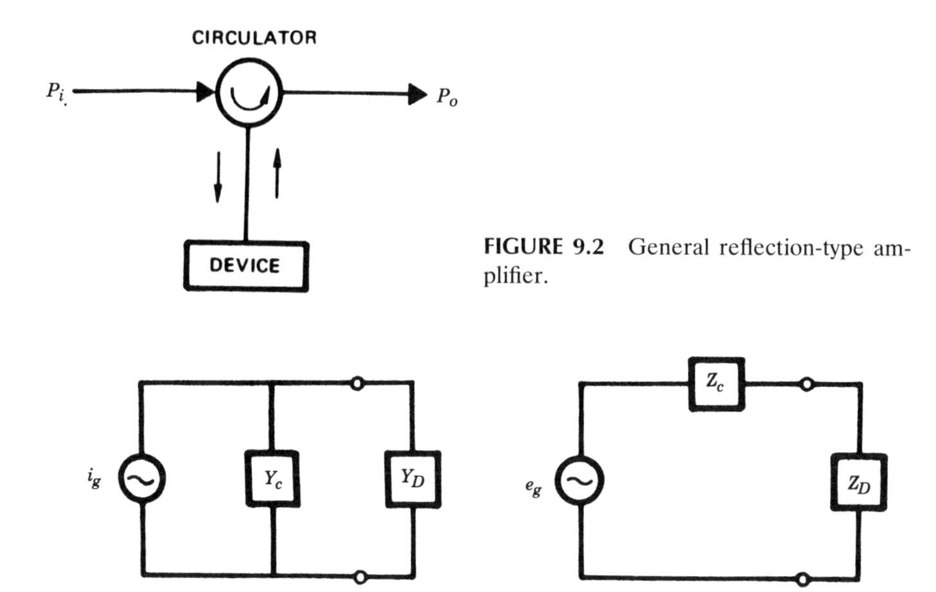

FIGURE 9.2 General reflection-type amplifier.

FIGURE 9.3 Equivalent circuits for the reflection amplifier.

9.3 GENERAL THEORY OF REFLECTION AMPLIFIERS

Two-terminal devices with negative resistance can be employed as reflection-type stable amplifiers. A circulator is used to separate the input and output ports. Figure 9.2 shows a general reflection-type amplifier. The equivalent circuit for such a system can be represented in terms of a Norton or Thévenin equivalent circuit as shown in Figure 9.3. Z_C and Y_C are the equivalent-circuit impedance and admittance seen at the device terminals and include device packaging and transforming circuits. The power gain for this type of amplifier is the power reflection coefficient at the plane of the device terminal.

The power gain is given by*

$$\text{power gain} = \frac{P_o}{P_i} = |\Gamma|^2 = \left| \frac{Y_C - Y_D^*}{Y_C + Y_D} \right|^2 \tag{9.9}$$

The power generation efficiency is defined by

$$\eta = \frac{P_o - P_i}{P_{\text{dc}} + P_i} \tag{9.10}$$

*For example, the derivation can be found in Ref. 3.

The added efficiency is

$$\eta_a = \frac{P_o - P_i}{P_{\text{dc}}} \tag{9.11}$$

Y_D and Y_C can be expressed as

$$Y_D = -G_D + jB_D \qquad \text{with } G_D > 0 \tag{9.12}$$

and

$$Y_C = G_C + jB_C \tag{9.13}$$

Substituting into Equation (9.9) and assuming that

$$B_C + B_D = 0 \tag{9.14}$$

Equation (9.9) becomes

$$\text{power gain} = \frac{(G_C + G_D)^2}{(G_C - G_D)^2} \tag{9.15}$$

Clearly, any amount of gain can be obtained by the proper choice of G_C. Note that when $G_C = G_D$, the gain becomes infinite and the device oscillates. Therefore, for stable amplification, $G_C \neq G_D$.

9.4 INJECTION-LOCKED AMPLIFIERS

Both stable amplifiers and injection-locked amplifiers can be built. For a stable amplifier, the gain is lower and the bandwidth is wider. There is no output signal (P_o) if the input signal $P_i = 0$. The injection-locked amplifier

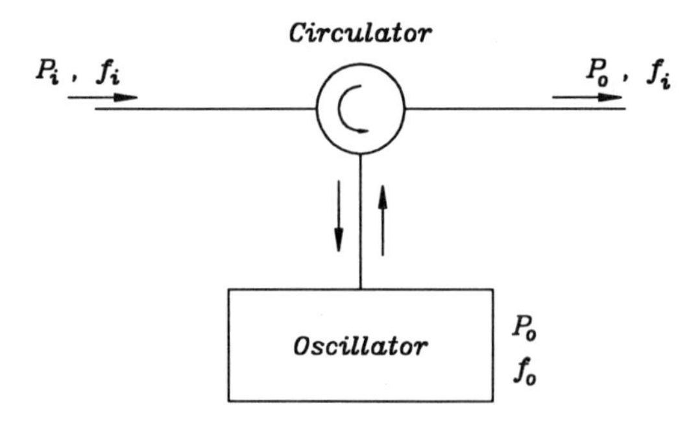

FIGURE 9.4 Injection-locked amplifier.

works differently and there is always an output signal. A detailed theory of injection-locked amplifiers is described by Kurokawa [4].

As shown in Figure 9.4, if an external signal at frequency f_i and of power P_i is injected into a free-running oscillator whose frequency is f_o with power output P_o, when f_i comes close to f_o the free-running oscillator is injection locked by this external input signal, and all the output power appears at f_i. The locking range ($2\Delta f$) depends on the external Q of the oscillator and the power gain of the system as given by Adler [5]:

$$\frac{2\Delta f}{f_o} = \frac{2}{Q_e}\left(\frac{P_o}{P_i}\right)^{-1/2} \tag{9.16}$$

where Δf is the one-side locking bandwidth and $2\Delta f$ is the total locking bandwidth. (Note that the left- and right-side locking bandwidth could be

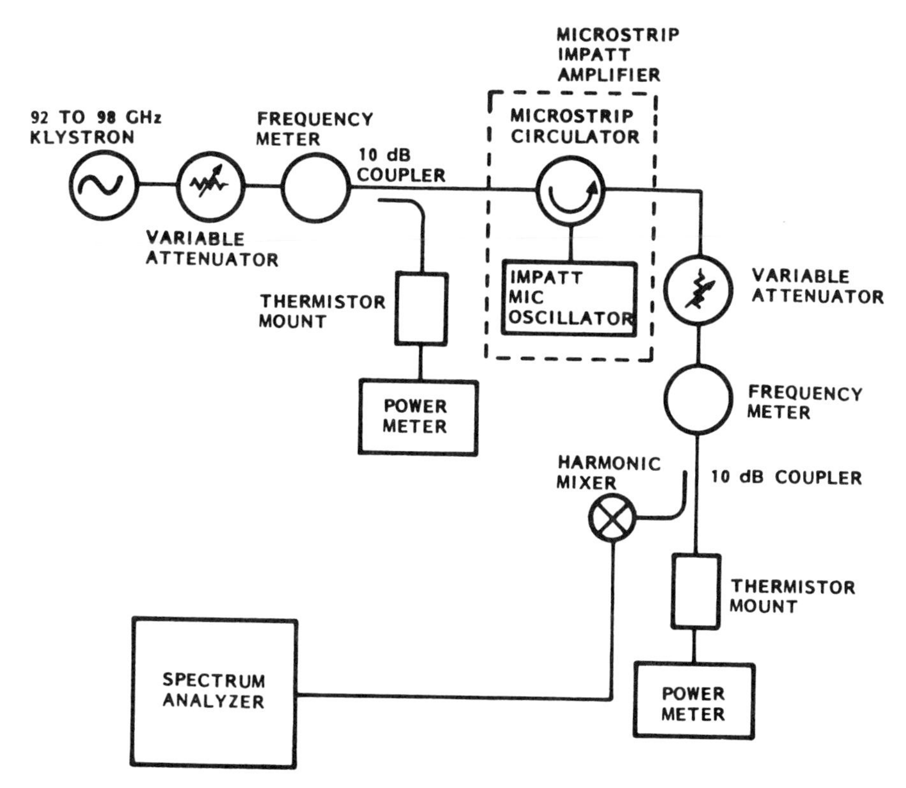

FIGURE 9.5 Injection-locking measurement system. (From Ref. 6 with permission from IEEE.)

slightly different; $2\Delta f$ is an approximation.) Q_e is the external Q of the oscillator circuit.

This phenomenon can be used to reduce the noise of an oscillator by locking it to an external low-noise source and can also be used as an amplifier, since within the locking range Δf, a small input signal at f_i appears as a large output signal at f_i. Of course, when power at f_i is removed, the output power will continue to appear at f_o. The oscillator works as an injection-locked amplifier with a gain given by

$$\text{gain} = \frac{P_o}{P_i} = \frac{4}{Q_e^2}\left(\frac{f_o}{2\Delta f}\right)^2 \tag{9.17}$$

The gain of this type of amplifier is generally high. Locking gains of 10 to 30 dB have been reported. Equation (9.16) can be used for the external-Q measurement of an oscillator since all other quantities in the equation can be measured.

Figure 9.5 shows a measurement setup for a 94-GHz IMPATT injection-locked amplifier. In this case, a stable tunable klystron source is used to lock a microstrip IMPATT oscillator. The input signal (P_i, f_i) and oscillator signal (P_o, f_o) are monitored by a spectrum analyzer and measured by power meters and frequency meters. The attenuators are used to adjust the gain. The input signal frequency f_i is varied until it is within the locking bandwidth

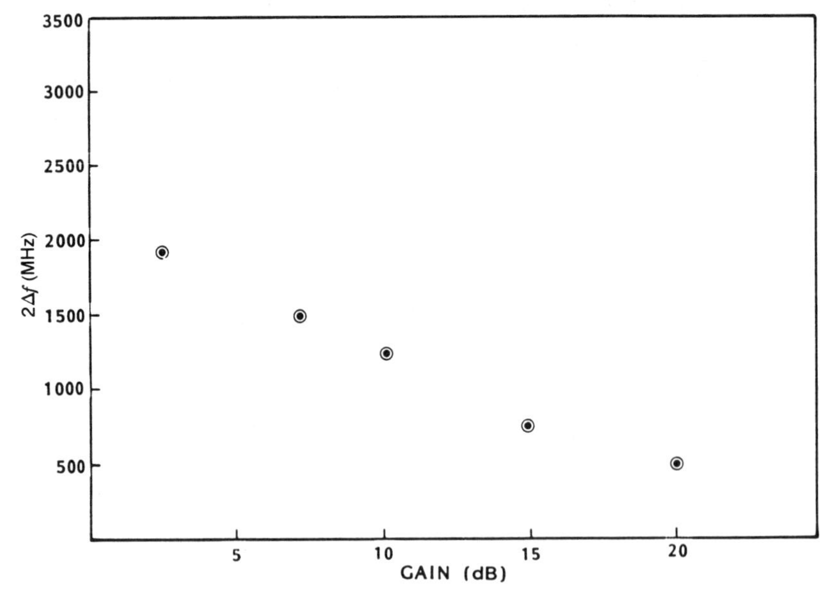

FIGURE 9.6 Injection-locking bandwidth ($2\Delta f$) as a function of power gain for a 94-GHz IMPATT amplifier. (From Ref. 6 with permission from IEEE.)

with respect to f_o. When this happens, the two signals snap together and become one signal at the frequency of f_i. By observing the spectrum analyzer, one can identify the locking bandwidth. From the readings of power meters, one can measure the power gain. A plot such as the one shown in Figure 9.6 can be obtained. It is interesting to note that the plot is almost linear, and the external Q can be found from the slope of the data line. For this example, the external Q is approximately 37.

It should be emphasized that the injection-locked amplifier has a higher gain and narrower bandwidth than those of a stable amplifier. Furthermore, a signal always appears in the output port of an injection-locked amplifier. Injection locking can be used to synchronize one or more oscillators to a lower power master or reference oscillator and also to reduce part of the FM noise.

REFERENCES

1. K. Kurokawa, "Some Basic Characteristics of Broadband Negative Resistance Oscillator Circuits," *Bell System Technical Journal*, Vol. 48, July–August 1969, pp. 1937–1955.
2. R. L. Eisenhart and P. J. Khan, "Some Tuning Characteristics and Oscillation Conditions of a Waveguide-Mounted Transferred-Electron Diode Oscillators," *IEEE Transactions on Electron Devices*, Vol. ED-19, September 1972, pp. 1050–1055.
3. K. Kurokawa, "Power Waves and Scattering Matrix," *IEEE Transactions on Microwave Theory and Techniques*, Vol. MTT-13, March 1965, pp. 194–202.
4. K. Kurokawa, "Injection-Locking of Microwave Solid-State Oscillators," *Proceedings of IEEE*, Vol. 61, October 1973, pp. 1386–1410.
5. R. Adler, "A Study of Locking Phenomena in Oscillator," *Proceedings of IRE*, Vol. 34, June 1946, pp. 351–357.
6. K. Chang et al., "W-Band (75–110 GHz) Microstrip Components," *IEEE Transactions on Microwave Theory and Techniques*, Vol. MTT-33, December 1985, pp. 1375–1382.

FURTHER READING

1. S. Yngvesson, *Microwave Semiconductor Devices*, Kluwer, Norwell, Mass., 1991, Chapter 6.
2. K. Chang and H. J. Kuno, "IMPATT and Related Transit-Time Diodes," in *Handbook of Microwave and Optical Components*, K. Chang, Editor, Wiley, New York, 1990, Vol. 2, Chap. 7.
3. E. Holzman and R. Robertson, *Solid-State Microwave Power Oscillator Design*, Artech House, Norwood, Mass., 1992.

CHAPTER 10

Transferred Electron Devices and Circuits

10.1 INTRODUCTION

The transferred electron effect was first discovered by J. B. Gunn [1], who in 1963 observed current oscillations as the applied voltage exceeded a certain threshold level on an *n*-type gallium arsenide (GaAs) specimen. The frequency of these oscillations could be made to lie in the microwave range by selecting the sample thickness and doping level. The carriers (electrons) are transferred from one end to the other in a bulk material by an electric field. The device is referred to as the Gunn diode, named after J. B. Gunn. The transferred electron effect (or Gunn effect) was later explained by Kroemer [2] to be the negative differential mobility mechanism, described in 1961 by Ridley and Watkins [3]. The theory was further refined by Hilsum [4].

Gunn devices can be fabricated in GaAs or InP or other III–V compound materials. The device exhibits negative differential resistance, which can be used for oscillator and amplifier applications. Gunn devices have been extensively used for low-noise local oscillators for mixers. They have also been used for low-power transmitters and wideband tunable sources.

Gunn oscillators have been produced from 2 to 140 GHz. For a single diode, continuous-wave (CW) power levels of up to several hundred milliwatts can be obtained in the X-, Ku-, and Ka-bands. A power output of 30 mW can be achieved from commercially available devices at 94 GHz. Higher power can be achieved by combining several devices in a power combiner. Gunn oscillators exhibit very low dc-to-RF efficiencies of 1 to 4%. The InP Gunn oscillator has a higher efficiency and output power than the GaAs oscillator, but it also requires a higher bias voltage and current.

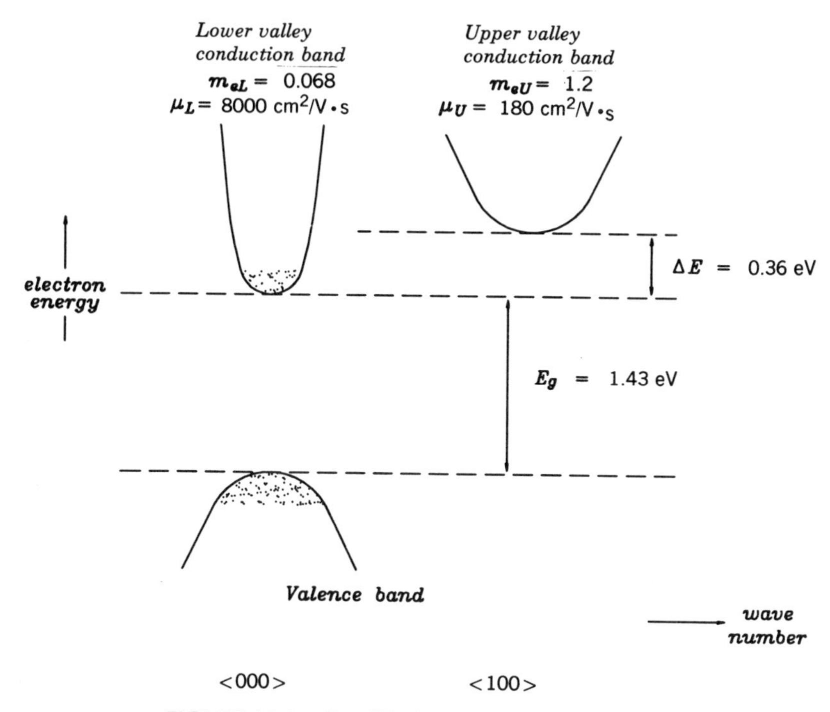

FIGURE 10.1 Simplified band structure of GaAs.

10.2 NEGATIVE DIFFERENTIAL RESISTANCE

The key to the use of a Gunn device as an oscillator is its negative resistance. The existence of the negative resistance was explained by Ridley and Watkins [3] and Hilsum [4], and is often called the Ridley–Watkins–Hilsum mechanism. For an n-type GaAs, this effect is attributed to its energy-band structure, which allows the transfer of conduction electrons from a high-mobility, lower-energy valley to a low-mobility, higher-energy satellite valley.

Figure 10.1 shows a two-valley band structure model of n-type GaAs. A high-mobility lower conduction band (i.e., lower valley) is separated by a low-mobility upper conduction band (i.e., higher valley). The electrons in the upper valley have heavier effective mass.

The I–V (or J–E) characteristics are shown in Figure 10.2. This I–V curve can be derived by considering the electron distributions in Figure 10.3. Consider the following three cases.

1. $0 \leq E < E_{th}$. At room temperature and low applied voltage, electrons are in the lower conduction band. When the bias is increased, the E

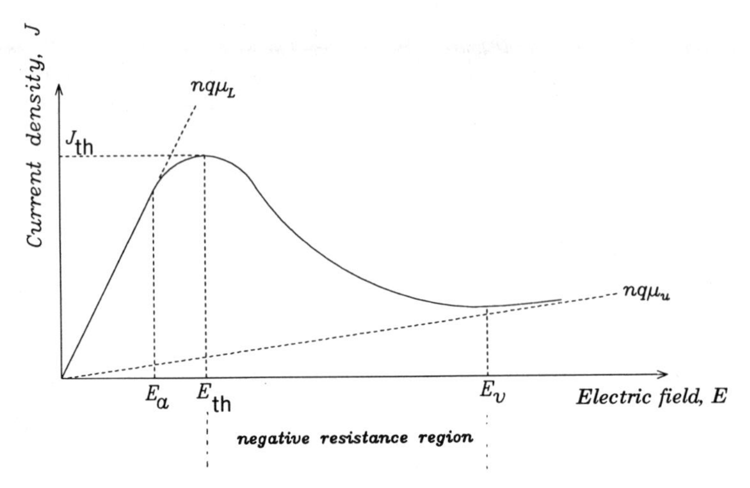

FIGURE 10.2 J–E (or I–V) characteristics.

field inside the device increases. Electrons gain energy and move fast. In this case, we have

$$n = n_L \qquad n_U = 0 \tag{10.1}$$

$$J = n_L q \mu_L E = \sigma_L E \tag{10.2}$$

$$\sigma_L = n_L q \mu_L \tag{10.3}$$

$$\mu_L = 8000 \text{ cm}^2/\text{V} \cdot \text{s} \qquad \text{for GaAs}$$

From Equation (10.2), it can be seen that the current density is linearly proportional to the electric field. The slope is $n_L q \mu_L$, which is

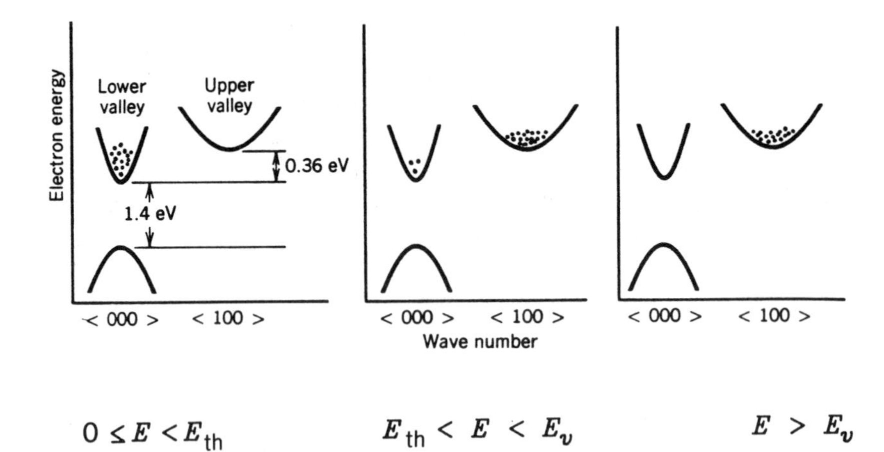

FIGURE 10.3 Electron distribution for different bias conditions.

constant. Since $I = JA$ and $V = EL$, A being the cross section and L being the length of the device, the I–V characteristics have the same shape as the J–E characteristics.

2. $E_{th} < E < E_v$. Further increase in the applied bias voltage will increase the E field and move some electrons from the lower conduction valley to the upper conduction valley. Since the electron mobility in the upper valley is much smaller than that in the lower valley, the total current density decreases as the E field increases. We have

$$n = n_L + n_U \tag{10.4}$$

$$J = n_L q \mu_L E + n_U q \mu_U E = \sigma E \tag{10.5}$$

$$\sigma = n_L q \mu_L + n_U q \mu_U \tag{10.6}$$

Since $\mu_L \gg \mu_U$, J and σ decrease as more electrons move from the lower valley to the upper valley. The decrease of J (or I) with E (or V) gives a negative resistance. In Figure 10.2, this region is labeled as the negative-resistance region.

3. $E > E_v$. When the E field is increased to E_v, all electrons move from the lower valley to the upper valley. Further increase in the E field will result in a linear increase in current density. The slope of the linearity is σ_U. We have the following equations for this region:

$$n = n_U \qquad n_L = 0 \tag{10.7}$$

$$J = n_U q \mu_U E = \sigma_U E \tag{10.8}$$

$$\sigma_U = n_U q \mu_U \tag{10.9}$$

In this region, the device exhibits a positive resistance.

From the two-valley theory above, the band structure of a semiconductor must satisfy the following criteria to exhibit negative resistance [4, 5].

1. The separation energy ΔE between the bottom of the lower valley and the bottom of the upper valley must be several times larger than the thermal energy (about 0.026 eV) at room temperature. This means that $\Delta E \gg kT = 0.026$ eV. Otherwise, the upper valley would be highly populated, due to thermal excitation.
2. Electrons in the lower valley must have higher mobility, smaller effective mass, and a lower density of state than the electrons in the upper valley. Otherwise, the device would not exhibit negative resistance.
3. The energy separation between the two valleys must be smaller than the gap energy between the conduction and the valence band. This

TABLE 10.1 Data for Two-Valley Semiconductor

Semiconductor	Gap Energy (at 300 K), E_g (eV)	Separation Energy between Two Valleys, ΔE (eV)	Threshold Field E_{th} (kV/cm)	Peak Velocity, v_p (10^7 cm/s)
Ge	0.80	0.18	2.3	1.4
GaAs	1.43	0.36	3.2	2.2
InP[a]	1.33	0.60	10.5	2.5
		0.80		
CdTe	1.44	0.51	13.0	1.5
InAs	0.33	1.28	1.60	3.6
InSb	0.16	0.41	0.6	5.0

Source: Ref. 5 with permission from Prentice-Hall.

[a]InP is a three-valley semiconductor: 0.60 eV is the separation energy between the middle and lower valleys, 0.8 eV that between the upper and lower valleys.

means that $E_g > \Delta E$. Otherwise, the semiconductor will break down and become highly conductive before the electrons begin to transfer from the lower to the upper valley.

Table 10.1 shows data for several two-valley semiconductors. The threshold E-field is 3.2 kV/cm for GaAs and 10.5 kV/cm for InP. Figure 10.4 shows the theoretical and experimental velocity-electric field characteristics of GaAs.

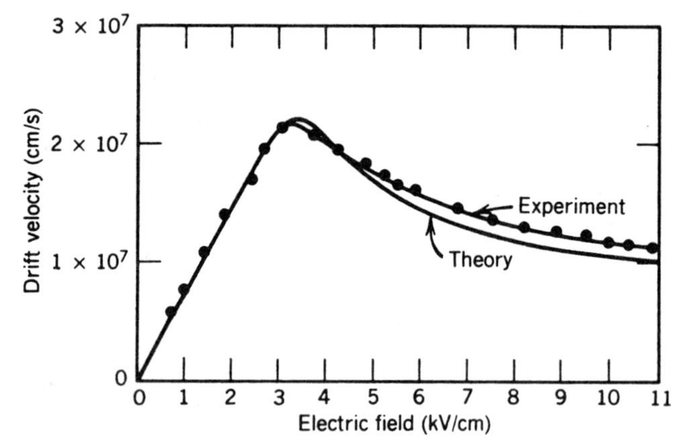

FIGURE 10.4 Theoretical and experimental velocity-electric field characteristic of GaAs. (From Ref. 6.)

10.3 MODES OF OPERATION

The common and simple mode of operation for a Gunn diode is the Gunn or dipole mode. In this mode of operation, to start the oscillation, a high-field domain or electron dipole is formed near the cathode. The formation of the domain is due to the noise or nonuniform doping near the cathode. This domain (charge spike) travels to the anode at the speed of v_s (saturated drift velocity, about 10^7 cm/s for GaAs as shown in Figure 10.4). Another spike is formed after the first one disappears. Figure 10.5 shows the formation of a charge spike in GaAs diode. The E field is obtained from the following equation:

$$\frac{\partial E}{\partial x} = -\frac{\Delta \rho}{\varepsilon_s} \tag{10.10}$$

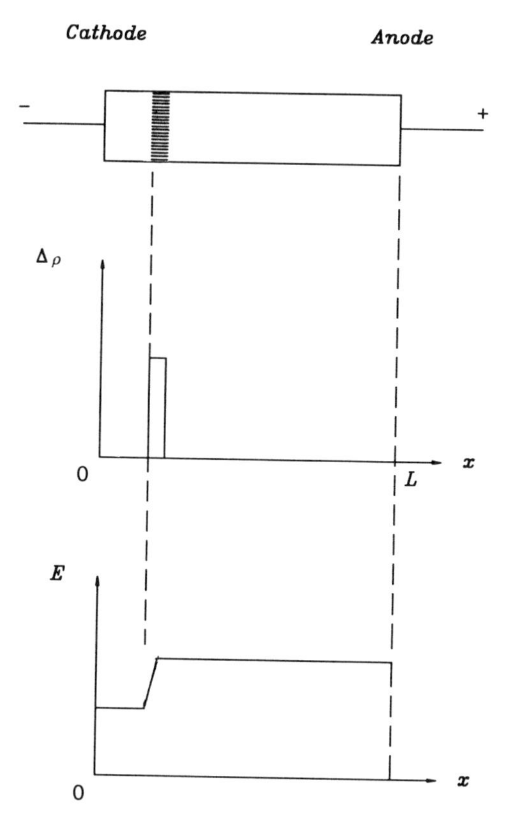

FIGURE 10.5 Formation of a domain (charge spike) inside the Gunn device. (From Ref. 7 with permission from IEEE.)

where $\Delta\rho$ is the charge density and ε_s is the GaAs dielectric constant. The time for the charge spike to travel from the cathode to the anode determines the frequency of operation. The time for electrons to be transferred over the device length L is

$$T = \frac{L}{v_s} \tag{10.11}$$

The frequency of operation is

$$f = \frac{1}{T} = \frac{v_s}{L} \tag{10.12}$$

For a GaAs Gunn device, we have

$$fL = v_s = 10^7 \text{ cm/s} \tag{10.13}$$

The Gunn mode of operation is simple to operate since only a dc bias greater than the threshold voltage is required. However, the efficiency is low, only a few percent.

Many other modes can be made to operate. These include the resonant Gunn mode, delayed mode, quenched domain mode, limited space-charge (LSA) mode, and hybrid mode [6]. The operation of these modes depends on the fL and n_0L products [8], as shown in Figure 10.6. Theoretical efficiencies

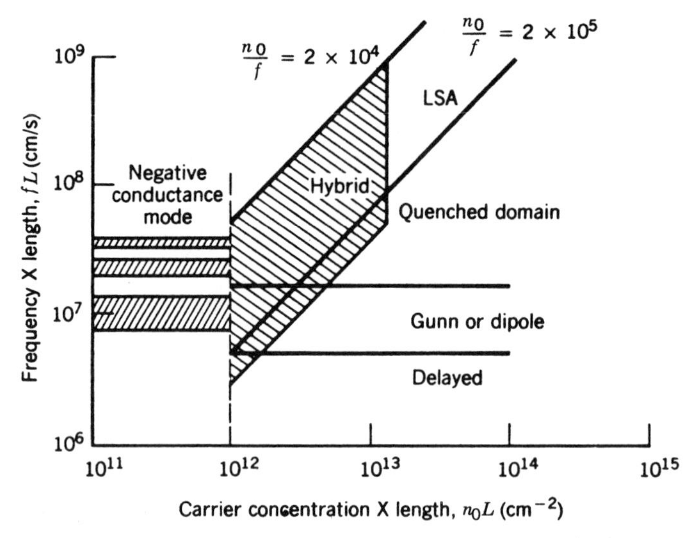

FIGURE 10.6 Mode chart for transferred electron devices. (From Ref. 8 with permission from *Journal of Applied Physics*.)

of over 20% have been reported for the quenched domain mode and LSA mode [6].

Example 10.1 Design a GaAs Gunn device to operate at 10 GHz in the Gunn mode. Assume that $n_0 = 2 \times 10^{15}$ cm^{-3}, $E_v = 20$ kV/cm, $E_{th} = 3$ kV/cm, and that the device produces 100 mW of output power with 3% dc-to-RF efficiency.

SOLUTION: $f = 10$ GHz and $v_s = 10^7$ cm/s. The device length is

$$L = \frac{v_s}{f} = 10^{-3} \text{ cm} \quad \text{or} \quad 10 \ \mu\text{m}$$

The threshold voltage is

$$V_{th} = E_{th} L = 3000 \times 10^{-3} \text{ V} = 3 \text{ V}$$

The bias or operating voltage is approximately three times the threshold voltage (see Figure 10.4). We have

$$V_{bias} = 3V_{th} = 9 \text{ V}$$

if $P_0 = 100$ mW and $\eta = P_0/P_{dc} = 3\%$, and

$$P_{dc} = \frac{P_0}{\eta} = 3.3 \text{ W}$$

The heat dissipated in the diode is

$$P_{diss} = P_{dc} - P_0 = 3.2 \text{ W}$$

$$I_{bias} = \frac{P_{dc}}{V_{bias}} = 0.367 \text{ A}$$

$$J = n_0 q v_s = (2 \times 10^{15})(1.6 \times 10^{-19})(10^7)$$

$$= 3.2 \times 10^3 \text{ A/cm}^2 = \frac{I_{bias}}{A}$$

The cross-sectional area of the diode is then

$$A = \pi r^2 = \frac{I_{bias}}{J} = 1.15 \times 10^{-4} \text{ cm}^2$$

and

$$r = 6 \times 10^{-3} \text{ cm} = 60 \ \mu\text{m}$$

10.4 DEVICE FABRICATION AND PACKAGING

Gunn devices can be fabricated using different epitaxial growth methods: vapor-phase epitaxial (VPE) growth, liquid-phase epitaxial (LPE) growth, or molecular beam epitaxy (MBE). The epitaxial layer with doping of n_0 is grown on the n^+ substrate as shown in Figure 10.7(a). A n^+ buffer epitaxial layer is grown on top of the n layer for contact purposes. Since the Gunn effect is a bulk effect, the epilayer is not required but is used to improve the ohmic contact and to control the precise length of the transit region.

The device is mounted on top of a heat sink. The device is normally packaged in a pill-type package as shown in Figure 10.7(b). For ease of integration, the entire package can be mounted on top of a screw. The heat

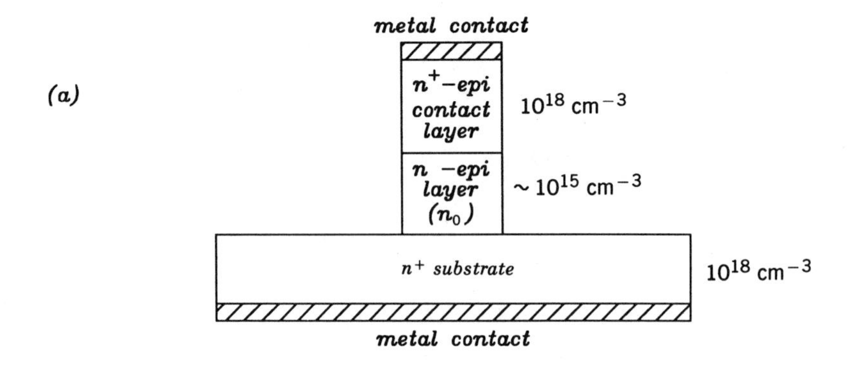

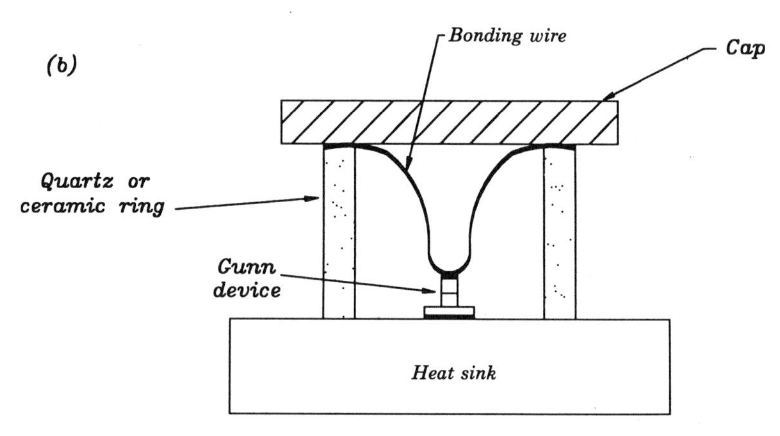

FIGURE 10.7 Gunn device and its package.

sink is important for heat dissipation due to the low efficiency of the Gunn diode.

Gunn diodes are commercially available from 4 to 94 GHz. For example, a M/A Com Gunn diode (Model *MA*-49107) operates at 10 GHz with a 100 mW of output power. The typical bias voltage is 8 V, and the typical bias current is 550 mA.

10.5 GUNN OSCILLATOR CIRCUITS

Gunn oscillators can be constructed in waveguide, coaxial, and microstrip circuits. For frequencies from 8 to 94 GHz, a common method of coupling a Gunn device into a waveguide circuit is to mount a package diode under a bias post as shown in Figure 10.8. The post, the coaxial section, and the reduced-height waveguide transformer are used to match the low device impedance to the high waveguide impedance. The cap resonator as shown in Figure 10.8(*e*) can be used to facilitate the impedance matching and to control the oscillating frequency. In all these circuits, a sliding short is used to tune the frequency and to maximize the output power. Equivalent circuits and analyses for these circuits are fairly complicated and can be found in Refs. 9 through 12.

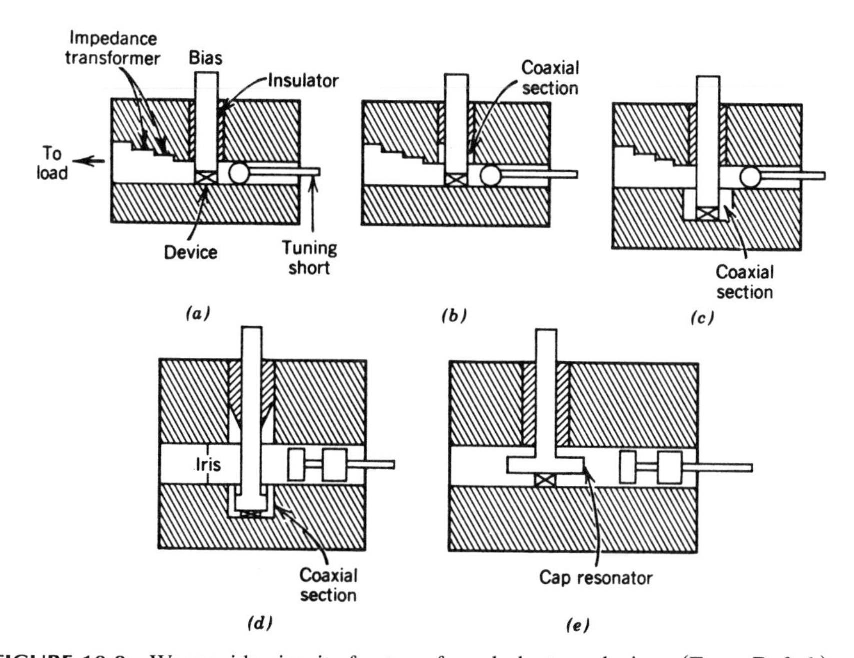

FIGURE 10.8 Waveguide circuits for transferred electron devices. (From Ref. 6.)

For oscillation to occur, the following conditions need to be satisfied, as discussed in Chapter 9.

$$\text{Im}(Z_D) + \text{Im}(Z_c) = 0 \tag{10.14}$$

and

$$|\text{Re}(Z_D)| \geq \text{Re}(Z_c) \tag{10.15}$$

where Z_D is the device impedance and Z_c is the transformed load impedance, as shown in Figure 10.9. For example, in a waveguide circuit, R_L is the impedance of the full-height waveguide, which is around 300 to 400 Ω. The matching circuit is a post-mounting network and a waveguide-height transformer. The circuit is designed to meet the conditions in Equations (10.14) and (10.15) at the desired frequency. The device impedance is a strong function of frequency and dc bias current and a weak function of the RF current and temperature. The circuit impedance is a function of frequency only. We have

$$Z_D = Z_D(f, I_0, I_{\text{RF}}, T) = -R_D + jX_D \tag{10.16}$$

$$Z_c(f) = R_c + jX_c \tag{10.17}$$

Equation (10.14) determines the frequency of oscillation. It can be seen that one can change the frequency by varying the dc bias current I_0. This is bias tuning. One can also change the frequency by moving the sliding short and thus varying Z_c. This is mechanical tuning.

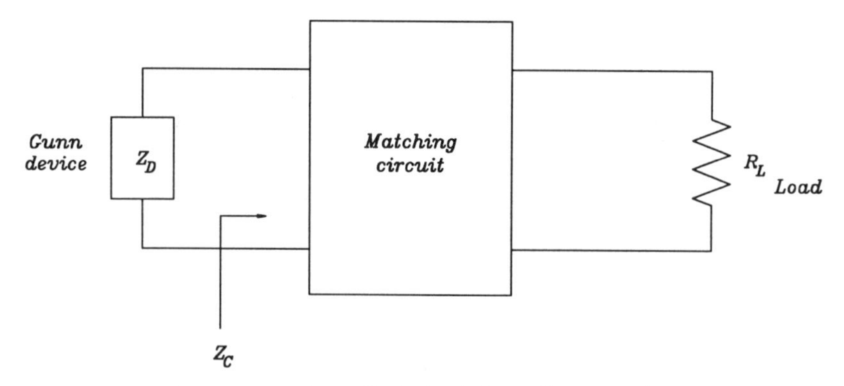

FIGURE 10.9 Simplified circuit for a Gunn oscillator.

10.6 VOLTAGE-CONTROLLED OSCILLATORS

Although bias tuning can be used to vary the oscillating frequency of a Gunn oscillator, the output power is not constant and varies over a wide range during tuning. Varactor-tuned circuits can be used to overcome this problem. Varactor-tuned oscillators provide fairly constant output power, wide tuning range, and fast response. Voltage-controlled oscillators (VCOs) can be used for FM sources, frequency-agile transmitters, tunable receiver local oscillators, and many other applications.

Varactor-tuned Gunn oscillators have been fabricated in waveguide, coaxial line, and microstrip. Figure 10.10 shows coaxial circuit arrangements, and Figure 10.11 shows the waveguide circuits [6]. Waveguide circuits offer higher Q but a lower tuning range than microstrip circuits.

Figure 10.12 shows the varactor-tuned microstrip Gunn oscillator developed by Rubin [13]. A tuning range in excess of 1 GHz was achieved in the 35-GHz region. Figure 10.13 shows a varactor-tuned Gunn oscillator built on coplanar waveguide (CPW) and slot line [14]. The VCO provides 16.3 ± 0.45 dBm output power throughout a 350-MHz tuning range centered at 10.37 GHz. Figure 10.13 shows the circuit configuration, and Figure 10.14 shows the power output as a function of frequency and varactor bias. The output power is fairly constant over the tuning range.

Example 10.2 A Gunn device is connected to an abrupt junction varactor diode in series. The circuit is coupled to a 50-Ω load through a quarter-wavelength transformer with a characteristic impedance of 10 Ω. The overall circuit is shown in Figure 10.15. The Gunn device has $R_D = 10\ \Omega$, $C_D = 1$ pF, and $L_p = 1$ nH. The varactor is an unpackaged chip with $C_j = 1$ pF at 0 V and $V_{\mathrm{bi}} = 1$ V. Determine if the circuit oscillates. What are the oscillating frequencies if the reverse biases of the varactor are (a) 0 V and (b) 20 V?

SOLUTION: (a)

$$C_j = C_{j0} = 1 \text{ pF}$$

$$R_c = \frac{Z_0^2}{R_L} = \frac{10^2}{50} = 2\ \Omega$$

Since $R_D > R_c$, it will oscillate. The oscillation frequency is calculated by

$$\omega = 2\pi f = \frac{1}{\sqrt{L_p C_T}}$$

$$C_T = \frac{C_D C_j}{C_D + C_j} = 0.5 \text{ pF}$$

$$\omega = 2\pi f = \frac{1}{\sqrt{1 \times 10^{-9} \times 0.5 \times 10^{-12}}} = 44.7 \times 10^9$$

$$f = 7.12 \text{ GHz}$$

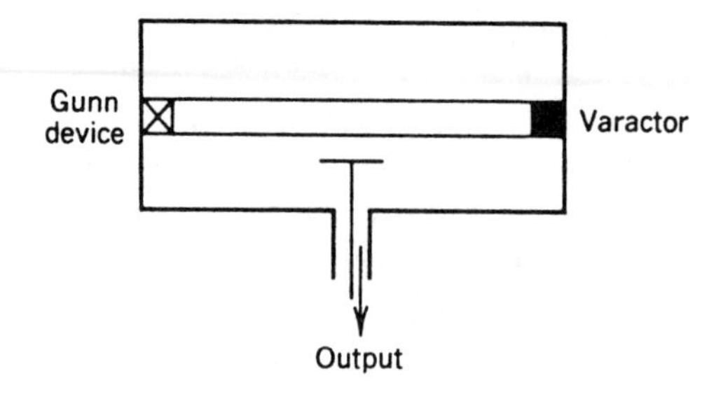

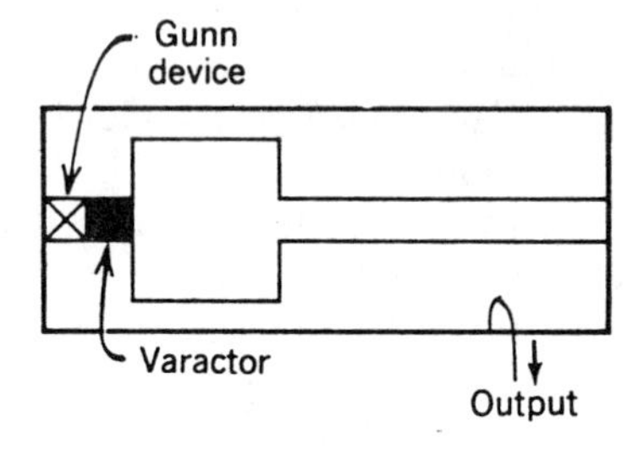

FIGURE 10.10 Varactor-tuned Gunn oscillator coaxial cavity circuits.

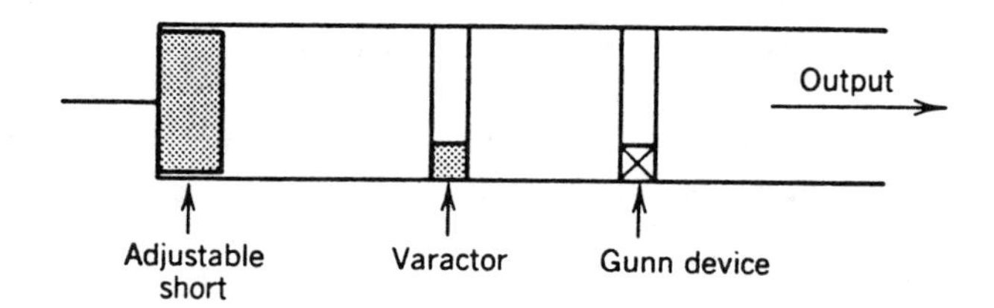

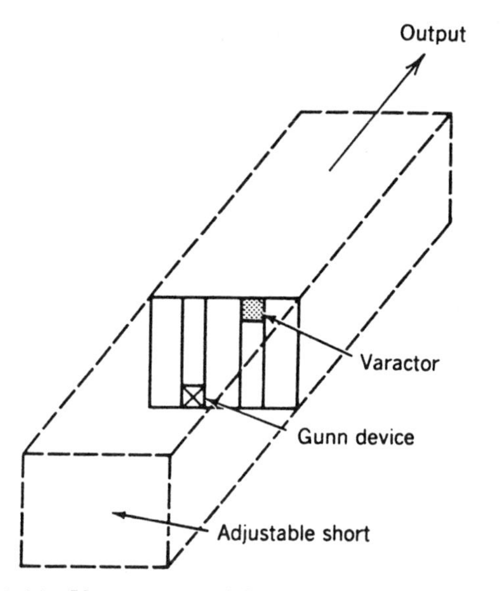

FIGURE 10.11 Varactor-tuned Gunn oscillator waveguide circuits.

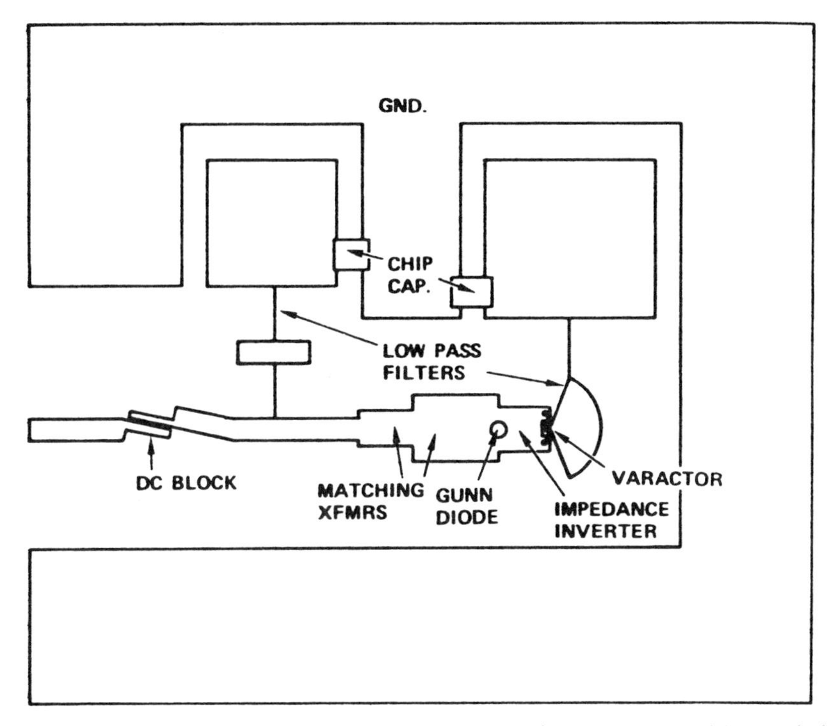

FIGURE 10.12 Varactor-tuned microstrip oscillator. (From Ref. 13 with permission from IEEE.)

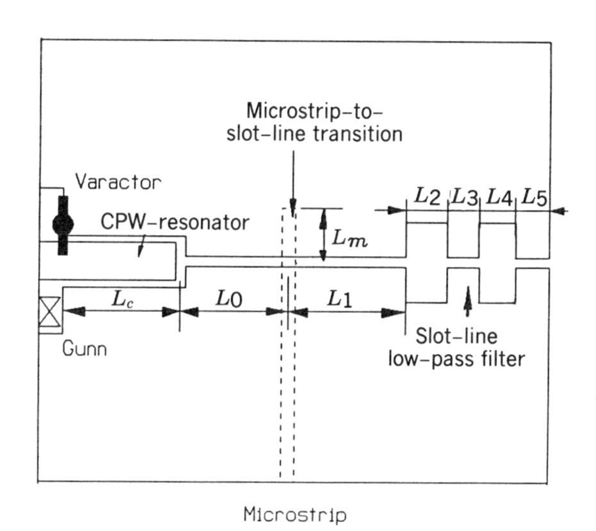

FIGURE 10.13 Varactor-tuned CPW/slot-line Gunn VCO configuration. (From Ref. 14 with permission from IEEE.)

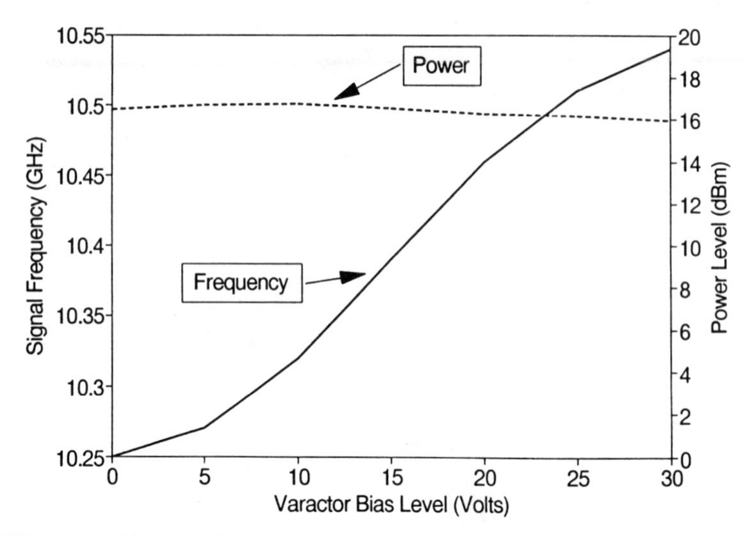

FIGURE 10.14 Varactor bias voltage versus frequency and power output. (From Ref. 14 with permission from IEEE.)

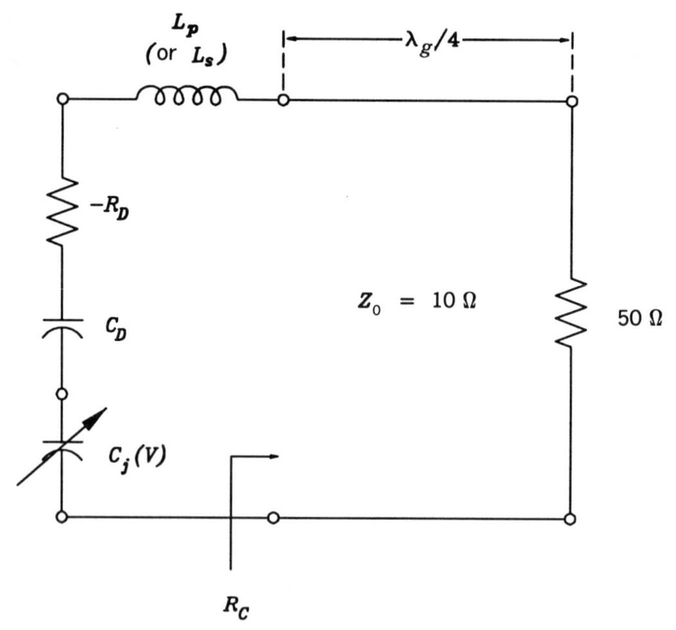

FIGURE 10.15 Gunn VCO circuit.

(b) $$C_j = C_{j0}\left(1 - \frac{V}{V_{\text{bi}}}\right)^{-1/2} = C_{j0}\left(1 + \frac{20}{1}\right)^{-1/2}$$

$$= \frac{1}{\sqrt{21}} = 0.218 \text{ pF}$$

$$C_T = \frac{1 \times 0.218}{1 + 0.218} = 0.179 \text{ pF}$$

$$\omega = 2\pi f = \frac{1}{\sqrt{0.179 \times 10^{-12} \times 1 \times 10^{-9}}} = 74.74 \times 10^9$$

$$f = 11.90 \text{ GHz}$$

Example 10.3 A 60-GHz microstrip Gunn VCO is shown in Figure 10.16. A varactor chip is mounted in shunt with the Gunn device. A two-section

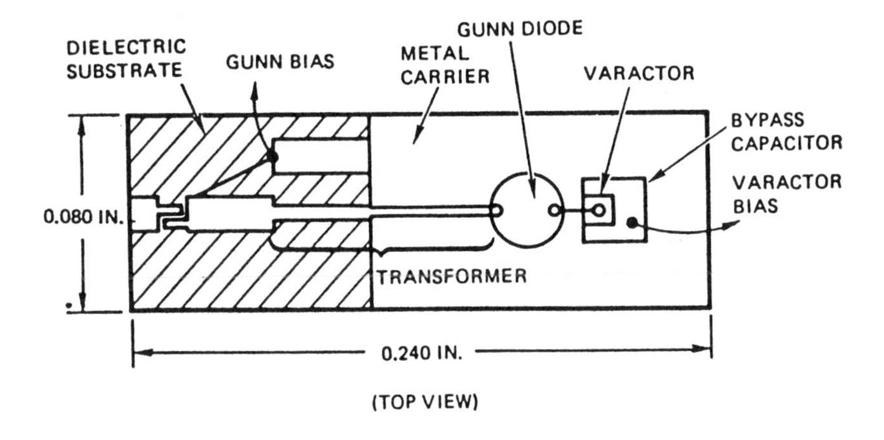

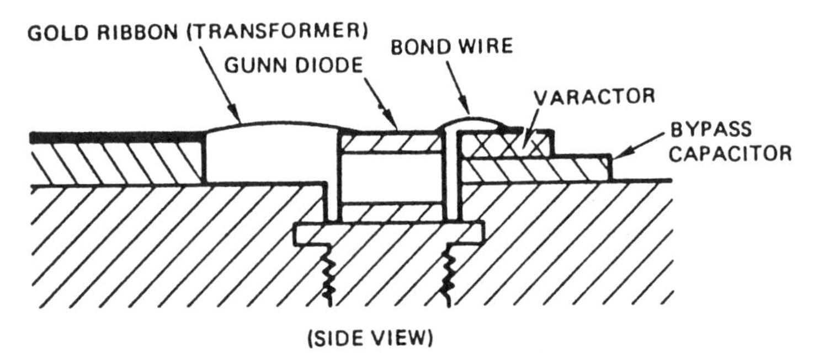

FIGURE 10.16 V-band microstrip Gunn VCO. (From Ref. 15 with permission from IEEE.)

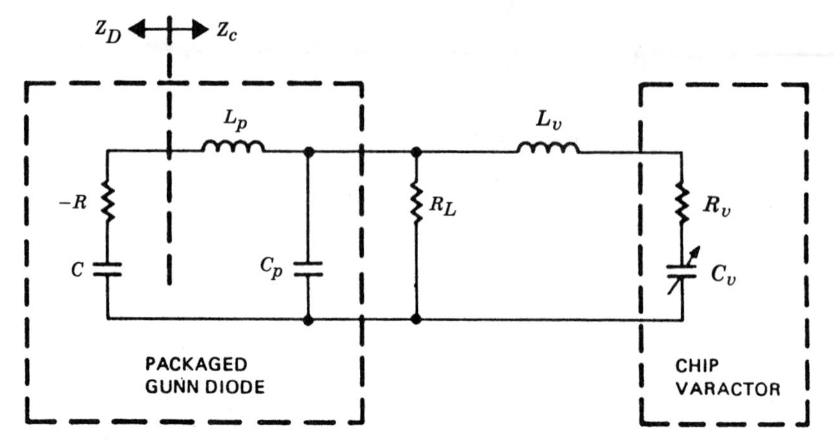

FIGURE 10.17 Equivalent circuit of the VCO shown in Figure 10.16.

transformer is used to accomplish impedance matching, and R_L is the transformed impedance seen by the Gunn device. Draw the equivalent circuit and give an expression for Z_c in terms of device and circuit parameters.

SOLUTION: The equivalent circuit is shown in Figure 10.17. $-R$ is the negative resistance of the Gunn diode; C is the capacitance of the Gunn diode; L_p is the package inductance of the Gunn diode ($\approx$ 0.15 nH); C_p is the package capacitance of the Gunn diode ($\approx$ 0.12 pF); R_L is the transformed load impedance; L_v is the bonding wire inductance; R_v is the series resistance of the varactor; and C_v is the junction capacitance of the varactor. The expression for Z_c can be derived as follows:

$$Z_c = j\omega L_p + Z_1$$

where Z_1 is the impedance looking into C_p.

$$Y_1 = j\omega C_p + \frac{1}{R_L} + \frac{1}{j\omega L_v + R_v - j(1/\omega C_v)}$$

$$= \frac{j\omega C_p R_L[j(\omega L_v - 1/\omega C_v) + R_v] + [R_v + j(\omega L_v - 1/\omega C_v)] + R_L}{R_L[R_v + j(\omega L_v - 1/\omega C_v)]}$$

$$Z_c = j\omega L_p$$

$$+ \frac{R_L[R_v + j(\omega L_v - 1/\omega C_v)]}{R_L + R_v - \omega^2 C_p R_L L_v + R_L(C_p/C_v) + j[\omega L_v - 1/\omega C_v + \omega C_p R_L R_v]}$$

$$Z_D = -R - j\frac{1}{\omega C}$$

Using the two equations above, the varactor tuning range can be computed by solving Equation (10.14) for different values of C_v.

PROBLEMS

P10.1 Design a Gunn device operating at 5 GHz. The electrons travel at a saturated speed of 10^7 cm/s.

(a) What is the length L of this device?

(b) What is the threshold voltage if $E_{th} = 3.2$ kV/cm?

P10.2 In Example 10.2, what are the oscillating frequencies if the reserve bias to the varactor is 20 V and Z_0 is **(a)** 20 Ω and **(b)** 30 Ω.

P10.3 A varactor diode has $C_j(V) = 20$ pF at $V = -10$ V and $C_j(V) = 10$ pF at $V = -20$ V. The varactor is used to tune an oscillator as shown in Figure P10.3. The active device has Z_D given by $Z_D = -5 + j\omega \times 10^{-10}$ Ω, where ω is the angular frequency. A quarter-wavelength transformer is used for impedance matching. Determine the oscillating frequency for the following conditions:

(a) $Z_1 = 10$ Ω, $V = -10$ V

(b) $Z_1 = 30$ Ω, $V = -20$ V

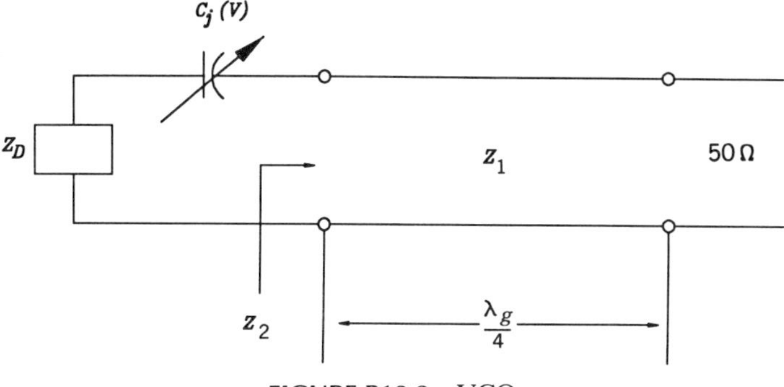

FIGURE P10.3 VCO.

P10.4 Draw the equivalent circuit for the coaxial VCO shown in Figure P10.4. Include all package parasitics of the devices, but neglect the mounting structure parasitics.

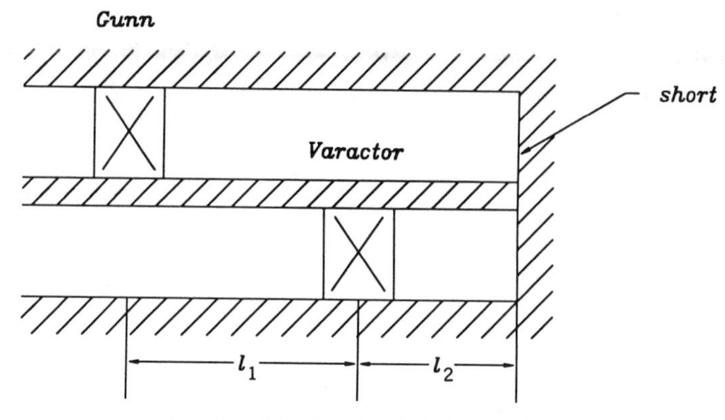

FIGURE P10.4 Coaxial Gunn VCO.

P10.5 Write a computer program to calculate the theoretical tuning range with the varactor junction capacitance varied from 0.1 to 0.8 pF in Example 10.3. The following parameters are used: $C = 0.14$ pF, $L_p = 0.15$ nH, $C_p = 0.12$ pF, $R_L = 250\ \Omega$, $L_v = 0.2$ nH, and $R_v = 2.3\ \Omega$. Plot the tuning curve. What is the tuning range in GHz? (Concentrate on the frequency range from 55 to 65 GHz.)

P10.6 Explain how the circuit shown in Figure 10.12 works.

P10.7 Explain how the circuit shown in Figure 10.13 works.

REFERENCES

1. J. B. Gunn, "Microwave Oscillations of Current in III–V Semiconductors," *Solid-State Communications*, Vol. 1, September 1963, pp. 89–91.
2. H. Kroemer, "Theory of the Gunn Effect," *Proceedings of the IEEE*, Vol. 52, December 1964, p. 1736.
3. B. K. Ridley and T. B. Watkins, "The Possibility of Negative Resistance Effects in Semiconductors," *Proceedings of the Physical Society* (*London*), August 1961, pp. 293–304.
4. C. Hilsum, "Transferred Electron Amplifiers and Oscillators," *Proceedings of the IEEE*, Vol. 50, February 1962, pp. 185–189.
5. S. Y. Liao, *Microwave Solid-State Devices*, Prentice Hall, Englewood Cliffs, N.J., 1985, Chapter 5.
6. C. Sun, "Transferred Electron Devices," in *Handbook of Microwave and Optical Components*, K. Chang, Editor, Wiley, New York, 1990, Vol. 2, Chapter 6.
7. A. Kroemer, "Negative Conductance in Semiconductors," *IEEE Spectrum*, Vol. 5, January 1968, p. 47.

8. J. A. Copeland, "LSA Oscillator Diode Theory," *Journal of Applied Physics*, Vol. 38, July 1967, pp. 3096–3101.
9. L. Lewin, "A Contribution to the Theory of Probes in Waveguides," *Proceedings of the IEEE*, Monograph 259R, October 1957, pp. 109–116.
10. R. L. Eisenhart and P. J. Khan, "Theoretical and Experimental Analysis of a Waveguide Mounting Structure," *IEEE Transactions on Microwave Theory and Techniques*, Vol. MTT-19, August 1971, pp. 706–719.
11. K. Chang and R. L. Ebert, "W-Band Power Combiner Design," *IEEE Transactions on Microwave Theory and Techniques*, Vol. MTT-28, April 1980, pp. 295–305.
12. M. E. Bialkowski, "Modeling of a Coaxial-Waveguide Power-Combining Structure," *IEEE Transactions on Microwave Theory and Techniques*, Vol. MTT-34, September 1986, pp. 937–942.
13. D. Rubin, "Varactor-Tuned Millimeter-Wave MIC Oscillator," *IEEE Transactions on Microwave Theory and Techniques*, Vol. MTT-24, November 1976, pp. 866–867.
14. J. A. Navarro, Y. Shu, and K. Chang, "A Novel Varactor Tunable Coplanar Waveguide-Slotline Gunn VCO," *1991 IEEE MTT Microwave Symposium Digest Technical Papers*, Boston, June 1991, pp. 1187–1190.
15. K. Chang et al., "V-Band Low-Noise Integrated Circuit Receiver," *IEEE Transactions on Microwave Theory and Techniques*, Vol. MTT-31, February 1983, pp. 146–154.

FURTHER READING

1. S. M. Sze, *Physics of Semiconductor Devices*, 2nd ed., Wiley, New York, 1981, Chap. 11.
2. C. Sun, "Transferred Electron Devices," in *Handbook of Microwave and Optical Components*, K. Chang, Editor, Wiley, New York, 1990, Vol. 2, Chap. 6.
3. W. A. Davis, *Microwave Semiconductor Circuit Design*, Van Nostrand Reinhold, New York, 1984, Chap. 15.
4. E. A. Wolff and R. Kaul, *Microwave Engineering and Systems Applications*, Wiley, New York, 1988, Chap. 13.

CHAPTER 11

IMPATT Devices and Circuits

11.1 INTRODUCTION

The term *IMPATT* stands for "*IMP*act Ionization *A*valanche *T*ransit *T*ime." The IMPATT device is one of the most powerful solid-state microwave sources. The IMPATT diode uses impact-ionization and transit-time properties to produce the negative resistance required for oscillation and amplification of microwave signals.

The concept of the IMPATT diode was first proposed by Read in 1958 for the relatively complex diode structure n^+pip^+, which is now commonly called the Read diode [1]. It was not until 1965 that the experimental observation of the IMPATT oscillation from a p–n junction diode was reported by Johnston et al. [2]. Since then a large amount of effort has been directed toward understanding and developing practical microwave and millimeter-wave IMPATT devices, oscillators, and amplifiers. IMPATT devices have been fabricated with Si, GaAs, and InP semiconductor materials. Silicon IMPATT devices have been operated at frequencies up to 394 GHz [3].

Although three-terminal devices are more commonly used at microwave frequencies now, the IMPATT diode is still very useful as a power source at millimeter-wave frequencies. At present, it generates the highest output power among all solid-state devices at millimeter-wave frequencies. At lower microwave frequencies, FETs and bipolar transistors, which are three-terminal devices, have replaced IMPATTs for most applications.

Due to avalanche breakdown and electron (hole) multiplication, the IMPATT device is inherently noisy. In general, IMPATTs have 10 dB higher AM noise than that of Gunn diodes. For this reason, the IMPATT diode is not suitable for use as a local oscillator in a receiver.

11.2 DEVICE PHYSICS [4, 5]

Microwave oscillation and amplification of an IMPATT diode are due to the frequency-dependent negative resistance arising from the phase delay between the current and voltage waveforms in the avalanche breakdown and transit-time processes. Since Read first proposed a negative resistance diode consisting of an n^+pip^+ (or p^+nin^+) structure, many other configurations have been developed. For silicon diodes, single-drift diodes (p^+nn^+) and double-drift diodes (p^+pnn^+) are the two most popular structures. For GaAs diodes, single-drift Read diodes ($p^+n^+nn^+$ or $p^+n^-n^+nn^+$), double-drift Read diodes ($p^+pp^+n^+nn^+$ or $p^+pp^+nn^+nn^+$), and hybrid Read diodes ($p^+pn^+nn^+$ or $p^+pnn^+nn^+$) have been fabricated. In this section, the physics of avalanche breakdown in a p–n junction is discussed. Next, a Read diode is used for an analytical description and understanding of operation theory of an IMPATT diode. For complicated structures, computer programs are generally used to generate small- and large-signal device parameters.

Avalanche breakdown normally imposes an upper limit on the reverse voltage applied to most p–n junction diodes. In IMPATT devices, the same avalanche breakdown, which is caused by avalanche multiplication (or impact ionization), can be used effectively to generate microwave power.

A p–n junction is shown in Figure 11.1 together with its I–V characteristics. As the reverse bias voltage approaches V_B, the electric field around the p–n junction reaches a very high value and avalanche breakdown occurs. The ionization rate α (the probability that an electron–hole pair will be generated per centimeter) for holes and electrons can be represented by

$$\alpha = Ae^{-(b/E)^m} \tag{11.1}$$

where m, b, and A are constants depending on the material, and E is the electric field. In general, the ionization rates for holes (α_p) and electrons (α_n) are functions of x, and they are not equal (i.e., $\alpha_n \neq \alpha_p$). For a p–n junction, the multiplication factor M_p of holes can be written as (see Appendix F for a derivation)

$$\frac{1}{M_p} = 1 - \int_0^w \alpha_p \exp\left[-\int_0^x (\alpha_p - \alpha_n)\, dx'\right] dx \tag{11.2}$$

where w is the width of the depletion region. A similar result applies to the multiplication factor M_n of electrons:

$$\frac{1}{M_n} = 1 - \int_0^w \alpha_n \exp\left[-\int_x^w (\alpha_n - \alpha_p)\, dx'\right] dx \tag{11.3}$$

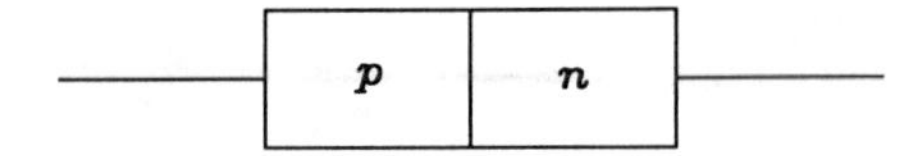

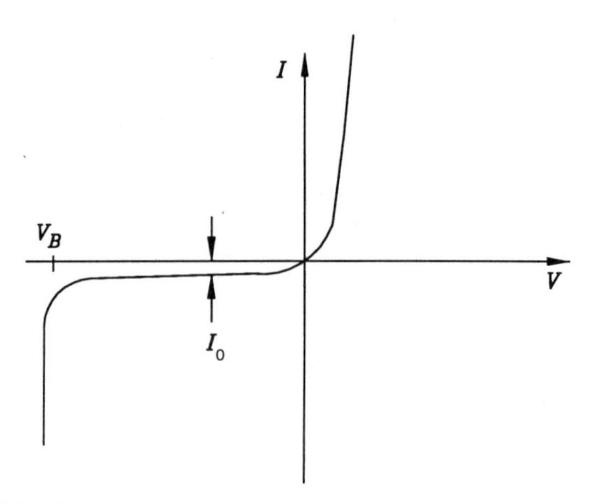

FIGURE 11.1 p-n junction and its I–V characteristics.

Avalanche breakdown occurs as M_p or M_n approaches infinity. The breakdown condition is thus given by

$$\int_0^W \alpha_p \exp\left[-\int_0^x (\alpha_p - \alpha_n)\, dx'\right] dx = 1 \tag{11.4}$$

or

$$\int_0^W \alpha_n \exp\left[-\int_x^W (\alpha_n - \alpha_p)\, dx'\right] dx = 1 \tag{11.5}$$

Solving Equations (11.4) and (11.5) gives the avalanche breakdown voltages. For a semiconductor with equal ionization rates for electrons and holes, Equations (11.4) and (11.5) reduce to

$$\int_0^W \alpha\, dx = 1 \tag{11.6}$$

IMPATT operation theory can best be understood by considering a Read diode as shown in Figure 11.2. The electric-field distribution inside the device and the ionization rates are derived from $dE/dx = qN/\varepsilon_s$, where N is the doping concentration. The device can be divided into two regions: the avalanche region and the drift region. The avalanche region is a small high-field region inside the n layer near the p–n junction, where impact ionization occurs and electrons and holes are generated. The drift region is the low-field region, where carriers drift at the saturation velocity. Due to the time delay in the avalanche and drift processes, there is a phase shift between the voltage and current. The total phase shift is equal to

$$\theta = \omega(\tau_a + \tau_d) \tag{11.7}$$

where τ_a is a time delay attributed to avalanche multiplication and τ_d is due

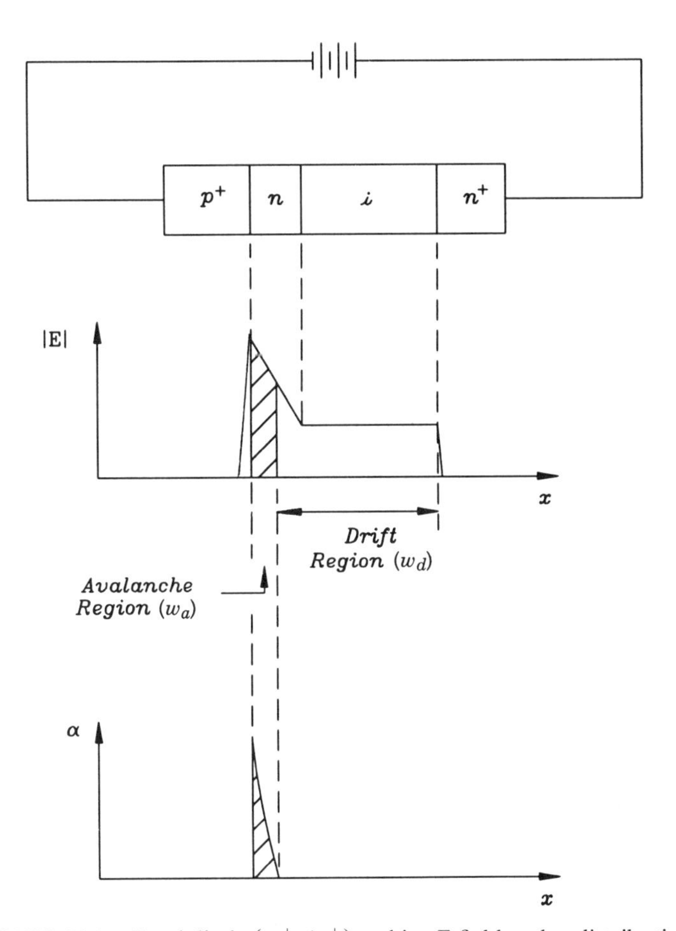

FIGURE 11.2 Read diode (p^+nin^+) and its E-field and α distributions.

to a finite drift time. For negative resistance to occur, θ needs to be greater than 90°. Consider the ionization rate as

$$\alpha = A \exp\left[-\left(\frac{b}{E_{dc} + E_{RF}}\right)^m\right] \tag{11.8}$$

Here the total field is equal to $E_{dc} + E_{RF}$, and let $E_{dc} \simeq$ breakdown field. Then the addition of an ac field will cause avalanche breakdown due to the exponential increase in electron–hole pairs. This additional ac field could be introduced due to a noise signal in the oscillation startup. The current generated by the avalanche process will have its maximum when the sinusoidal RF field passes through zero. As shown in Figure 11.3, the avalanche process will contribute a 90° inductive lag to the current generated in the avalanche region. This current will be injected into the drift region of the diode for an additional delay. The current induced in the external circuit by this change is shown in the bottom of Figure 11.3. It is obvious that the external current is more than 90° out of phase with respect to the RF voltage. A negative resistance is therefore created.

Example 11.1 For a single drift (p^+nn^+) IMPATT diode, draw the E-field distribution and identify the avalanche and drift regions.

SOLUTION:

$$\frac{dE}{dx} = \frac{qN}{\varepsilon_s}$$

In the p^+ region, $N = N_A^- = -N_A$, and the slope of the E vs. x curve is proportional to N_A. In the n region, the slope is proportional to N_D^+ (or $+N_D$). Note that $N_A \gg N_D$. The electric field distributions for two biases V_1 and V_B are shown in Figure 11.4. The areas under these curves are equal to their corresponding bias voltages. When the reverse bias is set at V_1, the E field inside the device is small, and no avalanche breakdown occurs. When the reverse bias is increased to V_B, in a small region near the p–n junction, the E field will be high enough to cause avalanche breakdown. This small region is the avalanche region inside which α is large. The drift region is the n region, where the generated electrons travel at the saturation velocity.

For the single-drift structure shown, only electrons contribute to the negative resistance since they travel through the drift region. The holes generated travel to the p^+ terminal without passing through a drift region. To use both electrons and holes effectively, double-drift devices (p^+pnn^+) are used to achieve higher efficiency and output power.

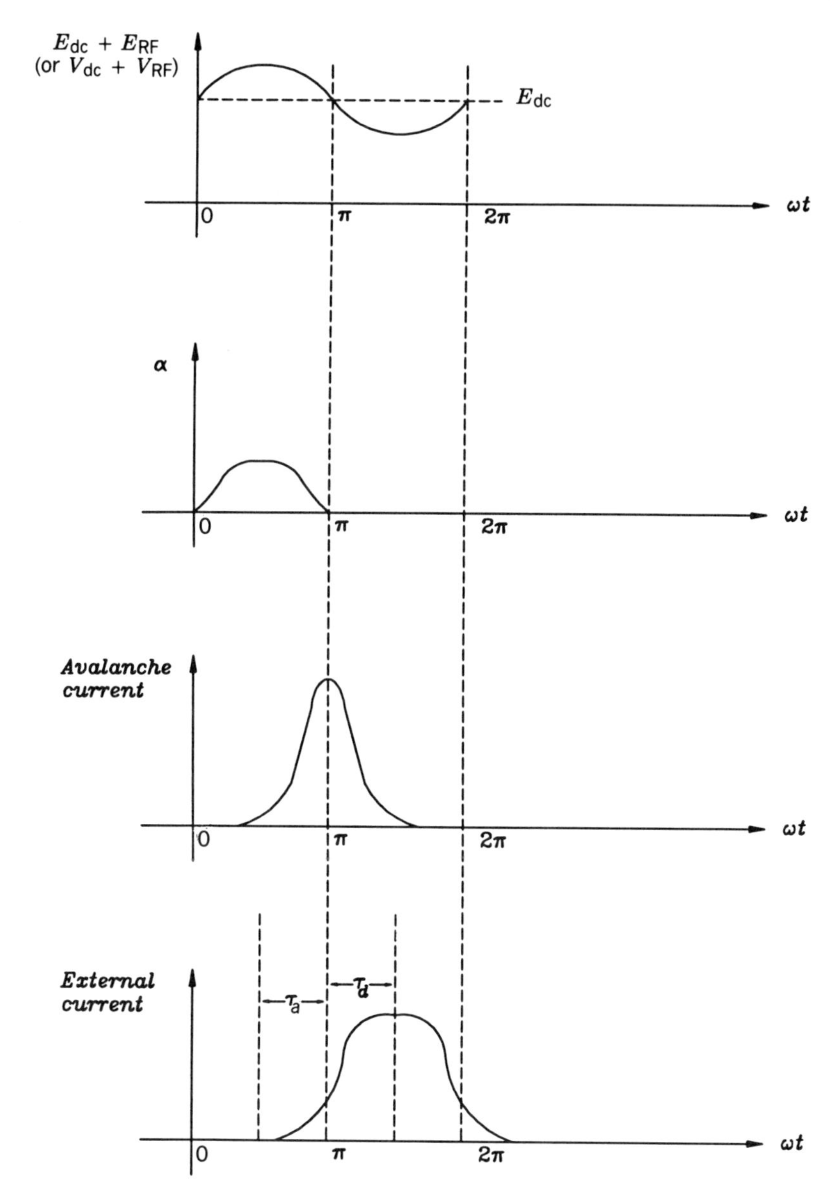

FIGURE 11.3 When the current is out of phase with the voltage, negative resistance is created.

11.3 SMALL-SIGNAL AND LARGE-SIGNAL ANALYSIS

Consider a p–n junction under reverse avalanche breakdown conditions as shown in Figure 11.5 for a single- or double-drift diode. Assuming a one-dimensional analysis and neglecting the diffusion effects, a set of equations

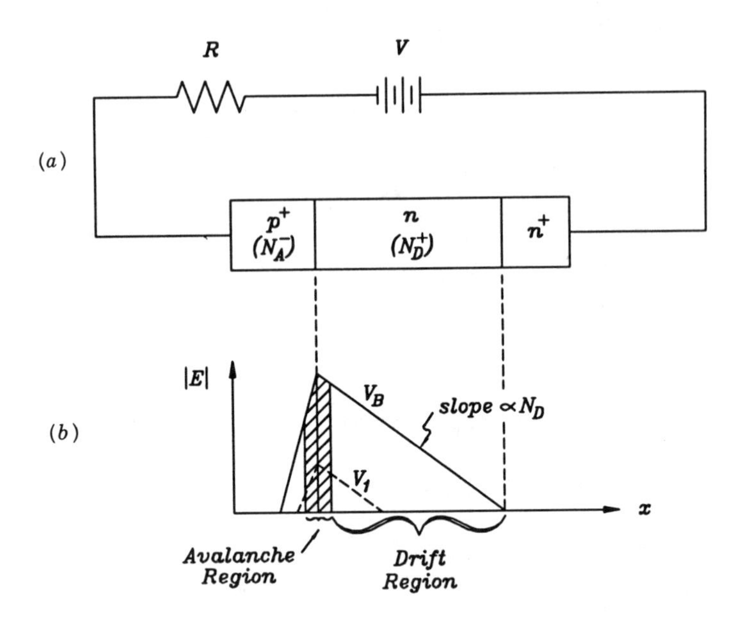

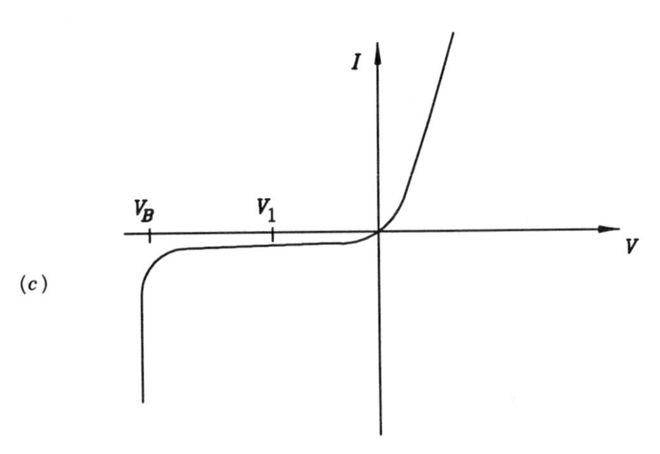

FIGURE 11.4 (*a*) Single-drift p^+nn^+ diode; (*b*) E-field distribution for reverse bias of V_1 and V_B (the E-field is maximum at the junction and minimum at the boundary of the space-charge region; see Section 5.2 for details); (*c*) I–V characteristics.

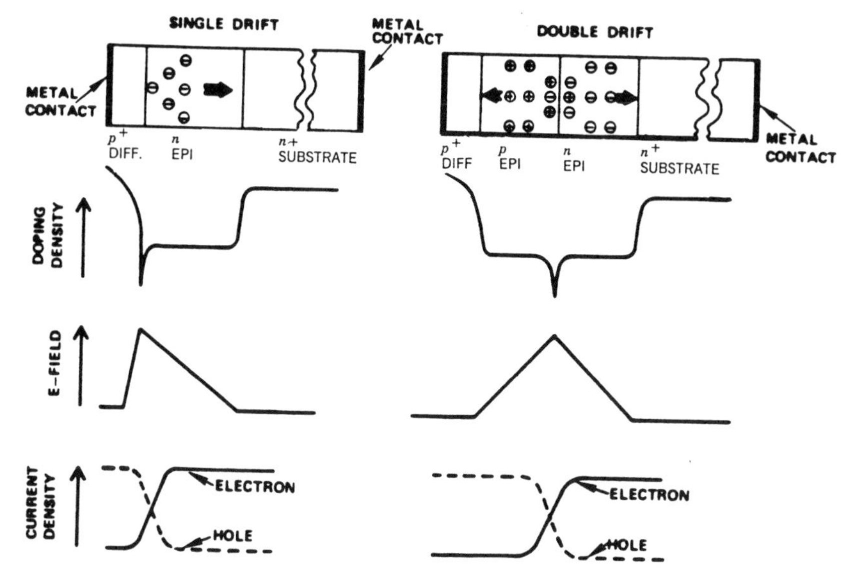

FIGURE 11.5 IMPATT diode model. (From Ref. 6 with permission from Academic Press.)

governing dynamics of electrons and holes are given by Poisson's equation [7]:

$$\frac{\partial E}{\partial x} = \frac{q}{\varepsilon_s}(N_D - N_A + p - n) \tag{11.9}$$

and by continuity equations for electrons and holes:

$$\frac{\partial n}{\partial t} = \frac{1}{q}\left(\frac{\partial J_n}{\partial x} + \alpha_n J_n + \alpha_p J_p\right) \tag{11.10}$$

$$\frac{\partial p}{\partial t} = \frac{1}{q}\left(\frac{-\partial J_p}{\partial x} + \alpha_n J_n + \alpha_p J_p\right) \tag{11.11}$$

where

N_D, N_A = doping concentration in the n and p materials
p, n = hole and electron carrier densities, respectively
ε_s = dielectric constant of semiconductor
J_p, J_n = hole and electron current densities
t = time
x = distance from the junction
q = electron charge

The total current density is

$$J = qnv_n + qpv_p + \varepsilon_s \frac{\partial E}{\partial t} \tag{11.12}$$

where v_p and v_n are drift velocities of holes and electrons.

A. Small-Signal Analysis

For the small-signal analysis, it is assumed that the ac signal is small compared with the dc component. The small-signal analysis provides useful information about the impedance and frequency responses of the device. The electric field and current density J can be written as ac signals superimposed upon the dc components, that is,

$$E = E_0 + E_1 e^{j\omega t} \tag{11.13}$$

$$J = J_0 + J_1 e^{j\omega t} \tag{11.14}$$

$$\Delta E = E_1 e^{j\omega t} \tag{11.15}$$

$$\alpha = \alpha_0 + \frac{\partial \alpha}{\partial E} \Delta E = \alpha_0 + \alpha' E_1 e^{j\omega t} \tag{11.16}$$

$$n = n_0 + n_1 e^{j\omega t} \tag{11.17}$$

$$p = p_0 + p_1 e^{j\omega t} \tag{11.18}$$

J_1 and E_1 represent ac components, which are smaller than the dc components (J_0, E_0). The higher-order terms of the ac components are neglected. ω is the angular frequency of operation, and $\alpha' = \partial\alpha/\partial E$.

Substituting Equations (11.13) through (11.18) into (11.9) through (11.12), we can separate the dc and ac terms (see Problem P11.4). The ac components are obtained

$$-\frac{\partial E_1}{\partial x} = j\left(\frac{\omega}{v_p}\right)E_1 + \frac{1}{\varepsilon_s}\left(\frac{1}{v_n} + \frac{1}{v_p}\right)J_{n1} - \frac{J_1}{\varepsilon_s v_p} \tag{11.19}$$

$$-\frac{\partial J_{n1}}{\partial x} = (\alpha'_n J_{n0} + \alpha'_p J_{p0} - j\omega\varepsilon_s \alpha_{p0})E_1 + \left(\alpha_{n0} - \alpha_{p0} - j\frac{\omega}{v_n}\right)J_{n1} + \alpha_{p0} J_1 \tag{11.20}$$

where the derivatives with respect to t are replaced by $j\omega$ and $J_n = J_{n0} + J_{n1}$, $J_1 = J_{n1} + J_{p1}$.

The dc terms are

$$\frac{\partial E_0}{\partial x} = -\frac{1}{\varepsilon_s}\left(\frac{1}{v_n} + \frac{1}{v_p}\right)J_{n0} + \left(\frac{q}{\varepsilon_s}\right)(N_D - N_A) + \frac{J_0}{\varepsilon_s v_p} \quad (11.21)$$

$$-\frac{\partial J_{n0}}{\partial x} = (\alpha_{n0} - \alpha_{p0})J_{n0} + \alpha_{p0}J_0 \quad (11.22)$$

These equations can be solved numerically (using the finite-difference method, for example) subject to boundary conditions. The ac impedance Z_1 of the devices can be obtained by

$$Z_1 = \frac{1}{Y_1} = \frac{V_1}{J_1 A} \quad (11.23)$$

where A is the cross-sectional area of the device. The ac voltage V_1 across the depletion region is given by

$$V_1 = \int E_1\, dx \quad (11.24)$$

From the ac solutions of J_1 and E_1, the ac impedance of the device as a function of frequency can be found from Equations (11.23) and (11.24). Note that N_A and N_D depend on the doping profile, which is a function of x. The parameters α_n, α_p, v_p, and v_n vary with temperature.

For a specified current density, junction temperature, and doping profile, a numerical analysis calculates and plots the dc electric field as a function of distance. The dc solution is used to calculate the device small-signal conductance and susceptance per unit area and the device Q factor as a function of frequency for a specified frequency range [8]. Figure 11.6 shows a typical computer output for a 35-GHz pulsed double-drift (p^+pnn^+) silicon IMPATT diode with $N_A = N_D = 3 \times 10^{16}$ cm^{-3}, $J = 350$ μA/μm^2, and a junction temperature of 250°C. The lengths for the n and p regions are 2.5 μm. It can be seen that the conductance of the diode is negative over a broad bandwidth. The susceptance goes through a resonance below which it is inductive and above which it is capacitive. The device would be designed to achieve a maximum $|Q|$ at the desired frequency.

Although the RF power limit, efficiency, and many other phenomena of an IMPATT device are determined by large-signal analysis, small-signal values provide a convenient basis for impedance information under various conditions. Small-signal analysis has been used extensively for designing devices and circuits.

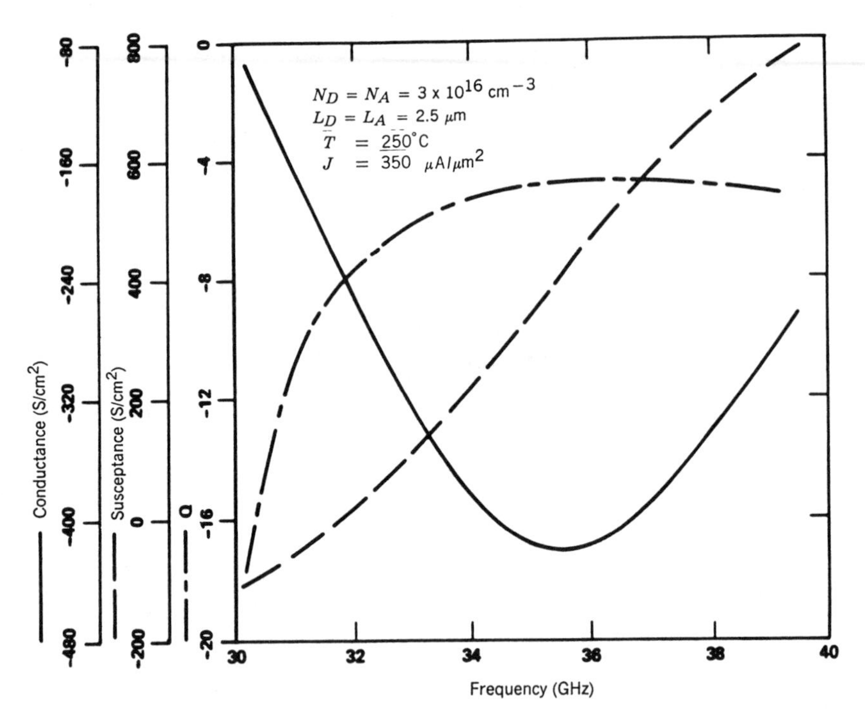

FIGURE 11.6 Calculated small-signal RF characteristics for 35-GHz pulsed IMPATTs. The negative Q value is due to the negative resistance.

B. Simplified Small-Signal Model

To illustrate the physical significance of the avalanche multiplication and the transit-time effects in an IMPATT diode, a simplified small-signal model based on the Read diode structure has been developed [1, 9]. In this model, the device is divided into three regions:

1. *Avalanche Region*. The thickness is very thin, the ionization rate is uniform, and the transit-time delay is negligible.
2. *Drift Region*. No carriers are generated and all carriers entering from the avalanche region move at saturated velocities.
3. *Inactive Region*. Additional parasitic resistance is introduced.

Furthermore, we assume that

$$\alpha_n = \alpha_p = \alpha \tag{11.25}$$

$$v_n = v_p = v_s \tag{11.26}$$

where v_s is the saturated velocity.

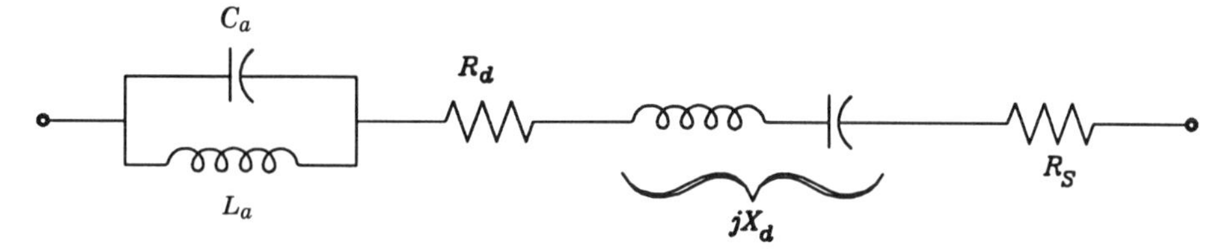

FIGURE 11.7 Equivalent circuit obtained from simplified small-signal analysis.

A detailed analysis is given in Appendix G. From this analysis, an equivalent circuit is derived as shown in Figure 11.7. C_a and L_a are due to the avalanche region. C_a and L_a form a resonant circuit with the resonant frequency given by

$$f_r = \frac{1}{2\pi} \frac{1}{\sqrt{L_a C_a}} \tag{11.27}$$

For the operating frequency $f > f_r$, R_d is negative. The negative resistance is the key for the IMPATT diode to be used as an oscillator or amplifier. R_d and jX_d are contributed by the drift region. R_s is the series resistance. Derivations and equations for L_a, C_a, R_d, and jX_d are provided in Appendix G.

C. Large-Signal Analysis

To understand nonlinear effects, transient response, power output, efficiency, and frequency tuning of the oscillator and amplifier, large-signal analysis must be used. The detailed large-signal performance can be obtained by solving Equations (11.9) through (11.11) with appropriate boundary conditions. Large-signal analyses and nonlinear effects have been studied by many researchers, as given in Refs. 10 through 14.

11.4 DOPING PROFILES AND DEVICE DESIGN

IMPATT devices can be made in many different configurations of doping profiles. The two most common semiconductor materials used are Si and GaAs, although InP and Ge can also be used. Silicon diodes normally have 5 to 10% efficiency, while GaAs diodes can operate with high efficiency (10 to 20%) at millimeter-wave frequencies.

Two simple and commonly used doping profiles are single-drift (SDR) and double-drift (DDR) profiles. The single-drift, uniform doping profile is the simplest structure. Its doping and electric field profiles are shown in Figure 11.8. Although the n^+pp^+ IMPATT has also been fabricated and shown

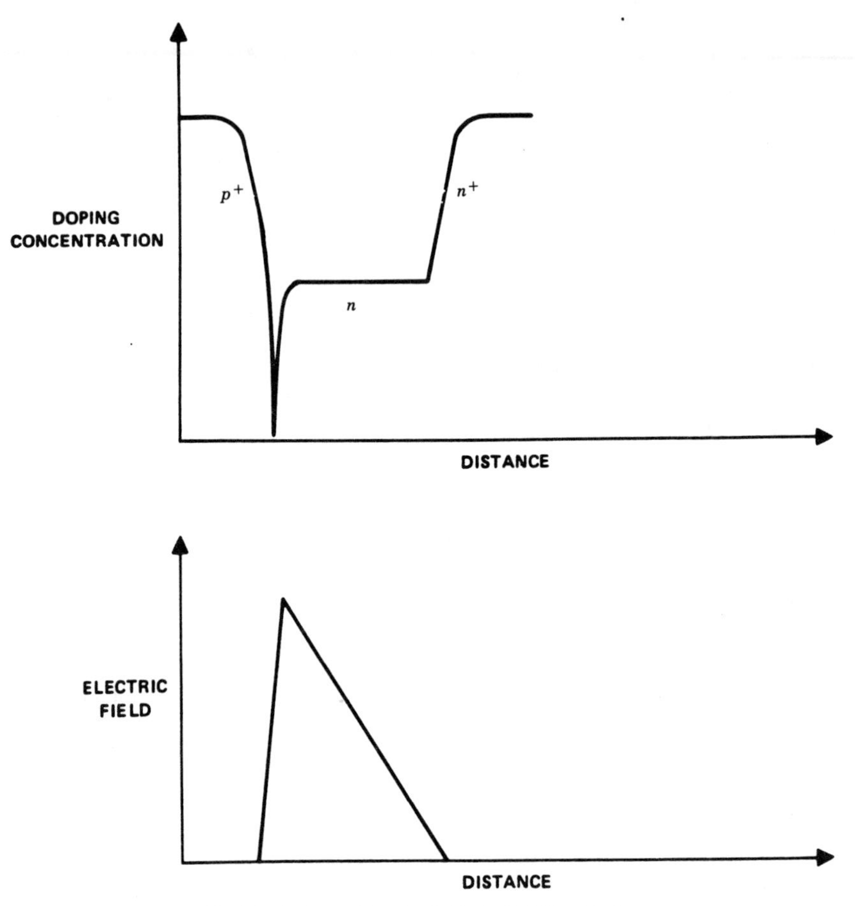

FIGURE 11.8 Single-drift IMPATT diode.

better efficiency, the p^+nn^+ type is easier to fabricate since the n^+ substrate is more common. The avalanche region voltage is approximately one-half of the total applied voltage.

The doping and electric-field profiles of the double-drift diode are shown in Figure 11.9. A double-drift diode consists of a p-type single-drift IMPATT and an n-type single-drift IMPATT connected in series. In the double-drift diode, however, both hole and electron drift regions share a common avalanche zone. The resultant improvement in the Q-factor enables the diode to operate at a higher efficiency. In addition, since a double-drift diode is equivalent to two single-drift diodes connected in series, a double-drift diode with twice as large a junction area as that of a single diode maintains the same impedance level as that of the single-drift diode. This results in four times as much output power as that of the single-drift diode.

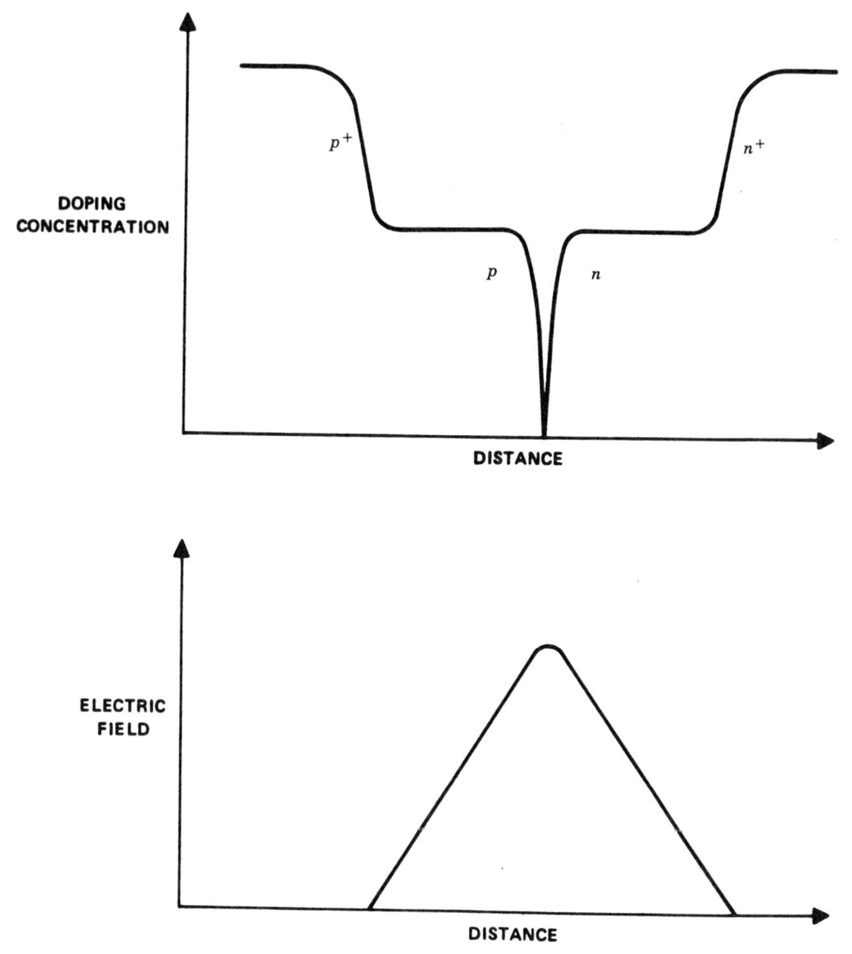

FIGURE 11.9 Double-drift IMPATT diode.

Other profiles, such as the single-drift Read profile ($p^+n^+nn^{++}$), double-drift Read profile ($p^{++}pp^+n^+\ nn^{++}$), and hybrid profile ($p^{++}pn^+nn^{++}$) have also been fabricated for high-power and high-efficiency applications, especially in GaAs substrates [4].

IMPATT device design is based on the iterative process shown in Figure 11.10. The theoretical values for the diode parameters such as doping densities and epilayer thickness are first obtained via a small-signal computer calculation. IMPATT diode wafers are then fabricated using these diode parameters. After diode fabrication, the diode profile is characterized by C–V measurement or SIMS (secondary-ion mass spectroscopy) analysis. Finally, RF testing of diodes yields information and correlates performance with device parameters.

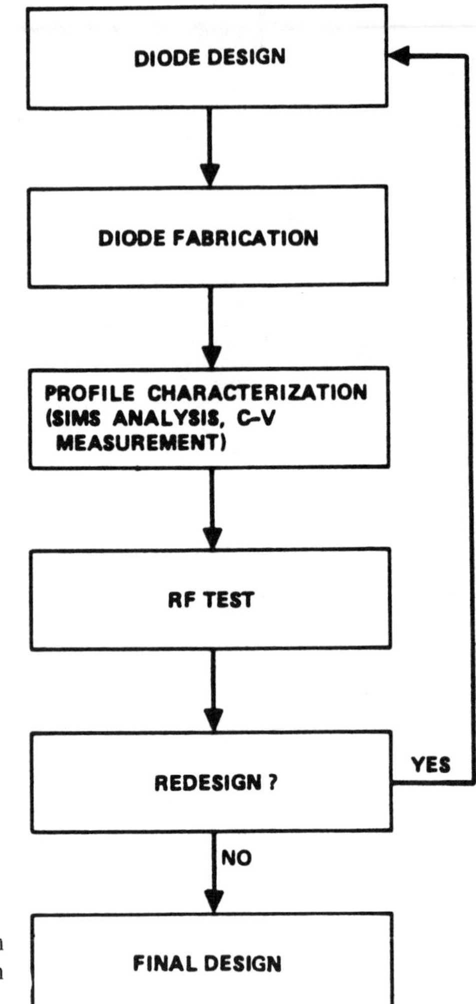

FIGURE 11.10 IMPATT device design procedure. (From Ref. 15. Permission from IEEE.)

Strictly speaking, optimum diode design requires knowledge of the large-signal characteristics of the device, and its performance is strongly dependent on the circuit parameters as well. Since exact analysis and accurate prediction of the circuit response are difficult at millimeter-wave frequencies, the diode design is based on the small-signal analysis, modifying the design subsequently subject to the experimental results. For a specified current density, junction temperature, and junction doping profile, a small-signal computer program calculates and plots the dc electric field as a function of distance. The computer program then uses this dc solution to calculate the small-signal RF conductance and susceptance per unit area, and the device Q, as a function of frequency for a specified frequency range. The designer runs the program iteratively for different values of input parameters, such as doping density, until a condition is reached for which the plot of device $|Q|$ versus frequency displays its maximum near the desired frequency of operation (see Figure 11.6 as an example). The parameters of the device that produce this are then taken as the design values. Note that the optimum frequencies cover a relatively wide range; that is, the device Q varies slowly with frequency around its maximum point.

11.5 DEVICE FABRICATION AND PACKAGING

Various methods can be used for IMPATT diode fabrication. Consider the fabrication of a silicon p^+nn^+ single-draft diode as an example. The fabrication procedure involves the following tasks: epitaxial growth, low-temperature diffusion, substrate thinning, metallization, and packaging.

The growth and evaluation of high-quality multilayer epitaxial silicon films on low-resistivity substrates is fundamental to the fabrication of high-performance millimeter-wave IMPATT diodes. Starting with an n^+ substrate, the growth of silicon epitaxial material (n layer) for IMPATT diodes is accomplished through the following simplified reaction:

$$SiH_4 \rightarrow Si + 2H_2$$

The reaction is complete at temperatures above 800°C. The growth temperature used is 1000°C because this temperature is optimum for obtaining good thickness control and reproducibility while limiting out-diffusion of dopant from the substrate. For a very thin layer, the use of molecular beam epitaxy (MBE) would be advantageous.

After one evaluates the epigrowth and is satisfied with the results, a shallow p^+ diffusion is followed to form a highly doped region for ohmic contact. The p^+nn^+ diode is formed. To ensure that the p dopant (boron) diffusion does not degrade the epitaxial n-type doping profile, the diffusion temperature must be kept low (ca. 1000°C). At this temperature, the boron diffusion is relatively slow, as is the out-diffusion of arsenic from the substrate. Because of the difference in diffusion coefficients, diffusion into the surface from a high-concentration boron source is typically more than five times faster than out-diffusion from the substrate. Instead of diffusion, the p^+ layer can also be obtained by ion implementation.

To reduce the series resistance, the substrate (n^+ layer) thickness needs to be reduced before metallization. A final thickness of 10 to 15 μm is generally desirable. Metallization on the substrate and p^+ sides allows for connection to the ohmic contact.

For reliable operation, IMPATT diode junction temperatures should be kept as low as possible. On the basis of a life test, it has been estimated that 250°C is safe for multiple-year operations. To keep the junction below 250°C and at the same time achieve the maximum output power, the heat dissipated in the diode must be removed efficiently. Heat removal can best be accomplished by placing a heat sink as close as possible to the junction where the heat is generated. To accomplish this goal in practice, the diode is thermal-compression bonded to a copper or silver heat sink with the p^+ side down. Due to its high thermal conductivity, metallized type IIA diamond is also used as a heat sink for millimeter-wave diodes.

Depending on the operating frequencies, various packaging techniques are used to achieve maximum performance. A good package should have the following features:

1. There should be low RF loss from the packaging material.
2. There should be low electrical parasitics so that the package self-resonance is well above the operating frequency.
3. There should be low thermal impedance between the diode chip and the remainder of the millimeter-wave circuit.
4. The package must be mechanically rugged.
5. The package must be hermetically sealable.
6. The package must be reproducible.

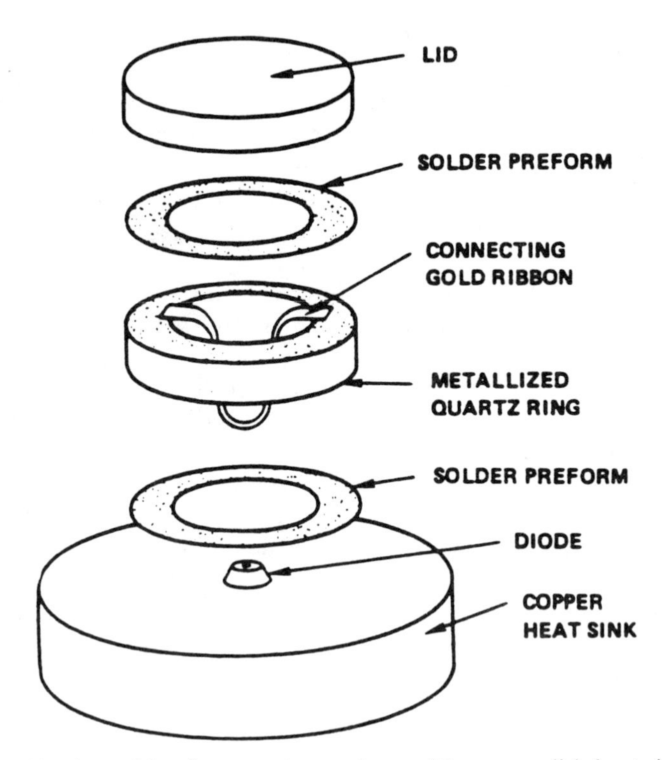

FIGURE 11.11 Assembly of quartz ring package with copper disk heat sink. (From Ref. 4.)

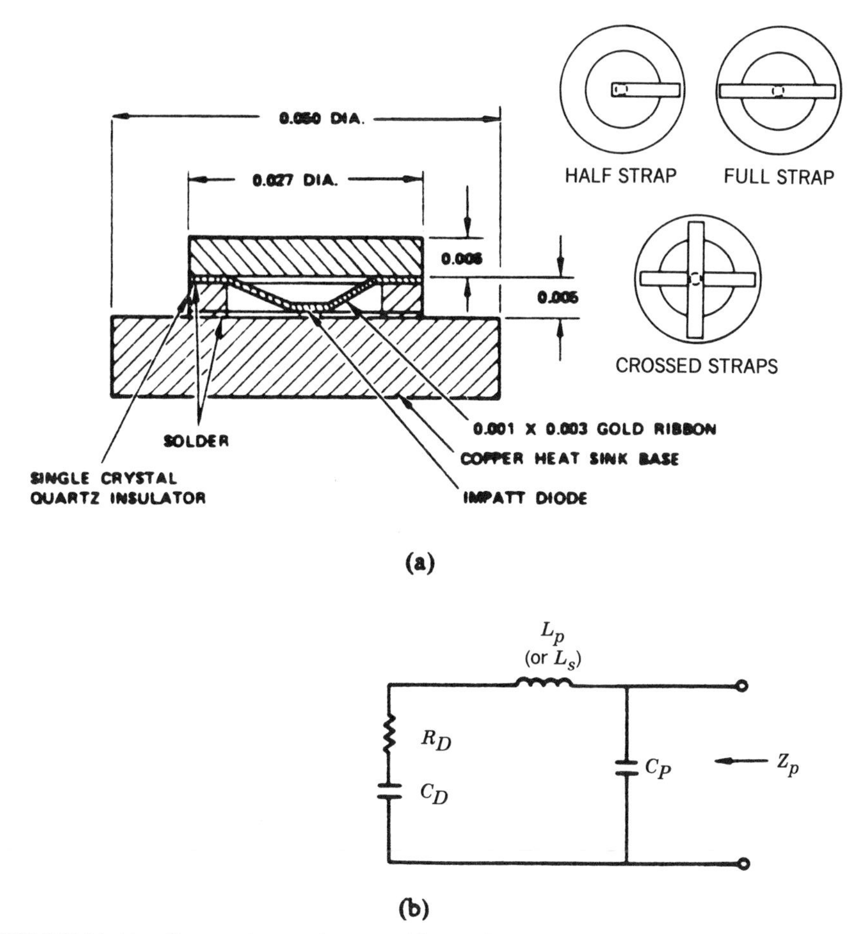

FIGURE 11.12 Quartz ring package and its equivalent circuit for a 94-GHz IMPATT diode. All measurements are in inches. (From Ref. 8 with permission from IEEE.)

At low frequencies, a ceramic pill type of package is sufficient. At high frequencies (above 60 GHz), a quartz-ring pill type of package is used to reduce the package capacitance. Figure 11.11 is an exploded view of the package, showing the method of assembly for a quartz ring package. The dimensions for a 94-GHz diode are shown in Figure 11.12. The package parasitics are modeled by an ideal inductance L_p (or L_s) representing the gold ribbon, and a capacitance C_p representing the package capacity, which is due largely to the quartz ring. The device impedance R_D and C_D will be transformed to an external impedance Z_p. A typical value for the package capacitance is 0.1 pF and the package inductance is 0.03, 0.04, and 0.075 nH for crossed-strap, full-strap, and half-strap configurations, respectively.

For frequencies above 100 GHz, an open package is used to further reduce the package parasitics. A double quartz-standoff package and direct-contacting method have been used for 140- and 217-GHz operation [4, 15].

11.6 OSCILLATOR CIRCUITS

The conditions for oscillation have been described in Chapter 9. At the oscillation frequency f_0, we have

$$R_c(f_0) < |R_D(f_0, I_0, I_{RF}, T)| \tag{11.28}$$

$$X_c(f_0) = -X_D(f_0, I_0, I_{RF}, T) \tag{11.29}$$

Equation (11.29) can be used to calculate the oscillation frequency. Equation (11.28) controls the output power. $Z_D = R_D + jX_D$ is the device impedance, and $Z_c = R_c + jX_c$ is the circuit impedance seen by the device. R_D is negative for the circuit to oscillate.

Many oscillator circuits have been developed in waveguide for IMPATT diodes. Examples of these circuits are depicted in Figure 11.13. They may be grouped into three basic types: reduced height, cap resonator, and cross-coupled coaxial waveguide cavity. In each circuit, a sliding short tuning element is used to achieve optimum performance at a specified frequency. The same

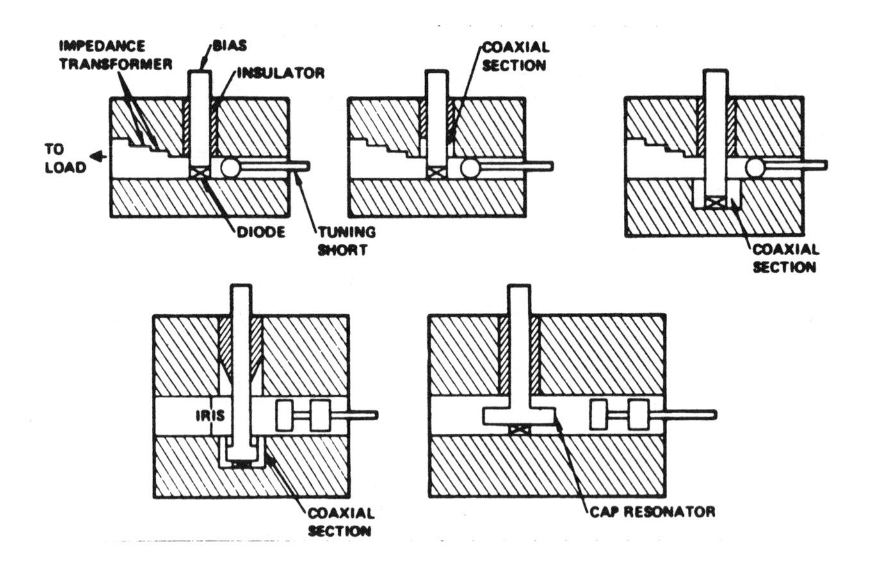

FIGURE 11.13 Examples of IMPATT oscillator circuits. (From reference 6 with permission by Academic Press.)

type of circuit can be used for CW or pulsed oscillators, but the dimensions would be optimized for each application.

Theoretical analyses are available for these circuits. For the reduced-height or full-height post-mounting circuit, analyses have been developed by Eisenhart and Khan [16] and Fong et al. [17]. The cross-coupled coaxial-waveguide circuit can be designed using a circuit model of the type developed

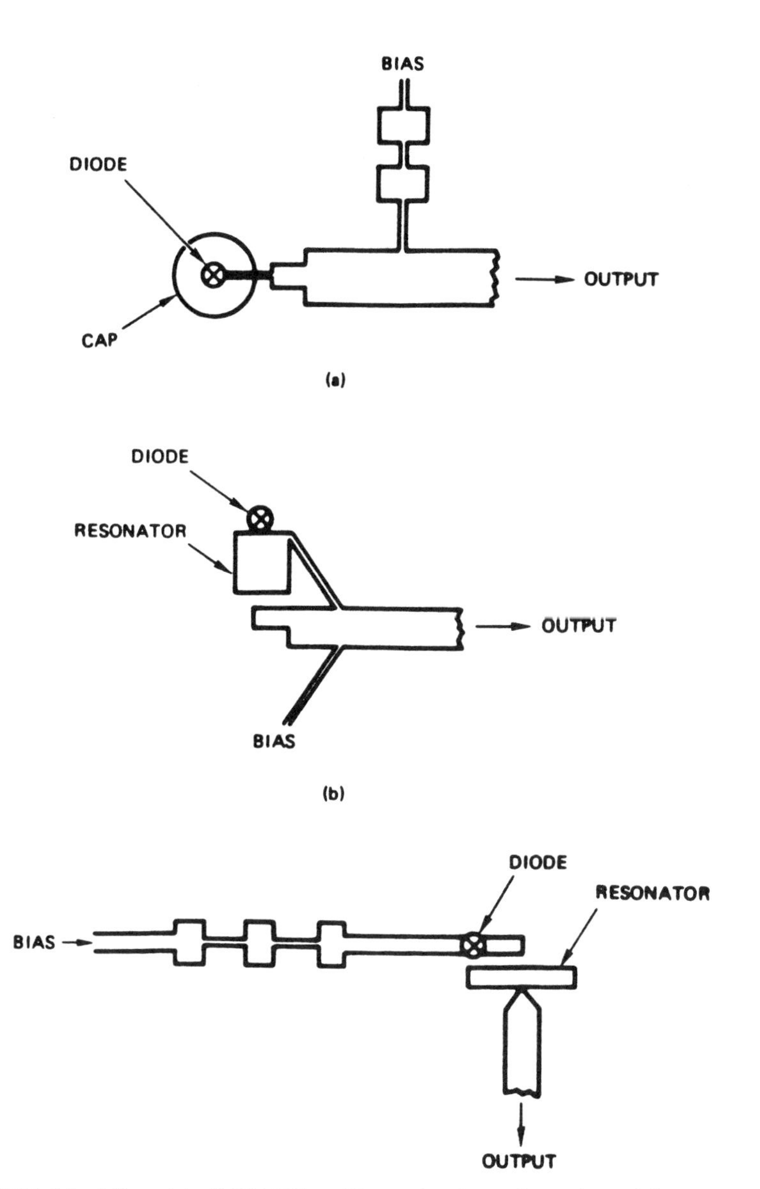

FIGURE 11.14 Microstrip IMPATT oscillator circuit configurations: (*a*) resonant-cap circuit; (*b*) and (*c*) resonant-line circuits.

by Chang and Ebert [8], Lewin [18], or Williamson [19]. The cap resonator circuit has been resolved by Bates [20].

Microstrip circuits have also been developed for IMPATT diodes, although the output power is generally less than that from a waveguide circuit. Microstrip oscillator circuits generally consist of an impedance-matching network and a resonant circuit for stabilizing the oscillation. The resonant circuit can be a metal resonant cap [21, 22], a resonant line section [23, 24], or a dielectric resonator [25]. Various circuit configurations are given in Figure 11.14.

CW (continuous-wave) IMPATT oscillators have been built from 7 to 400 GHz using silicon diodes and up to 130 GHz with GaAs diodes [4]. At 10 GHz, silicon IMPATT oscillators can produce 5 to 10 W of output power. At the millimeter-wave frequencies power levels of 2 to 3 W at 35 GHz, 1 to 2 W at 60 GHz, 500 mW to 1 W at 94 GHz, 100 mW at 140 GHz, 60 mW at 170 GHz, 78 mW at 185 GHz, 25 mW at 217 GHz, 7.5 mW at 285 GHz, and 200 μW at 361 GHz have been achieved. The power variation as a function of frequency roughly follows a relationship Pf = constant for frequencies below 100 GHz and Pf^2 = constant for frequencies above 100 GHz. This indicates that the power is determined by thermal limitation for frequencies below 100 GHz and by circuit impedance limits for frequencies above 100 GHz. The steep power falloff at high frequencies is due primarily to the increased adverse effects of diode, package, and mounting parasitics.

Since the IMPATT diode has a negative resistance over a very wide frequency band, the oscillator can be tuned over a wide frequency range using mechanical tuning or bias-current tuning techniques. Figure 11.15

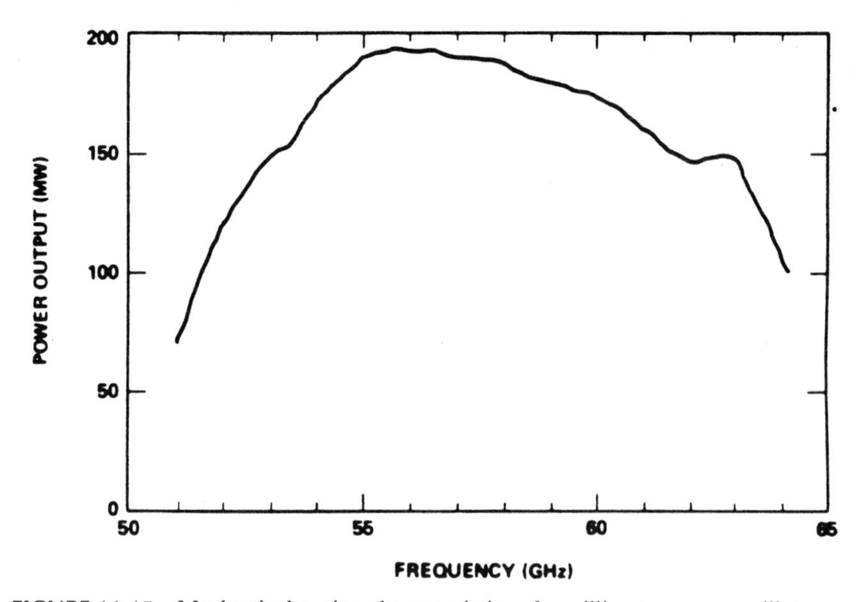

FIGURE 11.15 Mechanical tuning characteristics of a millimeter-wave oscillator.

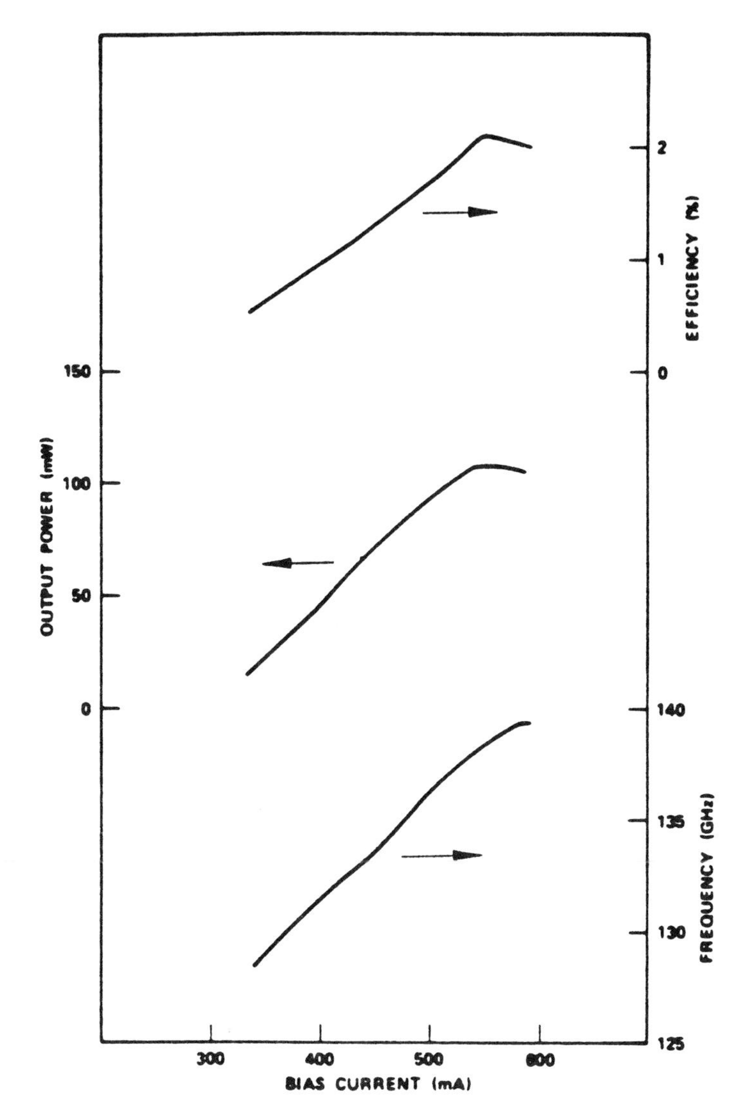

FIGURE 11.16 Output power, frequency, and efficiency for a CW IMPATT diode. (From Ref. 15 with permission from IEEE.)

shows the mechanical tuning characteristics of a 60-GHz IMPATT oscillator [6] as an example. It can be seen that the oscillator can be tuned over 10 GHz. Figure 11.16 shows the bias tuning characteristics of a 140-GHz oscillator.

The efficiency of a silicon IMPATT diode oscillator is normally below 10% for a junction temperature of 250°C. Some examples are: 10% at 40 GHz, 6% at 60 GHz, 5.8% at 94 GHz. For frequencies below 50 GHz, GaAs IMPATT diodes offer higher efficiency and power output compared with silicon diodes.

At 10 GHz, an output power of 20 to 30 W has been reported. At higher frequencies, power levels reported are 5 W at 20 GHz with 15 to 19% efficiency, 2.8 W at 44 GHz with 18% efficiency, and 5 mW at 130 GHz with 0.5% efficiency [4].

In many system applications, high-peak-power pulsed oscillators are required. IMPATT devices can be operated as pulsed power sources to achieve high peak power output over a relatively short pulse width (ca. 100 ns) with a low duty cycle.

In pulsed oscillators, the maximum pulse width as well as the pulse duty factor is one of the most important parameters that determines the achievable peak output power. Since IMPATT devices have small thermal time constants, the junction temperature rises rapidly within a pulse. To achieve high peak power output from IMPATT oscillators, the pulse width should be kept below 100 ns and the pulse duty factor below 1%. For a longer pulse width or a higher pulse duty factor, peak power output will decrease due to the requirement of keeping the peak junction temperature below the upper limit for reliable operation.

Another important property associated with pulsed operation of an IMPATT oscillator is the frequency chirping effect. As the junction temperature increases according to the transient thermal impedance change with a pulse cycle, the diode impedance (or admittance) changes. Typically, frequency chirping greater than 1 GHz can be obtained in the pulsed output signal. Noting that the frequency of oscillation is also dependent on the bias current, one can control the amount of frequency chirping to meet specific system requirements by shaping the bias pulse current waveform, as shown in Figure 11.17.

The peak power output reported from a silicon IMPATT oscillator is 50 W at 10 GHz, 30 W at 35 GHz, 13 W at 94 GHz, 3 W at 140 GHz, 1.3 W at 170 GHz, and 0.7 W at 217 GHz [4]. A pulse width of 100 ns with a 25- to 50-kHz pulse repetition rate is normally used. The efficiency is generally below 10% for silicon diodes.

11.7 AMPLIFIER CIRCUITS

As discussed in Chapter 9, reflection-type amplifiers can be constructed using two-terminal negative-resistance devices. IMPATT devices have been used effectively as both stabilized amplifiers and injection-locked amplifiers for microwave and millimeter-wave frequencies. While injection-locked amplifiers (or oscillators) are suited for high-gain (> 20 dB), narrow-bandwidth (< 1 GHz) operation, stabilized amplifiers are for low-gain (< 10 dB/stage) and broader-bandwidth (> 1 GHz) applications. The maximum power achievable from an amplifier is approximately the same as that obtainable from the same device operated as an oscillator. The operating theory for amplifiers using two-terminal devices is given in Chapter 9 and will not be repeated here.

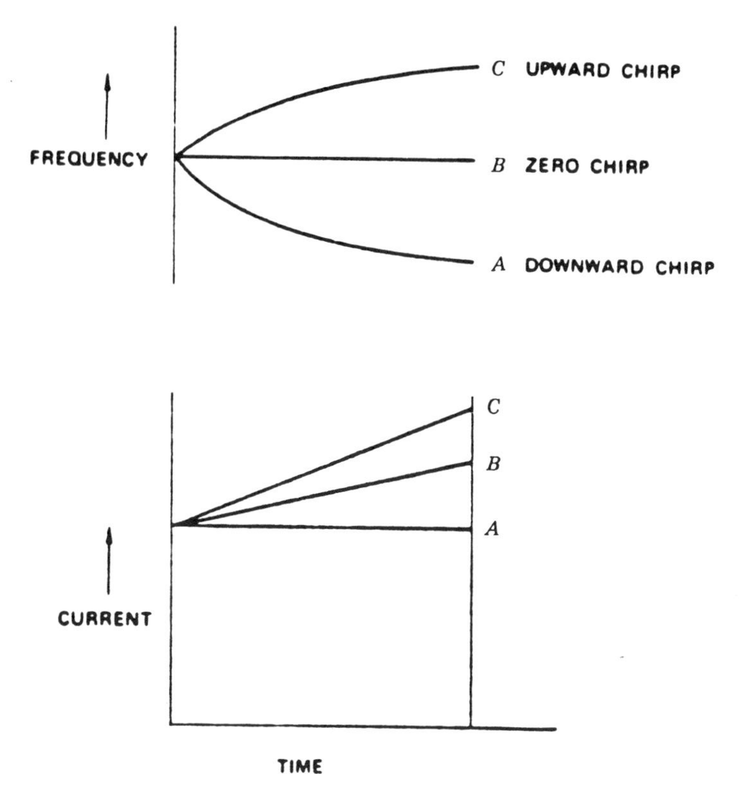

FIGURE 11.17 Frequency chirp characteristics of a double-drift IMPATT diode responding to a current ramp. (From Ref. 26 with permission from IEEE.)

Both injection-locked and stable amplifiers using IMPATT diodes have been demonstrated for CW and pulsed applications. The circuits are similar to oscillator circuits except that a circulator is needed and $|R_D| < R_c$. Some examples are given below.

A 50-GHz silicon IMPATT diode amplifier was reported in 1968 by Lee et al. [27]. Both injection-locked and stable amplifiers were reported. The injection-locked amplifier had 20 dB locking gain over a 500-MHz bandwidth, and the stable amplifier demonstrated 13 dB gain with a 3-dB bandwidth of 1 GHz. The circuit used a cap resonator circuit. Scherer reported a three-stage X-band injection-locked IMPATT amplifier with a total of 36 dB gain and a power output of 0.2 W over a 200-MHz bandwidth [28]. Kuno studied the nonlinear effects and large-signal effects of a stable or injection-locked amplifier [29, 30]. Since a circulator is always needed for IMPATT amplifier construction, the effects of a nonideal circulator on the amplifier performance are important information. The effects have been studied by Bates and Khan [31].

Microstrip IMPATT amplifiers have also been developed using microstrip circulators [24]. Figure 11.18 shows a W-band all-microstrip IMPATT amplifier [24]. At the top is the microstrip circulator, with the input and

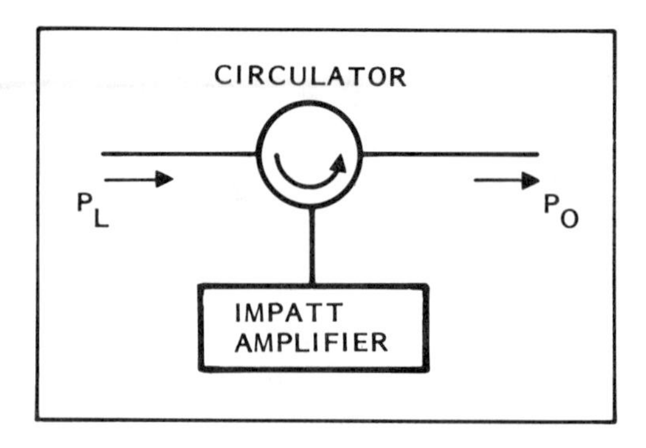

FIGURE 11.18 W-band microstrip IMPATT amplifier. (From Ref. 24 with permission from IEEE.)

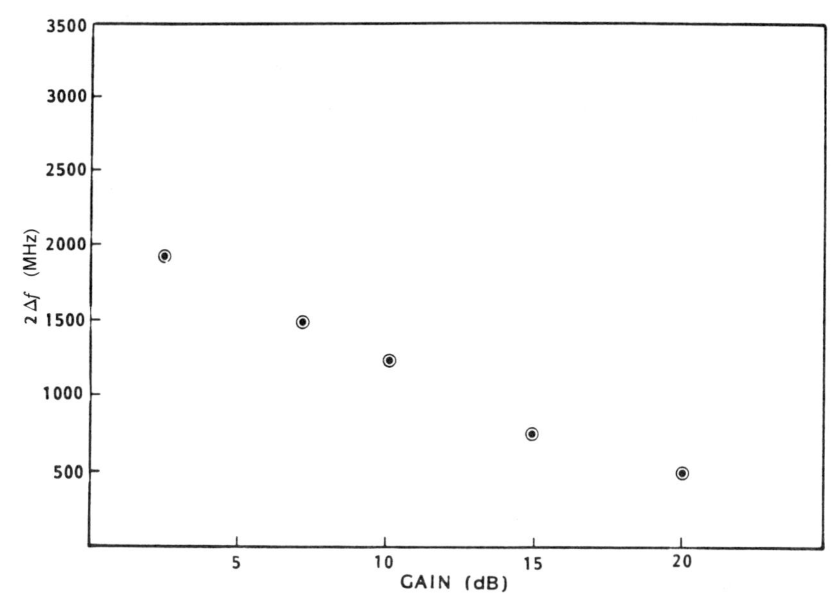

FIGURE 11.19 Injection-locking characteristics of a W-band microstrip IMPATT amplifier. (From Ref. 24 with permission from IEEE.)

output ports coupled to the transitions. At the bottom is the microstrip IMPATT oscillator. Using the measurement setup shown in Chapter 9, the injection-locking bandwidth as a function of power gain can be measured (Figure 11.19). The external Q_e can be calculated using Equation (9.17). For this particular circuit, the Q_e is approximately 37. Figure 11.20(*a*) shows the free-running spectrum of the IMPATT oscillator. After the application of the locking signal, the spectrum exhibits the low-noise characteristic shown in Figure 11.20(*b*).

11.8 POWER COMBINERS

Although the IMPATT diode is the most powerful millimeter-wave solid-state device, the output power from a single diode is limited by fundamental thermal and impedance problems. To meet many system requirements, it is necessary to combine several diodes to achieve high power levels. Many power-combining techniques have been developed in the microwave and millimeter-wave frequency range during the past 20 years. Reviews of these techniques are given by Russell [32] and by Chang and Sun [33]. The methods of power combining fall mainly into four categories, as shown in Figure 11.21 [33]: chip-level combiners, circuit-level combiners, spatial (or quasi-optical)

(*a*)

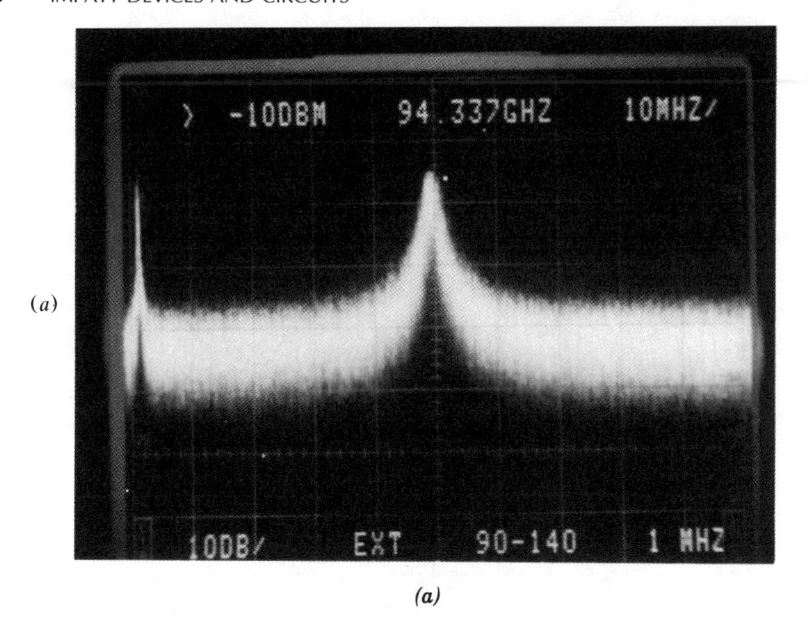

(a)

(*b*)

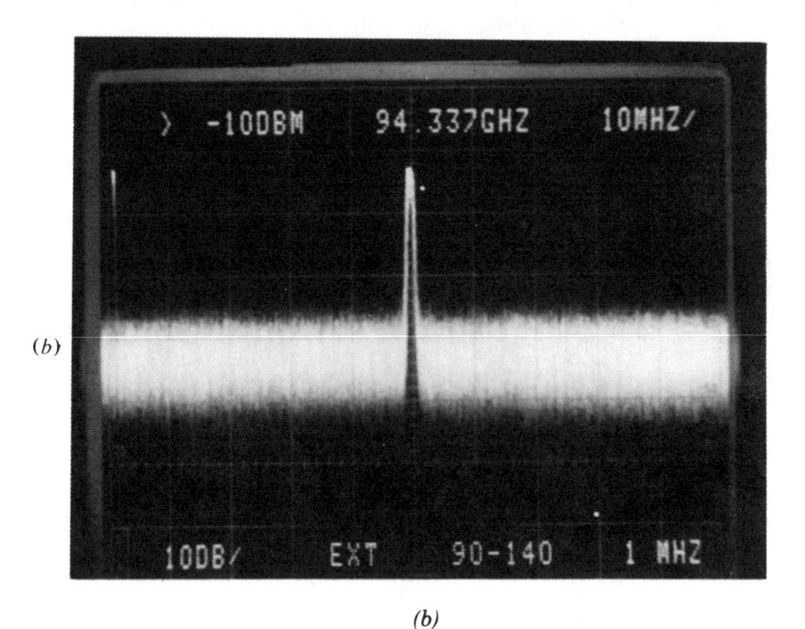

(b)

FIGURE 11.20 Spectrums of locked and unlocked IMPATT amplifier: (*a*) free-running spectrum; (*b*) injection-locked spectrum. (From Ref. 24 with permission from IEEE.)

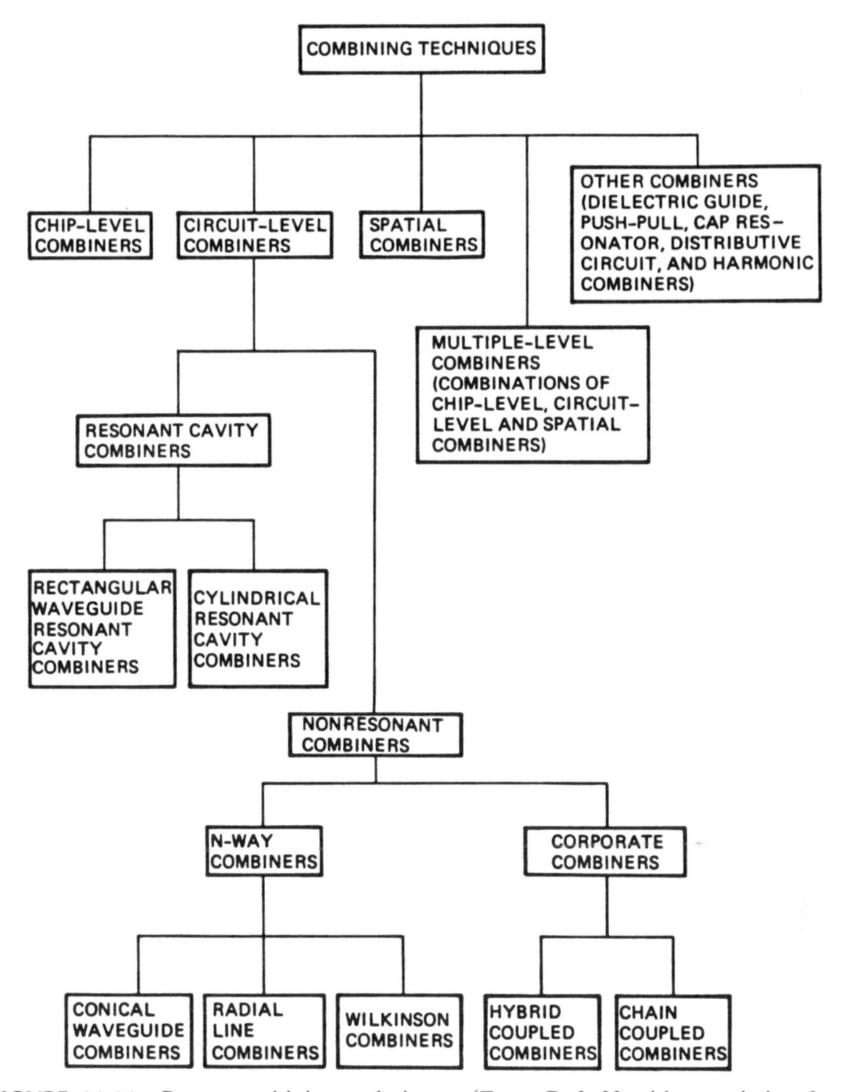

FIGURE 11.21 Power-combining techniques. (From Ref. 33 with permission from IEEE.)

combiners, and combinations of these three. The circuit-level combiners can be divided further into resonant and nonresonant combiners. Resonant combiners include rectangular- and cylindrical-waveguide resonant-cavity combining techniques. The nonresonant combiners include hybrid-coupled, conical waveguide, radial-line, and Wilkinson-type combiners. For IMPATT diodes, the three most commonly used combining techniques are rectangular resonator, cylindrical resonator, and hybrid-coupled combiners. IMPATT combiners have been built from 10 to 220 GHz. Figure 11.22 shows the

(a)

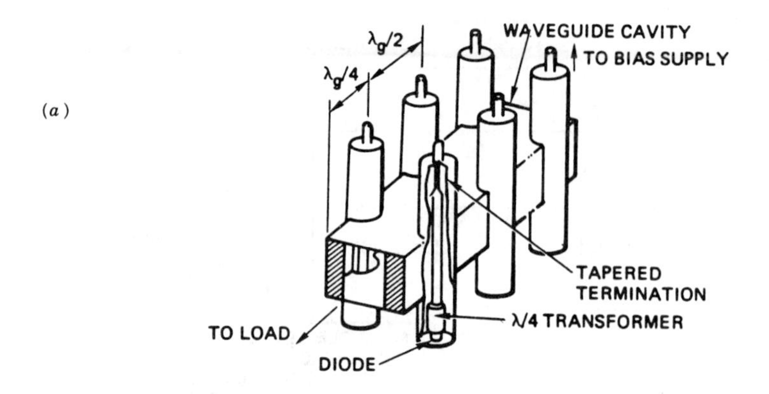

(b)

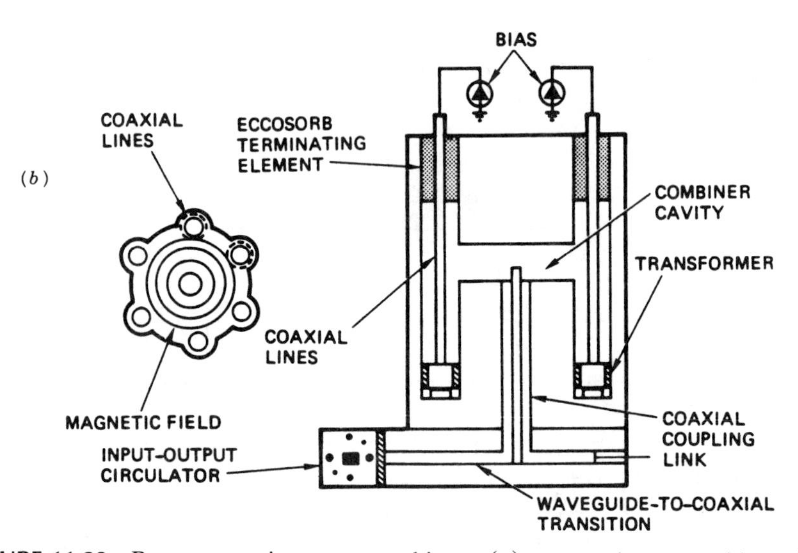

FIGURE 11.22 Resonant cavity power combiners: (*a*) rectangular waveguide cavity; (*b*) cylindrical waveguide cavity. (From Ref. 33 with permission from IEEE.)

resonant combiners using a rectangular waveguide cavity and a cylindrical waveguide cavity to combine power output from several diodes.

11.9 AM AND FM NOISE

Noise characteristics of oscillators are important properties for system applications. The AM noise of IMPATT oscillators is about 10 dB higher than

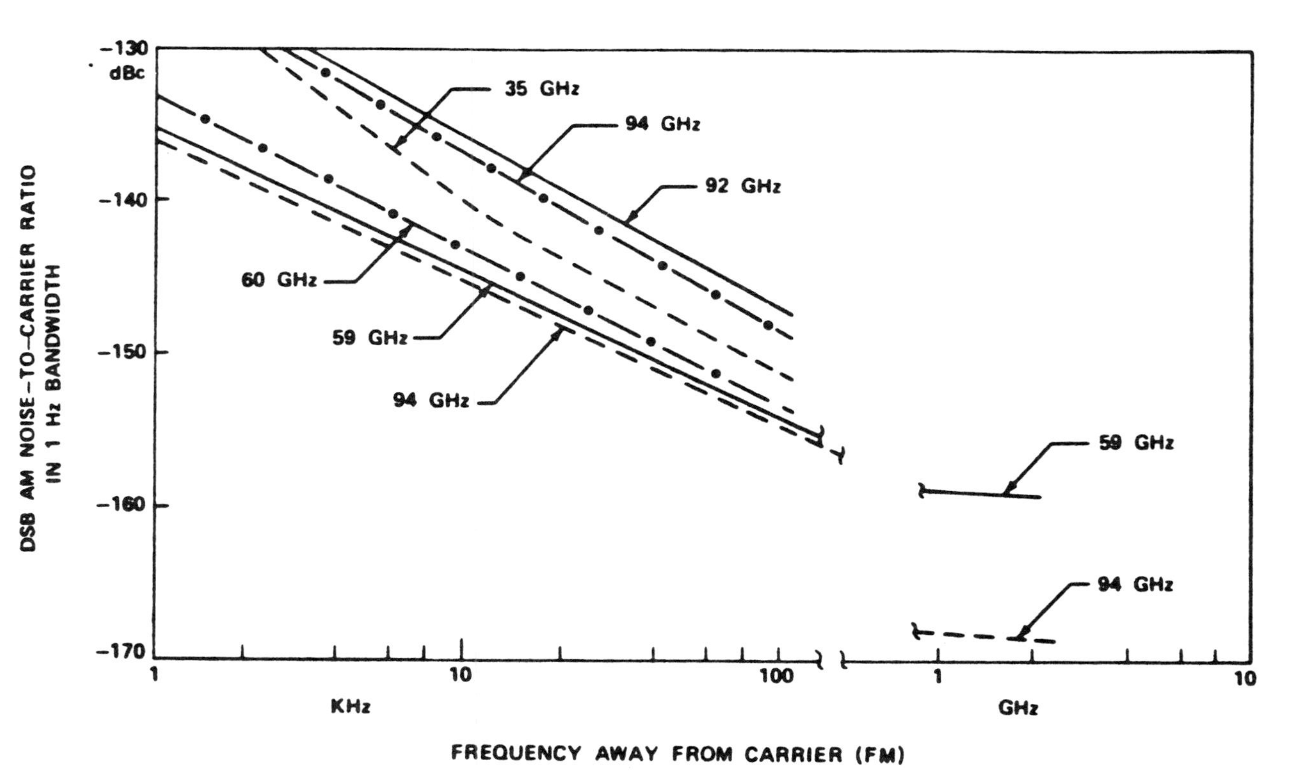

FIGURE 11.23 AM noise characteristics of millimeter-wave oscillators. ——, IMPATT; - - -, Gunn; - • - •, klystron. (From Ref. 6 with permission from Academic Press.)

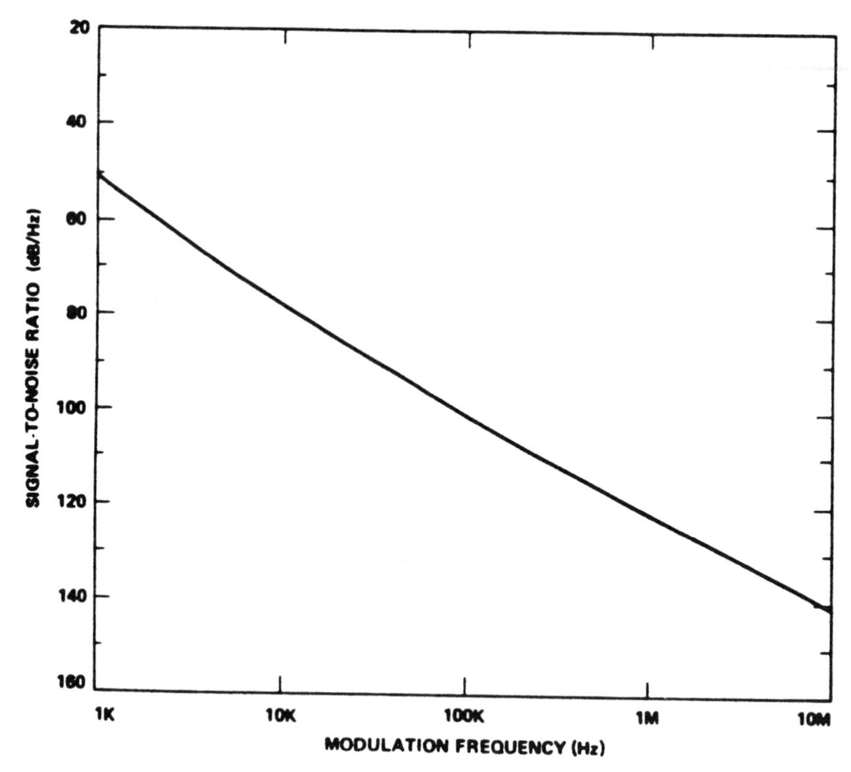

FIGURE 11.24 Measured FM noise characteristics of millimeter-wave IMPATT oscillator. (From Ref. 6 with permission from Academic Press.)

that for Gunn oscillators, making the Gunn diode more suitable than the IMPATT for local oscillator application. The IMPATT oscillator also suffers from parametric and bias oscillation, which eventually leads to diode burnout if care is not taken to prevent it.

Figure 11.23 shows the measured AM noise characteristics of typical millimeter-wave oscillators [4, 6]. For comparison, those of Gunn oscillators and klystrons are also shown. It is interesting to note that AM noise characteristics of IMPATT oscillators near the carrier frequency are similar to those of Gunn oscillators and klystrons. At higher modulation frequencies (higher than several hundred MHz), however, the IMPATT oscillator noise is higher than that of a Gunn oscillator or a klystron. For this reason IMPATT oscillators are in general difficult to use as local oscillators for mixers in receiver applications.

Shown in Figure 11.24 are measured FM noise characteristics of a millimeter-wave IMPATT oscillator [4, 6]. The FM noise characteristics of the oscillator are strongly dependent on the circuit Q. Values of the circuit-loaded Q for typical millimeter-wave IMPATT oscillators range between 20 and 100.

These values are based on injection-locking gain–bandwidth measurements.

PROBLEMS

P11.1 The doping structure shown in Figure P11.1 is called a double-drift Read low–high–low structure used for a GaAs IMPATT diode. Sketch the electric-field profile for this structure at a reverse bias near breakdown.

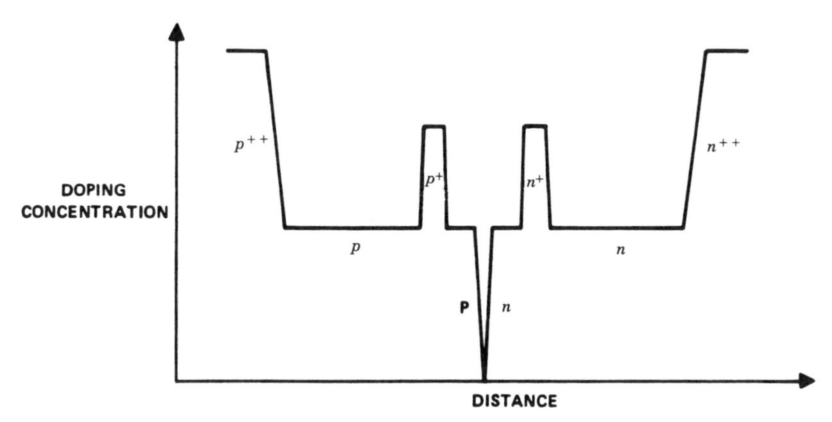

FIGURE P11.1 Double-drift Read low–high–low structure.

P11.2 The doping structure shown in Figure P11.2 is called a hybrid Read flat–high–low structure used for a GaAs IMPATT diode. Sketch the

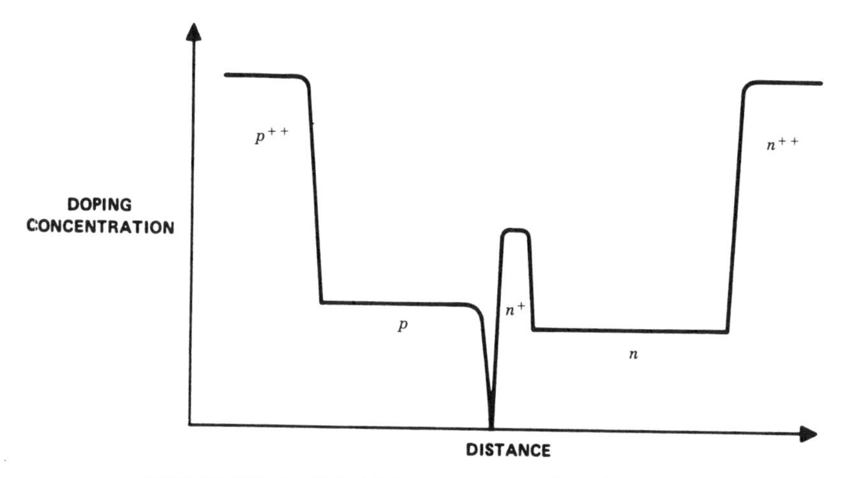

FIGURE P11.2 Hybrid Read flat–high–low structure.

electric-field profile for this structure at a reverse bias near breakdown.

P11.3 A p^+nn^+ device is shown in Figure P11.3. Assume that $N_D = 0$ and that the ionization rates of electrons and holes are equal and expressed as $\alpha_n = \alpha_p = \alpha = A\exp[-(b/E)^m]$. Derive an expression for the breakdown voltage V_B assuming that the entire n region is depleted.

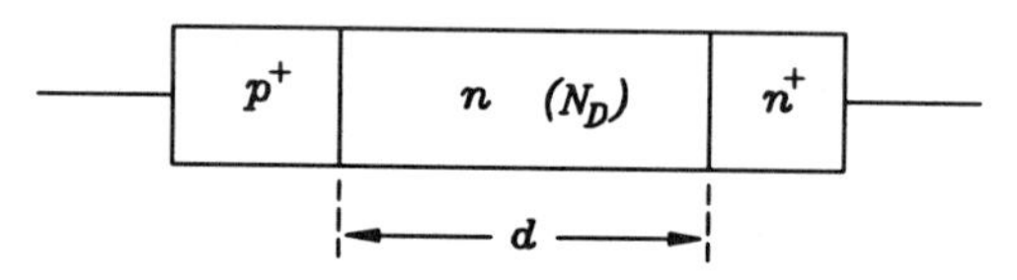

FIGURE P11.3 A p^+nn^+ device.

P11.4 Prove Equations (11.19) through (11.22) by substituting Equations (11.13) through (11.18) into Equations (11.9) through (11.12).

P11.5 A packaged IMPATT diode can be represented by the equivalent circuit shown in Figure P11.5.

(a) Derive an expression for the terminal impedance Z_p in terms of R_D, C_D, L_s, and C_p.

(b) Calculate the package resonant frequency if $L_s = 0.03$ nH and $C_p = 0.1$ pF.

(c) Calculate Z_p at 90 GHz if $R_D = -2\ \Omega$ and $C_D = 4$ pF.

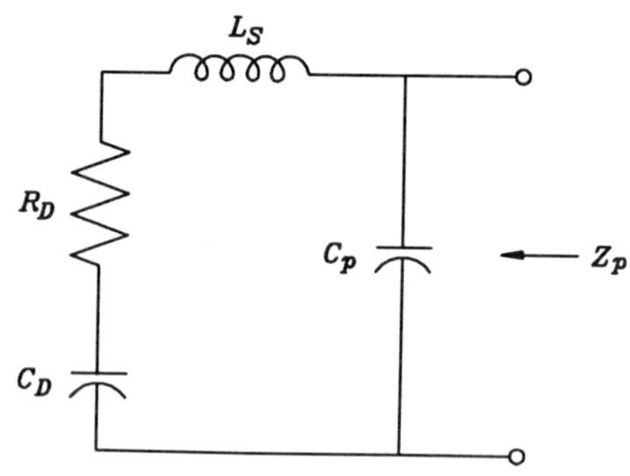

FIGURE P11.5 IMPATT equivalent circuit.

P11.6 Figure P11.6 shows the experimental data of an injection-locked IMPATT amplifier. Calculate the circuit Q from this chart.

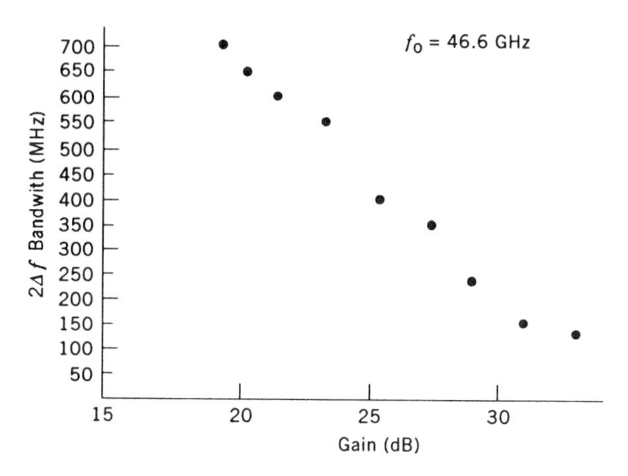

FIGURE P11.6 Injection-locked amplifier measurement.

P11.7 A GaAs linear p–n junction varactor is used to tune an IMPATT oscillator circuit as shown in Figure P11.7. At zero bias voltage, the varactor has a junction capacitance of 30 pF. $V_{bi} = 1.3$ V for GaAs. The IMPATT Z_D is given by $Z_D = -16 + j(2\omega \times 10^{-10})\ \Omega$. Calculate the oscillating frequency when the varactor is biased at **(a)** 0 V and **(b)** -20 V.

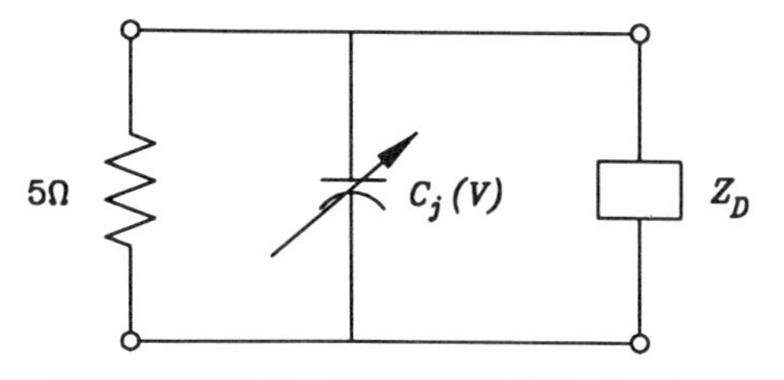

FIGURE P11.7 IMPATT VCO circuit.

P11.8 Draw the equivalent circuit for the circuit shown in Figure P11.8. Include the package parasitics for the diodes. The dc block can be

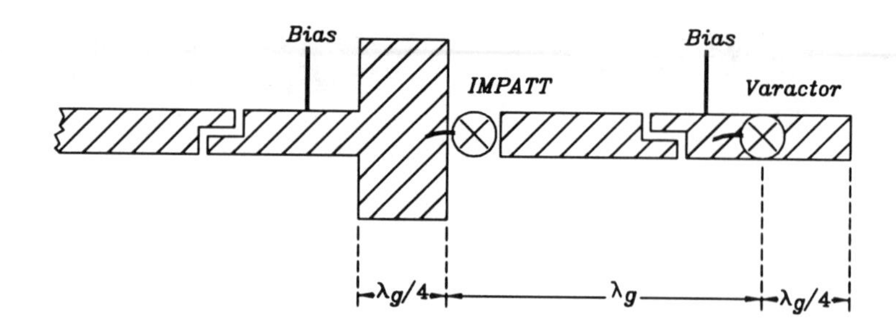

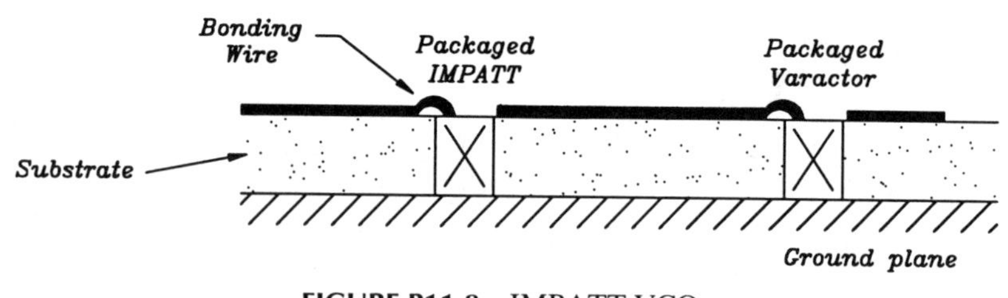

FIGURE P11.8 IMPATT VCO.

represented by a series capacitance. Should this be a large or a small capacitance? Explain.

P11.9 A silicon IMPATT diode has the following parameters:

dc bias current density	10 kA/cm^2
Device diameter	20 μm
Dielectric constant of silicon	$12\varepsilon_0$
Avalanche region width	0.1 μm
Depletion region width	1.5 μm
Saturated drift velocity	10^6 cm/s
Device series resistance (R_s)	1 Ω
Average electric field in avalanche region	5×10^5 V/cm

$$\alpha = 10^3 \exp\left(-\frac{500}{E}\right)^3 \text{ cm}^{-1} \qquad E \text{ in kV/cm}$$

Find

(a) the avalanche resonant frequency f_r.

(b) the equivalent-circuit elements (L_a, C_a, R_d, X_d, and R_s) at a frequency of f_r + 10 GHz.

(c) the equivalent-circuit elements at a frequency of $f_r - 2$ GHz.

(d) Plot the device impedance ($Z_D = R_D + jX_D$) as a function of frequency in the range from $f_r - 2$ GHz to $f_r + 10$ GHz. (Read Appendix G before doing this problem.)

P11.10 Explain why the inductance is reduced when one uses cross-strap instead of full-strap and full-strap instead of half-strap configurations.

REFERENCES

1. W. T. Read, "A Proposed High-Frequency Negative Resistance Diode," *Bell System Technical Journal*, Vol. 37, March 1958, pp. 401–446.
2. R. L. Johnston, B. C. DeLoach, Jr., and B. G. Cohen, "A Silicon Diode Microwave Oscillator," *Bell System Technical Journal*, Vol. 44, February 1965, pp. 369–372.
3. M. Ino, T. Ishibashi, and M. Ohmori, "CW Oscillation with p^+pn^+ Silicon IMPATT diodes in 200 GHz and 300 GHz Bands," *Electronics Letters*, Vol. 12, No. 6, March 18, 1976, pp. 148–149.
4. K. Chang and H. J. Kuno, "IMPATT and Related Transit-Time Devices," in *Handbook of Microwave and Optical Components*, K. Chang, Editor, Wiley, New York, 1990, Vol. 2, Chap. 7.
5. G. I. Haddad, P. T. Greiling, and W. E. Schroeder, "Basic Principles and Properties of Avalanche Transit-Time Devices," *IEEE Transactions on Microwave Theory and Techniques*, Vol. MTT-18, November 1970, pp. 752–772.
6. H. J. Kuno, "IMPATT Devices for Generation of Millimeter-Waves," in *Infrared and Millimeter-Waves*, K. Button, Editor, Academic Press, New York, 1979, Vol. 1, Chap. 2.
7. T. Misawa, "Negative Resistance in *p–n* Junctions Under Avalanche Breakdown Conditions, Parts I and II," *IEEE Transactions on Electron Devices*, Vol. ED-13, January 1966, pp. 137–151.
8. K. Chang and R. L. Ebert, "W-Band Power Combiner Design," *IEEE Transactions on Microwave Theory and Techniques*, Vol. MTT-28, April 1980, pp. 295–305.
9. M. Gilden and M. E. Hines, "Electronic Tuning Effects in the Read Microwave Avalanche Diode," *IEEE Transactions on Electron Devices*, Vol. ED-13, January 1966, pp. 169–175.
10. W. J. Evans and G. I. Haddad, "A Large-Signal Analysis of IMPATT Diodes," *IEEE Transactions on Electron Devices*, Vol. ED-15, October 1968, pp. 708–717.
11. I. L. Blue, "Approximate Large-Signal Analysis of IMPATT Oscillators," *Bell System Technical Journal*, Vol. 48, February 1969, pp. 383–396.
12. D. L. Scharfetter and H. K. Gummel, "Large-Signal Analysis of a Silicon Read Diode Oscillator," *IEEE Transactions on Electron Devices*, Vol. ED-16, January 1969, pp. 64–77.
13. K. Mouthaan, "Nonlinear Analysis of the Avalanche Transit-Time Oscillator," *IEEE Transactions on Electron Devices*, Vol. ED-16, November 1969, pp. 935–945.

14. P. T. Greiling and G. I. Haddad, "Large-Signal Equivalent Circuits of Avalanche Transit-Time Devices," *IEEE Transactions on Microwave Theory and Techniques*, Vol. MTT-18, November 1970, pp. 842–853.
15. K. Chang, W. F. Thrower, and G. M. Hayashibara, "Millimeter-Wave Silicon IMPATT Sources and Combiners for the 110–260 GHz Range," *IEEE Transactions on Microwave Theory and Techniques*, Vol. MTT-29, December 1981, pp. 1278–1284.
16. R. L. Eisenhart and P. J. Khan, "Theoretical and Experimental Analysis of a Waveguide Mounting Structure," *IEEE Transactions on Microwave Theory and Techniques*, Vol. MTT-19, August 1971, pp. 706–719.
17. T. T. Fong, K. P. Weller, and D. L. English, "Circuit Characterization of V-Band IMPATT Oscillators and Amplifiers," *IEEE Transactions on Microwave Theory and Techniques*, Vol. MTT-24, November 1976, pp. 752–758.
18. L. Lewin, "A Contribution to the Theory of Probes in Waveguides," *Proceedings of the IEE, Monograph*, 259R, October 1957, pp. 109–116.
19. A. G. Williamson, "Analysis and Modeling of Two-Gap Coaxial Line Rectangular Waveguide Junctions," *IEEE Transactions on Microwave Theory and Techniques*, Vol. MTT-31, March 1983, pp. 295–302.
20. B. D. Bates, "Analysis of Multiple-Step Radial-Resonator Waveguide Diode Mounts with Application to IMPATT Oscillator Circuits," in *IEEE MTT-S International Microwave Symposium Digest*, 1987, pp. 669–672.
21. G. B. Morgan, "Microstrip IMPATT-Diode Oscillator for 100 GHz," *Electron Letters*, Vol. 17, August 6, 1981, pp. 570–571.
22. P. Yen, D. English, C. Ito, and K. Chang, "Millimeter-Wave IMPATT Microstrip Oscillators," in *IEEE MTT-S International Microwave Symposium Digest*, 1983, pp. 139–140.
23. B. S. Glance and M. V. Schneider, "Millimeter-Wave Microstrip Oscillators," *IEEE Transactions on Microwave Theory and Techniques*, Vol. MTT-22, December 1974, pp. 1281–1283.
24. K. Chang, D. M. English, R. S. Tahim, A. J. Grote, T. Pham, C. Sun, G. M. Hayashibara, P. Yen, and W. Piotrowski, "W-Band (75–110 GHz) Microstrip Components," *IEEE Transactions on Microwave Theory and Techniques*, Vol. MTT-33, December 1985, pp. 1375–1382.
25. G. B. Morgan, "Stabilization of a W-Band Microstrip Oscillator by a Dielectric Resonator," *Electron Letters*, Vol. 18, June 24, 1982, pp. 556–558.
26. T. T. Fong and H. J. Kuno, "Millimeter-Wave Pulsed IMPATT Sources," *IEEE Transactions on Microwave Theory and Techniques*, Vol. MTT-27, May 1979, pp. 492–499.
27. T. P. Lee, R. D. Standley, and T. Misawa, "A 50 GHz Silicon IMPATT Diode Oscillator and Amplifier," *IEEE Transactions on Electron Devices*, Vol. ED-15, October 1968, pp. 741–747.
28. E. F. Scherer, "A Multistage High-Power Avalanche Amplifier at X-Band," *IEEE Journal of Solid-State Circuits*, Vol. SC-4, December 1969, pp. 396–399.
29. H. J. Kuno, "Analysis of Nonlinear Characteristics and Transient Response of IMPATT Amplifiers," *IEEE Transactions on Microwave Theory and Techniques*, Vol. MTT-21, November 1973, pp. 694–702.

30. H. J. Kuno and D. L. English, "Nonlinear and Large-Signal Characteristics of Millimeter-Wave IMPATT Amplifiers," *IEEE Transactions on Microwave Theory and Techniques*, Vol. MTT-21, November 1973, pp. 703–706.
31. B. D. Bates and P. J. Khan, "Influence of Non-ideal Circulator Effects on Negative-Resistance Amplifier Design," in *IEEE MTT-S International Microwave Symposium Digest*, 1980, pp. 174–176.
32. K. J. Russell, "Microwave Power Combining Techniques," *IEEE Transactions on Microwave Theory and Techniques*, Vol. MTT-27, May 1979, pp. 472–478.
33. K. Chang and C. Sun, "Millimeter-Wave Power Combining Techniques," *IEEE Transactions on Microwave Theory and Techniques*, Vol. MTT-31, February 1983, pp. 91–107.

FURTHER READING

1. S. M. Sze, *Physics of Semiconductor Devices*, 2nd ed., Wiley, New York, 1981, Chap. 10.
2. S. Yngvesson, *Microwave Semiconductor Devices*, Kluwer, Norwell, Mass., 1991, Chap. 3.
3. B. C. DeLoach, Jr., "Avalanche Transit-Time Microwave Diodes," in *Microwave Semiconductor Devices and Their Circuit Applications*, H. A. Watson, Editor, McGraw-Hill, New York, 1969, Chap. 15.
4. K. Chang and H. J. Kuno, "IMPATT and Related Transit-Time Devices" in *Handbook of Microwave and Optical Components*, K. Chang, Editor, Wiley, New York, 1990, Vol. 2, Chap. 7.

CHAPTER 12

Field-Effect Transistors

12.1 INTRODUCTION

After the invention of the transistor in 1948, Shockley proposed the junction field-effect transistor (JFET) in 1952 [1]. A JFET is basically a voltage-controlled transistor. Because its conduction process involves mainly the majority carrier, the JFET is a *unipolar transistor*. The conventional transistor with both the majority and the minority carriers is customarily called a *bipolar transistor*.

A family tree of different kinds of field-effect transistors is shown in Figure 12.1. There are three major types of FETs: the JFET, the MESFET, and the MOSFET. MOSFETs (metal–oxide–semiconductor FETs) use oxide as an insulator sandwiched between the metal and the semiconductor, making it a special case of the MISFET (metal–insulator–semiconductor FET). MOSFETs are widely used in digital integrated circuits for computer memories and microprocessors. The MESFET (metal–semiconductor FET) using a Schottky–barrier gate was first proposed by Mead in 1966 [3]. In the past two decades, the GaAs MESFET has become the dominant microwave solid-state device for both low noise and power applications and is the most important solid-state device employed in monolithic microwave integrated circuit (MMIC) technology [4]. The GaAs MESFET has the following features and advantages:

1. Only majority carriers flow.
2. The conductivity of a layer of semiconductor is modulated (controlled) by an applied electric field: thus the name *field-effect transistor*.
3. The operation is controlled by a voltage at the third terminal rather than by a current as in the bipolar transistor.

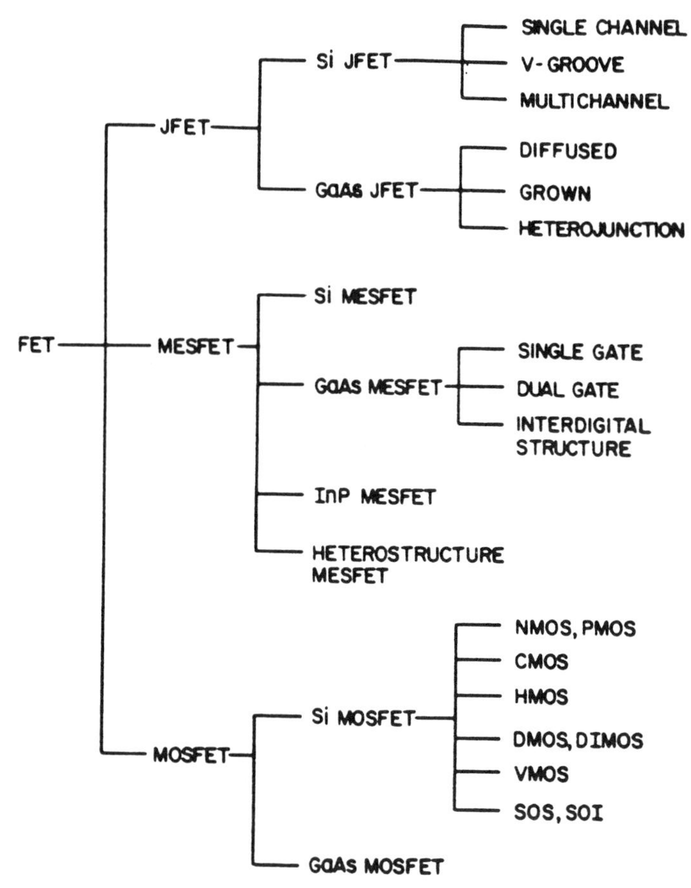

FIGURE 12.1 Family tree of field-effect transistors. (From Ref. 2 with permission from Wiley.)

4. It has voltage gain in addition to current gain.
5. It can operate at high frequencies, up to millimeter-wave frequencies.
6. It has low noise and high efficiency.
7. It has a high input resistance, up to several megaohms.
8. Compared to a bipolar transistor, it is relatively immune to radiation.

MESFETs have been used for many applications, including as low-noise amplifiers, power amplifiers, oscillators, mixers, high-speed logic circuits, switches, and phase shifters. In this chapter the device structures, operating principles, equivalent circuits, and dc and RF characteristics are introduced. The circuit applications are discussed in later chapters.

12.2 JUNCTION FIELD-EFFECT TRANSISTOR

Figure 12.2 shows a schematic diagram of an n-channel junction field-effect transistor (JFET) and its circuit symbol. The structure consists of a conductive channel controlled by a p-type gate. The reverse bias between gate and source creates a voltage-controlled space-charge (or depletion) region, as discussed in Chapter 5. The channel opening depends on the applied voltage V_{gs} (note that V_{gs} is negative). When $V_{gs} = 0$, the channel is wide and electrons can move freely from source to drain. When $|V_{gs}|$ is increased, I_{ds} decreases due to the narrowing of the channel opening. When $V_{gs} = V_p =$ pinch-off voltage, the channel is shut off and $I_{ds} = 0$. Figure 12.3 shows the dc I–V characteristics. The characteristics can be divided into three regions. In the linear region, the drain voltage is small and I_{ds} is proportional to V_{ds}. In the saturation region, the drain current I_{ds} is constant and is independent of V_{ds}. In the breakdown region, the drain current increases very rapidly,

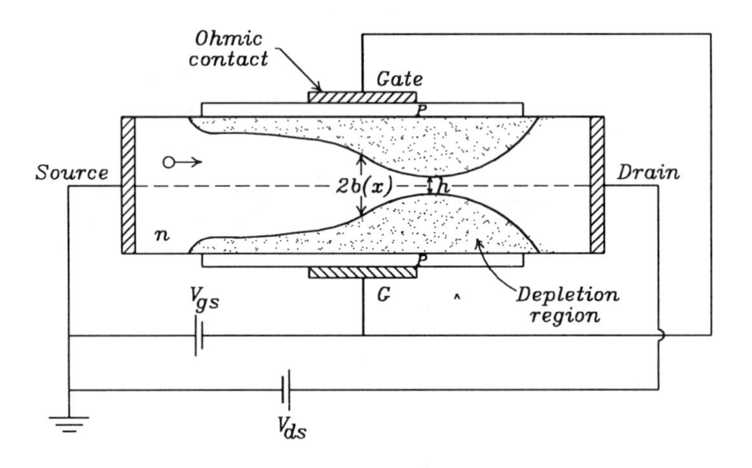

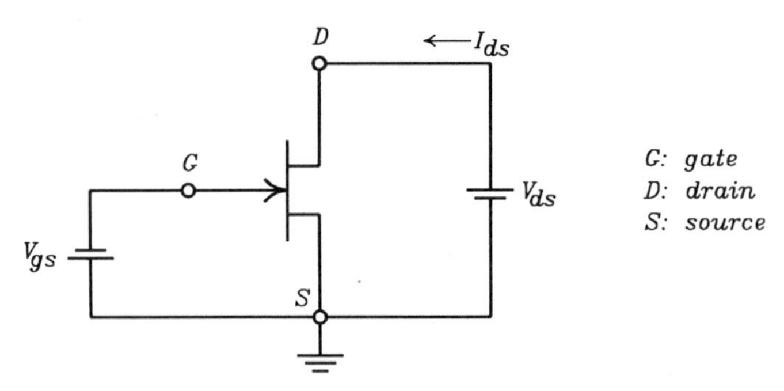

FIGURE 12.2 JFET basic structure and its circuit symbol.

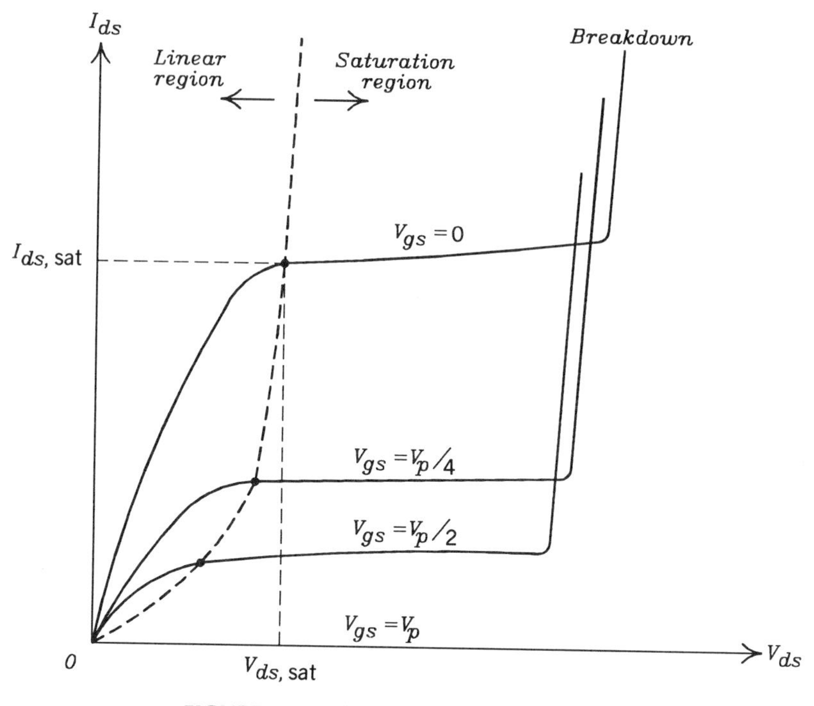

FIGURE 12.3 *I–V* characteristics of JFET.

with a slight increase in V_{ds}. The saturation drain current $I_{ds,\text{sat}}$ is larger for a lower reverse gate bias V_{gs} since the channel opening is wider.

12.3 MESFET OPERATION PRINCIPLES AND DC CHARACTERISTICS

Figure 12.4 shows a schematic diagram and circuit symbol of a MESFET. The structure consists of a semi-insulating GaAs substrate, a buffer layer, and an *n*-type layer. The *n*-type layer has doping between 8×10^{16} and 2×10^{17} cm^{-3}. The electron mobility is in the range 3000 to 4500 cm^2/V · s. The source and drain contacts are ohmic contacts using Au-Ge, for example. The gate contact is a Schottky-barrier contact using evaporated aluminum [5].

The reverse bias between the gate and the source modulates the depletion region, and the forward bias between the drain and source accelerates the electrons. As the reverse bias increases, the channel opening height decreases and the drain current decreases accordingly. The drain current is thus modulated by the gate voltage. *I–V* characteristics similar to those shown in Figure 12.3 are obtained.

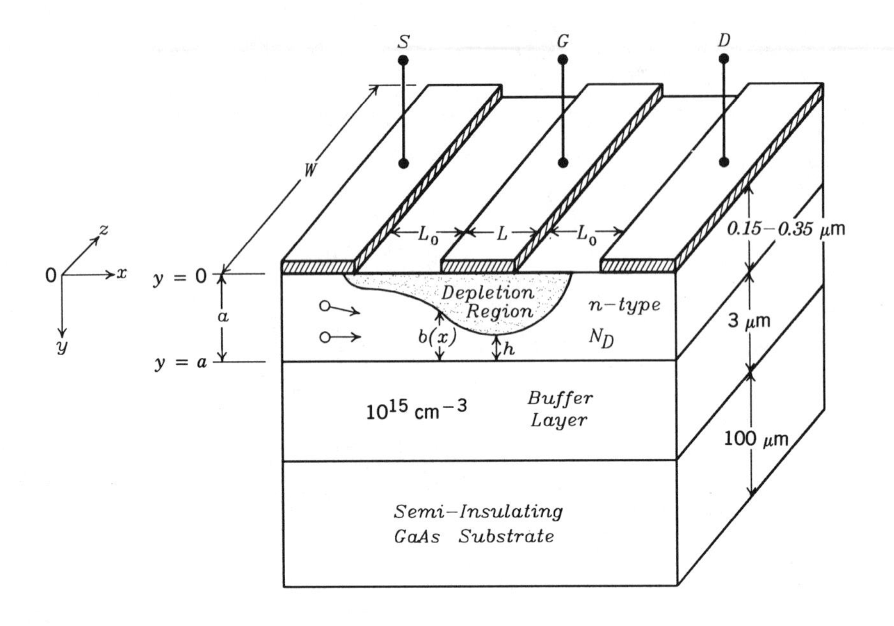

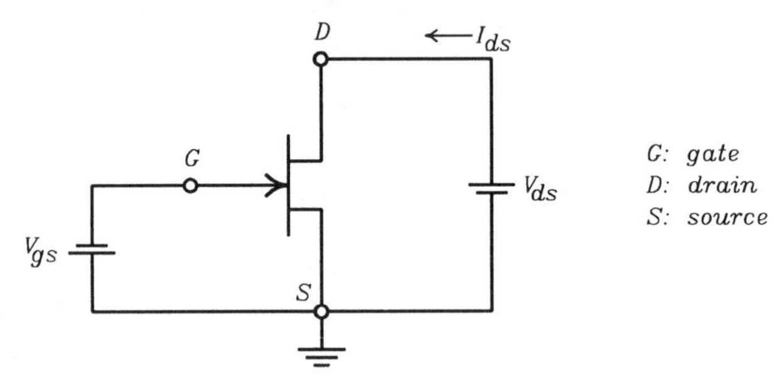

FIGURE 12.4 MESFET basic structure and its circuit symbol.

When the reverse bias is increased to the pinch-off voltage, the channel opening is shut off. We have

$$V_{gs} = V_p \qquad I_{ds} = 0 \qquad \text{and} \qquad h = 0 \tag{12.1}$$

From the Poisson's equation for the voltage, we have

$$\frac{d^2V}{dy^2} = -\frac{\rho}{\varepsilon} = -\frac{qN_D}{\varepsilon_r \varepsilon_0} \tag{12.2}$$

where N_D is the doping concentration. The boundary conditions are:

$$\text{At } y = 0: \quad V = 0$$

$$\text{At } y = a: \quad E_y = -\frac{dV}{dy} = 0 \quad \text{for pinch-off case}$$

$$\text{At } y = a: \quad V = |V_p|$$

Integrating Equation (12.2) once and using the boundary $E_y = 0$ at $y = a$, we have

$$\frac{dV}{dy} = -\frac{qN_D}{\varepsilon_r \varepsilon_0}(y - a) \tag{12.3}$$

Integrating (12.3) again and using $V = 0$ at $y = 0$ gives

$$V(y) = -\frac{qN_D}{2\varepsilon_r \varepsilon_0}(y^2 - 2ay) \tag{12.4}$$

Using the boundary condition $V = V_p$ at $y = a$ results in

$$V(a) = |V_p| = \frac{qN_D a^2}{2\varepsilon_r \varepsilon_0} \tag{12.5}$$

In practice, the saturation drain current depends on V_{gs} and can be approximated by

$$I_{ds,\text{sat}} = I_{dss}\left(1 - \frac{V_{gs}}{V_p}\right)^2 \tag{12.6}$$

and $I_{ds} = I_{ds,\text{sat}}$ for $V_{ds} > V_{ds,\text{sat}}$ (saturation region). I_{dss} is defined as

$$\begin{aligned} I_{dss} &= \text{saturation drain current at } V_{gs} = 0 \\ &= \frac{qN_D \mu h W V_p}{3L} \end{aligned} \tag{12.7}$$

where L is the gate length, W the channel width, and μ the electron mobility. Note that $I_{ds,\text{sat}} = 0$ when $V_{gs} = V_p$ and $I_{ds,\text{sat}} = I_{dss}$ when $V_{gs} = 0$ from Equation (12.6). The transconductance is defined as

$$g_m = \left.\frac{dI_{ds}}{dV_{gs}}\right|_{V_{ds}=\text{constant}} \quad \text{S} \tag{12.8}$$

From Equation (12.6) we have

$$g_m = -\frac{2I_{dss}}{V_p}\left(1 - \frac{V_{gs}}{V_p}\right) \tag{12.9}$$

For small V_{ds}, I_{ds} is approximately proportional to V_{ds} (linear region), given by

$$I_{ds} = A\left[1 - \left(\frac{V_{gs}}{V_p}\right)^{1/2}\right]V_{ds} \tag{12.10}$$

A is a constant depending on channel dimensions and the doping concentration.

12.4 SMALL-SIGNAL EQUIVALENT CIRCUIT

The small-signal equivalent circuit is useful for circuit designs at a low power level. Figure 12.5 shows the small-signal equivalent circuit and the location of the circuit element in the FET structure [2, 4–6]. The various components in the model are defined in the following:

Intrinsic Elements

R_i	input (channel) resistance
C_{gs}	gate-to-channel capacitance
C_{dg}	gate–drain feedback capacitance
R_{ds}	drain–source resistance
g_m	low-frequency transconductance
C_{dc}	drain-to-channel capacitance
τ_0	phase delay due to carrier transit in channel where carrier velocity is saturated ($E > E_p$, threshold electric field)

Extrinsic Elements

C_{ds}	drain–source capacitance
R_d	drain-to-channel resistance, including contact resistance
R_s	source-to-channel resistance, including contact resistance
R_g	gate–metal resistance

Parasitic inductances L_d, L_s, and L_g could be added in series with R_d, R_s, and R_g in the equivalent circuit to account for the effects of bonding wires. The values of these elements depend on the channel doping, channel

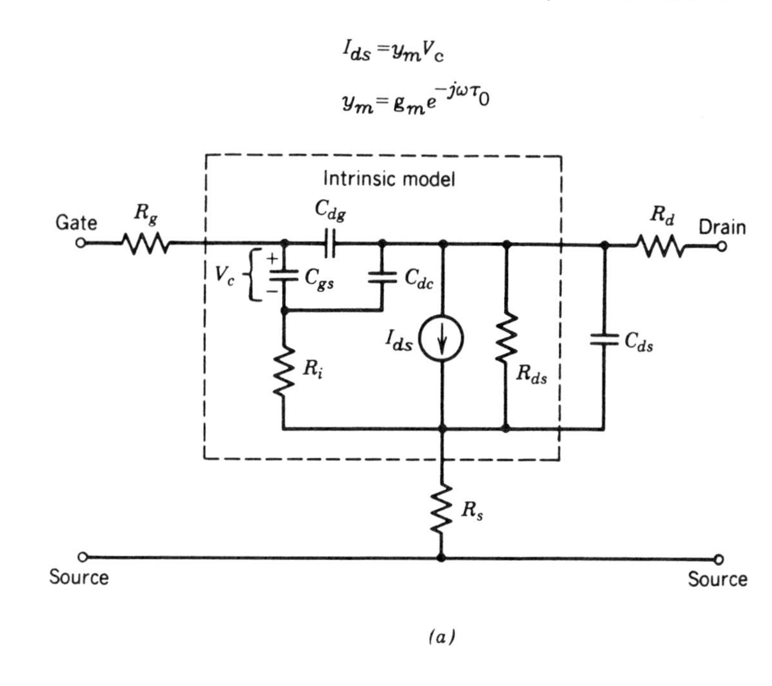

(a)

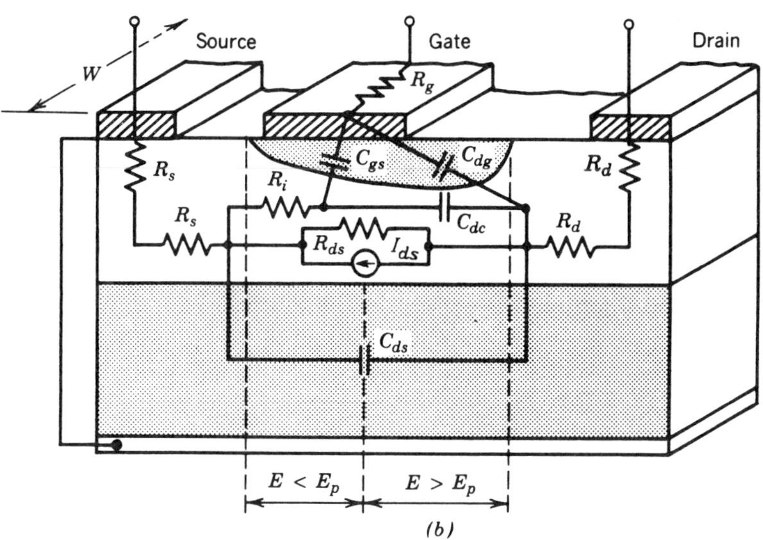

(b)

FIGURE 12.5 Small-signal equivalent circuit of an MESFET and the physical origin of the circuit elements. E_p is the threshold electric field. (From Ref. 6 with permission from IEEE.)

TABLE 12.1 Equivalent-Circuit Parameters of a Low-Noise GaAs MESFET with a 1- × 500-μm Gate (HP Experimental, $N_D = 1 \times 10^{17}$ cm^{-3})

Intrinsic Elements	dc Bias	Extrinsic Elements
$g_m = 53$ S		$C_{ds} = 0.12$ pF
$\tau_0 = 5.0$ ps		$R_g = 2.9$ Ω
$C_{gs} = 0.62$ pF		$R_d = 3$ Ω
$C_{dg} = 0.014$ pF		$R_s = 2.0$ Ω
$C_{dc} = 0.02$ pF		$L_g = 0.05$ nH[a]
$R_i = 2.6$ Ω		$L_d = 0.05$ nH[a]
$R_{ds} = 400$ Ω		$L_s = 0.04$ nH[a]
	$V_{ds} = 5$ V	
	$V_{gs} = 0$	
	$I_{ds} = 70$ mA	

Source: Ref. 6 with permission from IEEE.

[a]Contacting inductances of the test fixture in series with R_g, R_d, and R_s, respectively.

type, material, and dimensions. The large values of the extrinsic resistances will seriously decrease power gain and efficiency and increase the noise figure of the device. Typical element values for a GaAs MESFET with a 1-μm gate length and a 500-μm gate width are listed in Table 12.1.

The transit time for the saturated velocity case is given by

$$\tau = \frac{L}{v_s} \tag{12.11}$$

For a 1-μm gate length, the transit time in a GaAs FET is on the order of 10 ps. The cutoff frequency determined by the transit time is

$$f_{co} = \frac{1}{2\pi\tau} = \frac{v_s}{2\pi L} \tag{12.12}$$

From Equation (12.12), a GaAs MESFET has a higher cutoff frequency than that of a Si MESFET since the saturation drift velocity v_s is 2×10^7 cm/s for GaAs at an electric field of 3 kV/cm (see Figure 10.4) and 8×10^6 cm/s for silicon at 15 kV/cm.

The maximum frequency of oscillation, f_{max}, is defined as the frequency at which the unilateral power gain is reduced to unity and is given by [7]

$$f_{max} = \frac{f_{co}}{2}\sqrt{\frac{R_{ds}}{R_g}} \tag{12.13}$$

12.5 DEVICE OPTIMIZATION FOR LOW-NOISE APPLICATIONS

To optimize the MESFET device for low-noise applications, several critical equivalent-circuit elements and device parameters should be optimized, as follows [4]:

1. Reduce the gate length L.
2. Increase the transconductance g_m.
3. Reduce the gate-to-source capacitance C_{gs}.
4. Minimize the parasitic resistances R_g and R_s.

Equations (12.12) and (12.13) show that reducing the gate length increases the maximum operating frequency $f_{\max}$. Gate lengths of 0.2 to 0.5 μm are routinely fabricated using electron beam lithographics. Gate lengths as short as 0.1 μm have been demonstrated. Commercially available MESFETs have a 1-μm gate length operating up to 10 GHz, a 0.5-μm gate length up to 18 GHz, and a 0.2-μm gate length up to 30 GHz and millimeter-wave frequencies. Reducing the gate length effectively increases g_m, since g_m varies approximately as $L^{-1/3}$ [8]. A short gate length also contributes to a low gate-to-source capacitance.

To minimize the source resistance, one can recess the gate to increase the cross-sectional area of the channel between the source and gate, use the n^+ contact layer under the ohmic metal to reduce the contact resistance, offset the gate toward the source, and use a selectively implanted source contact. A figure showing these techniques is given in Figure 12.6.

To minimize the gate resistance, one can increase the gate metal thickness. A T- or mushroom-shaped gate permits a short gate while keeping the resistance low. Another approach is to use multiple gate fingers or feeds. Figure 12.7 shows these techniques.

12.6 DEVICE FABRICATION AND PACKAGING

The fabrication of a MESFET includes three major steps: (1) the growth and characterization of the semi-insulating substrate; (2) the growth and characterization of the active-device region, which consists of one or more thin n-type layers; and (3) device packaging and testing.

An active-device region can be grown by epitaxial growth or ion implantation. Epitaxial growth techniques include vapor-phase epitaxy (VPE), molecular beam epitaxy (MBE), metal organic chemical vapor deposition (MOCVD), and several hybrid versions of MBE and MOCVD [4]. Figure 12.8 summarizes the process flow sequence.

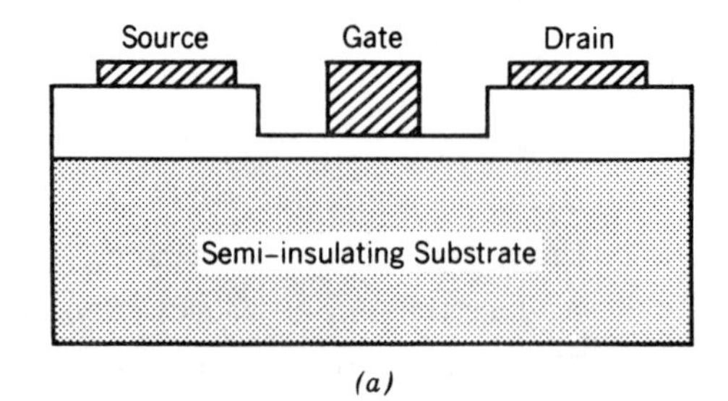

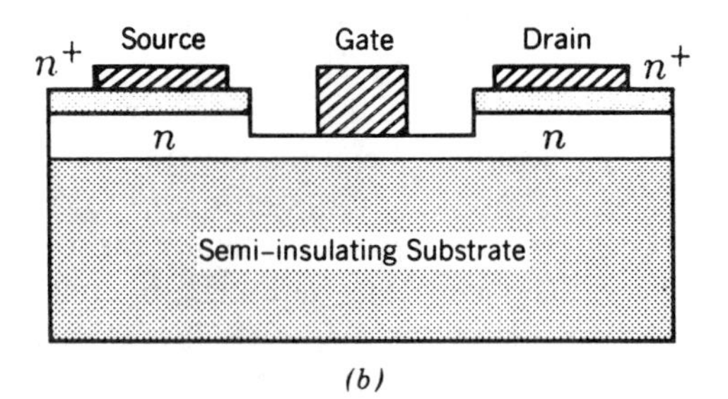

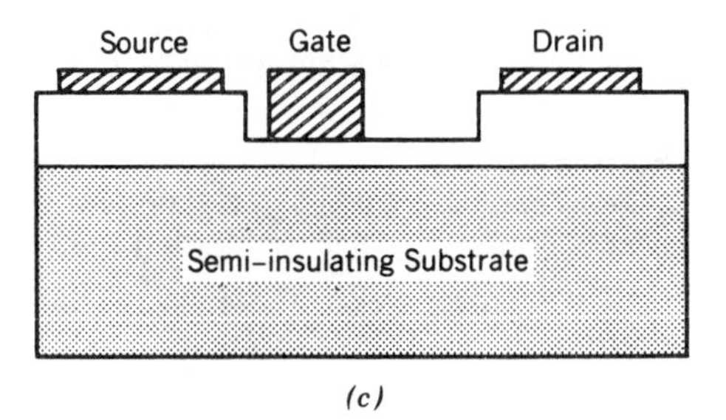

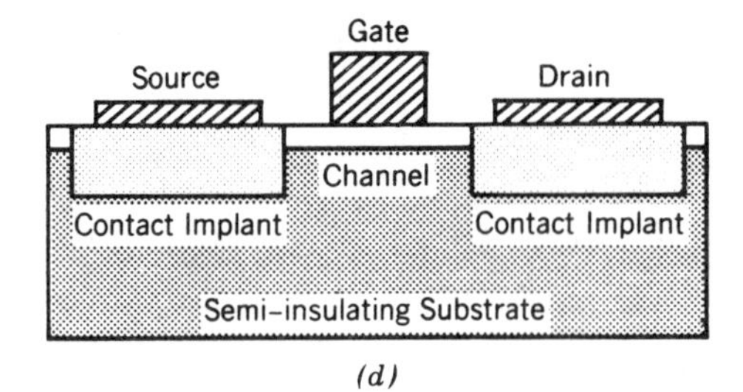

FIGURE 12.6 Techniques for reducing MESFET source resistance: (*a*) recessing the gate; (*b*) an n^+ contact under the ohmic metal reduces the contact resistance and hence the source resistance; (*c*) offsetting the gate toward the source; (*d*) selectively implanting ohmic contact layers. (From Ref. 4 with permission from Wiley.)

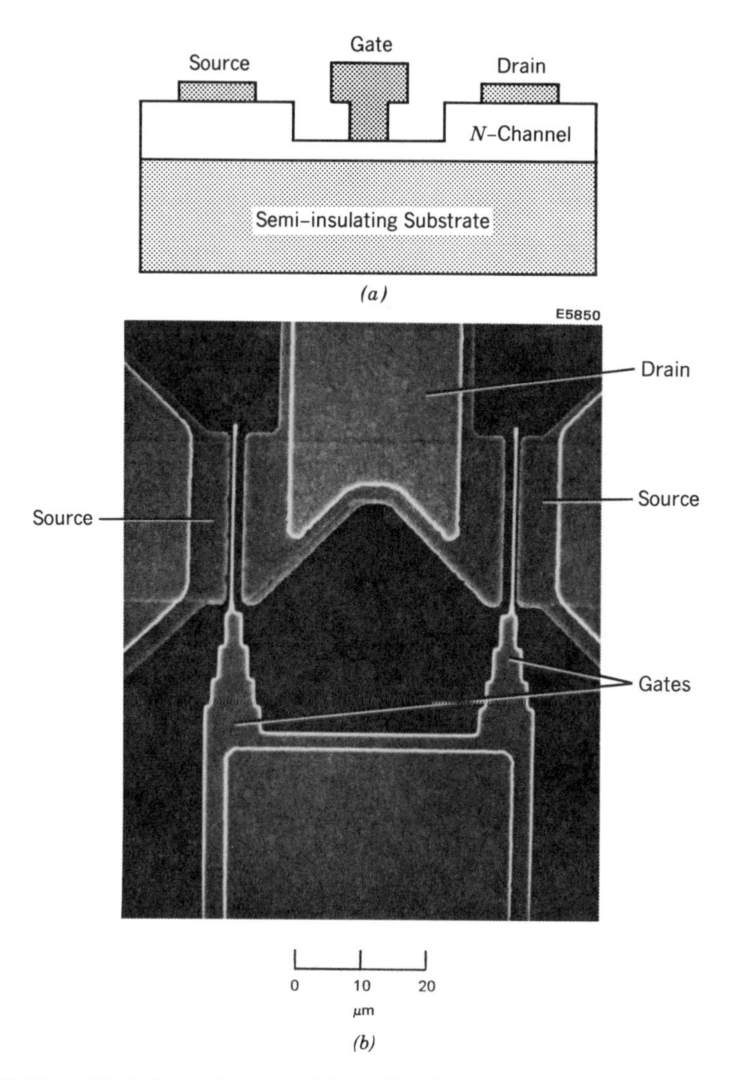

FIGURE 12.7 Techniques for minimizing MESFET gate resistance: (*a*) T- or mushroom-shaped gate metal cross sections; (*b*) multiple gate fingers and gate feeds. (From Ref. 4 with permission from Wiley.)

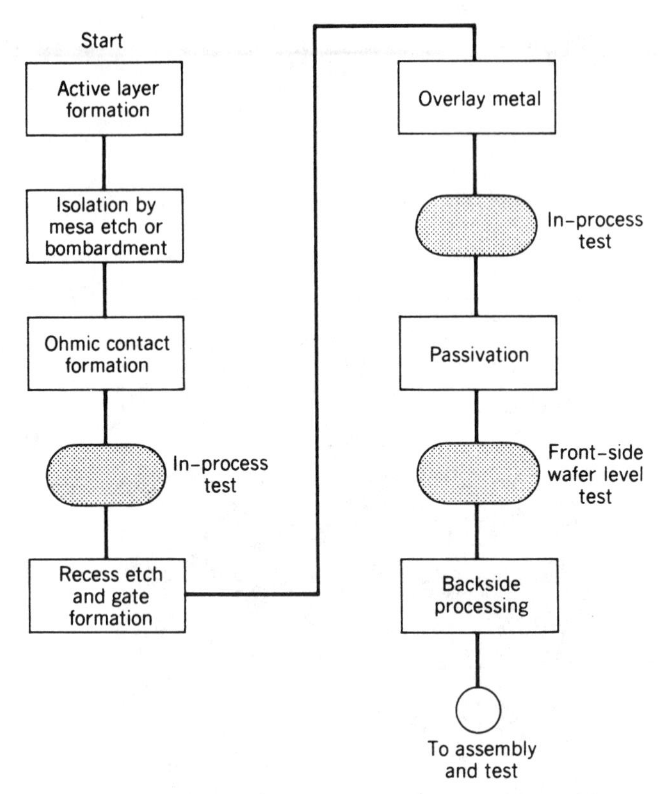

FIGURE 12.8 MESFET process flow sequence. (From Ref. 4 with permission from Wiley.)

After metallization, the device is ready for packaging if it is used as a discrete device. In a monolithic circuit, the individual device occupies a small part of a chip and will not be packaged. The top-view geometries of a low-noise and a power MESFET are shown in Figure 12.9. The power FET has a geometry consisting of many parallel gate fingers arranged so as to increase gate width. A large gate width is important for high power output and heat dissipation. Figure 12.10(*a*) shows a scanning electron micrograph of the top view of a low-noise FET, and Figure 12.10(*b*) shows a power FET.

The package should have a minimal effect on the overall device performance. Figure 12.11 shows various packaging configurations. The package in Figure 12.11(*a*) is typical for low-noise FETs and Figure 12.11(*b*) for power FETs. Figure 12.11(*c*) shows a chip carrier with a grounded source and 50-Ω input and output lines.

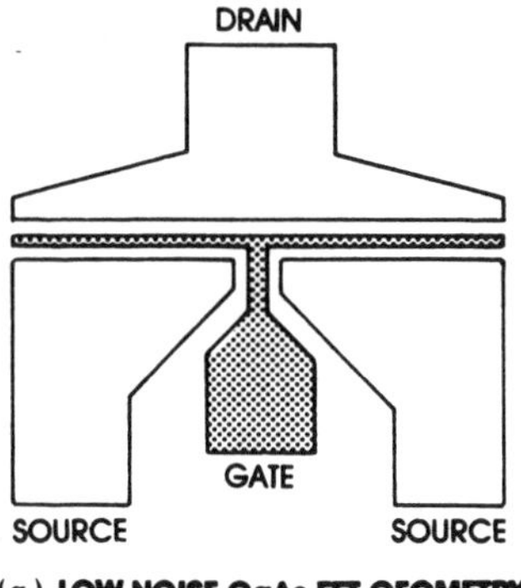

(*a*) LOW-NOISE GaAs FET GEOMETRY

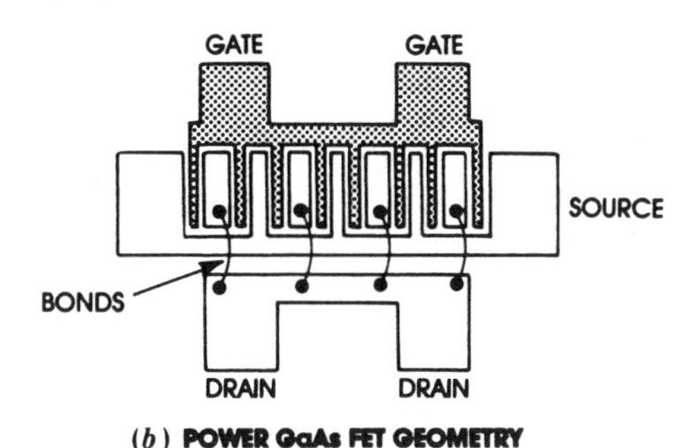

(*b*) POWER GaAs FET GEOMETRY

FIGURE 12.9 MESFET geometry. (From Ref. 9 with permission from Artech House.)

12.7 RF CHARACTERIZATION OF MESFETs [4]

DC characterization can be carried out using a curve tracer, and a typical display will look as shown in Figure 12.3. The RF characterization is more involved and usually carried out by two-port S-parameter measurements. This characterization is preferred because (1) a standard impedance (50 Ω) is easily provided at the input and output; (2) such terminations usually result in stable operation at all frequencies; (3) extensive computer software packages for device modeling and amplifier design are readily available; and (4) instrumentation and on-wafer probing systems are available for rapid RF characterization of devices by measuring their two-port S parameters.

Two-port common-source S parameters are usually measured for low-noise amplifier applications. This configuration, shown in Figure 12.12, also provides the maximum small-signal gain. Multistage low-noise amplifiers usually employ several gain stages after the first or second stage of low-noise amplification.

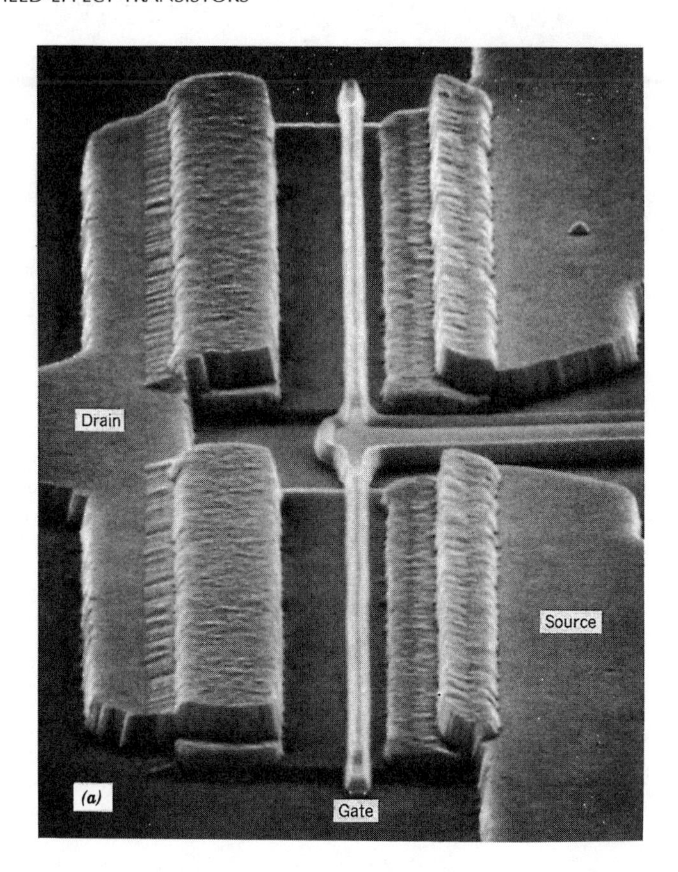

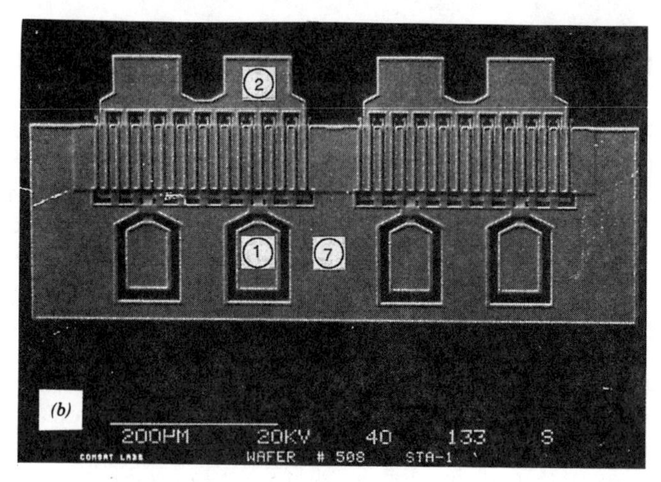

FIGURE 12.10 (*a*) Top view of a low-noise FET with a T-gate length of approximately 0.2 μm; (*b*) Top view of a power FET. [(*a*) From Ref. 4; (*b*) from Ref. 10 with permission from Wiley.]

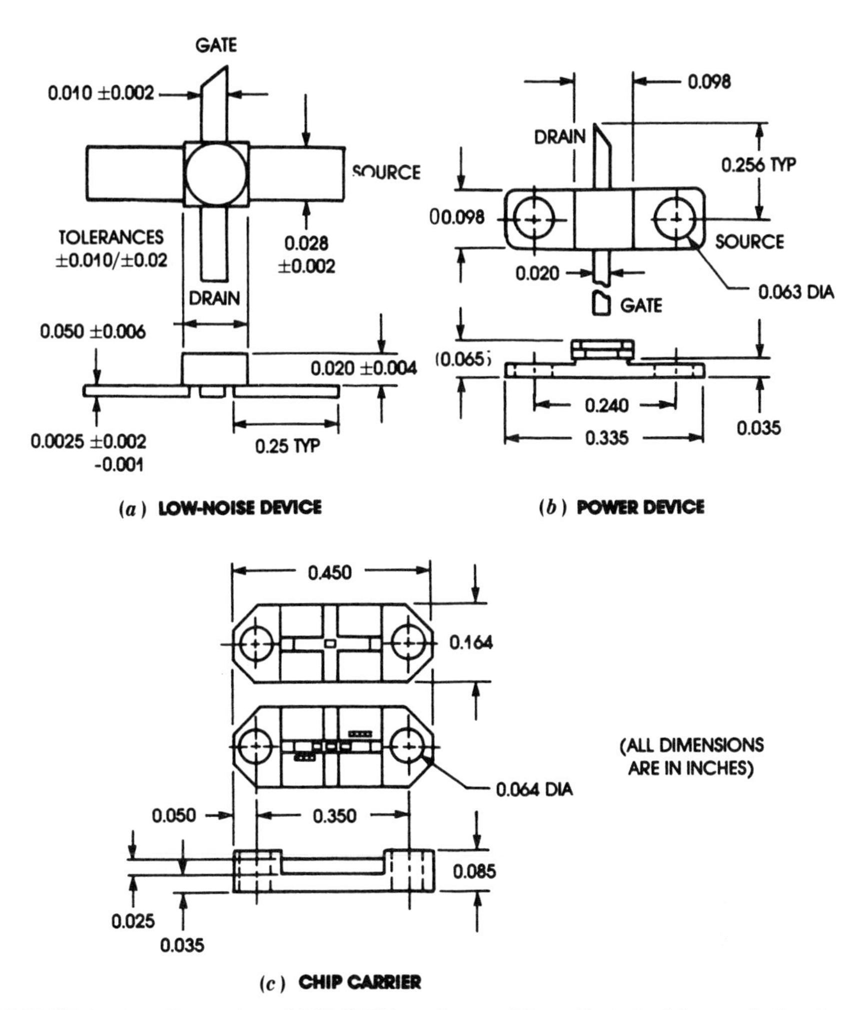

FIGURE 12.11 Examples of MESFET packages. (From Ref. 9 with permission from Artech House.)

With the exception of on-wafer measurements, S parameters are normally measured in a 50-Ω transmission line, usually microstrip. This configuration is shown in Figure 12.12. In this geometry, the position of the reference planes that define input gate inductance and the output drain inductance must be determined accurately by through-line or short-circuit measurements. In this way, the measured S parameters include the parasitic inductances of the bonding leads, which are also present when the chip is bonded into an amplifier. Bonding leads are kept as short as possible to minimize inductance so as to retain circuit bandwidth.

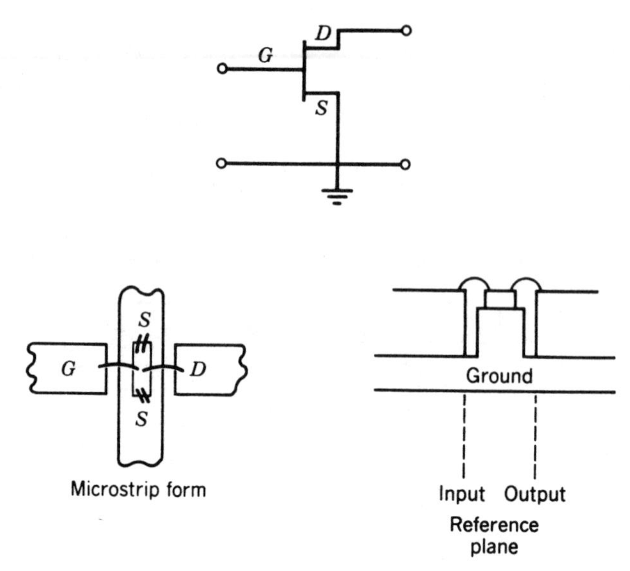

FIGURE 12.12 Common-source MESFET configuration. (From Ref. 4 with permission from Wiley.)

Recently, on-wafer probing systems capable of accurate, small-signal S-parameter characterization have become available. These probing systems are based on a tapered coplanar transmission-line system, which is used to make signal and ground contact to transistors at the wafer level. This technique offers several advantages:

1. It allows device evaluation preceding backside wafer processing.
2. It eliminates the need for device mounting.
3. It allows nondestructive device evaluation.
4. It greatly improves device evaluation throughput so that 100% RF evaluation is possible.

Figure 12.13 shows an RF wafer probe together with a typical probe footprint layout. To ensure accurate measurements, the device pad layout must be matched to the probe. Shown in the figure is a ground–signal–ground layout, which requires three contacts at both the input and output of the transistor. This configuration is favored in most applications.

To account for wafer probe loss, an accurate S-parameter characterization of the probe must be obtained. The wafer probe is a noninsertable device that has a miniature coaxial connector at one end and a coplanar footprint on the other. The probe's S parameters are obtained using standard network

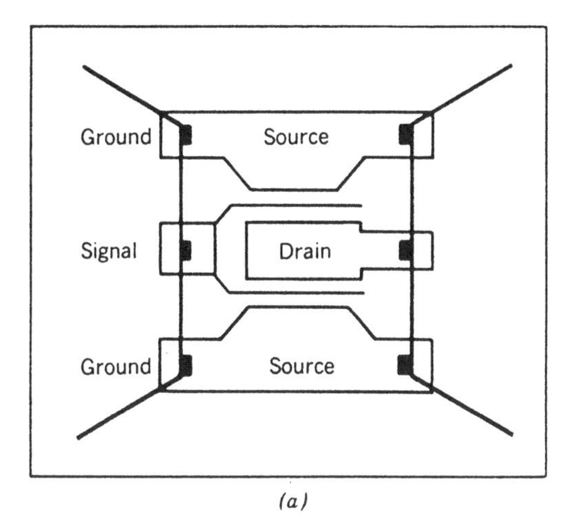

(a)

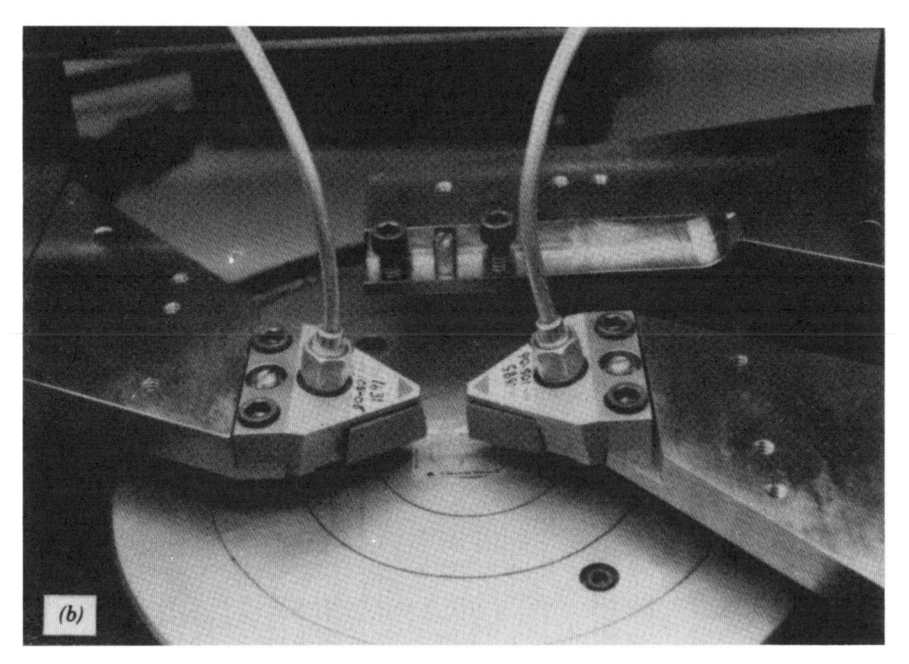

FIGURE 12.13 On-wafer RF probe: (*a*) ground–signal–ground probe footprint; (*b*) probe hardware. (From Ref. 4 with permission from Wiley.)

analyzer analysis two-port calibration procedures in conjunction with sets of on-wafer coplanar and coaxial standards (open, short, and load). Although the setup and calibration times for these measurements are significant, once completed, large numbers of devices can be characterized quickly, with little human labor and with computer-formatted output data.

REFERENCES

1. W. Shockley, "A Unipolar Field-Effect Transistor," *Proceedings of the IRE*, Vol. 40, November 1952, pp. 1365–1376.
2. S. M. Sze, *Physics of Semiconductor Devices*, 2nd ed., Wiley, New York, 1981, Chap. 6.
3. C. A. Mead, "Schottky-Barrier Gate Field-Effect Transistor," *Proceedings of the IEEE*, Vol. 54, February 1966, pp. 307–308.
4. T. A. Midford, "FETs: Low-Noise Applications," in *Handbook of Microwave and Optical Components*, K. Chang, Editor, Wiley, New York, 1990, Vol. 2, Chap. 11.
5. S. Y. Liao, *Microwave Solid-State Devices*, Prentice Hall, Englewood Cliffs, N.J., 1985, Chap. 4.
6. C. A. Liechti, "Microwave Field-Effect Transistor—1976," *IEEE Transactions on Microwave Theory and Techniques*, Vol. MTT-24, June 1976, pp. 279–300.
7. I. Bahl and P. Bhartia, *Microwave Solid-State Circuit Design*, Wiley, New York, 1988, Chap. 7.
8. H. Fukui, "Design of Microwave GaAs MESFETs for Broad-Band Low-Noise Amplifiers," *IEEE Transactions on Microwave Theory and Techniques*, Vol. MTT-27, July 1979, pp. 643–650.
9. T. S. Laverghetta, *Solid-State Microwave Devices*, Artech House, Norwood, Mass., 1987, Chap. 5.
10. H. A. Hung, T. Smith, and H. Huang, "FETs: Power Applications," in *Handbook of Microwave and Optical Components*, K. Chang, Editor, Wiley, New York, 1990, Vol. 2, Chap. 10.

CHAPTER 13

Bipolar Transistors, HEMTs, and HBTs

13.1 INTRODUCTION

Although the GaAs MESFET has become the most popular solid-state device, other types of transistors are also frequently used. In this chapter we describe some of these transistors: bipolar transistor, high electron mobility transistor (HEMT), and heterojunction bipolar transistor (HBT).

13.2 MICROWAVE SILICON BIPOLAR TRANSISTOR

A junction transistor consists of a silicon crystal in which a layer of p-type silicon is sandwiched between two layers of n-type silicon. This is called an *n-p-n transistor*. Alternatively, a *p-n-p transistor* can be formed. Figure 13.1 shows the schematics of both n-p-n and p-n-p transistors. Since the electron mobility in silicon is higher than the hole mobility, all microwave silicon transistors are of n-p-n type. The bipolar transistor can be considered as two p–n junctions cascaded together in which both majority and minority carriers are present. Both electron and hole movement contributes to the operation of a bipolar transistor. The microwave bipolar transistor can be considered as a scaled-up version of the low-frequency transistor.

Since the bipolar transistor was invented by Schockley in 1948, the technology has been improved steadily in low-frequency and digital applications. Significant microwave applications began in the 1960s with the commercially available discrete n-p-n silicon transistors having maximum frequencies of oscillation (f_{max}) in excess of 10 GHz [1, 2]. The advances in GaAs MESFET technology in the past two decades have, however, caused

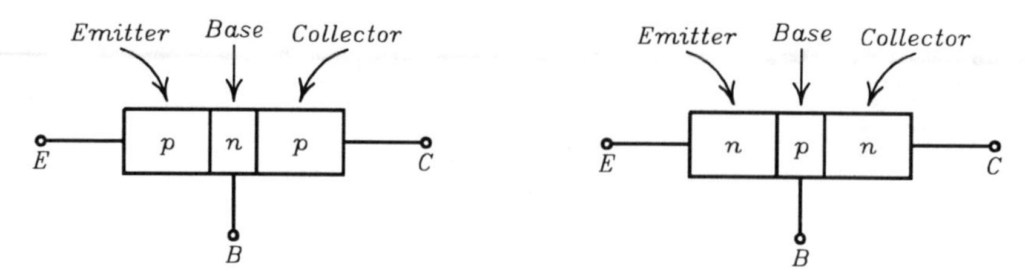

FIGURE 13.1 The *p-n-p* and *n-p-n* bipolar transistors.

the MESFET to take the lead in microwave applications due to its superior noise, gain, and power performance at high frequencies. The GaAs MESFET also has the advantages of higher electron mobility and effective saturated velocity, easier isolation of devices using the semi-insulating nature of GaAs, and high impedance. The use of bipolar silicon transistors is limited to frequencies below 4 GHz.

Figure 13.2 shows the step-by-step fabrication process starting with an *n*-type epitaxially grown silicon layer (2 to 5 μm thick). The *p*-type material (10^{17} to 10^{18} cm^{-3}) is diffused into the layer through window openings to form the base region. A shallow, heavily doped *n*-type material (10^{20} cm^{-3}) is diffused into the emitter opening. To achieve good microwave performance, the depths of the diffusions are kept very small. For example, the total junction depth of the base of a 2-GHz transistor is only on the order of 0.3 μm, and the depth for emitter is about 0.2 μm [3]. These small depths keep the transit delay time small.

The equivalent hybrid-π model of the intrinsic bipolar transistor is shown in Figure 13.3(*a*). Since r_{ce} is very large at microwave frequencies, the simplified model shown in Figure 13.3(*b*) can be used. Additional parasitic resistances, capacitances, and inductances should be included for an actual packaged transistor. Figure 13.3(*c*) shows the extrinsic circuits including the parasitic elements R_e, L_e, L_b, L_c, C_{be}, and C_{ce} [4].

The maximum frequency of oscillation ($f_{\max}$) is given by [4]

$$f_{\max} = \sqrt{\frac{f_T}{8\pi r_{b'e} C_{b'c}}} \tag{13.1}$$

where f_T is the gain–bandwidth frequency or cutoff frequency, which is the frequency where the short-circuit gain approximates unity. f_T can be expressed as [4]

$$f_T \approx \frac{g_m}{2\pi C_{b'e}} \tag{13.2}$$

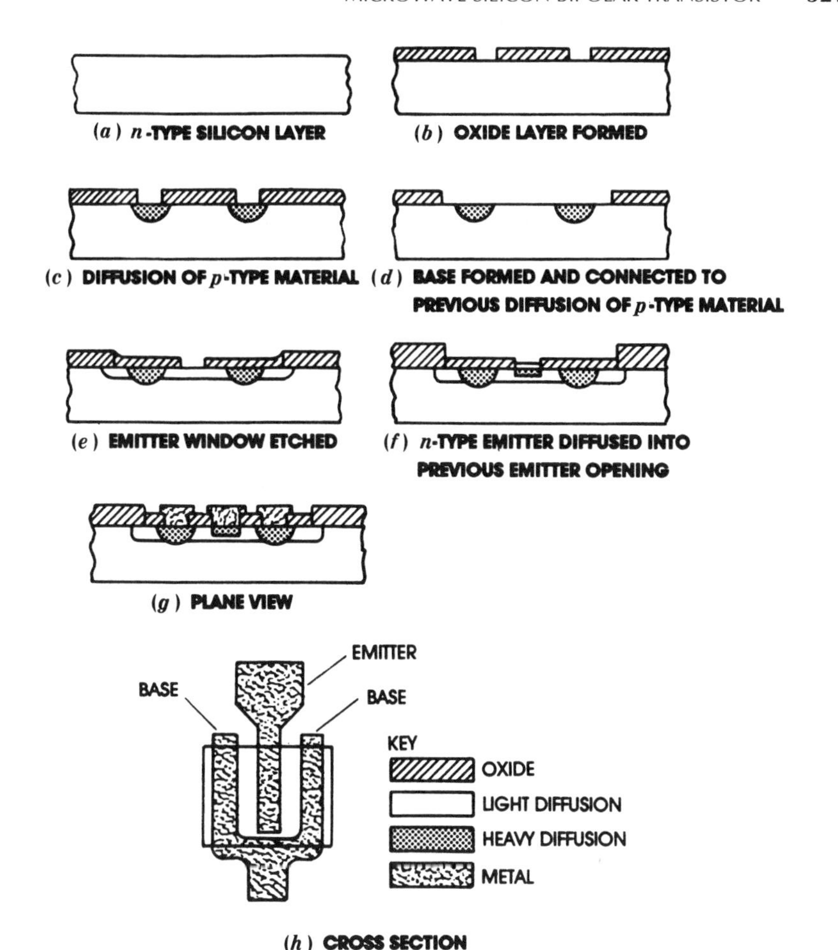

FIGURE 13.2 Transistor fabrication process. (From Ref. 3 with permission from Artech House.)

or expressed in terms of total signal time delay from the emitter to collector by [5]

$$f_T = \frac{1}{2\pi\tau_{ec}} \tag{13.3}$$

τ_{ec} can be represented approximately by the summation of the base delay time τ_b and the base-to-collector depletion-layer delay time τ_c [4]. This is written as

$$\tau_{ec} \approx \tau_b + \tau_c \tag{13.4}$$

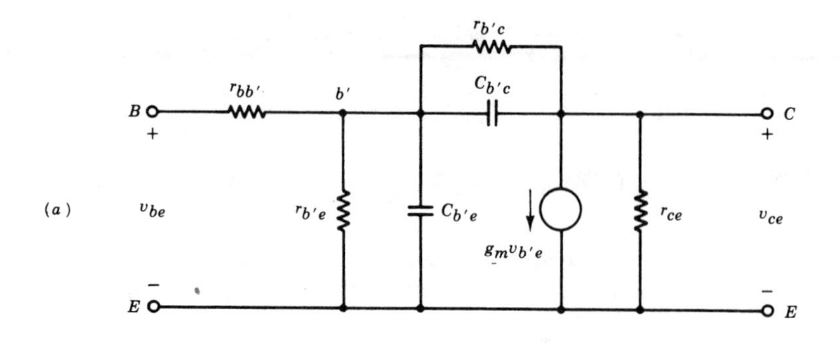

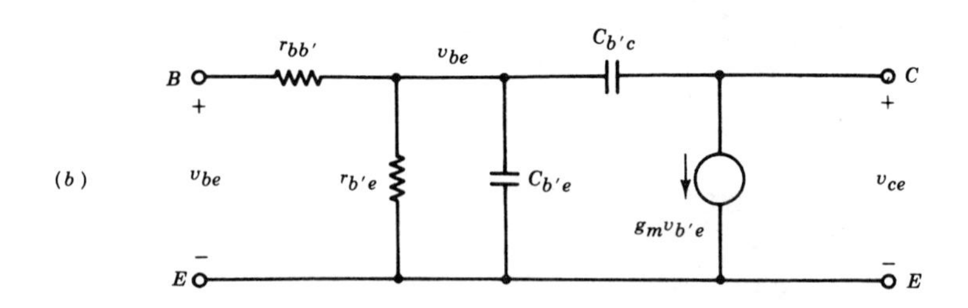

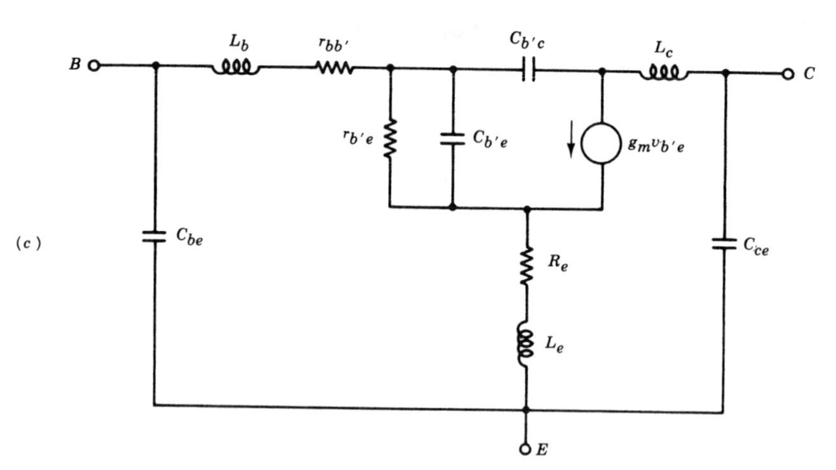

FIGURE 13.3 Equivalent circuit of a microwave bipolar transistor: (*a*) hybrid-π model of a common-emitter transistor; (*b*) simplified hybrid-π model; (*c*) model including packaged parasitics. (From Ref. 4 with permission from Prentice Hall.)

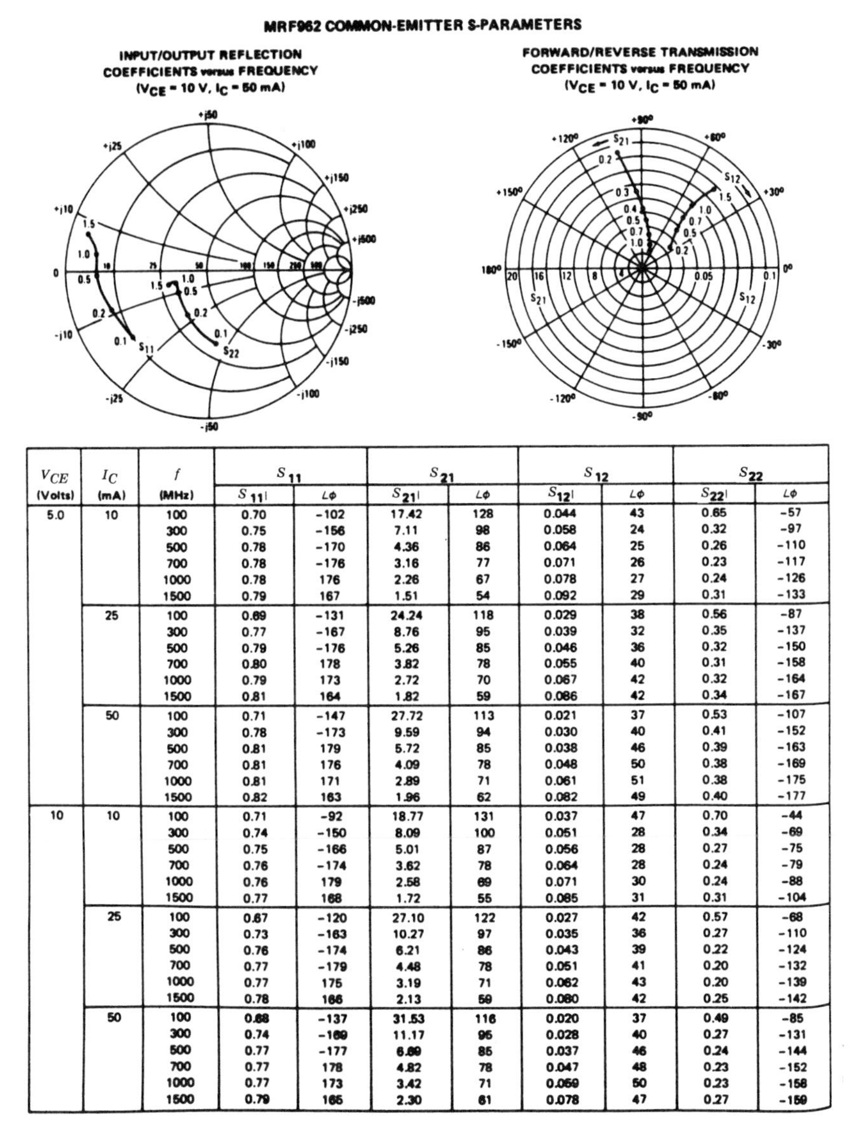

V_{CE} (Volts)	I_C (mA)	f (MHz)	S_{11} $\|S_{11}\|$	S_{11} $\angle\phi$	S_{21} $\|S_{21}\|$	S_{21} $\angle\phi$	S_{12} $\|S_{12}\|$	S_{12} $\angle\phi$	S_{22} $\|S_{22}\|$	S_{22} $\angle\phi$
5.0	10	100	0.70	-102	17.42	128	0.044	43	0.65	-57
		300	0.75	-156	7.11	98	0.058	24	0.32	-97
		500	0.78	-170	4.36	86	0.064	25	0.26	-110
		700	0.78	-176	3.16	77	0.071	26	0.23	-117
		1000	0.78	176	2.26	67	0.078	27	0.24	-126
		1500	0.79	167	1.51	54	0.092	29	0.31	-133
	25	100	0.69	-131	24.24	118	0.029	38	0.56	-87
		300	0.77	-167	8.76	95	0.039	32	0.35	-137
		500	0.79	-176	5.26	85	0.046	36	0.32	-150
		700	0.80	178	3.82	78	0.055	40	0.31	-158
		1000	0.79	173	2.72	70	0.067	42	0.32	-164
		1500	0.81	164	1.82	59	0.086	42	0.34	-167
	50	100	0.71	-147	27.72	113	0.021	37	0.53	-107
		300	0.78	-173	9.59	94	0.030	40	0.41	-152
		500	0.81	179	5.72	85	0.038	46	0.39	-163
		700	0.81	176	4.09	78	0.048	50	0.38	-169
		1000	0.81	171	2.89	71	0.061	51	0.38	-175
		1500	0.82	163	1.96	62	0.082	49	0.40	-177
10	10	100	0.71	-92	18.77	131	0.037	47	0.70	-44
		300	0.74	-150	8.09	100	0.051	28	0.34	-69
		500	0.75	-166	5.01	87	0.056	28	0.27	-75
		700	0.76	-174	3.62	78	0.064	28	0.24	-79
		1000	0.76	179	2.58	69	0.071	30	0.24	-88
		1500	0.77	168	1.72	55	0.085	31	0.31	-104
	25	100	0.67	-120	27.10	122	0.027	42	0.57	-68
		300	0.73	-163	10.27	97	0.035	36	0.27	-110
		500	0.76	-174	6.21	86	0.043	39	0.22	-124
		700	0.77	-179	4.48	78	0.051	41	0.20	-132
		1000	0.77	175	3.19	71	0.062	43	0.20	-139
		1500	0.78	166	2.13	59	0.080	42	0.25	-142
	50	100	0.68	-137	31.53	116	0.020	37	0.49	-85
		300	0.74	-169	11.17	96	0.028	40	0.27	-131
		500	0.77	-177	6.69	85	0.037	46	0.24	-144
		700	0.77	178	4.82	78	0.047	48	0.23	-152
		1000	0.77	173	3.42	71	0.059	50	0.23	-158
		1500	0.79	166	2.30	61	0.078	47	0.27	-159

FIGURE 13.4 S-parameter data for the Motorola MRF962 transistor. (From *Motorola RF Data Manual*; reproduced with permission of Motorola, Inc.)

The RF characterization and S-parameter measurements can be carried out using the methods described in Section 12.7 for the GaAs MESFET. Figure 13.4 shows an example of the S-parameter data for a bipolar transistor.

TABLE 13.1 Heterojunction FET Names and Origin

Acronym	Name	Device Aspect	Origin
HEMT	High electron mobility transistor	Electron mobility	Fujitsu
MODFET	Modulation doped FET	Epitaxial layer doping	Cornell, University of Illinois, Rockwell
TEGFET	Two-dimensional electron Gas FET	Electron distribution	Thomson CSF
SDHT	Selectively doped hetero-structure transistor	Epitaxial layer doping	AT & T Bell Labs

Source: Ref. 7 with permission from Artech House.

13.3 HIGH ELECTRON MOBILITY TRANSISTOR

High electron mobility transistors (HEMTs) and heterojunction bipolar transistors belong to a new class of III–V semiconductor devices using heterojunctions for their operation. Unlike conventional devices, which use junctions between like materials, heterojunction devices utilize junctions formed between semiconductors of different compositions, such as GaAs/AlGaAs and InGaAs/InP. Heterojunction device design has an additional degree of freedom since one can vary the band structure as well as the doping level.

With the discovery of two-dimensional electron gas (2-DEG) at the *n*-AlGaAs/GaAs heterojunction in 1978 [6] and with the advances in the molecular beam epitaxy (MBE) growth technique for III–V compound materials, great progress has been made in HEMT devices in the United States, Japan, and Europe. Various names have been used, including modulation-doped FET (MODFET), two-dimensional electron gas FET (TEGFET), selectively doped heterojunction transistor (SDHT), and heterojunction FET (HFET). Table 13.1 lists the origins of these names used interchangeably for this device.

13.4 OPERATION PRINCIPLES OF HEMTs

In a normal homojunction MESFET, the electrons move in the active layer, which is an *n*-type GaAs layer with a doping of 10^{17} cm^{-3}. As a result, the electron mobility is degraded due to strong ionized impurity scattering [8]. This fact becomes exceedingly important in submicrometer structures with a gate length of less than 1 μm. As the gate length decreases, the doped layer must be made thinner and the doping density made higher in such a way that good control by the gate can be preserved and that the threshold voltage can be maintained constant. Therefore, the scattering problem becomes more severe for submicron devices. Use of one or more heterojunctions to separate

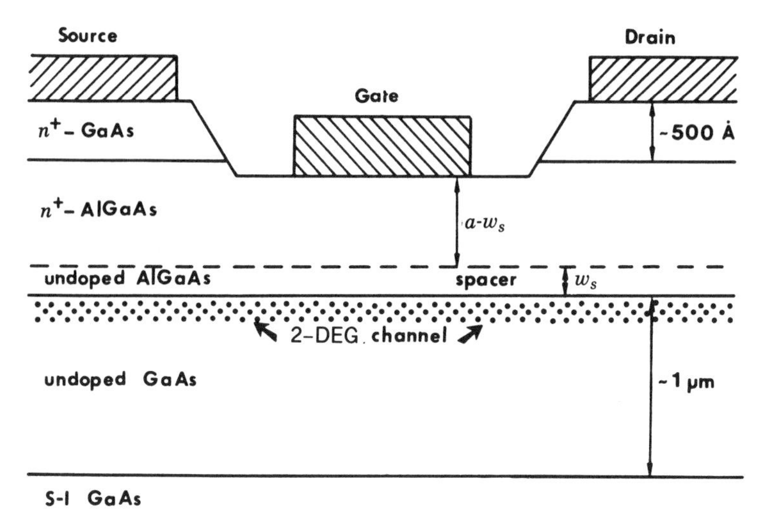

FIGURE 13.5 Typical structure of HEMT. (From Ref. 8 with permission from Wiley.)

the moving carriers from their parent impurities was proposed as a solution for achieving high mobility in III–V compounds by doping only the large-band-gap material (i.e., selective doping or modulated doping).

Figure 13.5 shows a typical structure of a HEMT device. The main difference from the conventional MESFET is the presence of a heterojunction formed at the interface of an AlGaAs doped alloy [$Al_{x_{Al}}Ga_{1-x_{Al}}As = x_{Al}(AlAs) + (1 - x_{Al})GaAs$] grown on an undoped GaAs layer. AlAs and GaAs have nearly the same lattice constant; therefore, any combination of them will provide a good lattice match at the interface. The ratio of Al to Ga in the AlGaAs is typically 25% Al and 75% Ga, although other compositions are also widely used [7]. This can be written as $Al_{0.25}Ga_{0.75}As$. Due to the higher band gap of AlGaAs compared to that in the adjacent GaAs region, free electrons diffuse from the AlGaAs into GaAs and form a two-dimensional electron gas (2-DEG) at the heterojunction. The transport properties of this (2-DEG) region are considerably superior to those of free electrons in a conventional MESFET, where the channel region must be doped to obtain the charge carriers. Due to the absence of ionized donors in the channel of a HEMT, electrons forming the 2-DEG suffer little Coulomb scattering and enjoy high mobility. This is an improvement over the MESFET, for which electron mobility is limited by scattering with ionized donors present in the channel [7]. Possible advantages of HEMTs are given in Table 13.2. Recent advancements in sub-quarter-micron lithography and materials technology have opened new avenues for further improvement over the basic HEMT structure. Table 13.3 lists some important variations of basic HEMT structures.

TABLE 13.2 Advantages of HEMTs[a]

- High electron mobility
- Small source resistance
- High f_T due to high electron velocity in large electric fields
- High transconductance due to small gate-to-channel separation
- High output resistance
- Higher Schottky barrier height due to deposition of Schottky metal on AlGaAs instead of on GaAs

Source: Ref. 7 with permission from Artech House.

[a]Many of these advantages are significantly enhanced as the device is cooled.

TABLE 13.3 Variations of Basic HEMT Structures

Salient Features or Name	Description
Graded (Al,Ga)As layer	(Al,Ga)As donor layer graded in composition from (Al,Ga)As at 2-DEG interface to GaAs at cap interface to facilitate ohmic contact formation
Inverted HEMT structure	Undoped GaAs drift layer grown on top of doped (Al,Ga)As layer
Multiple-heterojunction HEMT	Epitaxial layer structure with more than one heterointerface providing electrons to the 2-DEG
Planar doped layer	Doping of the donor layer confined to a single atomic layer
Quantum well drift layer	Thin (300–400 Å) low band-gap drift region confined on both sides by potential barrier formed by hetero-junction with higher band-gap materials
Superlattice buffer layer	Buffer layer composed of undoped (Al,Ga)As/GaAs layers [e.g., 30 periods of AlAs (15 Å)/GaAs (25 Å)]
Superlattice donor layer	Superlattice donor layer [e.g., (Al,Ga)As/GaAs] with doping confined to GaAs
Pseudomorphic HEMT	Drift region comprising low-band-gap, high-mobility layer, *not* lattice matched to donor layer (e.g., $Al_{0.25}Ga_{0.75}As/In_{0.53}Ga_{0.47}As/GaAs$ layers)
(Al,In)As/(Ga,In)As /InP HEMT	Lattice-matched HEMT epitaxial layer with (Al,In)As donor layer and (Ga,In)As drift layer on semi-insulating InP substrate
HEMT on silicon substrate	HEMT epitaxial layer grown on a silicon substrate
Heterostructure insulated gate FET (HIGFET)	No donor layer, 2-DEG induced in channel by varying gate electrode potential, similar to MOSFETs
p-Channel HEMT	Donor layer doped with Be to provide a 2-DEG hole gas at interface

Source: Ref. 7 with permission from Artech House.

HEMTs have been used up to 94 GHz for low-noise and power amplifier applications. A device noise figure of 1.8 dB at 60 GHz was reported using AlGaAs/GaAs HEMTs [9]. Using InAlAs/InGaAs devices, minimum noise figures of 0.5, 1.2, and 2.1 dB have been demonstrated at 18, 60, and 94 GHz, respectively. [10].

13.5 PSEUDOMORPHIC HEMTs

The pseudomorphic HEMTs use InGaAs as the two-dimensional electron gas channel material instead of GaAs as shown in Figure 13.6. The potential advantages of using a thin InGaAs layer as the pseudomorphic channel are:

1. Enhanced electron transport in InGaAs compared to GaAs
2. Improved confinement of carriers in the channel
3. Layer conduction band discontinuity at the AlGaAs/InGaAs heterojunction, which allows higher sheet charge density, higher current density, and transconductance

The higher transconductance results in higher forward transducer gain, and thus higher power gain and larger gain–bandwidth product. The pseudomor-

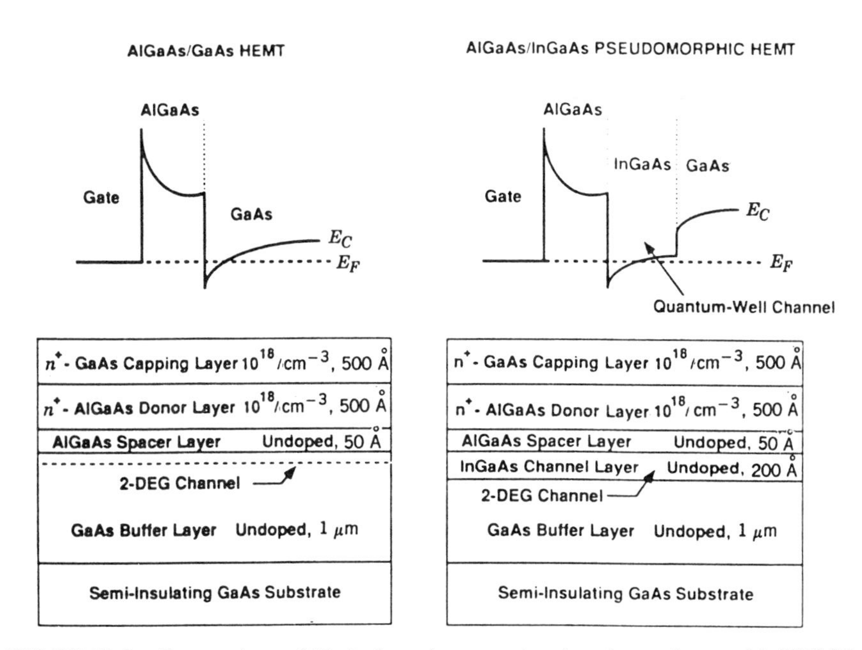

FIGURE 13.6 Comparison of GaAs-based conventional and pseudomorphic HEMTs. (From Ref. 7 with permission from Artech House.)

phic HEMTs are suitable for higher-frequency and broadband applications. A maximum frequency of oscillation (f_{max}) of 350 GHz and a unity current-gain cutoff frequency (f_T) of 150 GHz have been reported [11, 12].

13.6 HETEROJUNCTION BIPOLAR TRANSISTOR

The heterojunction bipolar transistor (HBT) is a modified bipolar transistor that relies on the use of heterojunctions for its operation. The HBT was first proposed by Kroemer in 1957 [13]. In the HBT, the emitter and base are formed in a semiconductor with different band gaps. The emitter is in the wider-band-gap material. The structure provides a potential barrier for the injection of holes, in an *n-p-n* transistor, into the emitter while facilitating electron injection into the base [7]. A high emitter injection efficiency can be maintained even with heavy base doping (to reduce base spreading resistance) and light emitter doping (to reduce base–emitter junction capacitance). Parasitics resistances and capacitances in the HBT can be significantly lower than those in its silicon counterpart, with resultant improvements in performance at high frequencies.

Figure 13.7 shows a schematic cross section of a typical *n-p-n* HBT structure. An *n*-type emitter is formed in the wide-band-gap AlGaAs, while the *p*-type base is formed in the lower-band-gap GaAs. The *n*-type collector is also formed in GaAs. A heavily doped n^+ GaAs layer is used between the emitter contact and the AlGaAs layer to form an ohmic contact.

HBTs can be used for the same applications as silicon bipolar transistors except that they offer higher speed and higher operating frequency. Applica-

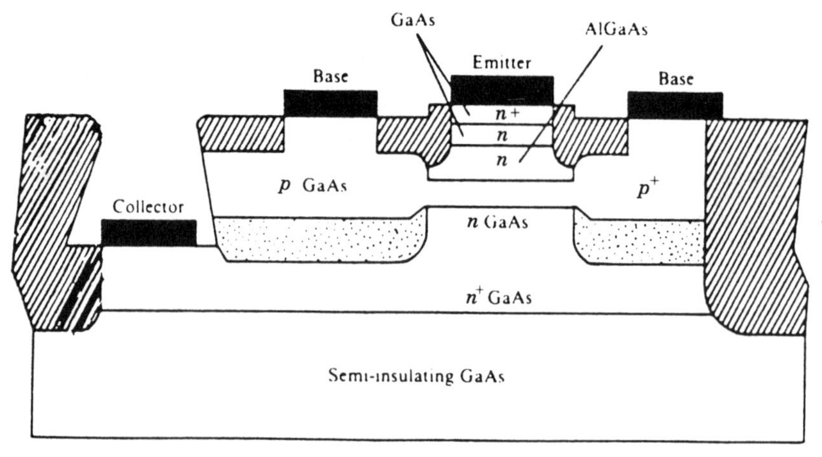

FIGURE 13.7 HBT device structure. (From Ref. 14 with permission from TAB Books.)

TABLE 13.4 Advantages of HBTs over Silicon Bipolar Transistors

- Due to the wide-band-gap emitter, a much higher base doping concentration can be used, decreasing base resistance.
- Emitter doping can be lowered and minority-carrier storage in the emitter can be made negligible, reducing base–emitter capacitance.
- High electron mobility, built-in drift fields, and velocity overshoot combine to reduce electron transit time.
- Semi-insulating substrates help reduce pad parasitics and allow convenient integration of devices.
- Early voltages are higher and high injection effects are negligible due to high base doping.

Source: Ref. 7 with permission from Artech House.

TABLE 13.5 Advantages of HBTs over MESFETs

- The key distances that govern electron transit time are established by epitaxial growth, not by lithography, which allows high f_T with modest processing requirements.
- The entire emitter area conducts current, leading to high current-handling capability and high transconductance.
- The "control region" of the device, the base–emitter junction, is very well shielded from the output voltage, leading to low output conductance; taken together with the high transconductance, enormous values of voltage amplification factor g_m/g_0 are attainable.
- Breakdown voltage is controllable directly by the epitaxial structure of the device.
- The threshold voltage for output current flow is governed by the built-in potential of the base–emitter junction, leading to well matched characteristics.
- The device is well shielded from traps in the bulk and surface regions, contributing to low $1/f$ noise and absence of trap-induced frequency dependence of output resistance or current lag phenomena.

Source: Ref. 7 with permission from Artech House.

tions in microwaves include power amplifiers, oscillators, and mixers. Tables 13.4 and 13.5 compare HBTs with silicon bipolar transistors and MESFETs [7, 15].

13.7 CONCLUSIONS

The technology in transistors has been advancing very rapidly in the past 20 years. This has created many new applications in microwave circuits. The state-of-the-art performance of these devices is changing every day. Comparisons of various transistors can be found in many literatures [7, 16]. For a circuit designer, however, the fundamental design theory applies regardless of what type of transistor is used. In subsequent chapters, applications of transistors to circuit design are presented.

REFERENCES

1. H. F. Cooke, "Microwave Transistors: Theory and Design," *Proceedings of the IEEE*, Vol. 59, August 1971, pp. 1163–1181.
2. C. P. Snapp, "Microwave Silicon Bipolar Transistors and Monolithic Integrated Circuits," in *Handbook of Microwave and Optical Components*, K. Chang, Editor, Wiley, New York, 1990, Vol. 2, Chap. 8.
3. T. S. Laverghetta, *Solid-State Microwave Devices*, Artech House, Norwood, Mass., 1987, Chap. 5.
4. G. Gonzalez, *Microwave Transistor Amplifiers*, Prentice Hall, Englewood Cliffs, N.J., 1984, Chap. 1.
5. S. M. Sze, *Physics of Semiconductor Devices*, 2nd ed., Wiley, New York, 1981, Chap. 3.
6. R. Dingle, H. L. Stormer, A. C. Gossard, and W. Wiexmann, "Electron Mobility in Modulation-Doped Semiconductor Superlattices," *Applied Physics Letters*, Vol. 33, 1978, pp. 665–667.
7. F. Ali and A. Gupta, Editors, *HEMTs & HBTs: Devices, Fabrication, and Circuits*, Artech House, Norwood, Mass., 1991.
8. J. Zimmermann and G. Salmer, "High-Electron-Mobility Transistors: Principles and Applications," in *Handbook of Microwave and Optical Components*, K. Chang, Editor, Wiley, New York, 1990, Vol. 2, Chap. 9.
9. K. H. G. Duh, P. C. Chao, P. M. Smith, L. F. Lester, B. R. Lee, J. M. Ballingall, and Y. M. Kao, "Millimeter-Wave Low Noise Amplifiers," *IEEE MTT-S International Microwave Symposium Digest*, 1988, pp. 923–926.
10. K. H. G. Duh, P. C. Chao, P. Ho, M. Y. Kao, P. M. Smith, J. M. Ballingall, and A. A. Jabra, "High Performance InP-Based HEMT Millimeter-Wave Low Noise Amplifiers," *IEEE MTT-S International Microwave Symposium Digest*, 1989, pp. 805–808.
11. L. D. Nguyen, P. J. Tasker, D. C. Radulescu, and L. F. Eastman, "Characterization of Ultra-High-Speed Pseudomorphic AlGaAs/InGaAs (on GaAs) MODFETs," *IEEE Transactions on Electron Devices*, Vol. ED-36, October 1989, pp. 2243–2248.
12. L. F. Lester, P. M. Smith, P. Ho, P. C. Chao, R. C. Tiberio, K. H. G. Duh, and E. D. Wolf, "0.15 μm Gate-Length Double Recess Pseudomorphic HEMT with f_{max} of 350 GHz," *IEEE International Electron Devices Meeting Technical Digest*, 1988, p. 172.
13. H. Kroemer, "Theory of a Wide-Gap Emitter for Transistors," *Proceedings of the IRE*, Vol. 45, No. 11, November 1957, pp. 1535–1537.
14. N. Sclater, *Galliam Arsenide IC Technology*, TAB Books, Blue Ridge Summit, Pa., 1988.

15. P. M. Asbeck, M. F. Chang, K. C. Wang, D. C. Miller, G. J. Sullivan, N. H. Sheng, E. Sovero, and J. A. Higgins, "Heterojunction Bipolar Transistors for Microwave and Millimeter-Wave Integrated Circuits," *IEEE Transactions on Microwave Theory and Techniques*, Vol. MTT-35, No. 12, December 1987, pp. 1462–1470.

16. J. Berenz and B. Dunbridge, "MMIC Device Technology for Microwave Signal Processing Systems," *Microwave Journal*, Vol. 31, No. 4, April 1988, pp. 115–131.

CHAPTER 14

Transistor Amplifiers

14.1 INTRODUCTION

In this chapter we introduce some basic design principles used for microwave transistor amplifiers. The design procedure is based on the S parameters of the transistor. The transistor could be a bipolar transistor, a FET, a HEMT, or a HBT. Although modern design procedures are usually carried out using computer-aided design tools, the use of equations and graphs will help in understanding some fundamental design principles.

The most important design considerations for a microwave transistor amplifier are power gain, noise, bandwidth, stability, and bias arrangements. The design normally starts with a set of specifications and proper selection of the transistor. The input- and output-matching networks are designed to achieve the required stability, amplifier gain, noise, and bandwidth. DC bias circuits are designed not to disturb the RF performance.

An amplifier should not oscillate in the operating bandwidth. An unconditionally stable transistor amplifier will not oscillate under any passive termination of the input or output circuits. A potentially unstable transistor could oscillate under certain passive terminations and requires careful design to avoid oscillation.

A microwave transistor amplifier can be classified as a low-noise amplifier or a power amplifier. The major consideration in a low-noise amplifier design is to achieve low noise and in a power amplifier design is to achieve high output power and efficiency.

14.2 POWER GAIN

Figure 14.1 shows a microwave transistor amplifier. The circuit consists of an input-matching network, a transistor, and an output-matching network. The

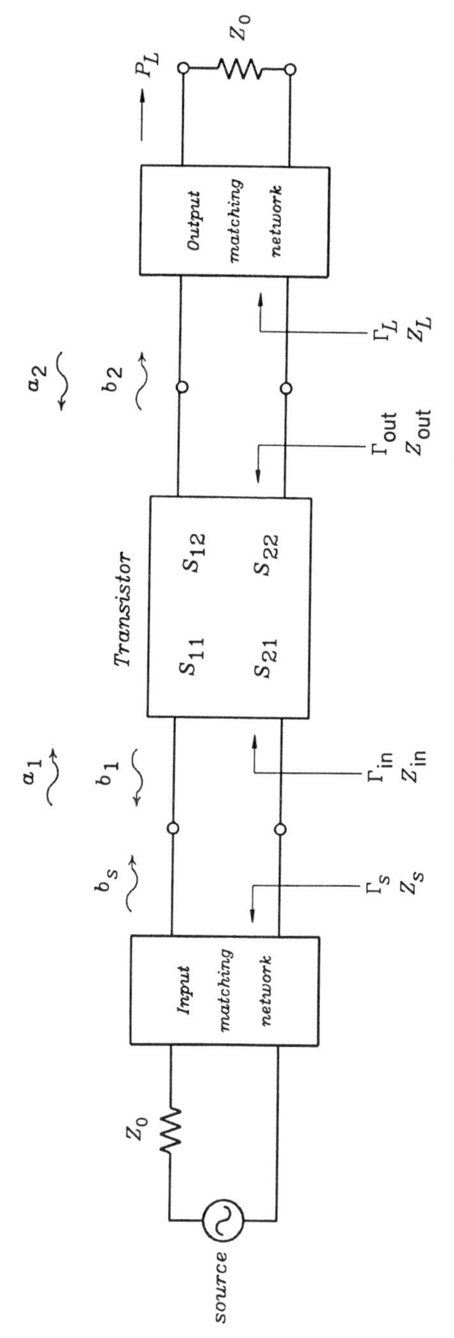

FIGURE 14.1 Transistor amplifier circuit.

amplifier is connected to a source with a characteristic impedance of Z_0 and a load with the same characteristic impedance.

Four equations can be written governing this circuit:

$$b_1 = S_{11}a_1 + S_{12}a_2 \tag{14.1}$$

$$b_2 = S_{21}a_1 + S_{22}a_2 \tag{14.2}$$

$$a_1 = b_s + \Gamma_s b_1 \tag{14.3}$$

$$a_2 = b_2\Gamma_L \tag{14.4}$$

Equations (14.1) and (14.2) are obtained from the definition of S parameters of the transistor, which is a two-port network. Γ_s and Γ_L are the reflection coefficients looking into the source and the load, respectively, from the transistor. From Equations (14.1) through (14.4) we can derive:

From (14.3): $b_s = a_1 - \Gamma_s b_1$

From (14.1): $= a_1 - \Gamma_s S_{11}a_1 - \Gamma_s S_{12}a_2$

From (14.4): $= a_1 - \Gamma_s S_{11}a_1 - \Gamma_s S_{12}b_2\Gamma_L$

From (14.2): $= (1 - \Gamma_s S_{11})\dfrac{b_2 - S_{22}a_2}{S_{21}} - \Gamma_s S_{12}b_2\Gamma_L$

From (14.4): $= (1 - \Gamma_s S_{11})\dfrac{b_2 - S_{22}b_2\Gamma_L}{S_{21}} - \Gamma_s S_{12}b_2\Gamma_L$

$$= b_2\frac{(1 - \Gamma_s S_{11}) - S_{22}(1 - \Gamma_s S_{11})\Gamma_L - \Gamma_s S_{12}\Gamma_L S_{21}}{S_{21}}$$

Therefore,

$$\frac{b_2}{b_s} = \frac{S_{21}}{1 - \Gamma_s S_{11} - S_{22}\Gamma_L + S_{11}S_{22}\Gamma_s\Gamma_L - S_{12}S_{21}\Gamma_s\Gamma_L} \tag{14.5}$$

The transducer power gain G_T is defined as

$$G_T = \frac{\text{power delivered to the load}}{\text{power available from the source}} = \frac{P_L}{P_{\text{avs}}} \tag{14.6}$$

The power delivered to the load is

$$P_L = |b_2|^2 - |a_2|^2$$

From (14.4):

$$P_L = |b_2|^2 - |b_2|^2|\Gamma_L|^2$$

$$= |b_2|^2\left(1 - |\Gamma_L|^2\right) \tag{14.7}$$

The power available (P_{avs}) from a source is defined as the power delivered by the source to a conjugately matched load. From Figure 14.1, this implies that $\Gamma_{\text{in}} = \Gamma_s^*$, so

$$|b_s|^2 = P_{\text{avs}} - P_{\text{avs}}|\Gamma_{\text{in}}|^2$$

$$= P_{\text{avs}} - P_{\text{avs}}|\Gamma_s^*|^2$$

$$= P_{\text{avs}} - P_{\text{avs}}|\Gamma_s|^2$$

$$P_{\text{avs}} = \frac{|b_s|^2}{1 - |\Gamma_s|^2} \tag{14.8}$$

Substituting (14.7) and (14.8) into (14.6) gives

$$G_T = \frac{|b_2|^2}{|b_s|^2}\left(1 - |\Gamma_L|^2\right)\left(1 - |\Gamma_s|^2\right) \tag{14.9}$$

Substituting (14.5) into (14.9), we have

$$G_T = \frac{|S_{21}|^2\left(1 - |\Gamma_s|^2\right)\left(1 - |\Gamma_L|^2\right)}{|(1 - \Gamma_s S_{11})(1 - S_{22}\Gamma_L) - S_{12}S_{21}\Gamma_s\Gamma_L|^2} \tag{14.10}$$

The transducer power gain is a function of the transistor S parameters and Γ_s and Γ_L.

From Equations (14.1) through (14.4), Γ_{in} and Γ_{out} in Figure 14.1 can be written as (see Problem P14.1)

$$\Gamma_{\text{in}} = \frac{b_1}{a_1} = S_{11} + \frac{S_{12}S_{21}\Gamma_L}{1 - S_{22}\Gamma_L} \tag{14.11}$$

and

$$\Gamma_{\text{out}}\big|_{b_s=0} = \left.\frac{b_2}{a_2}\right|_{b_s=0} = S_{22} + \frac{S_{12}S_{21}\Gamma_s}{1 - S_{11}\Gamma_s} \tag{14.12}$$

Note that if $\Gamma_L = 0$ (i.e., matched load) or $S_{12} = 0$ (i.e., unilateral transistor), then $\Gamma_{\text{in}} = S_{11}$. Using Equations (14.11) and (14.12), (14.10) becomes

$$G_T = \frac{1 - |\Gamma_s|^2}{|1 - \Gamma_{\text{in}}\Gamma_s|^2}|S_{21}|^2 \frac{1 - |\Gamma_L|^2}{|1 - S_{22}\Gamma_L|^2} \tag{14.13}$$

or

$$G_T = \frac{1 - |\Gamma_s|^2}{|1 - S_{11}\Gamma_s|^2}|S_{21}|^2 \frac{1 - |\Gamma_L|^2}{|1 - \Gamma_{\text{out}}\Gamma_L|^2} \tag{14.14}$$

14.3 POWER GAIN FOR A UNILATERAL TRANSISTOR

For a unilateral transistor, $S_{12} = 0$. The signal only flows from the input port to the output port of the transistor. Most transistors have small S_{12} values and can be considered as unilateral. In this case, Equations (14.10), (14.11), and (14.12) become

$$G_T = G_{Tu} = \frac{1 - |\Gamma_s|^2}{|1 - S_{11}\Gamma_s|^2}|S_{21}|^2 \frac{1 - |\Gamma_L|^2}{|1 - S_{22}\Gamma_L|^2} \tag{14.15}$$

$$\Gamma_{\text{in}} = S_{11} \tag{14.16}$$

$$\Gamma_{\text{out}} = S_{22} \tag{14.17}$$

Equation (14.15) can be written in the form

$$G_{Tu} = G_s G_0 G_L \tag{14.18}$$

where

$$G_s = \frac{1 - |\Gamma_s|^2}{|1 - S_{11}\Gamma_s|^2} \tag{14.19}$$

$$G_0 = |S_{21}|^2 \tag{14.20}$$

$$G_L = \frac{1 - |\Gamma_L|^2}{|1 - S_{22}\Gamma_L|^2} \tag{14.21}$$

The first term in (14.18) depends on S_{11} and the source reflection coefficient Γ_s. The second term depends on S_{21}. The third term depends on S_{22} and the load reflection coefficient Γ_L. The microwave transistor amplifier can be considered to consist of three different gain (or loss) blocks, as shown in Figure 14.2. The terms G_s and G_L are the gain or loss produced by the matching or mismatching of the input or output circuits [1]. The term G_s affects the degree of matching and mismatching between Γ_s and S_{11}. Al-

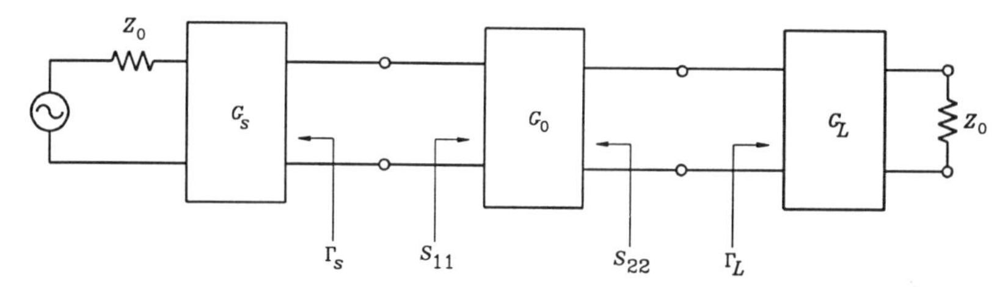

FIGURE 14.2 Unilateral transducer power gain.

though the G_s block consists of passive components, it can have either a gain or a loss. G_s depends on the mismatch between Z_0, the input-matching network, and S_{11}. Decreasing the mismatch loss can be treated as providing gain. Equation (14.18) can be written in terms of decibels as

$$G_{Tu}(\text{dB}) = G_s(\text{dB}) + G_0(\text{dB}) + G_L(\text{dB}) \tag{14.22}$$

The transducer power gain can be maximized by optimizing the input- and output-matching networks. The maximum gain occurs when (see Problem P14.2)

$$\Gamma_s = S_{11}^* = \Gamma_{\text{in}}^* \tag{14.23}$$

$$\Gamma_L = S_{22}^* = \Gamma_{\text{out}}^* \tag{14.24}$$

We have $G_s = G_{s,\max}$ and $G_L = G_{L,\max}$. G_{Tu} is then maximized and given by

$$G_{\text{Tu,max}} = \frac{1}{1 - |S_{11}|^2}|S_{21}|^2\frac{1}{1 - |S_{22}|^2} \tag{14.25}$$

Given a unilateral transistor with known S parameters, an amplifier can be designed with a maximum gain following the procedure given below.

1. Check the stability (to be discussed later).
2. Calculate $G_{\text{Tu,max}}$.
3. Find $\Gamma_s = S_{11}^*$ and $\Gamma_L = S_{22}^*$.
4. From Γ_s and Γ_L, find Z_s and Z_L from the Smith chart or

$$Z_s = Z_0\frac{1 + \Gamma_s}{1 - \Gamma_s} \qquad Z_L = Z_0\frac{1 + \Gamma_L}{1 - \Gamma_L}$$

5. Design the matching networks to match from Z_0 to Z_s and Z_0 to Z_L using the methods described in Chapter 2.

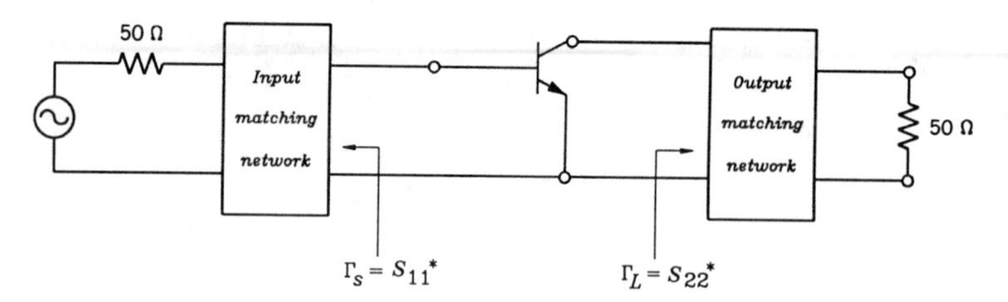

FIGURE 14.3 Transistor amplifier for maximum gain.

Example 14.1 Design a microwave transistor amplifier (shown in Figure 14.3) for maximum gain using LC matching elements for a transistor with the following S parameters at 1 GHz:

$$S_{11} = 0.706 \angle -160^\circ \qquad S_{12} = 0$$
$$S_{21} = 5.01 \angle 85^\circ \qquad S_{22} = 0.508 \angle -20^\circ$$

SOLUTION: For maximum gain, $\Gamma_s = S_{11}^*$ and $\Gamma_L = S_{22}^*$.

$$\begin{aligned} G_{\text{Tu, max}} &= \frac{1}{1 - |S_{11}|^2} |S_{21}|^2 \frac{1}{1 - |S_{22}|^2} \\ &= \frac{1}{1 - (0.706)^2} (5.01)^2 \frac{1}{1 - (0.508)^2} \\ &= 67.45 \text{ or } 18.3 \text{ dB} \end{aligned}$$

To design the input-matching network using LC elements and the Smith chart, first note that

$$\Gamma_s = S_{11}^* = 0.706 \angle 160^\circ$$

From the Smith chart (Z chart), we have

$$Z_s = 50(0.178 + j0.171)$$

Z_s can also be obtained from

$$Z_s = Z_0 \frac{1 + \Gamma_s}{1 - \Gamma_s}$$

From the Y–Z chart in the Γ_s plane shown in Figure 14.4, one can use a shunt L_1 and a series C_1 to match the 50-Ω line to Z_s. The circuit is shown in Figure 14.5(a).

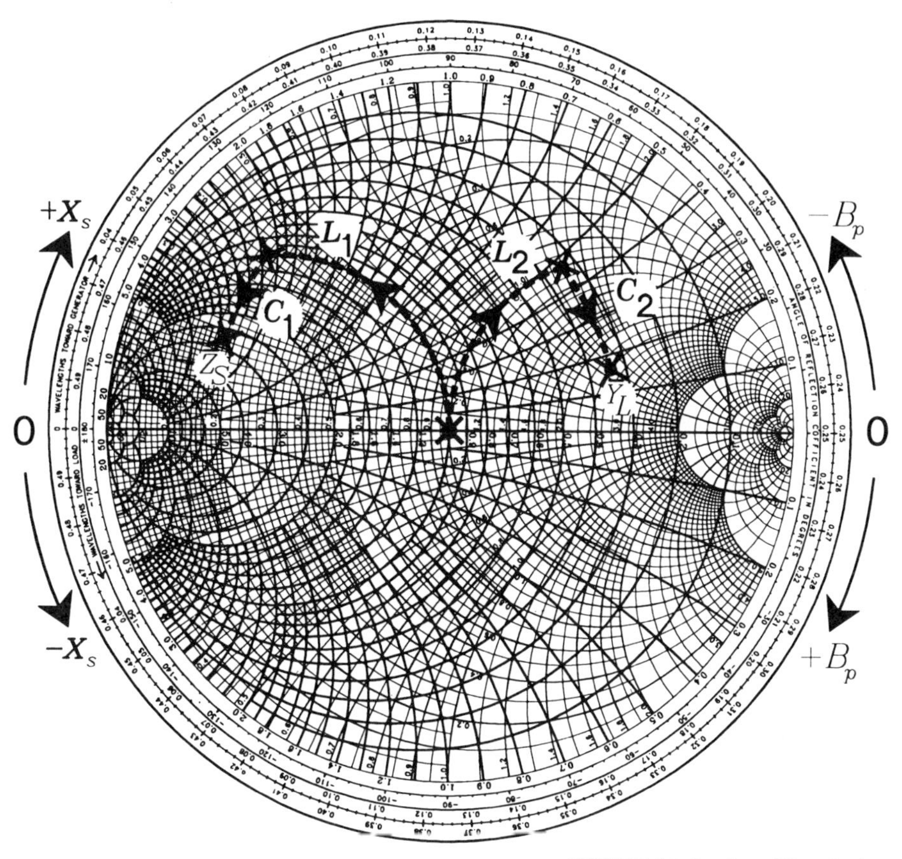

FIGURE 14.4 Amplifier design.

From the Y chart

$$\overline{Y}_{L1} = -2.2j - 0 = -2.2j$$

$$Z_{L1} = \frac{1}{\overline{Y}_{L1}} \times 50 = j22.73 = j\omega L_1$$

$$L_1 = 3.62 \text{ nH}$$

From the Z chart

$$\overline{Z}_{c1} = 0.18j - 0.375j = -0.195j$$

$$Z_{c1} = 50\overline{Z}_{c1} = -9.75j = -j\frac{1}{\omega C_1}$$

$$C_1 = 16.3 \text{ pF}$$

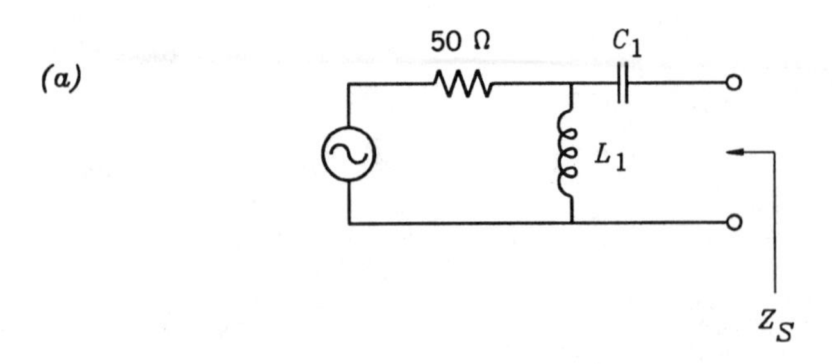

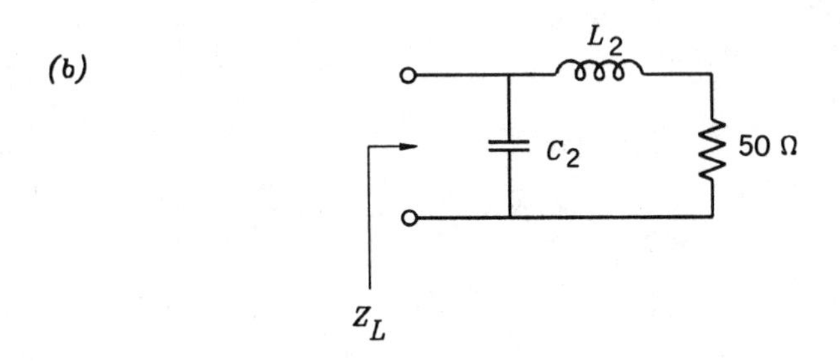

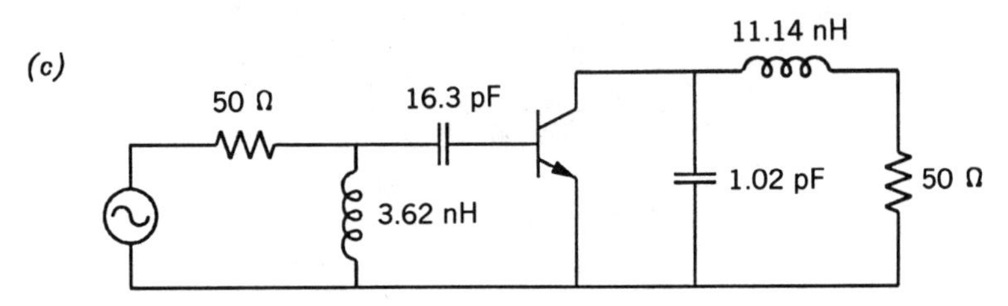

FIGURE 14.5 Amplifier design: (*a*) input-matching network; (*b*) output-matching network; (*c*) overall circuit.

The output-matching network can be designed using a Y–Z chart in the Γ_L plane (Figure 14.4). A series L_2 and a shunt C_2 are used to match a 50-Ω load to Z_L as shown in Figure 14.5(*b*):

$$\Gamma_L = S_{22}^* = 0.508 \angle 20^\circ$$

$$\bar{Y}_L = 0.335 - j0.157$$

From the Z chart,

$$Z_{L2} = 50(1.4j - 0) = 70j = j\omega L_2$$

$$L_2 = 11.14 \text{ nH}$$

From the Y chart,

$$\bar{Y}_{c2} = -0.16j - (-0.48j) = 0.32j$$

$$Z_{c2} = \frac{1}{\bar{Y}_{c2}} \times 50 = -156.3j = -j\frac{1}{\omega C_2}$$

$$C_2 = 1.02 \text{ pF}$$

The overall circuit is shown in Figure 14.5(c).

Example 14.2 Design a transistor amplifier with a maximum gain at 4 GHz using microstrip shunt stubs in a 50-Ω system. The FET has the following S

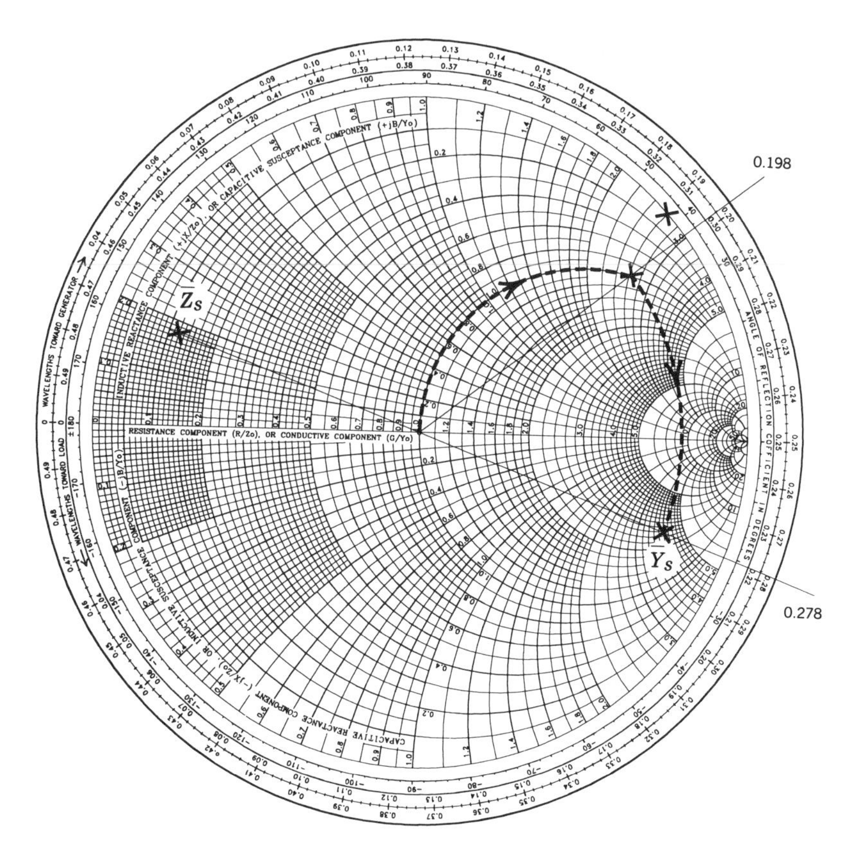

FIGURE 14.6 Γ_s plane.

parameters:

$$S_{11} = 0.8 \angle -160^\circ \qquad S_{12} = 0$$

$$S_{21} = 4.2 \angle 75^\circ \qquad S_{22} = 0.7 \angle -20^\circ$$

SOLUTION:

$$G_{Tu,\max} = \frac{1}{1 - |S_{11}|^2} |S_{21}|^2 \frac{1}{1 - |S_{22}|^2}$$

$$= \frac{1}{1 - (0.8)^2} (4.2)^2 \frac{1}{1 - (0.7)^2}$$

$$= 96 \text{ or } 19.8 \text{ dB}$$

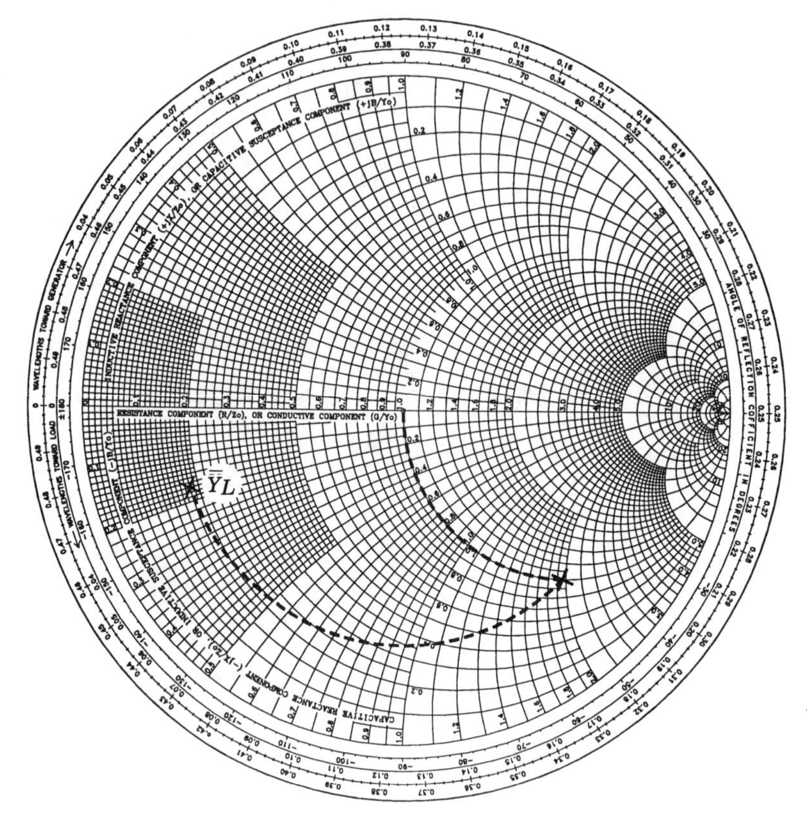

FIGURE 14.7 Γ_L plane.

For maximum gain

$$\Gamma_s = S_{11}^* = 0.8 \angle 160^\circ$$
$$\Gamma_L = S_{22}^* = 0.7 \angle 20^\circ$$

For the input-matching circuit, using a Smith chart (Figure 14.6), we have

$$\bar{Z}_s = 0.115 + j0.174$$
$$\bar{Y}_s = 2.64 - j4.00$$

From the design method given in Chapter 2, we have a transmission line length $= 0.278\lambda_g - 0.198\lambda_g = 0.080\lambda_g$. The open stub length $= 0.192\lambda_g$.

For the output-matching circuit, using a Smith chart (Figure 14.7), we have

$$\bar{Y}_L = 0.18 - j0.17$$

A short stub of $0.074\lambda_g$ is used, followed by a $0.158\lambda_g$ transmission line to match the 50-Ω load to Z_L. The complete design with bias circuits is shown in Figure 14.8.

14.4 STABILITY CONSIDERATIONS

To design an amplifier, it is important to avoid oscillation. The stability of an amplifier (or its resistance to oscillation) is an important design consideration. Stability depends on the S parameters of the transistor, the matching networks, and the terminations.

For the amplifier shown in Figure 14.9, the circuit is said to be unconditionally stable if the real parts of Z_{in} and Z_{out} are greater than zero for all

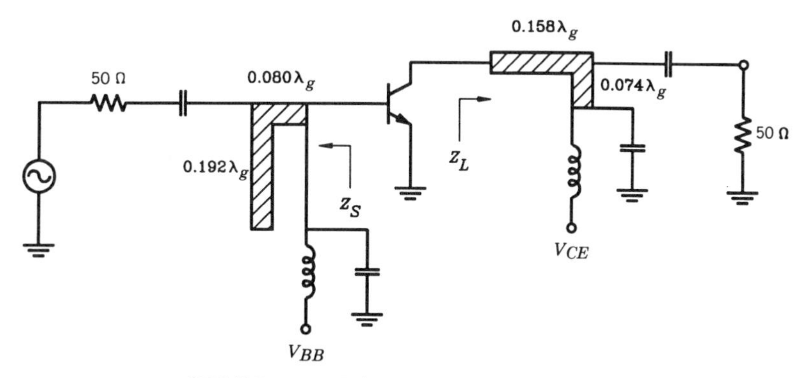

FIGURE 14.8 Microwave amplifier design.

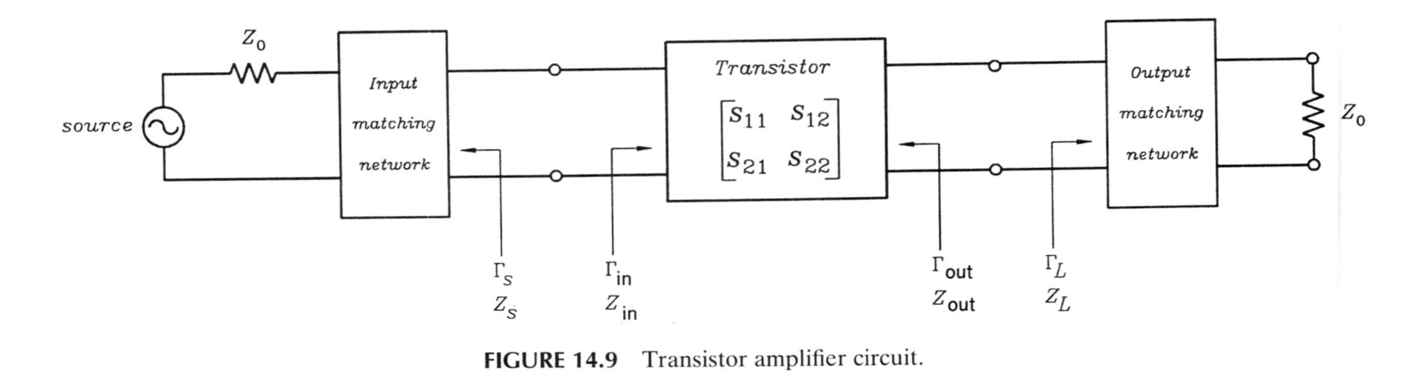

FIGURE 14.9 Transistor amplifier circuit.

passive load and source impedances. The circuit is potentially unstable if the real parts of Z_{in} or Z_{out} are negative for some passive load and source impedances. The conditions for unconditional stability can be written in terms of reflection coefficients as

$$|\Gamma_s| < 1 \tag{14.26}$$

$$|\Gamma_L| < 1 \tag{14.27}$$

$$|\Gamma_{\text{in}}| = \left| S_{11} + \frac{S_{12}S_{21}\Gamma_L}{1 - S_{22}\Gamma_L} \right| < 1 \tag{14.28}$$

$$|\Gamma_{\text{out}}| = \left| S_{22} + \frac{S_{12}S_{21}\Gamma_s}{1 - S_{11}\Gamma_s} \right| < 1 \tag{14.29}$$

Potentially unstable conditions can be written as

$$|\Gamma_{\text{in}}| > 1 \quad \text{or} \quad |\Gamma_{\text{out}}| > 1 \tag{14.30}$$

for some passive Z_s or Z_L.

For a matched load and source, $\Gamma_s = 0$ or $\Gamma_L = 0$. Equation (14.30) is then equivalent to

$$|S_{11}| > 1 \quad \text{or} \quad |S_{22}| > 1 \tag{14.31}$$

Therefore, if $|S_{11}| > 1$ or $|S_{22}| > 1$, the network cannot be unconditionally stable since for $\Gamma_s = 0$ or $\Gamma_L = 0$ terminations, $|\Gamma_{\text{in}}| > 1$ or $|\Gamma_{\text{out}}| > 1$.

From Equations (14.26) through (14.29), the following conditions can be derived for the unconditionally stable case [1, 2]. Note that these conditions depend on the S parameters of the transistor.

$$|S_{11}| < 1 \tag{14.32}$$

$$|S_{22}| < 1 \tag{14.33}$$

$$K = \frac{1 - |S_{11}|^2 - |S_{22}|^2 + |\Delta|^2}{2|S_{12}S_{21}|} > 1 \tag{14.34}$$

$$|\Delta| = |S_{11}S_{22} - S_{12}S_{21}| < 1 \tag{14.35}$$

For a unilateral transistor, $S_{12} = 0$. Equation (14.34) is automatically satisfied since $K = \infty$. The conditions for the unconditionally stable case become

$$|S_{11}| < 1 \tag{14.36}$$

$$|S_{22}| < 1 \tag{14.37}$$

For a potentially unstable transistor, it is important to design Z_s and Z_L (or Γ_s and Γ_L) for which $|\Gamma_{in}| < 1$ and $|\Gamma_{out}| < 1$. The regions of Γ_s and Γ_L (or Z_s and Z_L) that produce $|\Gamma_{in}| < 1$ and $|\Gamma_{out}| < 1$ can be determined in the Smith chart using the graphical method outlined below.

To find the regions of Z_s and Z_L for which the amplifier is stable, we start with the boundaries of the regions by setting $|\Gamma_{in}| = 1$ and $|\Gamma_{out}| = 1$. $|\Gamma_{in}| = 1$ corresponds to a circle in the Γ_L plane, and $|\Gamma_{out}| = 1$ corresponds to a circle in the Γ_s plane. Second, we determine which side of the $|\Gamma_{in}| = 1$ circle is stable (i.e., which side gives $|\Gamma_{in}| < 1$). A similar procedure applies to $|\Gamma_{out}|$.

The circles obtained are called *stability circles*. The circles are obtained by setting

$$|\Gamma_{in}| = \left| S_{11} + \frac{S_{12}S_{21}\Gamma_L}{1 - S_{22}\Gamma_L} \right| = 1 \tag{14.38}$$

$$|\Gamma_{out}| = \left| S_{22} + \frac{S_{12}S_{21}\Gamma_s}{1 - S_{11}\Gamma_s} \right| = 1 \tag{14.39}$$

From Equations (14.38) and (14.39) we can derive (see Problem P14.3)

$$\left| \Gamma_L - \frac{(S_{22} - \Delta S_{11}^*)^*}{|S_{22}|^2 - |\Delta|^2} \right| = \left| \frac{S_{12}S_{21}}{|S_{22}|^2 - |\Delta|^2} \right| \tag{14.40}$$

and

$$\left| \Gamma_s - \frac{(S_{11} - \Delta S_{22}^*)^*}{|S_{11}|^2 - |\Delta|^2} \right| = \left| \frac{S_{12}S_{21}}{|S_{11}|^2 - |\Delta|^2} \right| \tag{14.41}$$

where $\Delta = S_{11}S_{22} - S_{12}S_{21}$.

The radii and centers of the circles for $|\Gamma_{in}| = 1$ and $|\Gamma_{out}| = 1$ in the Γ_L and Γ_s planes are given below.

1. Input stability circle (i.e., Γ_s value for $|\Gamma_{out}| = 1$):

$$\text{radius} = r_s = \left| \frac{S_{12}S_{21}}{|S_{11}|^2 - |\Delta|^2} \right| \tag{14.42}$$

$$\text{center} = c_s = \frac{(S_{11} - \Delta S_{22}^*)^*}{|S_{11}|^2 - |\Delta|^2} \tag{14.43}$$

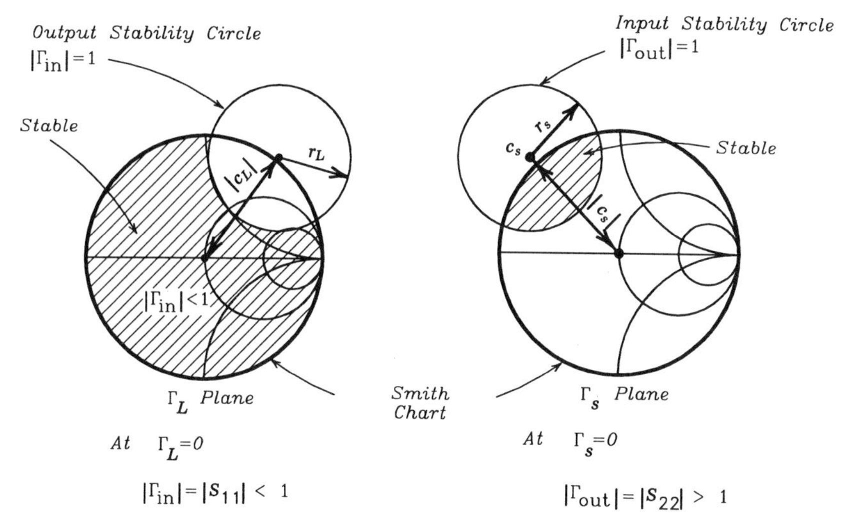

FIGURE 14.10 Example of stability circles in potentially unstable case.

2. Output stability circle (i.e., Γ_L value for $|\Gamma_{in}| = 1$):

$$\text{radius} = r_L = \left| \frac{S_{12}S_{21}}{|S_{22}|^2 - |\Delta|^2} \right| \tag{14.44}$$

$$\text{center} = c_L = \frac{(S_{22} - \Delta S_{11}^*)^*}{|S_{22}|^2 - |\Delta|^2} \tag{14.45}$$

For a transistor with S parameters at a specified frequency, one can plot the input and output stability circles on a Smith chart. Figure 14.10 shows an example of these circles. To determine which side of the circle in the Smith chart is stable, one can check the $\Gamma_L = 0$ or $\Gamma_s = 0$ point. $\Gamma_L = 0$ corresponds to the center of the Smith chart in the Γ_L plane. At $\Gamma_L = 0$, $|\Gamma_{in}| = |S_{11}|$ from Equation (14.38).

If $|S_{11}| > 1$, the side of the stability circle including the $\Gamma_L = 0$ point is unstable. This can be seen by examining Equations (14.30) and (14.31). If $|S_{11}| < 1$, the side of the stability circle including the $\Gamma_L = 0$ point is stable, as shown in Figure 14.10. A similar procedure applies to the Γ_s plane and the input stability circle.

For the unconditionally stable case, the input and output stability circles should fall completely outside the Smith chart. Furthermore, $|S_{11}| < 1$ and $|S_{22}| < 1$. Figure 14.11 shows the input and output stability circles for this case. Mathematically, this requires that

$$||c_L| - r_L| > 1 \quad \text{and} \quad |S_{11}| < 1 \tag{14.46}$$

$$||c_s| - r_s| > 1 \quad \text{and} \quad |S_{22}| < 1 \tag{14.47}$$

Example 14.3 Determine the stability for the following transistors and draw the input and output stability circles for the potentially unstable case.

(a) $S_{11} = 0.674 \angle -152°$
$S_{12} = 0.075 \angle 6.2°$
$S_{21} = 1.74 \angle 36.4°$
$S_{22} = 0.6 \angle -92.6°$

(b) $S_{11} = 0.7 \angle -50°$
$S_{12} = 0.27 \angle 75°$
$S_{21} = 5 \angle 120°$
$S_{22} = 0.6 \angle 80°$

SOLUTION:

(a) $|S_{11}| = 0.674 \qquad |S_{12}| = 0.075$
$|S_{21}| = 1.74 \qquad |S_{22}| = 0.6$

$$\Delta = S_{11}S_{22} - S_{12}S_{21} = 0.386 \angle 134°$$

$$|\Delta| = 0.386 < 1$$

$$K = \frac{1 - |S_{11}|^2 - |S_{22}|^2 + |\Delta|^2}{2|S_{12}S_{21}|} = 1.282 > 1$$

$$|S_{11}| < 1$$

$$|S_{22}| < 1$$

It is unconditionally stable.

(b) $|S_{11}| = 0.7 \qquad |S_{12}| = 0.27$
$|S_{21}| = 5 \qquad |S_{22}| = 0.6$

$$\Delta = S_{11}S_{22} - S_{12}S_{21} = 1.76 \angle 18.5°$$

$$|\Delta| > 1$$

$$K = \frac{1 - |S_{11}|^2 - |S_{22}|^2 + |\Delta|^2}{2|S_{12}S_{21}|} = 1.202 > 1$$

Since $|\Delta| > 1$, it is potentially unstable.

From Equations (14.42) through (14.45), we have the input stability circle given by

$$r_s = 0.518$$

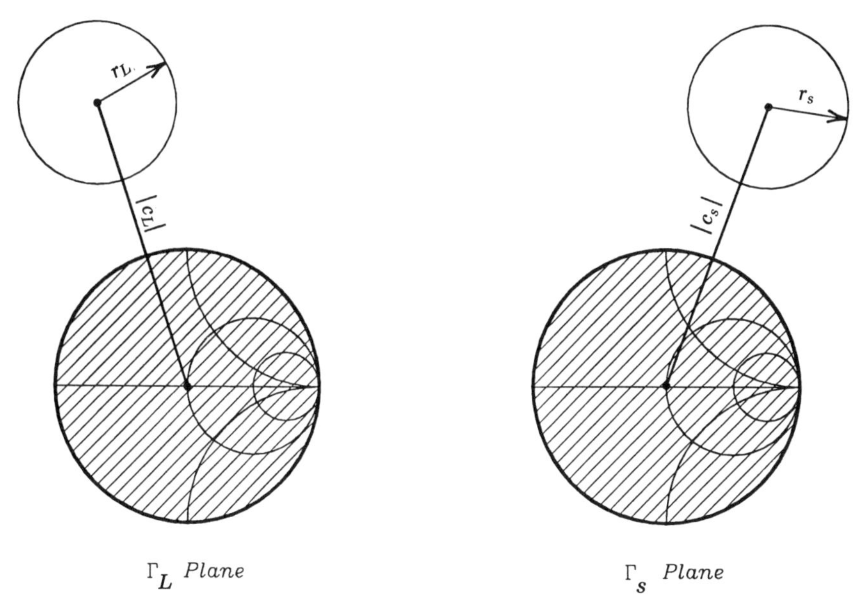

FIGURE 14.11 Example of stability circles in unconditionally stable case.

and

$$c_s = 0.152 \angle 82°$$

Since $|S_{22}| < 1$, the center of the Smith chart is in the stable region. For the output stability circle, we have

$$r_L = 0.494$$

and

$$c_L = 0.239 \angle -58°$$

The circles are plotted in Figure 14.12.

14.5 CONSTANT-GAIN CIRCLES FOR THE UNILATERAL CASE

In many design cases, maximum gain may not be desirable due to other considerations, such as noise figure and bandwidth. The amplifier is designed with the input and output matching sections to achieve a specified gain and noise figure over the bandwidth. The design is accomplished by plotting the constant-gain circles and constant-noise figure circles on the Smith chart and finding values of Z_s and Z_L that meet both the gain and noise requirements. The constant-gain circles represent loci of Γ_s and Γ_L that give fixed values of gain G_s and G_L.

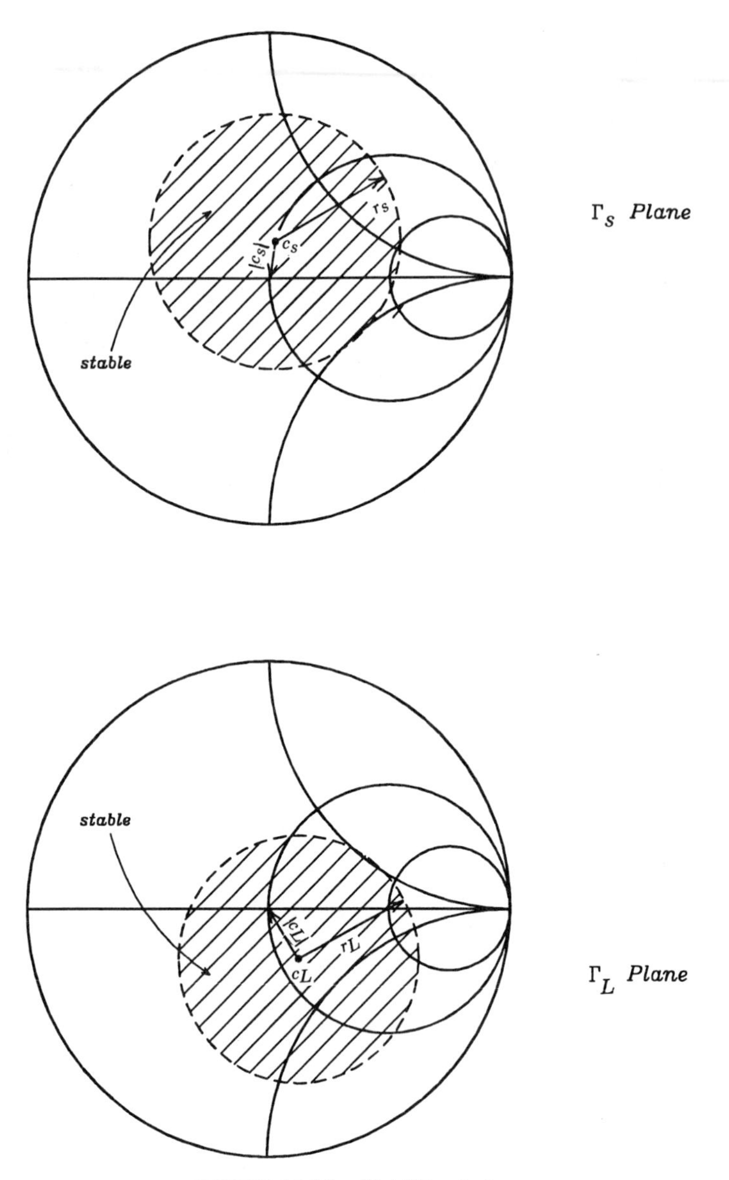

FIGURE 14.12 Stability circles.

For simplicity, only the unilateral cases are considered here. Most practical transistors have small values of $|S_{12}|$ and can be treated as unilateral devices. For the bilateral case (where S_{12} cannot be ignored), the constant-gain circles based on transducer power gain are more difficult to obtain and a simpler design procedure is to draw constant-gain circles for operating power gain or available power gain [1].

Now, considering the unilateral case, the transducer power gain is

$$G_{\mathrm{Tu}} = \frac{1 - |\Gamma_s|^2}{|1 - S_{11}\Gamma_s|^2}|S_{21}|^2 \frac{1 - |\Gamma_L|^2}{|1 - S_{22}\Gamma_L|^2}$$

$$= G_s \cdot G_0 \cdot G_L \tag{14.48}$$

Assuming the unconditionally stable case, $|S_{11}| < 1$ and $|S_{22}| < 1$, the constant-gain circles are in the Smith chart. From Equation (14.48),

$$G_s = \frac{1 - |\Gamma_s|^2}{|1 - S_{11}\Gamma_s|^2} \tag{14.49}$$

and

$$G_L = \frac{1 - |\Gamma_L|^2}{|1 - S_{22}\Gamma_L|^2} \tag{14.50}$$

The values of G_s and G_L are maximized when $\Gamma_s = S_{11}^*$ and $\Gamma_L = S_{22}^*$ with

$$G_{s,\max} = \frac{1}{1 - |S_{11}|^2} \tag{14.51}$$

$$G_{L,\max} = \frac{1}{1 - |S_{22}|^2} \tag{14.52}$$

Define normalized gain factors g_s and g_L as

$$g_s = \frac{G_s}{G_{s,\max}} = \frac{1 - |\Gamma_s|^2}{|1 - S_{11}\Gamma_s|^2}\left(1 - |S_{11}|^2\right) \tag{14.53}$$

$$g_L = \frac{G_L}{G_{L,\max}} = \frac{1 - |\Gamma_L|^2}{|1 - S_{22}\Gamma_L|^2}\left(1 - |S_{22}|^2\right) \tag{14.54}$$

Note that $0 \leq g_s \leq 1$ and $0 \leq g_L \leq 1$.

To derive the constant-gain circle in the Γ_s plane, we expand Equation (14.53) to give [3]

$$g_s|1 - S_{11}\Gamma_s|^2 = \left(1 - |\Gamma_s|^2\right)\left(1 - |S_{11}|^2\right) \tag{14.55}$$

$$\left(g_s|S_{11}|^2 + 1 - |S_{11}|^2\right)|\Gamma_s|^2 - g_s(S_{11}\Gamma_s + S_{11}^*\Gamma_s^*) = 1 - |S_{11}|^2 - g_s$$

$$\Gamma_s\Gamma_s^* - \frac{g_s(S_{11}\Gamma_s + S_{11}^*\Gamma_s^*)}{1 - (1 - g_s)|S_{11}|^2} = \frac{1 - |S_{11}|^2 - g_s}{1 - (1 - g_s)|S_{11}|^2} \tag{14.56}$$

Now adding $(g_s^2|S_{11}|^2)/[1-(1-g_s)|S_{11}|^2]^2$ to both sides gives

$$\left|\Gamma_s - \frac{g_s S_{11}^*}{1-(1-g_s)|S_{11}|^2}\right|^2 = \frac{\left(1-|S_{11}|^2-g_s\right)\left[1-(1-g_s)|S_{11}|^2\right]+g_s^2|S_{11}|^2}{\left[1-(1-g_s)|S_{11}|^2\right]^2}$$

Finally, (Problem P14.13)

$$\left|\Gamma_s - \frac{g_s S_{11}^*}{1-(1-g_s)|S_{11}|^2}\right| = \frac{\sqrt{1-g_s}\left(1-|S_{11}|^2\right)}{1-(1-g_s)|S_{11}|^2} \tag{14.57}$$

Equation (14.57) represents a circle in the Γ_s plane if g_s is fixed. The center and radius of the circle are

$$c_s = \frac{g_s S_{11}^*}{1-(1-g_s)|S_{11}|^2} \tag{14.58}$$

$$r_s = \frac{\sqrt{1-g_s}\left(1-|S_{11}|^2\right)}{1-(1-g_s)|S_{11}|^2} \tag{14.59}$$

Similarly, the constant-gain circles for the output network have

$$c_L = \frac{g_L S_{22}^*}{1-(1-g_L)|S_{22}|^2} \tag{14.60}$$

and

$$r_L = \frac{\sqrt{1-g_L}\left(1-|S_{22}|^2\right)}{1-(1-g_L)|S_{22}|^2} \tag{14.61}$$

From Equations (14.58) and (14.60), it can be seen that the centers of each family of circles lie along straight lines connecting the center of the Smith chart and S_{11}^* or S_{22}^*. The centers have the same angles determined by S_{11}^* or S_{22}^*. For two special cases: (1) $g_s = 1$ (maximum gain), $c_s = S_{11}^*$ and $r_s = 0$, the constant-gain circle for maximum gain is a point located at S_{11}^*. (2) For $G_s = 1$ (or 0 dB), then $g_s = 1/G_{s,\max} = 1-|S_{11}|^2$, $|c_s| = r_s = |S_{11}|/(1+|S_{11}|^2)$. The circle passes through the center of the Smith chart. Similar results apply to the constant-output-gain circles.

The procedure for drawing constant-gain circles in the Γ_s plane is as follows:

1. In the Γ_s plane, locate S_{11}^* and draw a line from the center of the Smith chart to S_{11}^*. At S_{11}^* the gain is

$$G_{s,\max} = \frac{1}{1 - |S_{11}|^2}$$

2. Determine the values of G_s to be drawn. Note that $0 \leq G_s \leq G_{s,\max}$. For each G_s, calculate the corresponding $g_s = G_s/G_{s,\max}$.
3. Calculate $|c_s|$ and r_s for each g_s.
4. Draw a circle for each g_s with a center at c_s and a radius of r_s.

The procedure is repeated for constant-output-gain circles in the Γ_L plane with S_{11}, G_s, c_s, and r_s replaced by S_{22}, G_L, c_L, and r_L.

Example 14.4 A transistor has the following S parameters at 10 GHz. Draw the constant-gain circles for G_s values of (a) $G_{s,\max}$, (b) 2 dB, (c) 1 dB, (d) 0 dB, and (e) -1 dB.

$$S_{11} = 0.706 \angle -160^\circ \qquad S_{12} = 0$$

$$S_{21} = 5.01 \angle 85^\circ \qquad S_{22} = 0.508 \angle -20^\circ$$

SOLUTION: (a) $G_{s,\max} = \dfrac{1}{1 - |S_{11}|^2} = 1.994 = 3 \text{ dB}$

$$c_s = S_{11}^* = 0.706 \angle +160^\circ$$

(b)

$$G_s = 2 \text{ dB} = 1.585$$

$$g_s = \frac{G_s}{G_{s,\max}} = 0.795$$

$$|c_s| = \frac{g_s |S_{11}^*|}{1 - |S_{11}|^2 (1 - g_s)} = 0.625$$

$$r_s = \frac{\sqrt{1 - g_s}\left(1 - |S_{11}|^2\right)}{1 - |S_{11}|^2 (1 - g_s)} = 0.253$$

(c)

$$G_s = 1 \text{ dB} = 1.259$$

$$g_s = \frac{G_s}{G_{s,\max}} = 0.631$$

$$|c_s| = 0.546$$

$$r_s = 0.373$$

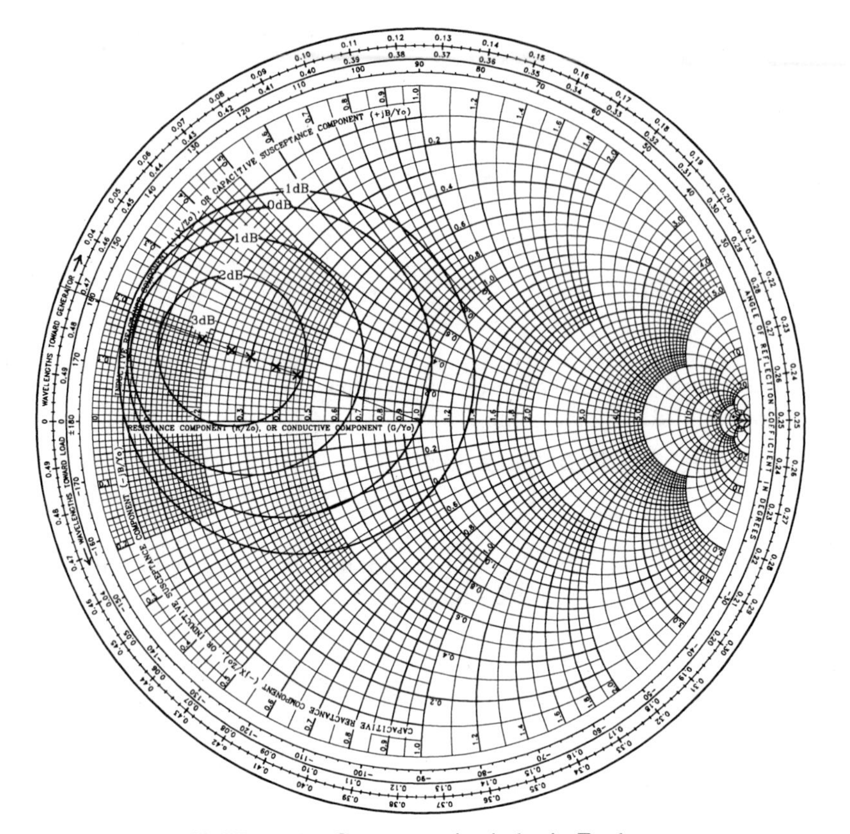

FIGURE 14.13 Constant-gain circles in Γ_s plane.

(d)

$$G_s = 0 \text{ dB} = 1$$
$$g_s = 0.501$$
$$|c_s| = 0.471$$
$$r_s = 0.472$$

(e)

$$G_s = -1 \text{ dB} = 0.794$$
$$g_s = 0.398$$
$$|c_s| = 0.401$$
$$r_s = 0.556$$

The circles are plotted in Figure 14.13.

14.6 CONSTANT-NOISE-FIGURE CIRCLES

For low-noise amplifier (LNA) design, the noise figure is an important design parameter. The noise figure of a two-port amplifier is given by [4]

$$F = F_{\min} + \frac{r_n}{\mathrm{Re}(\overline{Y}_s)} |\overline{Y}_s - \overline{Y}_{sm}|^2 \tag{14.62}$$

where

$F_{\min}$: minimum noise figure
$r_n = R_N/Z_0$: equivalent normalized noise resistance
$\bar{Y}_s$: normalized source admittance
$\bar{Y}_{sm}$: normalized source admittance resulting in minimum noise figure $F_{\min}$

Note that if $\bar{Y}_s = \bar{Y}_{sm}$, then $F = F_{\min}$ and $\Gamma_s = \Gamma_{sm}$ (a reflection coefficient corresponding to $\bar{Y}_{sm}$). r_n, $F_{\min}$, and Γ_{sm} can be measured and normally are provided by the manufacturer.

$\bar{Y}_s$ and $\bar{Y}_{sm}$ are related to Γ_s and Γ_{sm} by

$$\bar{Y}_s = \frac{1 - \Gamma_s}{1 + \Gamma_s}$$

and

$$\bar{Y}_{sm} = \frac{1 - \Gamma_{sm}}{1 + \Gamma_{sm}}$$

Substituting the two equations above into Equation (14.62) gives (Problem P14.14)

$$F = F_{\min} + \frac{4r_n|\Gamma_s - \Gamma_{sm}|^2}{\left(1 - |\Gamma_s|^2\right)|1 + \Gamma_{sm}|^2} \tag{14.63}$$

Equation (14.63) is used to calculate F for any Γ_s or Z_s if r_n, Γ_{sm}, and $F_{\min}$ are known.

Define a noise-figure parameter N_i, which corresponds to F_i, a given noise figure.

$$N_i = \frac{|\Gamma_s - \Gamma_{sm}|^2}{1 - |\Gamma_s|^2} = \frac{F_i - F_{\min}}{4r_n}|1 + \Gamma_{sm}|^2 \tag{14.64}$$

Then Equation (14.64) can be rearranged as

$$(\Gamma_s - \Gamma_{sm})(\Gamma_s^* - \Gamma_{sm}^*) = N_i - N_i|\Gamma_s|^2$$

$$|\Gamma_s|^2 - \Gamma_{sm}\Gamma_s^* + |\Gamma_{sm}|^2 - \Gamma_{sm}^*\Gamma_s = N_i - N_i|\Gamma_s|^2$$

$$|\Gamma_s|^2(1 + N_i) + |\Gamma_{sm}|^2 - 2\,\mathrm{Re}(\Gamma_s\Gamma_{sm}^*) = N_i \tag{14.65}$$

Multiplying (14.65) by $(1 + N_i)$ gives

$$|\Gamma_s|^2(1 + N_i)^2 + |\Gamma_{sm}|^2 - 2(1 + N_i)\mathrm{Re}(\Gamma_s\Gamma_{sm}^*) = N_i^2 + N_i\left(1 - |\Gamma_{sm}|^2\right)$$

Dividing by $(1 + N_i)^2$ gives

$$|\Gamma_s|^2 + \frac{|\Gamma_{sm}|^2}{(1+N_i)^2} - \frac{2\,\mathrm{Re}(\Gamma_s \Gamma_{sm}^*)}{1+N_i} = \frac{N_i^2 + N_i(1 - |\Gamma_{sm}|^2)}{(1+N_i)^2}$$

$$\left|\Gamma_s - \frac{\Gamma_{sm}}{1+N_i}\right|^2 = \frac{N_i^2 + N_i(1 - |\Gamma_{sm}|^2)}{(1+N_i)^2} \tag{14.66}$$

Equation (14.66) represents a family of circles with N_i as a parameter. These circles are called the constant-noise-figure circles, with centers and radii given by

$$c_{Fi} = \frac{\Gamma_{sm}}{1+N_i} \tag{14.67}$$

$$r_{Fi} = \frac{1}{1+N_i}\sqrt{N_i^2 + N_i(1 - |\Gamma_{sm}|)^2} \tag{14.68}$$

For the special case of minimum noise figure, $F_i = F_{\min}$; then $N_i = 0$, $c_{Fi} = \Gamma_{sm}$, and $r_{Fi} = 0$. This corresponds to a point located at Γ_{sm}, as expected.

A typical set of constant-noise-figure circles is shown in Figure 14.14. The following parameters are used: $F_{\min} = 3$ dB and $\Gamma_{sm} = 0.58 \angle 138°$. At point A, $\Gamma_s = 0.38 \angle 119°$ gives a noise figure of 4 dB.

In a unilateral case, a set of input constant-gain circles can be drawn together with the constant-noise-figure circles. An example for a GaAs FET is shown in Figure 14.15. Trade-offs in design need to be made between gain and noise figure. Maximum gain and minimum noise figure cannot, in general, be obtained simultaneously. In Figure 14.15, the maximum gain of $G_s = 3$ dB, obtained with $\Gamma_s = 0.7 \angle 110°$, corresponds to a noise figure of 4 dB. The minimum noise figure $F_{\min} = 0.8$ dB, obtained with $\Gamma_s = \Gamma_{sm} = 0.6 \angle 40°$, corresponds to a gain of $G_s \approx -1$ dB. A compromise can be drawn, for example, with $G_s = 2$ dB and $F_i = 2.5$ dB as Γ_s is represented by point A in Figure 14.15.

14.7 BANDWIDTH CONSIDERATION AND BROADBAND AMPLIFIERS

Broadband amplifiers are needed for many electronic warfare (EW), communication, and measurement applications. Since the S parameters of a three-terminal device vary with frequency, the design of a constant-gain amplifier over a broad bandwidth requires the matching input and output networks to compensate for the $|S_{21}|$ variations with frequency, and in the meantime, to keep a low noise figure. Several methods can be used for broadband amplifier design: (1) the use of compensated matching networks, (2) the use of

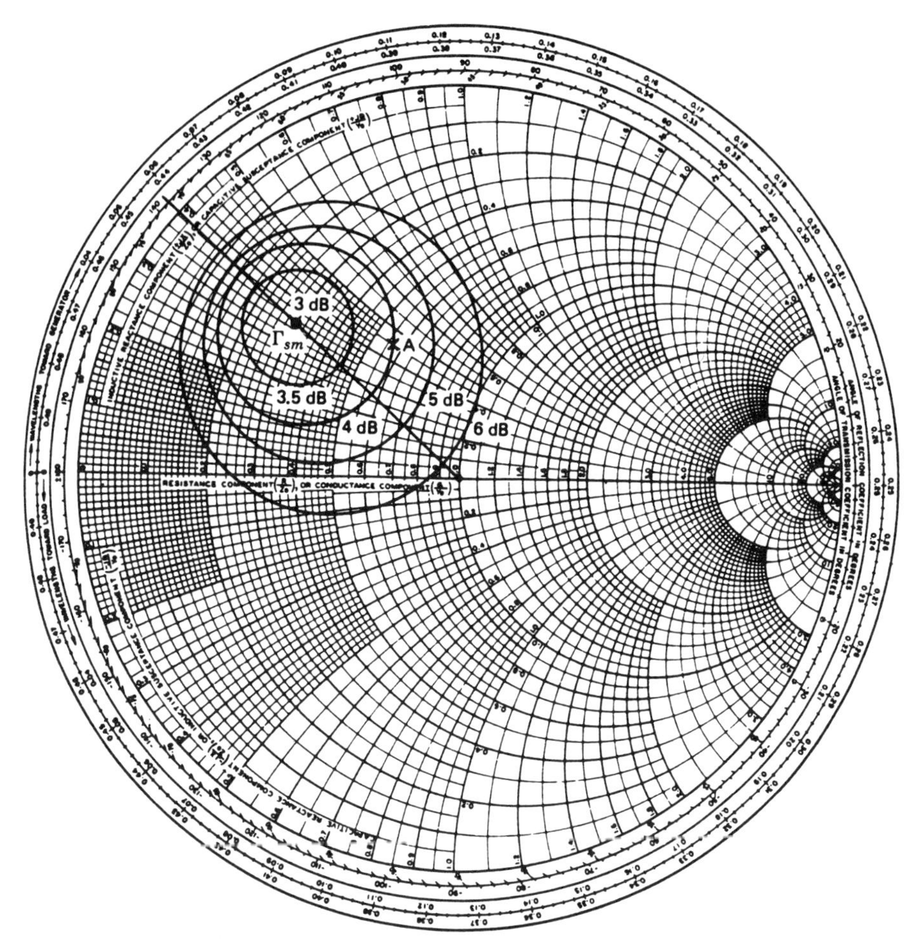

FIGURE 14.14 Typical constant-noise-figure circles in the Γ_s plane. (From Ref. 1 with permission from Prentice Hall.)

balanced structures, (3) the use of negative feedback circuits, and (4) the use of the distributed or traveling-wave design. Due to the complexity of the design, computer-aided design (CAD) tools such as SuperCompact or Touchstone are generally needed to facilitate the design.

Example 14.5 Design a broadband amplifier with a gain of 10 dB over the range 300 to 600 MHz. The transistor has the following S parameters:

f (MHz)	S_{11}	S_{21}	S_{12}	S_{22}
300	$0.25\angle -35°$	$4.5\angle 30°$	$0.01\angle 10°$	$0.8\angle -10°$
450	$0.32\angle -78°$	$3.2\angle 45°$	$0.05\angle -5°$	$0.9\angle -15°$
600	$0.2\angle -85°$	$2.0\angle 35°$	$0.02\angle -12°$	$0.85\angle -20°$

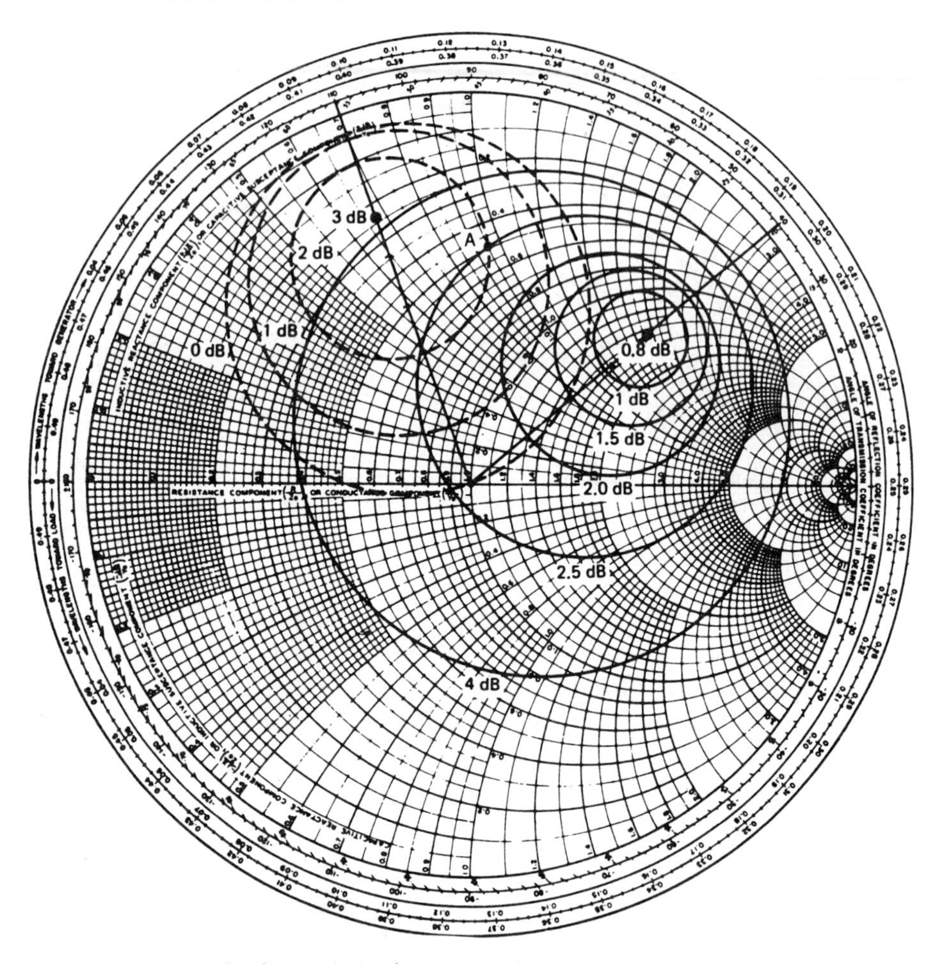

FIGURE 14.15 Noise figure circles (solid curves) and G_s constant-gain circles (dashed curves). The transistor is a GaAs FET with $V_{DS} = 4$ V, $I_{DS} = 12$ mA, and $f = 6$ GHz. (From Ref. 1 with permission from Prentice Hall.)

SOLUTION: Since $|S_{12}|$ is small, the transistor can be considered as unilateral.

$$G_{\text{Tu, max}} = G_{L,\text{max}} \cdot G_0 \cdot G_{s,\text{max}}$$

$$G_0 = |S_{21}|^2 = \begin{cases} 13 \text{ dB} & \text{at 300 MHz} \\ 10 \text{ dB} & \text{at 450 MHz} \\ 6 \text{ dB} & \text{at 600 MHz} \end{cases}$$

Using matching networks to compensate for gain and to achieve a constant 10-dB gain over the bandwidth, the gain for G_s and G_L should be

$$\begin{array}{rl} -3 \text{ dB} & \text{at 300 MHz} \\ 0 \text{ dB} & \text{at 450 MHz} \\ +4 \text{ dB} & \text{at 600 MHz} \end{array}$$

For the input-matching network, we have

$$G_{s,\max} = \frac{1}{1 - |S_{11}|^2} = \begin{cases} 0.280 \text{ dB} & \text{at } 300 \text{ MHz} \\ 0.469 \text{ dB} & \text{at } 450 \text{ MHz} \\ 0.177 \text{ dB} & \text{at } 600 \text{ MHz} \end{cases}$$

Obviously, little can be done about $G_{s,\max}$ since the maximum gain is small (from 0.177 to 0.469 dB) over the frequency range. The output-matching network gain is

$$G_{L,\max} = \frac{1}{1 - |S_{22}|^2} = \begin{cases} 4.437 \text{ dB} & \text{at } 300 \text{ MHz} \\ 7.212 \text{ dB} & \text{at } 450 \text{ MHz} \\ 5.567 \text{ dB} & \text{at } 600 \text{ MHz} \end{cases}$$

Neglecting the small effects due to G_s, one can design G_L as follows to compensate for the $|S_{21}|$ variations over the frequency range:

$$G_L = \begin{cases} -3 \text{ dB} & \text{at } 300 \text{ MHz} \\ 0 \text{ dB} & \text{at } 450 \text{ MHz} \\ +4 \text{ dB} & \text{at } 600 \text{ MHz} \end{cases}$$

This example illustrates the use of a compensated matching network to achieve constant-gain performance of an amplifier over a wide frequency range.

The use of compensated matching networks to achieve constant-gain flatness has the drawback of high input and output VSWR. Use of a balanced structure can overcome this problem. The balanced amplifier has the advantages of flat gain and good input and output VSWR. The price paid is that it requires two amplifiers and two couplers.

Figure 14.16 shows the circuit arrangement of a balanced amplifier. Two hybrid couplers are used to divide and combine the signal. Every time the signal crosses the centerline, it experiences a 90° phase shift. The input signal at port 1 splits into two signals at ports 2 and 3 with equal amplitude but 90° out of phase. When these signals reflect back to port 1, the reflected waves are 180° out of phase at port 1. Therefore,

$$|S_{11}| = 0.5|S_{11a} - S_{11b}| \tag{14.69}$$

where S_{11} is for the overall circuit and S_{11a} and S_{11b} are for amplifiers A and B, respectively. The reflected waves at port 4, which are in phase and add together, are dissipated in a resistor. Similarly, we can write

$$|S_{21}| = 0.5|S_{21a} + S_{21b}| \tag{14.70}$$

$$|S_{12}| = 0.5|S_{12a} + S_{12b}| \tag{14.71}$$

$$|S_{22}| = 0.5|S_{22a} - S_{22b}| \tag{14.72}$$

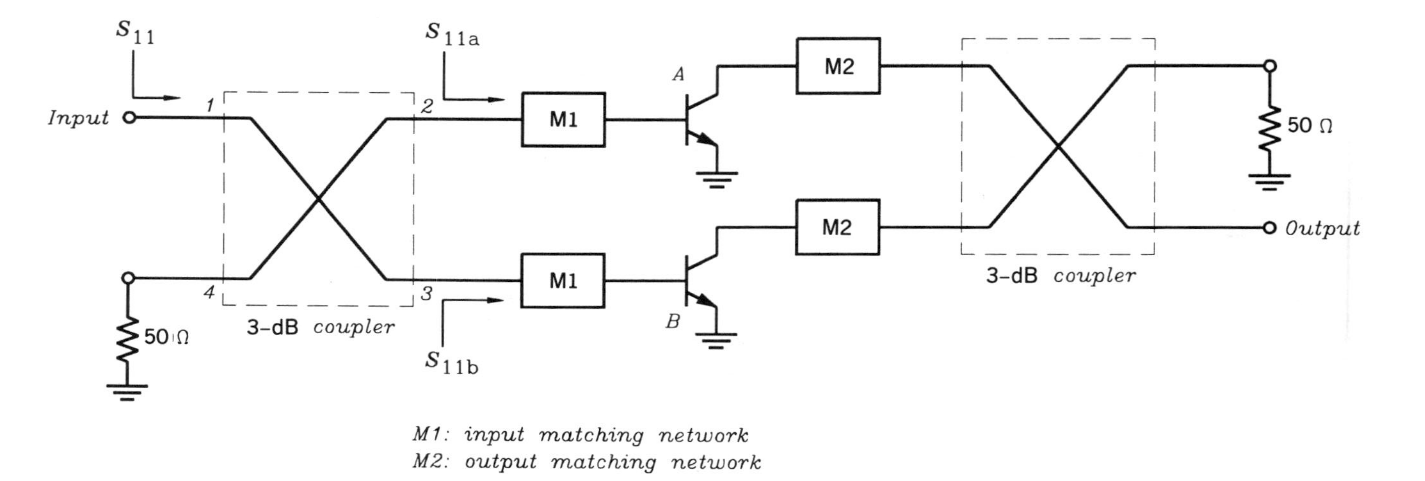

FIGURE 14.16 Balanced amplifier.

If the two amplifiers are nearly identical, we have

$$S_{11a} \approx S_{11b} \qquad S_{21a} \approx S_{21b}$$
$$S_{22a} \approx S_{22b} \qquad S_{12a} \approx S_{12b}$$

Substituting these into Equations (14.69) through (14.72) gives

$$S_{11} \approx 0 \qquad S_{22} \approx 0 \tag{14.73}$$

and the power gain

$$\begin{aligned} |S_{21}|^2 &= \tfrac{1}{4}|S_{21a} + S_{21b}|^2 \\ &= |S_{21a}|^2 \end{aligned} \tag{14.74}$$

The total gain of the balanced amplifier is equal to the gain of each individual amplifier. If $|S_{21a}|^2$ is designed with flat gain using the compensated matching networks described earlier, the overall gain is flat. Furthermore, the input and output VSWR are low, as given by Equation (14.73).

The use of negative feedback circuits for broadband amplifier design can be demonstrated by an example [1]. Figure 14.17(a) shows an FET amplifier with series and shunt feedback resistors R_1 and R_2. The simplified equivalent circuit is given in Figure 14.17(b). From Figure 14.17(b), the admittance (y) matrix can be written as (see Problem P14.15)

$$\begin{bmatrix} i_1 \\ i_2 \end{bmatrix} = \begin{bmatrix} \dfrac{1}{R_2} & -\dfrac{1}{R_2} \\ \dfrac{g_m}{1 + g_m R_1} - \dfrac{1}{R_2} & \dfrac{1}{R_2} \end{bmatrix} \begin{bmatrix} v_1 \\ v_2 \end{bmatrix} \tag{14.75}$$

From the y matrix, the S matrix can be derived as (use Table 3.1)

$$S_{11} = S_{22} = \frac{1}{D}\left[1 - \frac{g_m Z_0^2}{R_2(1 + g_m R_1)}\right] \tag{14.76}$$

$$S_{21} = \frac{1}{D}\left(\frac{-2g_m Z_0}{1 + g_m R_1} + \frac{2Z_0}{R_2}\right) \tag{14.77}$$

$$S_{12} = \frac{2Z_0}{DR_2} \tag{14.78}$$

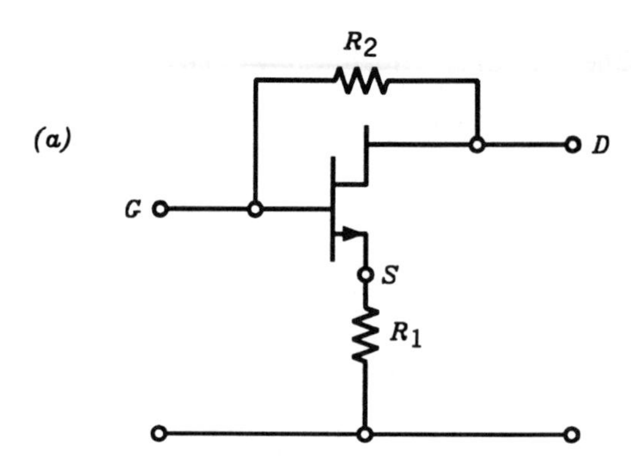

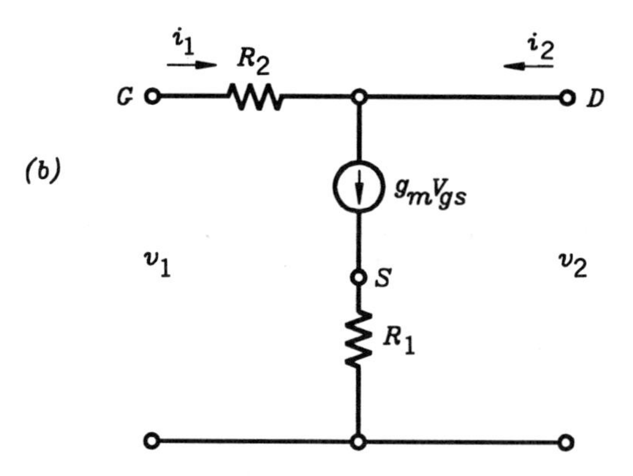

FIGURE 14.17 Amplifier with negative feedback circuits.

where

$$D = 1 + \frac{2Z_0}{R_2} + \frac{g_m Z_0^2}{R_2(1 + g_m R_1)}$$

If $S_{11} = S_{22} = 0$, we have

$$\frac{g_m Z_0^2}{R_2(1 + g_m R_1)} = 1$$

or

$$R_1 = \frac{Z_0^2}{R_2} - \frac{1}{g_m} \tag{14.79}$$

Substitution into Equation (14.77) gives

$$S_{21} = \frac{Z_0 - R_2}{Z_0} \tag{14.80}$$

Therefore, we have the following equations for R_1 and R_2:

$$R_1 = \frac{Z_0^2}{R_2} - \frac{1}{g_m} \tag{14.81}$$

$$R_2 = Z_0(1 + |S_{21}|) \tag{14.82}$$

Note that S_{21} is for the overall circuit, not just for the transistor. From Equation (14.81) it can be seen that when both R_1 and R_2 are used and g_m has a high value, the minimum input and output VSWR are obtained when $R_1 R_2 \approx Z_0^2$. Equation (14.80) indicates that S_{21} is only dependent on R_2.

For multioctave bandwidth, a distributed or traveling-wave amplifier is used. The technique has been used for vacuum tubes [5] and was applied to FETs in 1969 [6]. In this approach, the input and output capacitances of the active device, which are the bandwidth-limiting factors, are integrated with lumped elements to create distributed transmission lines, where they are rendered less harmful [7, 8]. Figure 14.18 shows a simplified circuit model of the traveling-wave monolithic amplifier. The circuit operates as follows. An RF signal applied at the input end of the gate line travels down the line to the other end, where it is absorbed by the terminating impedance. However, a significant portion of the signal is absorbed by the gate circuits of the individual FETs along the way. The input signal, sampled by the gate circuits at different phases (and generally, at different amplitudes) is transferred to the drain line via the gain mechanism of the FET. If the phase velocity of the signal in the drain line is identical to the phase velocity of the gate line, the various components of the amplified signal add constructively. This condition is obtained only for the forward-traveling wave. Any signal traveling back-

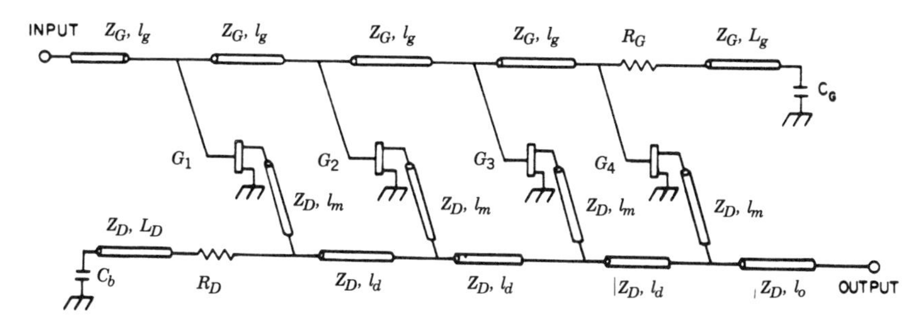

FIGURE 14.18 Schematic representation of a four-cell FET traveling-wave amplifier. (From Ref. 7 with permission from Howard W. Sams.)

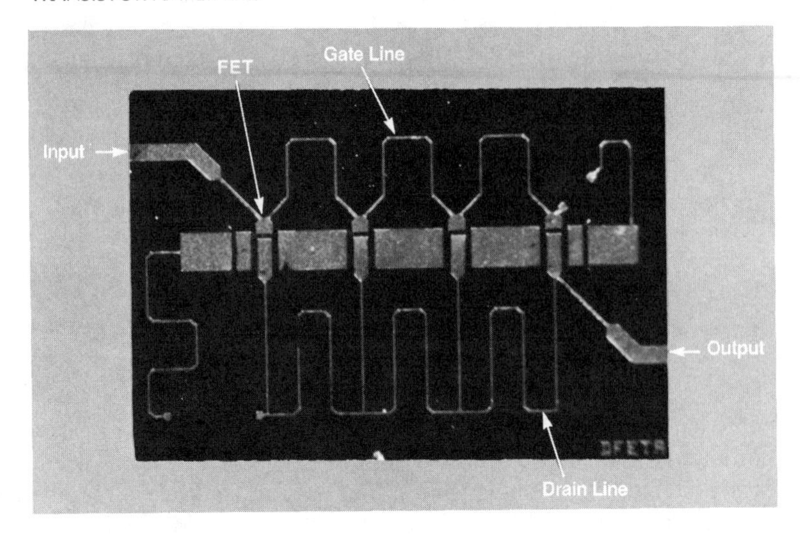

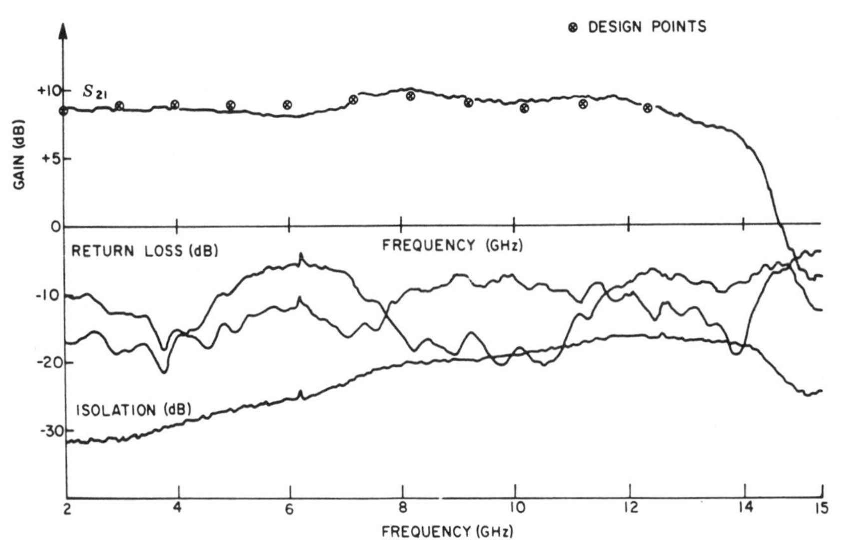

FIGURE 14.19 Four-FET monolithic distributed amplifier and its performance. (From Ref. 7 courtesy of Raytheon Co.)

ward on the drain line, which is not completely canceled, is absorbed by the drain line termination.

Figure 14.19 shows a 4-FET distributed amplifier and its performance. The monolithic chip has dimensions of $2.5 \times 1.65 \times 0.1$ mm. The performance shows a gain of approximately 9 dB over a very wide bandwidth.

14.8 HIGH-POWER AMPLIFIERS AND POWER COMBINERS

For high-power-amplifier applications, the device is operating in nonlinear regions with large-signal conditions. The large-signal S parameters are needed to design the transistor for power applications. The measurement of large-signal S parameters is generally difficult.

In the power amplifier, the 1-dB compression point, dynamic range, intermodulation, and intercept point described in Chapter 7 are useful parameters to characterize the amplifier. One would like to have high output power, high dc-to-RF efficiency, wide operating bandwidth, high 1-dB compression point, high dynamic range, and low intermodulation. A summary of the status of power FET performance in hybrid microwave integrated circuits (HMICs) and monolithic microwave integrated circuits (MMICs) is given in Figures 14.20 and 14.21, respectively.

The output power of a single transistor is directly proportional to the device size (total gate width). However, the size of a device is limited by thermal impedance, matching circuit bandwidth, and yield considerations. When more power is required than that can be provided by a single device, power-combining techniques are used [10, 11]. Many power-combining techniques have been demonstrated, as summarized in Figure 11.21.

Chip-level combining can be accomplished by designing the complete matching circuits of two individual FET cells using elements of characteristic impedance values that are half of that needed for the input or output impedance [9]. Chip-level combining uses small chip capacitors and matching circuits to match the 50-Ω impedance.

The use of planar 3-dB couplers (Wilkinson, Lange, branch-line, or rat-race couplers) is a straightforward method in building a two-device power combiner. The 3-dB coupler has the characteristics of wide bandwidth and high port-to-port isolation, which minimizes interactions among devices. However, the number of devices that can be combined is limited due to circuit loss of the couplers. Figure 14.22 shows the Lange coupler and branch-line coupler used for combining circuits [9].

Perhaps the most successful combining technique in FET power combining is the N-way multiport combiner using radial lines [12–14]. Figure 14.23 shows the configuration of a 30-way radial-line power combiner. A power output of 26 W was achieved at 11.3 GHz with a 0.5-dB bandwidth of

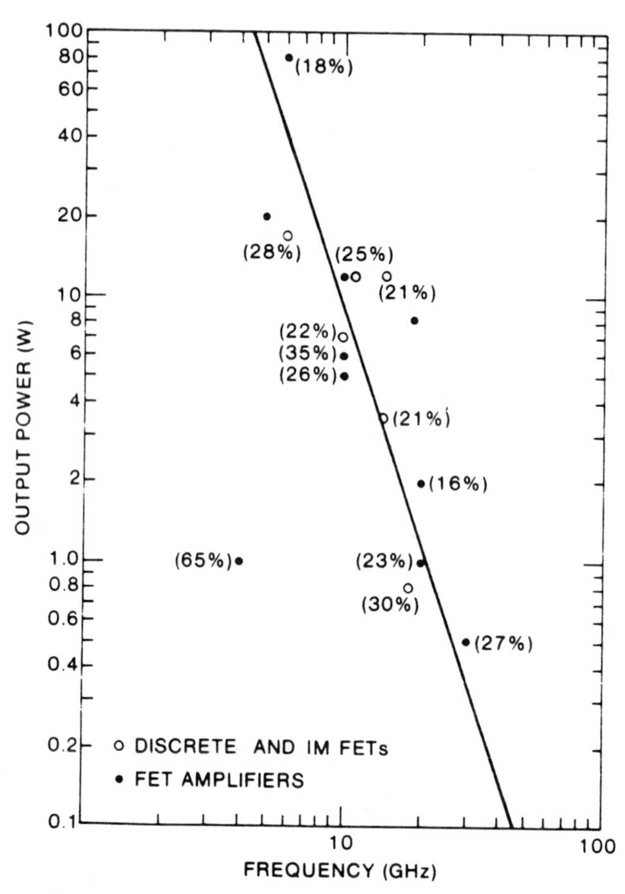

FIGURE 14.20 Status of discrete power FET and HMIC amplifier performance. Numbers in parentheses are power-added efficiencies. (From Ref. 9 with permission from Wiley.)

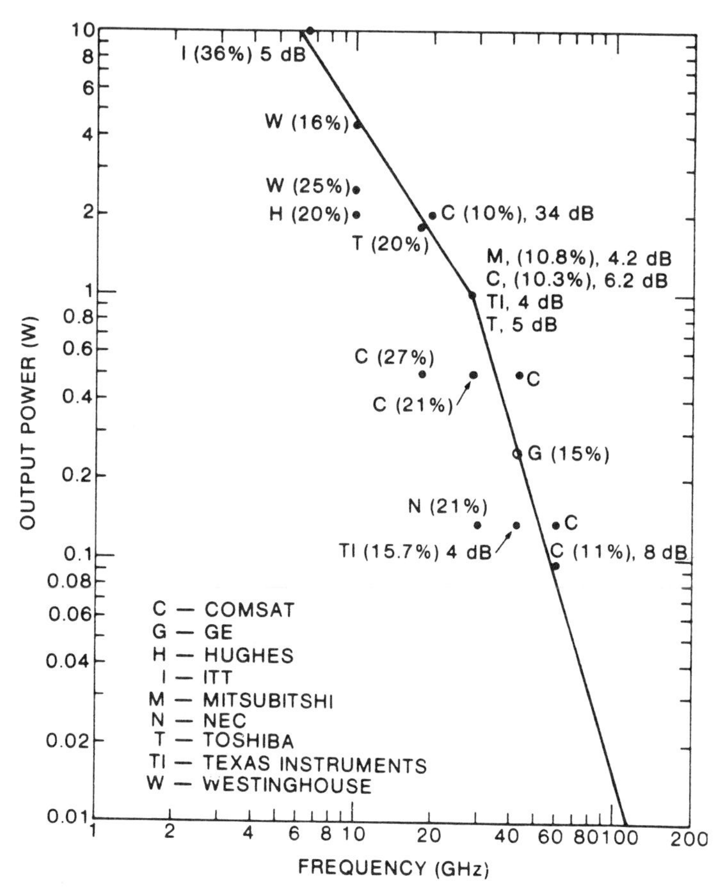

FIGURE 14.21 Status of MMIC power FET amplifier performance. Numbers in parentheses are power-added efficiencies. (From Ref. 9 with permission from Wiley.)

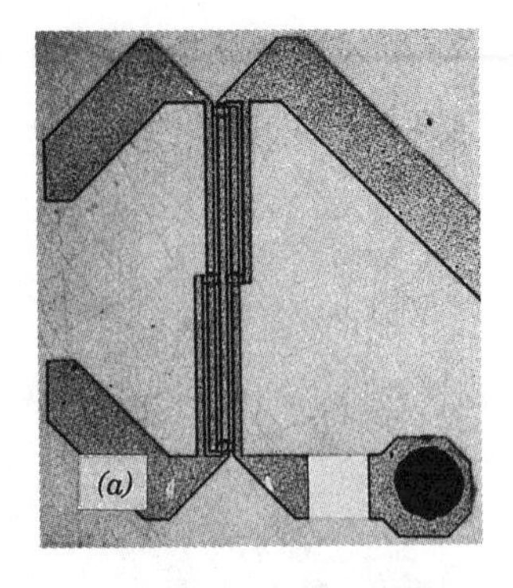

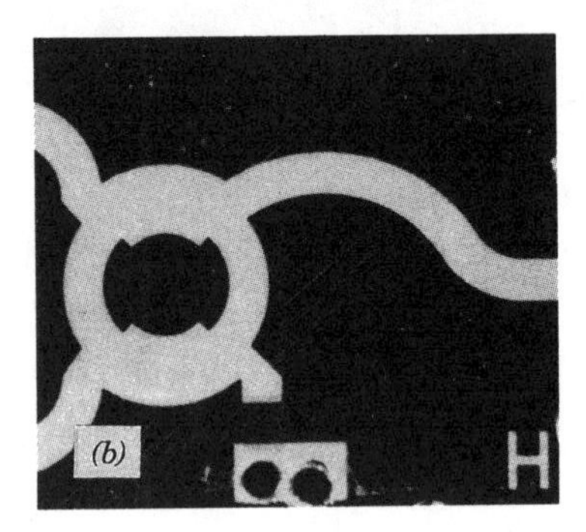

FIGURE 14.22 3-dB couplers; (*a*) Lange coupler; (*b*) branch-line coupler. (From Ref. 9.)

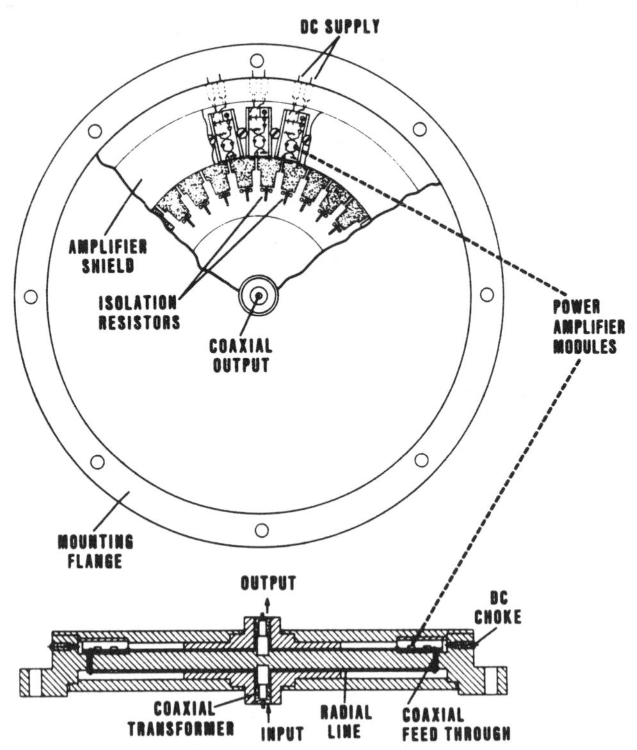

FIGURE 14.23 Radial-line power combiner. (From Ref. 13 with permission from

600 MHz. The power from a single amplifier is 1 W with approximately 5 dB gain.

Quasi-optical (or spatial) power combining is a feasible approach to combining the power of many radiating elements in free space [11, 15]. Quasi-optical power combining relies on arrays of individual oscillating elements, each of which radiates into free space. The oscillating elements are placed close to each other so that they will couple and will all oscillate at the same frequency through injection locking. Injection locking can be accomplished by mutual coupling or an external source. The total power radiated is the sum of the power radiated from all the individual oscillating elements. The individual element is an active antenna element obtained by integrating a FET oscillator with a planar antenna [16]. Quasi-optical power combiners using FET oscillators have been demonstrated with good combining efficiency [17, 18].

14.9 DC BIAS TECHNIQUES

DC biasing is an important design consideration for proper operation of amplifiers. The ideal biasing arrangement should select the proper quiescent point and hold the quiescent point constant over variations in transistor parameters and temperature. The dc and RF circuits should be isolated so that no RF signal leaks into the dc biasing circuit and the biasing circuit does not disturb RF performance.

Biasing techniques for two-terminal devices were discussed briefly in Section 8.6. DC biasing for three-terminal devices is more complicated. Since the GaAs FET is the most used three-terminal device, let us concentrate our discussion on this device. Figure 14.24 shows five basic dc biasing networks for GaAs FET amplifiers. The column "How" indicates the sequence in which the voltage must be applied to avoid transient burnout of the GaAs FET device during turn-on. For example, if the drain is biased positive before the gate is biased negative in Figure 14.24(*a*), the FET will operate momentarily beyond its safe operating region and the device will burn out.

A source resistor is used in Figure 14.24(*d*) and (*e*) to provide automatic transient protection. However, the resistor may degrade the noise figure of the amplifier and cause low-frequency oscillation. A zener diode can also be used to protect the amplifier.

Selection of the dc operating point depends on the application [1]. Four quiescent points, located at I, II, III, IV, are given for the typical GaAs FET I–V characteristics shown in Figure 14.25. For example, quiescent point I is recommended for low-noise, low-power applications. At point I, the FET operates at a low current level of $I_{dc} = 0.15 I_{DSS}$. For low noise and higher power gain, point II is chosen. The output power level can be increased by

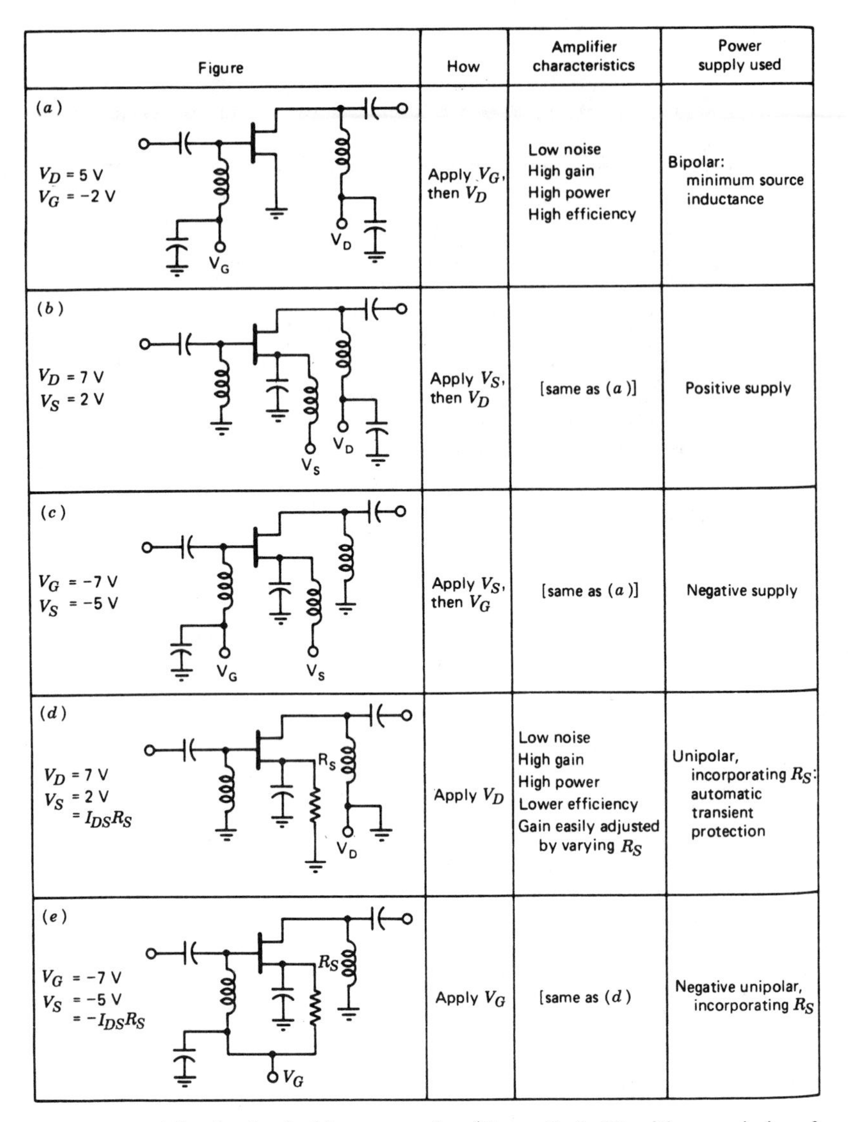

Figure	How	Amplifier characteristics	Power supply used
(*a*) $V_D = 5$ V, $V_G = -2$ V; V_G, V_D	Apply V_G, then V_D	Low noise High gain High power High efficiency	Bipolar: minimum source inductance
(*b*) $V_D = 7$ V, $V_S = 2$ V; V_S, V_D	Apply V_S, then V_D	[same as (*a*)]	Positive supply
(*c*) $V_G = -7$ V, $V_S = -5$ V; V_G, V_S	Apply V_S, then V_G	[same as (*a*)]	Negative supply
(*d*) $V_D = 7$ V, $V_S = 2$ V $= I_{DS}R_S$; R_S, V_D	Apply V_D	Low noise High gain High power Lower efficiency Gain easily adjusted by varying R_S	Unipolar, incorporating R_S: automatic transient protection
(*e*) $V_G = -7$ V, $V_S = -5$ V $= -I_{DS}R_S$; R_S, V_G	Apply V_G	[same as (*d*)	Negative unipolar, incorporating R_S

FIGURE 14.24 Five basic dc bias networks. (From Ref. 19 with permission from *Microwaves & RF*.)

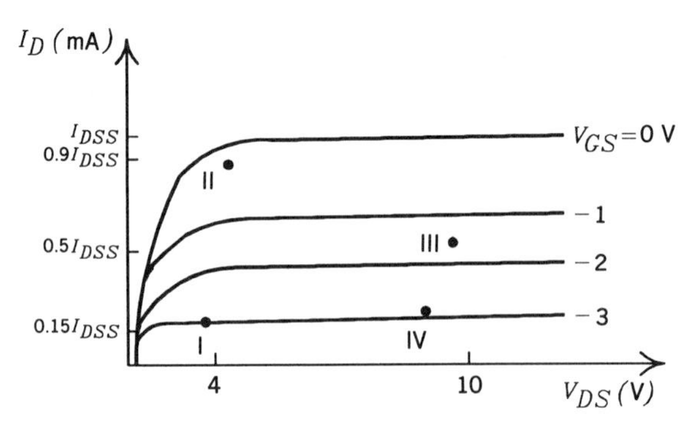

FIGURE 14.25 Typical GaAs FET characteristics and recommended quiescent points. (From Ref. 1 with permission from Prentice Hall.)

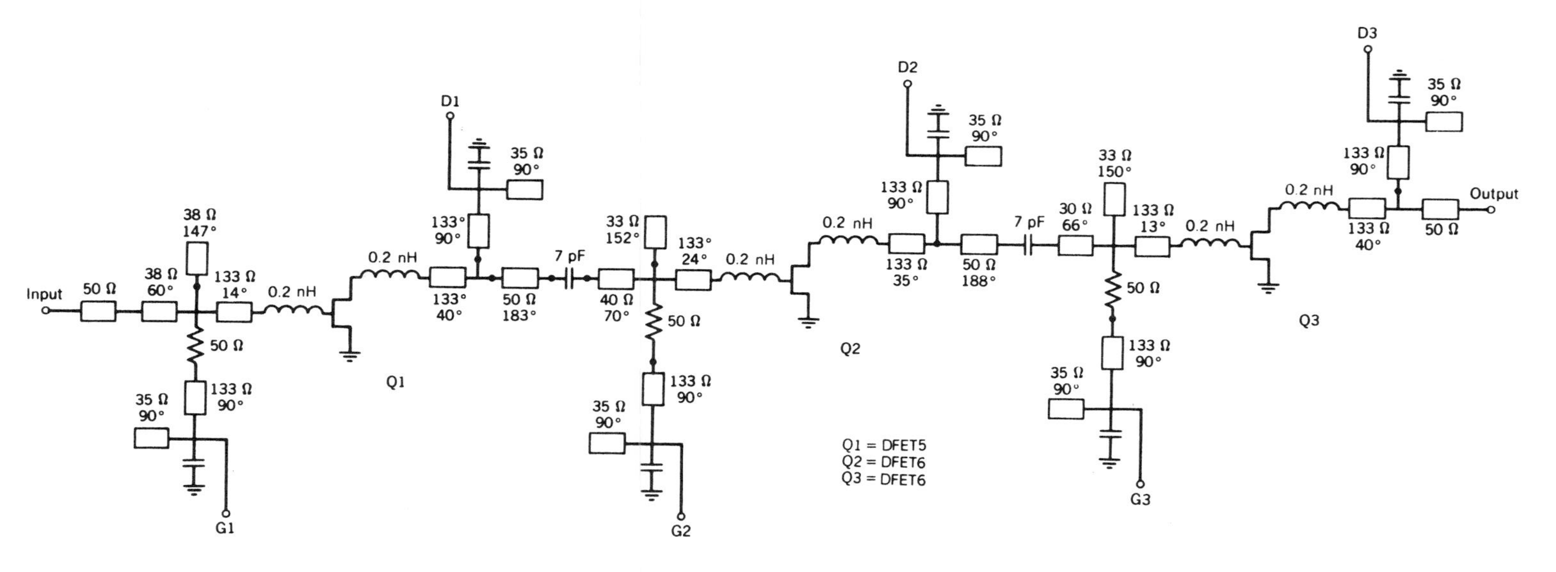

FIGURE 14.26 K-band MESFET amplifier design. (From Ref. 20.)

selecting quiescent point III with $I_{DS} = 0.5I_{DSS}$. Points I, II, and III are maintained in class A operation.* For class AB or B operation, the drain-to-source current must be decreased from point III. For high efficiency, point IV is selected for AB or B operation.

14.10 PRACTICAL CIRCUITS

In this section we present several practical circuit examples for amplifiers. Figure 14.26 shows a K-band (18 to 26 GHz) hybrid integrated circuit amplifier. The amplifier consists of three stages fabricated in a microstrip on multiple 0.025-in.-thick quartz substrates. Rectangular waveguides are used at the input and output ports. Low-loss microstrip-to-waveguide transitions are employed to couple the power from waveguide to microstrip, and vice versa. The entire set of cascaded amplifier circuits is mounted in a carrier located in a channel operating below waveguide cutoff. The circuit was designed using a commercial CAD software package. The circuit is shown in Figure 14.27. The performance of the amplifier is shown in Figure 14.28. Good agreement was achieved between the measurement and simulation.

Figure 14.29 shows the first monolithic low-noise amplifier reported by NEC. The circuit is a one-stage amplifier built on a 2.75 × 1.95 × 0.15 mm GaAs chip. The large pads are used for the source RF grounds. Figure 14.30 shows a two-stage X-band power FET amplifier. The circuit provides a saturated output power of 1.6 W over a 2-GHz bandwidth centered at 9 GHz with a gain of approximately 10 dB.

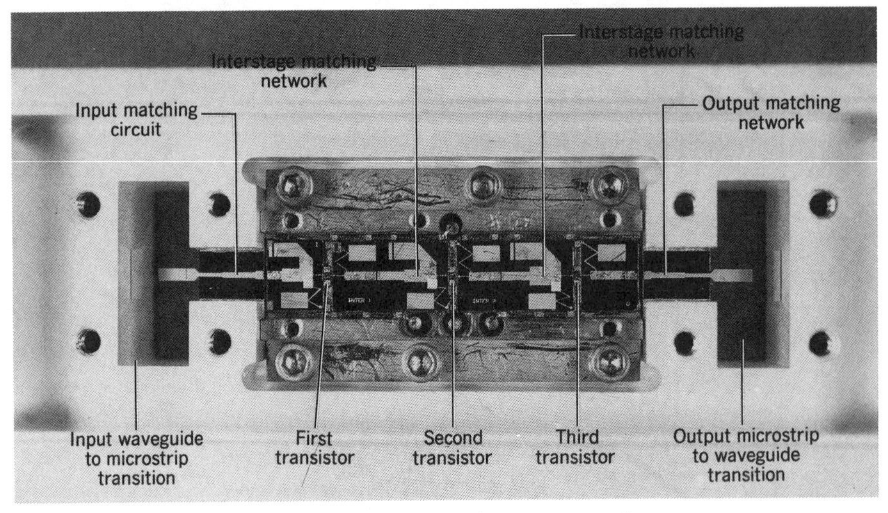

FIGURE 14.27 K-band MESFET amplifier. (From Ref. 20.)

*For the definitions of class A, AB, or B, refer to IEEE Standard Dictionary of Electrical and Electronics Terms or any other electronics books. The classification was originally used for vacuum tube amplifiers and then for bipolar transistors and FETs.

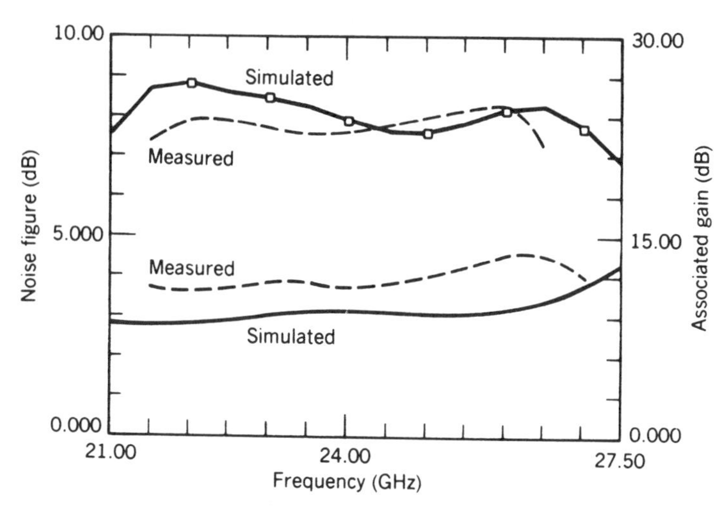

FIGURE 14.28 Comparison of the measured and simulated performance of the K-band MESFET amplifier. (From Ref. 20.)

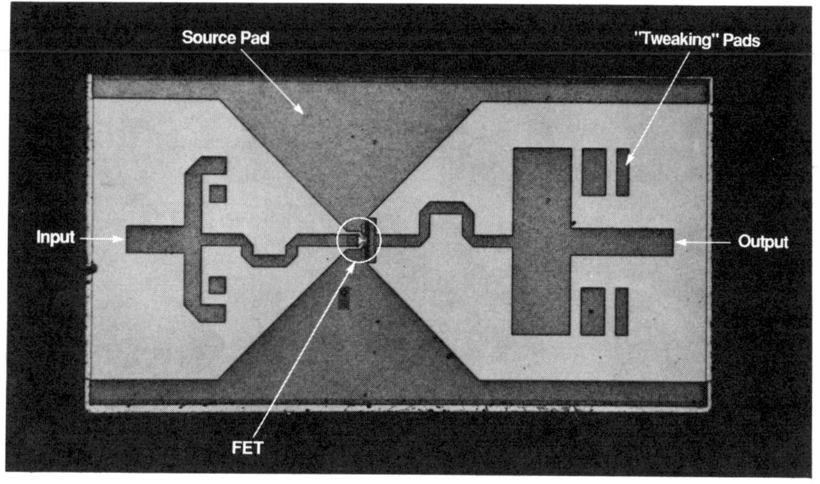

FIGURE 14.29 First monolithic low-noise amplifier. (From Refs. 7 and 21 with permission from IEEE.)

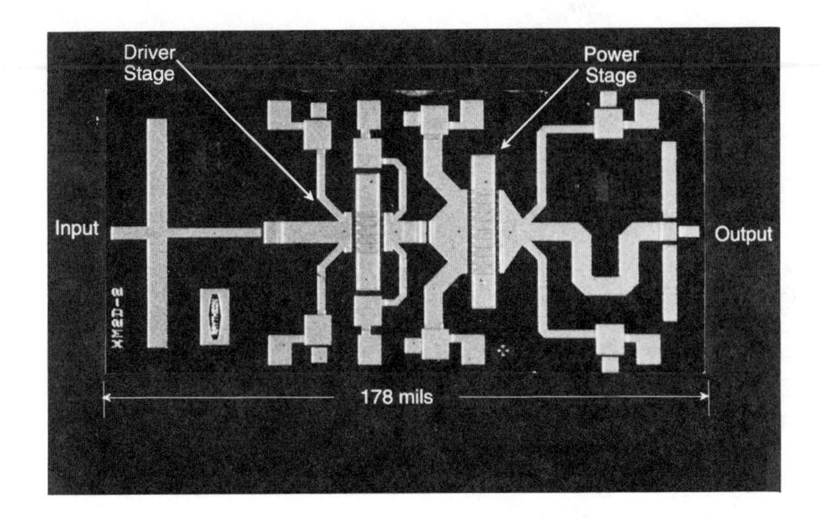

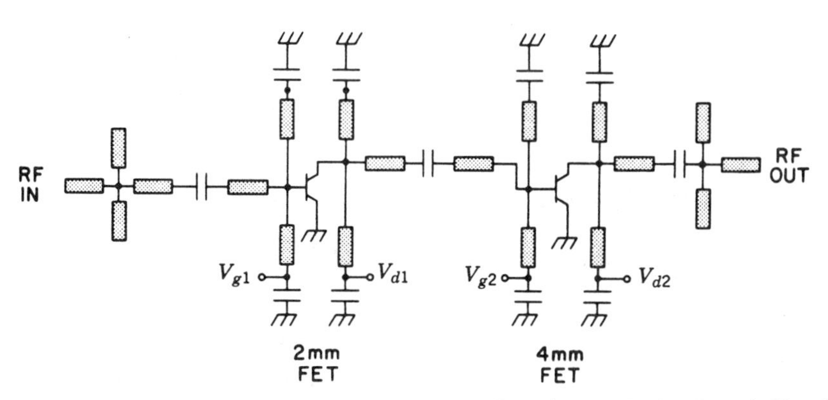

FIGURE 14.30 Two-stage 9-GHz power amplifier. (From Refs. 7 and 22 with permissions from IEEE and Raytheon Co.)

PROBLEMS

P14.1 Derive Equations (14.11) and (14.12).

P14.2 Prove that the maximum transducer gain occurs when $\Gamma_s = S_{11}^*$ and $\Gamma_L = S_{22}^*$.

P14.3 Derive Equations (14.40) and (14.41) for stability circles.

P14.4 In each of the stability circles shown in Figure P14.4, indicate clearly the stable region for Γ_s (or Z_s) and Γ_L (or Z_L). The solid circles are Smith charts.

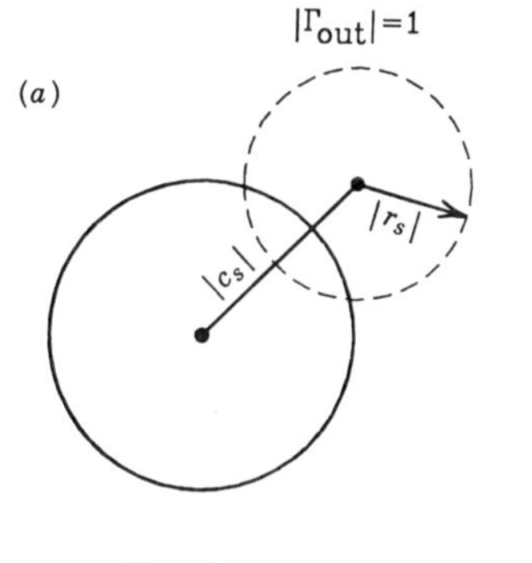

$K < 1$

$|S_{11}| < 1$

$|S_{22}| > 1$

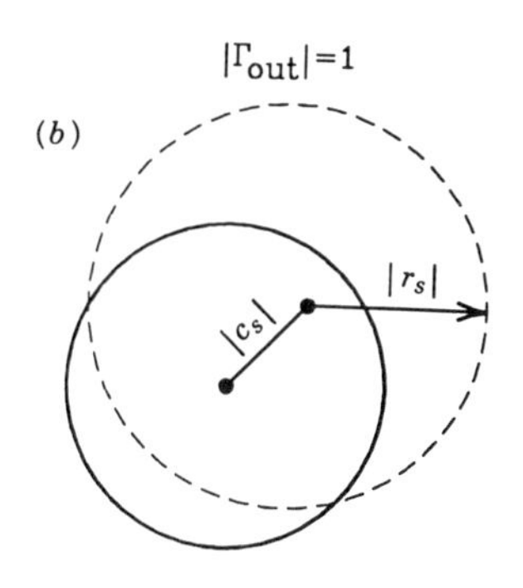

$K < 1$

$|S_{11}| > 1$

$|S_{22}| < 1$

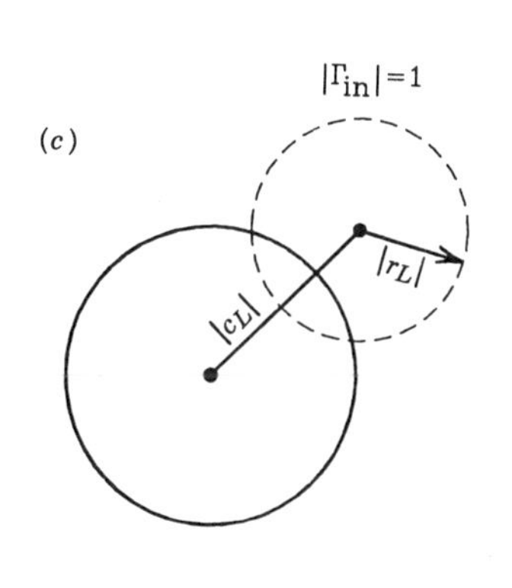

$K < 1$

$|S_{11}| < 1$

$|S_{22}| > 1$

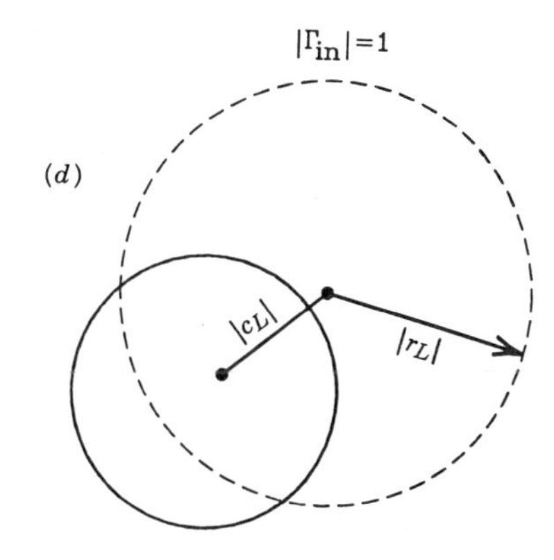

$K < 1$

$|S_{11}| > 1$

$|S_{22}| < 1$

FIGURE P14.4

P14.5 A FET amplifier has the following S parameters for the device: $S_{11} = 0.5\ \angle 180°$, $S_{21} = 4.0\ \angle 90°$, $S_{12} = 0$, and $S_{22} = 0.5\ \angle 45°$.

(a) Is the amplifier stable?

(b) What is the maximum gain in dB?

(c) What is the input impedance Z_{in}? ($Z_0 = 50\ \Omega$.)

(d) What is the load impedance Z_L for the maximum-gain case?

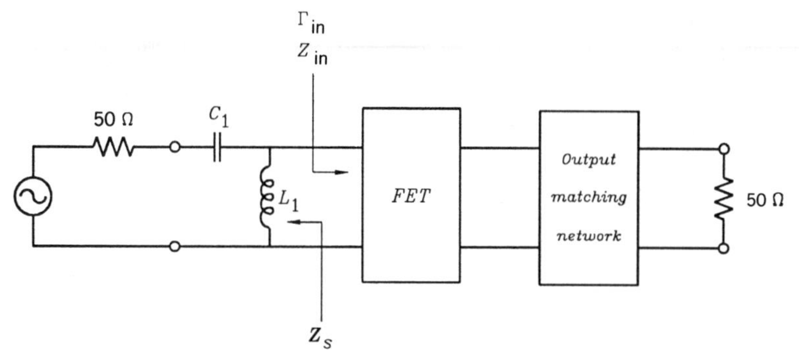

FIGURE P14.6

P14.6 An amplifier is operating at 10 GHz using an FET device with the following S parameters:
$S_{11} = 0.5 \angle -45°$, $S_{12} = 0$, $S_{21} = 5 \angle 30°$, and $S_{22} = 0.8 \angle -160°$. Design the amplifier for maximum gain using a 50-Ω input and output transmission lines as shown in Figure P14.6.

(**a**) Is the amplifier stable?

(**b**) What is the maximum gain in dB?

(**c**) Design the input matching network for maximum gain using series C_1 and shunt L_1 elements to match the 50-Ω line to Z_s. Give L_1 in nH and C_1 in pF. (Use a Z–Y Smith chart.)

(**d**) What are the values of Γ_{in} and its corresponding Z_{in}?

P14.7 A FET device has the following S parameters at 3 GHz: $S_{11} = 0.3 \angle -60°$, $S_{12} = 0$, $S_{21} = 2 \angle 45°$, and $S_{22} = 0.8 \angle -30°$. Design an amplifier (Figure P14.7) for maximum gain using this transistor and 50-Ω input and output transmission lines.

(**a**) Check the stability.

(**b**) What is the maximum gain in dB?

(**c**) Design an input-matching network for maximum gain using series L and shunt C elements to match a 50-Ω line to Z_s. Give L in nH and C in pF.

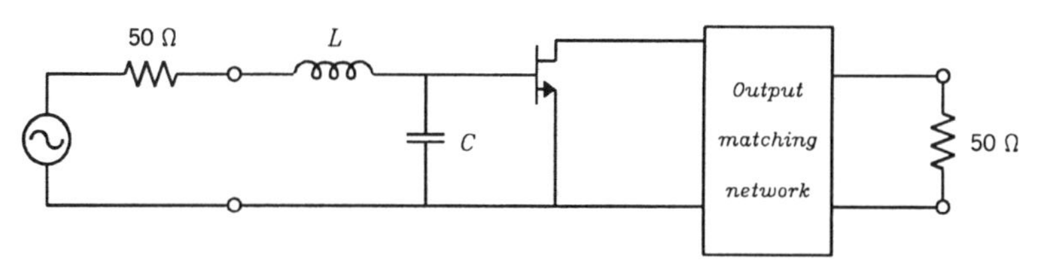

FIGURE P14.7

(d) Design an output-matching network for maximum gain using series C and shunt L elements to match a 50-Ω line to Z_L. Give L in nH and C in pF.

(e) What is the VSWR value seen by the source and by the load?

P14.8 At 10 GHz, a transistor has the constant-gain and constant-noise circles as shown in Figure P14.8. Determine the $\overline{Z}_s$ value required to achieve a noise figure of 3 dB and a unilateral gain of 15.2 dB. The transistor has $S_{21} = 4.45 \angle 65°$, $S_{22} = 0.21 \angle -80°$, and $S_{12} = 0$.

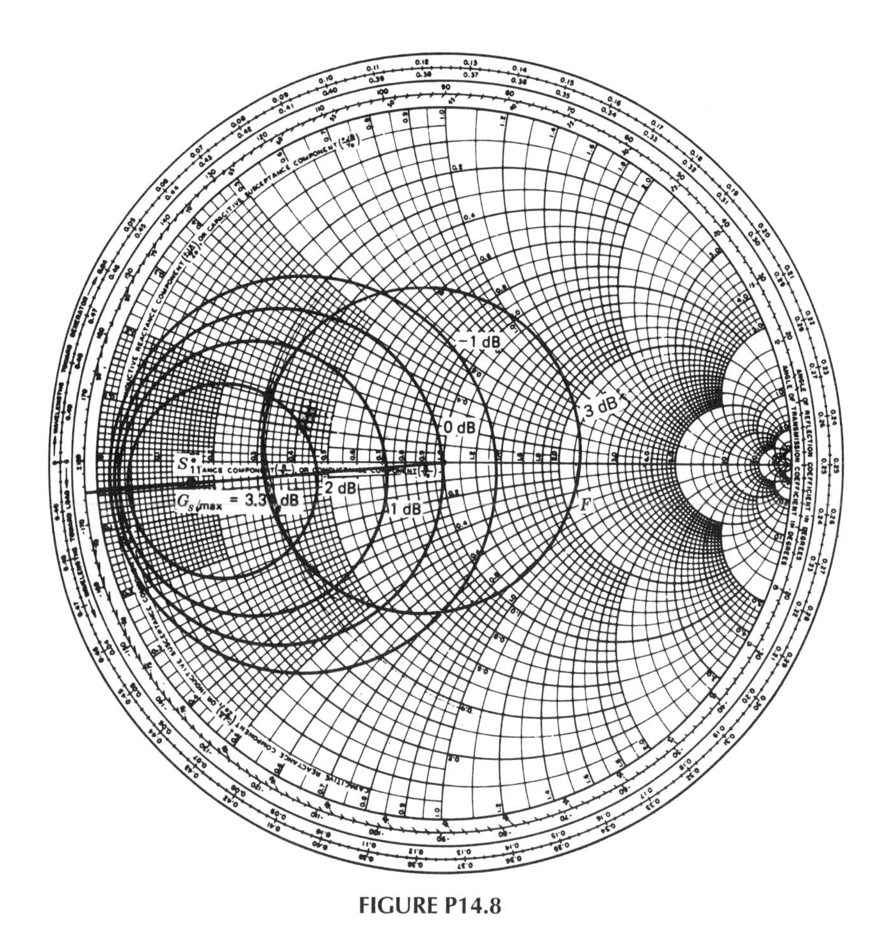

FIGURE P14.8

P14.9 In Figure P14.9, indicate the possible locations for a stable source impedance (Z_s) and stable load impedance (Z_L). The solid circle is the Smith chart.

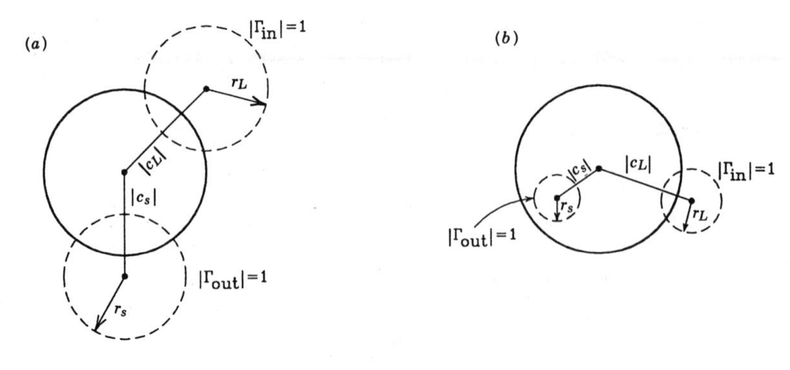

$K < 1,\ |\Delta| < 1,\ |S_{11}| < 1,\ |S_{22}| < 1$

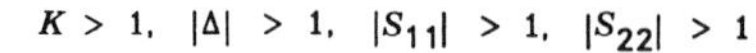

$K > 1,\ |\Delta| > 1,\ |S_{11}| > 1,\ |S_{22}| > 1$

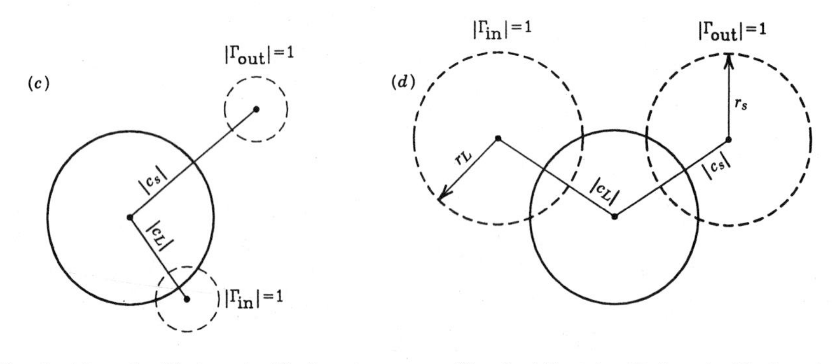

$K > 1,\ |\Delta| < 1,\ |S_{11}| > 1,\ |S_{22}| < 1$

$K > 1,\ |\Delta| < 1,\ |S_{11}| > 1,\ |S_{22}| < 1$

FIGURE P14.9

P14.10 The FET amplifier shown in Figure P14.10 has the following S parameters: $S_{11} = 0.5\ \angle 180^\circ$, $S_{21} = 3.0\ \angle 90^\circ$, $S_{12} = 0$, and $S_{22} = 0.5\ \angle -90^\circ$. The circuit is terminated by $R_s = 50\ \Omega$ and $R_L = 100\ \Omega$. The characteristic impedance of the system is 50 Ω. Find **(a)** Z_{in} in

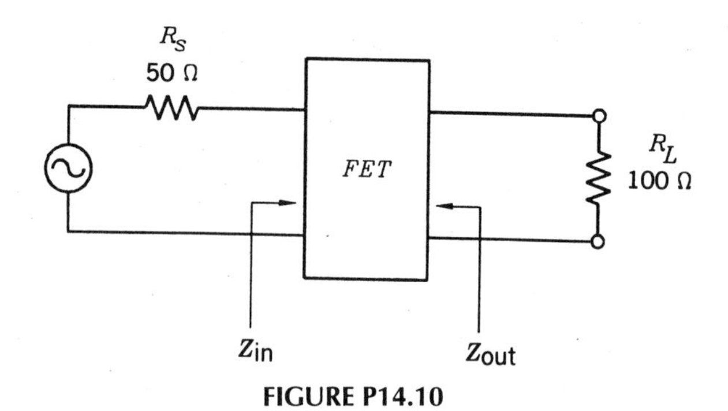

FIGURE P14.10

ohms, **(b)** Z_{out} in ohms, **(c)** the unilateral power gain in dB, and **(d)** the maximum unilateral power gain if the matching networks are used.

P14.11 A FET has the following parameters at 10 GHz:

$$S_{11} = 0.55 \angle 170^\circ \qquad F_{min} = 2.5 \text{ dB}$$
$$S_{12} = 0.05 \angle 23^\circ \qquad \Gamma_{sm} = 0.475 \angle 165^\circ$$
$$S_{21} = 1.7 \angle 25^\circ \qquad R_N = 3.5 \ \Omega$$
$$S_{22} = 0.84 \angle -67^\circ$$

Since $|S_{12}|$ is small, use the unilateral approximation.

(a) Draw the constant-gain circles for $G_s = 0$ and $+2$ dB.

(b) Draw the constant-noise circle for $F = 2$ dB.

(c) What is Γ_s if one wants to have $G_s = 2$ dB and $F = 2$ dB?

(d) What is Γ_s if one wants to have $G_s = 0$ dB and $F = 2$ dB?

P14.12 Design a FET amplifier for a minimum-noise figure and as high a gain value as possible. Use open stubs and quarter-wavelength transformers for matching. The circuit will be fabricated in microstrip on a Duroid substrate with $\varepsilon_r = 10$ and thickness $h = 0.050$ in. The FET device has the following S parameters at 3 GHz over a 100-MHz bandwidth:

$$S_{11} = 0.9 \angle -90^\circ \qquad \Gamma_{sm} = 0.5 \angle 135^\circ$$
$$S_{12} = 0 \qquad F_{min} = 3 \text{ dB}$$
$$S_{21} = 2 \angle 90^\circ$$
$$S_{22} = 0.5 \angle -45^\circ \qquad R_N = 4 \ \Omega$$

The amplifier is operated with a bandwidth of 100 MHz.

(a) What is the total gain in dB with the minimum-noise figure?

(b) Design the input-matching network using a quarter-wavelength transformer and 50-Ω open stub to match a 50-Ω line to Z_s.

(c) Design the output matching network using a quarter-wavelength transformer and 50-Ω open stud to match a 50-Ω line to Z_L.

(d) In part (b), give the actual length and width of the matching elements in inches. (Use charts or formulas to obtain the answer.)

(e) In part (c), give the actual length and width of the matching elements in inches.

(f) If an input signal of -20 dBm is received by the amplifier, what is the output signal power in dBm and the output noise power in dBm?

P14.13 Derive Equation (14.57).

P14.14 Derive Equation (14.63) from (14.62).

P14.15 Prove Equation (14.75).

REFERENCES

1. G. Gonzalez, *Microwave Transistor Amplifiers*, Prentice-Hall, Englewood Cliffs, N.J., 1984.
2. K. Kurokawa, "Power Waves and Scattering Matrix," *IEEE Transactions on Microwave Theory and Techniques*, Vol. MTT-13, March 1965, pp. 194–202.
3. D. M. Pozar, *Microwave Engineering*, Addison-Wesley, Reading, Mass., 1990, Chap. 11.
4. H. Rothe and W. Dahlke, "Theory of Noisy Fourpoles," *Proceedings of the IRE*, Vol. 44, June 1956, pp. 811–818.
5. E. L. Gintzon, W. R. Hewlett, J. H. Jasburg, and J. D. Noe, "Distributed Amplification," *Proceedings of the IRE*, Vol. 36, 1948, pp. 456–459.
6. W. Jutzi, "A MESFET Distributed Amplifier with 2 GHz Bandwidth," *Proceedings of the IEEE*, Vol. 57, 1969, pp. 1195–1196.
7. R. A. Pucel, "Applications of Monolithic Microwave Integrated Circuit," in *Gallium Arsenide Technology*, D. K. Ferry, Editor, Howard W. Sams, Indianapolis, Ind., 1985, pp. 249–301.
8. Y. Ayasli, L. Mozzi, J. L. Vorhause, L. D. Reynolds, and R. A. Pucel, "A Monolithic GaAs 1–13 GHz Traveling-Wave Amplifier," *IEEE Transactions on Microwave Theory and Techniques*, Vol. MTT-30, July 1982, pp. 976–981.
9. H. A. Hung, T. Smith, and H. Huang, "FETs: Power Applications," in *Handbook of Microwave and Optical Components*, K. Chang, Editor, Wiley, New York, 1990, Vol. 2, Chap. 10.
10. K. J. Russell, "Microwave Power Combining Techniques," *IEEE Transactions on Microwave Theory and Techniques*, Vol. MTT-27, May 1979, pp. 472–478.
11. K. Chang and C. Sun, "Millimeter-Wave Power Combining Techniques," *IEEE Transactions on Microwave Theory and Techniques*, Vol. MTT-31, February 1983, pp. 91–107.
12. J. Goel, "A K-Band GaAs FET Amplifier with 8.2 W Output Power," *IEEE Transactions on Microwave Theory and Techniques*, Vol. MTT-32, March 1984, pp. 317–324.
13. E. Belohoubeck et al., "30-Way Radial Power Combiner for Miniature GaAs FET Power Amplifiers," *IEEE-MTT-S International Microwave Symposium Digest*, June 1986, pp. 515–519.
14. G. W. Swift and D. I. Stones, "A Comprehensive Design Technique for Radial Wave Power Combiner," *IEEE-MTT-S International Microwave Symposium Digest*, June 1988, pp. 279–281.
15. J. W. Mink, "Quasi-optical Power Combining of Solid-State Millimeter-Wave Sources," *IEEE Transactions on Microwave Theory and Techniques*, Vol. MTT-34, February 1986, pp. 273–279.

16. K. Chang, K. A. Hummer, and G. K. Gopalakrishnan, "Active Radiating Element Using FET Source Integrated with Microstrip Patch Antenna," *Electronics Letters*, Vol. 24, October 13, 1988, pp. 1347–1348.
17. Z. B. Popovic, R. M. Weikle II, M. Kim, K. A. Potter, and D. B. Rutledge, "Bar-Grid Oscillators," *IEEE Transactions on Microwave Theory and Techniques*, Vol. MTT-38, March 1990, pp. 225–230.
18. R. A. York and R. C. Compton, "Quasi-optical Power Combining Using Mutually Synchronized Oscillator Arrays," *IEEE Transactions on Microwave Theory and Techniques*, Vol. MTT-39, June 1991, pp. 1000–1009.
19. G. D. Vendelin, "Five Basic Bias Designs for GaAs FET Amplifiers," *Microwaves & RF*, February 1978, pp. 40–42.
20. T. A. Midford, "FETs: Low Noise Applications," in *Handbook of Microwave and Optical Components*, K. Chang, Editor, Wiley, New York, 1990, Vol. 2, Chap. 11.
21. A. Higashisaka and T. Mizuta, "20-GHz Band Monolithic GaAs FET Low-Noise Amplifier," *IEEE Transactions on Microwave Theory and Techniques*, Vol. MTT-29, January 1981, pp. 1–6.
22. A. Platzker, M. S. Durschlag, and J. Vorhaus, "Monolithic Broadband Power Amplifier at X-Band," *IEEE Monolithic Circuit Symposium Digest*, 1983, pp. 59–61.

CHAPTER 15

Transistor Oscillators

15.1 INTRODUCTION

The design of transistor oscillators is very similar to that of transistor amplifiers. The same transistors, dc bias circuits, and S parameters are used for the oscillator design except that the circuit is designed to be unstable. The small- and large-signal S parameters provide all the information necessary for designing an oscillator.

Resonators are normally incorporated into the oscillator design to achieve low noise and high-frequency stability. The resonator can be a lumped element, a distributed transmission line, a cavity, or a dielectric disk. A varactor diode or ferrimagnetic material can be used to electronically tune the oscillating frequency.

15.2 TWO-PORT TRANSISTOR OSCILLATORS

Figure 15.1 shows a two-port transistor oscillator. Z_L is the load impedance and Z_T is the terminating impedance seen by the transistor. The terminating network is used to provide $|\Gamma_{out}| > 1$ which is necessary for oscillation to start. The load network determines the oscillation frequency and power delivered to the 50-Ω load. The operating principle is similar to that of the one-port oscillator using a two-terminal device described in Chapter 9. Two approaches are generally used to design an oscillator: the small-signal approach and the large-signal approach.

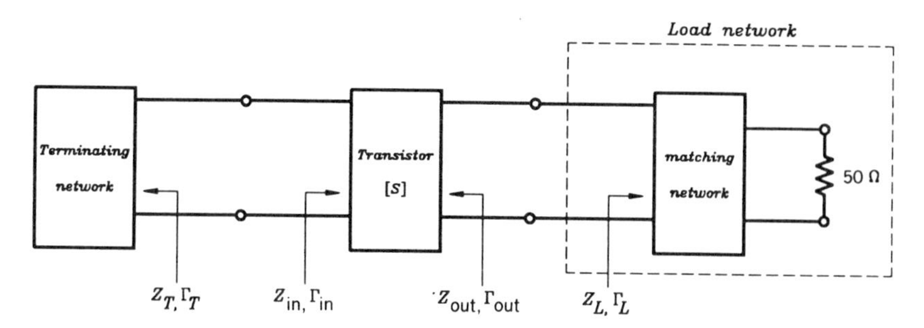

FIGURE 15.1 Two-port transistor oscillator circuit.

15.3 SMALL-SIGNAL DESIGN APPROACH USING SMALL-SIGNAL S PARAMETERS

For oscillation to occur at a frequency of f_0, the following conditions need to be satisfied:

$$|R_{out}(V, f_0)| > R_L(f_0) \tag{15.1}$$

$$X_{out}(V, f_0) + X_L(f_0) = 0 \tag{15.2}$$

where R_{out} is negative. The first equation ensures that $|\Gamma_{out}| > 1$, and the second equation determines the oscillation frequency. As long as $|R_{out}(V, f_0)|$ at a certain voltage V is greater than R_L, the network has the potential for oscillation. In other words, when the power supply voltages (V_{ds}, V_{gs}) of a MESFET oscillator are turned on, the oscillations start to build from the noise level. The output amplitude (voltage V, current I, and power output P) continues to grow until it is limited by the saturation effects of the device. Therefore, the ability of a transistor to begin oscillations in a given circuit is determined by the noise level and thus can be analyzed by small-signal techniques. The negative resistance R_{out} is a function of voltage, and as the oscillation power is increased, the negative-resistance value is reduced. If the negative resistance is decreased to a value lower than the load resistance, oscillation will cease. This problem can be eliminated by designing the magnitude of the negative resistance at $V = 0$ to be much larger than the load. For example, a design factor of $|R_{out}(0, f_0)| = 3R_L(f_0)$ is used in practice to ensure that the oscillation does not stop as it approaches steady-state conditions [1].

The conditions for oscillation can also be expressed as [2]

$$K < 1 \tag{15.3}$$

$$\Gamma_{in}\Gamma_T = 1 \tag{15.4}$$

$$\Gamma_{out}\Gamma_L = 1 \tag{15.5}$$

Since $|\Gamma_L|$ and $|\Gamma_T|$ are less than 1, Equations (15.4) and (15.5) imply that $|\Gamma_{in}| > 1$ and $|\Gamma_{out}| > 1$.

It can be proved that if the output port oscillates, the terminating port also oscillates. The output port oscillates if

$$\Gamma_{out}\Gamma_L = 1 \tag{15.6}$$

Equations (14.11) and (14.12) are rewritten as

$$\Gamma_{in} = S_{11} + \frac{S_{12}S_{21}\Gamma_L}{1 - S_{22}\Gamma_L} \tag{15.7}$$

$$\Gamma_{out} = S_{22} + \frac{S_{12}S_{21}\Gamma_T}{1 - S_{11}\Gamma_T} \tag{15.8}$$

From Equations (15.6) and (15.8), we have

$$\Gamma_L = \frac{1}{\Gamma_{out}} = \frac{1 - S_{11}\Gamma_T}{S_{22} - \Delta\Gamma_T} \tag{15.9}$$

where $\Delta = S_{11}S_{22} - S_{12}S_{21}$. Equation (15.9) can be rearranged to give

$$\Gamma_T = \frac{1 - S_{22}\Gamma_L}{S_{11} - \Delta\Gamma_L} \tag{15.10}$$

Equation (15.7) can be written as

$$\Gamma_{in} = \frac{S_{11} - \Delta\Gamma_L}{1 - S_{22}\Gamma_L} \tag{15.11}$$

Multiplying Equation (15.10) by (15.11) gives

$$\Gamma_T\Gamma_{in} = 1 \tag{15.12}$$

Therefore, the terminating port is also oscillating. The results can be generalized to an n-port oscillator by showing that the oscillator is oscillating simultaneously at each port.

The following steps can be used to design a transistor oscillator:

1. Select a potentially unstable transistor at the frequency of oscillation. If it is not potentially unstable, use feedback elements to make it unstable.
2. Design the terminating network, Z_T or Γ_T, to make $|\Gamma_{out}| > 1$ by choosing Γ_T or Z_T in the unstable zone of the stability circle.

3. From Z_T and the transistor small-signal S parameters, calculate Γ_{out} and confirm $|\Gamma_{out}| > 1$.

$$\Gamma_{out} = S_{22} + \frac{S_{12}S_{21}\Gamma_T}{1 - S_{22}\Gamma_T}$$

$$|\Gamma_{out}| > 1$$

4. Choose the load according to the oscillation conditions as follows:

$$X_L(f_0) = -X_{out}(f_0)$$
$$R_L(f_0) = \tfrac{1}{3}|R_{out}(0, f_0)|$$

The value chosen for Z_L usually produces a working oscillator. The measured oscillation frequency will be shifted from the design value since $X_{out}(f_0)$, used in determining f_0, is assumed to be independent of the amplitude V.

5. Design the load-matching network to transform a 50-Ω load to Z_L.

Example 15.1 The S parameters of a GaAs MESFET at 8 GHz are $S_{11} = 0.980\ \angle 163°$, $S_{12} = 0.390\ \angle -54°$, $S_{21} = 0.675\ \angle -161°$, and $S_{22} = 0.465\ \angle 120°$. Design an 8-GHz oscillator using this device [1].

SOLUTION: (1) Check the stability at 8 GHz.

$$K = \frac{1 - |S_{11}|^? - |S_{22}|^2 + |\Delta|^2}{2|S_{12}S_{21}|} = 0.529 < 1 \qquad \text{potentially unstable}$$

$$\Delta = S_{11}S_{22} - S_{12}S_{21}$$

(2) Choose Z_T or Γ_T to make $|\Gamma_{out}| > 1$ by drawing a stability circle in the input (terminating) plane. The input stability circle has a radius and center given by

$$c_s = \frac{(S_{11} - \Delta S_{22}^*)^*}{|S_{11}|^2 - |\Delta|^2} = 1.35\angle -156°$$

$$r_s = \left| \frac{S_{12}S_{21}}{|S_{11}|^2 - |\Delta|^2} \right| = 0.521$$

The circle is drawn in Figure 15.2. If Z_T is chosen at A shown in Figure 15.2, then

$$Z_T = -j7.5\ \Omega$$

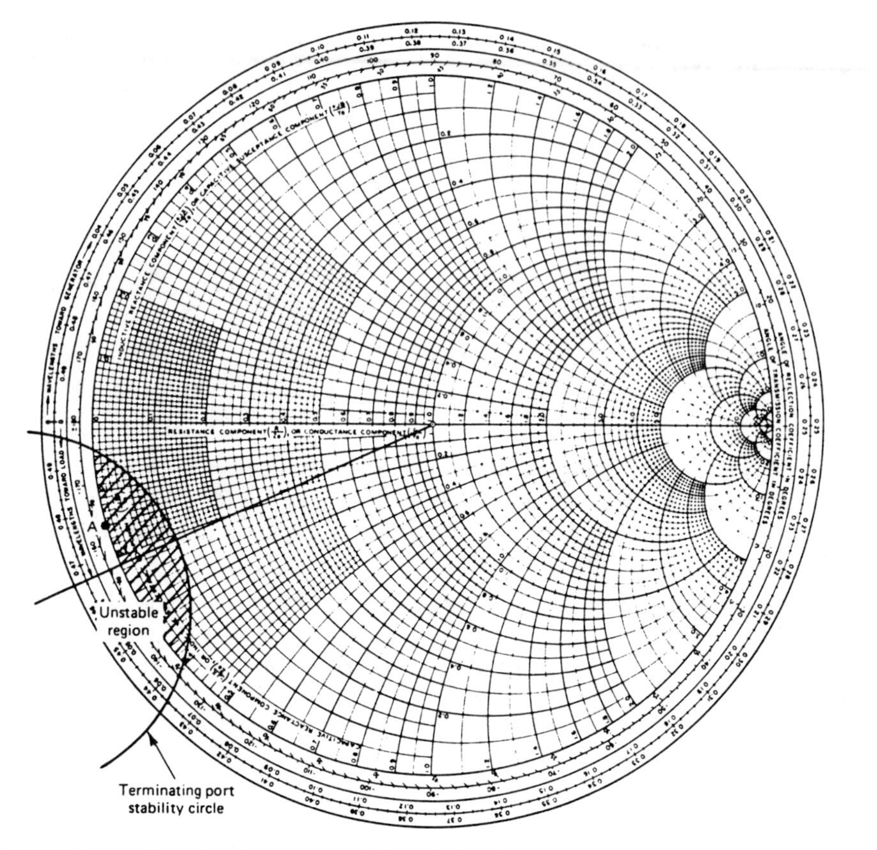

FIGURE 15.2 Terminating port stability circle. (From Ref. 1 with permission from Prentice Hall.)

Z_T can be obtained by using an open-circuited stub of $Z_0 = 50\ \Omega$ and $l = 0.226\lambda_g$.

$$Z_T = -j7.5 = -jZ_0 \cot \frac{2\pi l}{\lambda_g}$$

(3) With Z_T connected,

$$\Gamma_{\text{out}} = S_{22} + \frac{S_{12}S_{21}\Gamma_T}{1 - S_{11}\Gamma_T} = 12.8 \angle -16.6^\circ$$

which corresponds to $Z_{\text{out}} = -58.0 - j2.61\ \Omega$.

(4) Design $R_L = \frac{1}{3}|R_{\text{out}}| = 19.3\ \Omega$ and $X_L = -X_{\text{out}} = j2.61\ \Omega$.

15.4 LARGE-SIGNAL DESIGN APPROACH

The large-signal design approach is the same as that of the small-signal technique. The main difference is the determination of the device large-signal S parameters. The large-signal S parameters can be determined either by measurements or by using a large-signal device model. From the large-signal S parameters of the device, one can do the following:

1. Select Γ_T to produce $|\Gamma_{out}| > 1$.
2. Evaluate Γ_{out} and Z_{out}.

$$Z_{out}(V_0, f_0) = R_{out}(V_0, f_0) + jX_{out}(V_0, f_0)$$

   $R_{out}(V_0, f_0)$ and $X_{out}(V_0, f_0)$ are large-signal parameters at voltage V_0 and oscillating frequency f_0.
3. Choose Z_L as

$$R_L(f_0) = -R_{out}(V_0, f_0)$$

   and

$$X_L(f_0) = -X_{out}(V_0, f_0)$$

   The two conditions above will ensure that $\Gamma_{out}\Gamma_L = 1$ (see Problem P15.2).

It should be noted that R_{out} and X_{out} do not change with time since a steady state has been reached. Only in the large-signal approach can one evaluate the generated power output and efficiency of the oscillator using a large-signal analysis such as the harmonic balance technique. The frequency of oscillation can also be predicted more accurately.

15.5 STABLE OSCILLATORS USING RESONANT CIRCUITS

For fixed-frequency operation, one would like to have the output frequency stay stable. Better stability corresponds to lower FM and phase noise. To accomplish frequency stability, series or parallel resonant circuits are generally employed in the terminating network. The most common resonators are lumped elements, distributed transmission lines (stubs), cavities, rings, and dielectric resonators [3]. These resonators are used for fixed-frequency or mechanically tuned oscillators. For electronically tunable oscillators, a YIG or varactor is used.

A requirement for all resonators is high Q or low loss. For a high-Q resonator (> 50), the reflection coefficient is simply the outer boundary of the Smith chart, with the phase depending on the transmission-line length between the resonator and the active device [2]. Lumped-element resonators

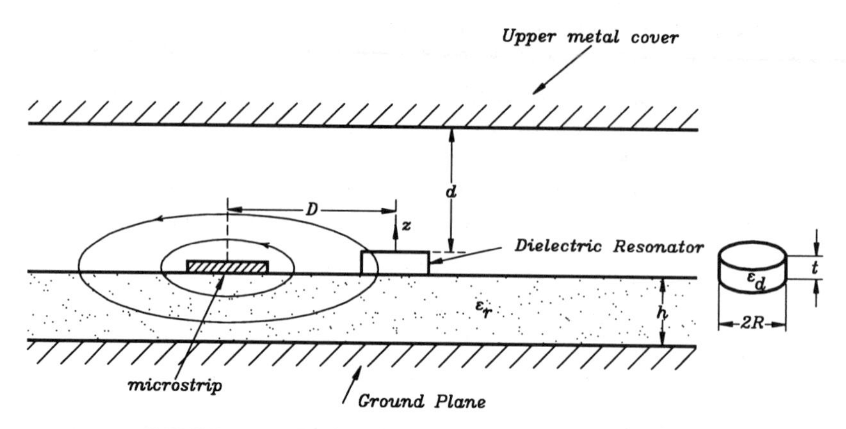

FIGURE 15.3 Dielectric resonator and microstrip line.

are formed from capacitors and inductors with associated parasitics. Distributed-element resonators are usually open or short transmission lines. Other possible distributed resonant elements include rings, disks, stubs, and so on. Cavity resonators are normally made of coaxial line or waveguide. The most popular low-cost, high-Q resonator for transistor circuits is a dielectric resonator (dielectric puck) coupled to a microstrip line as shown in Figure 15.3. A dielectric resonator (DR) is mounted close to a microstrip line. The DR is a cylindrical puck with a height of t and a radius of R. A dielectric spacer may be added under the puck to improve the loaded Q value by optimizing the coupling. The distance D determines the coupling between the resonator and the microstrip line.

A dielectric resonator is a piece of unmetallized ceramic with a high dielectric constant in which the electromagnetic fields are confined to the dielectric region and its immediate vicinity [3]. A good dielectric resonator should have low loss, a high unloaded Q factor, a low-temperature coefficient, and a high relative dielectric constant. A dielectric constant between 30 and 40 is generally used. A commonly used resonant mode in a cylindrical resonator is the dominant mode, denoted as $TE_{01\delta}$. When the relative dielectric constant is about 40, most of the electric and magnetic energy of the dominant mode is located inside the cylinder.

A complete solution for the resonant frequency of the microstrip dielectric resonator can be obtained either through solution of the boundary value problem or by using the transverse resonance procedure. Some design equations can be found in reference 3. A typical example using barium tetratitanate on a Cu-Flon substrate with various h/R and d/R values is shown in Figure 15.4. Dielectric resonators have no basic frequency limitations and can be used in millimeter-wave frequencies. Table 15.1 shows some

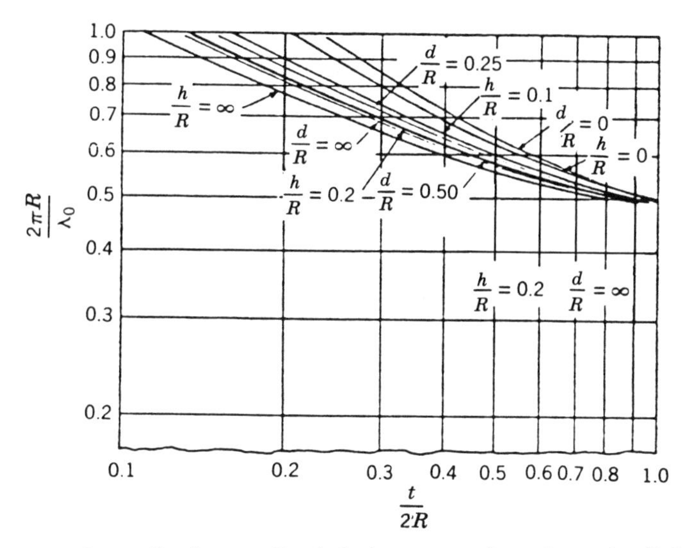

FIGURE 15.4 Generalized normalized design curves for microstrip dielectric resonators with $\varepsilon_r = 2.1$ and $\varepsilon_d = 37$. (From Ref. 3 with permission from Wiley.)

dielectric resonator parameters and performance data at millimeter-wave frequencies.

Figure 15.5 shows a transistor oscillator built using a lumped-element resonator and a matching network. The feedback is accomplished with a base inductor. C and L_E form the resonant circuit, and L_1, L_2, and C_1 form the matching network. To use the dielectric resonator, the circuit can be modified into that shown in Figure 15.6(a) with the equivalent circuit shown in Figure 15.6(b).

A dielectric resonator can be placed in the feedback circuit between the input and output circuits as shown in Figure 15.7. An oscillator can be

TABLE 15.1 Dielectric Resonator Parameters and Performance in Millimeter-Wave Frequencies

Material	Resonant Frequency (GHz)		t	$2R$	
	Design	Measured	(mils)	(mils)	Q_0
Barium tetratitanate	94	93.49	7.5	28	346
Barium tetratitanate	53	52.25	15.5	50	700
Alumina	94	92.75	19.7	49	1100

Source: Ref. 3 with permission from Wiley.

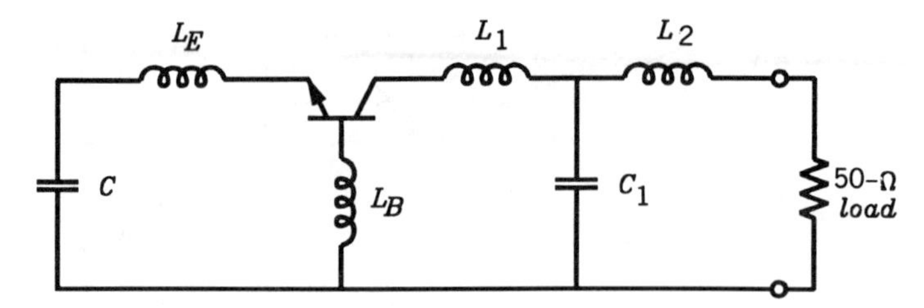

FIGURE 15.5 Transistor oscillator using lumped elements.

considered as an amplifier with positive feedback. The gain of a feedback amplifier is given by (see Problem P15.1)

$$A = \frac{\mu}{1 - \mu B} \tag{15.13}$$

where μ is the gain of the amplifier and B is the feedback circuit gain (or loss), as shown in Figure 15.8. The gain of the feedback amplifier becomes infinite when the loop gain μB equals unity and the phase shift is 360°. Therefore, the total phase shift at f_0 in Figure 15.7 is $\phi_A + \phi_R + \phi_c = 2n\pi$, where $n = 1, 2, \ldots$. ϕ_A, ϕ_R, and ϕ_c are insertion phases of the amplifier, resonator, and the remaining part of the feedback circuit. A small portion of the output signal is constantly fed back to the input in phase with the input

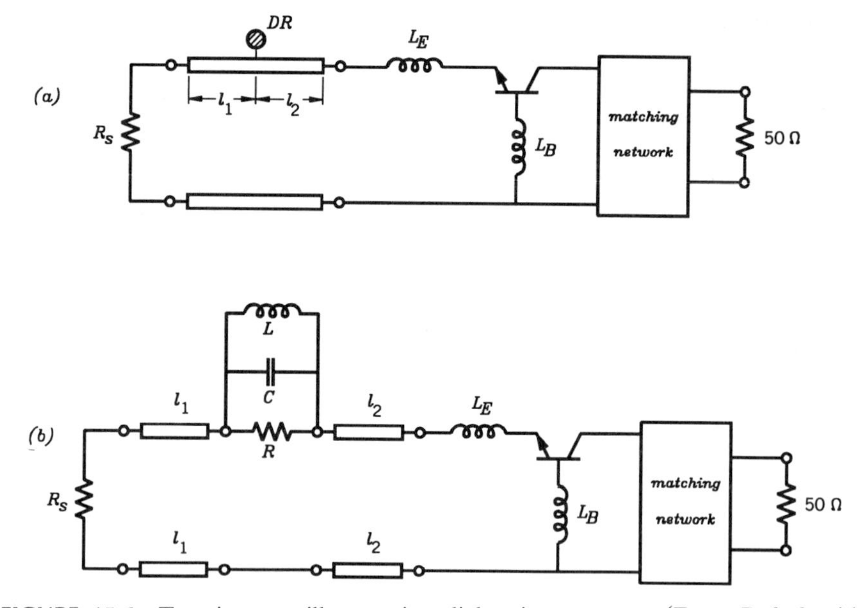

FIGURE 15.6 Transistor oscillator using dielectric resonator. (From Ref. 2 with permission from Wiley.)

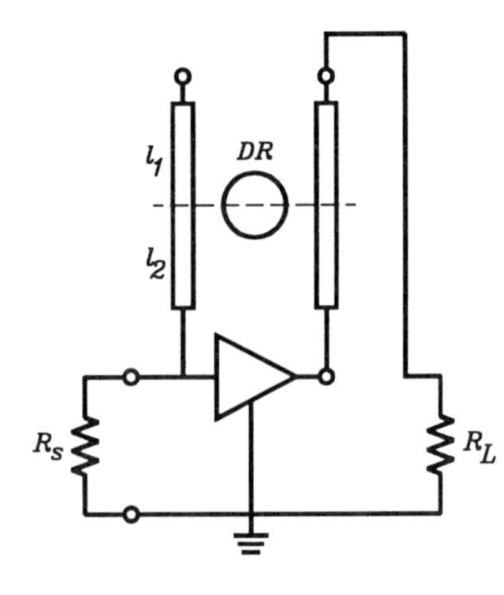

FIGURE 15.7 Dielectric resonator in the feedback circuit of an oscillator. (From Ref. 2 with permission from Wiley.)

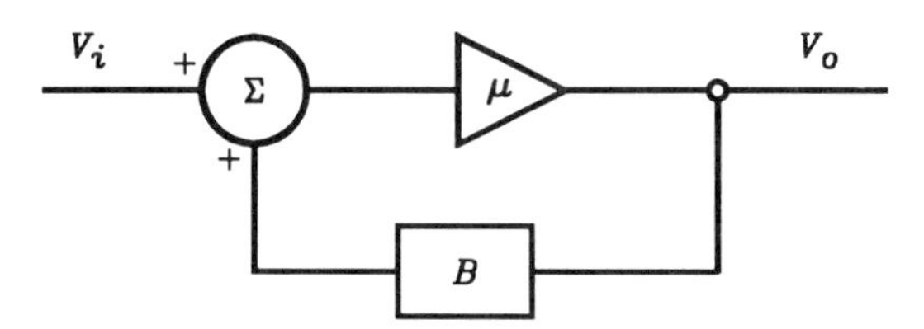

FIGURE 15.8 Feedback amplifier.

signal. The initial input signal is generated by noise and the energy source is the dc power supply. For a stable transistor, feedback circuits can be used to make it unstable.

15.6 VOLTAGE-TUNABLE OSCILLATORS

Similar to oscillators using two-terminal devices discussed in earlier chapters, voltage-tunable or voltage-controlled oscillators (VTOs or VCOs) can be obtained using the following methods:

1. Change the bias to the active device.
2. Incorporate a YIG resonator in the circuit.
3. Incorporate a varactor in the circuit.

The use of bias tuning is the simplest method, but it gives a limited tuning range with a large variation in power output. The use of YIG [yttrium iron garnet, $Y_2Fe_2(FeO_4)_3$] resonators provides very wide tuning ranges. A YIG resonator consists of a ferrimagnetic material that can be modeled by a parallel *RLC* circuit as shown in Figure 15.9. The YIG sphere couples

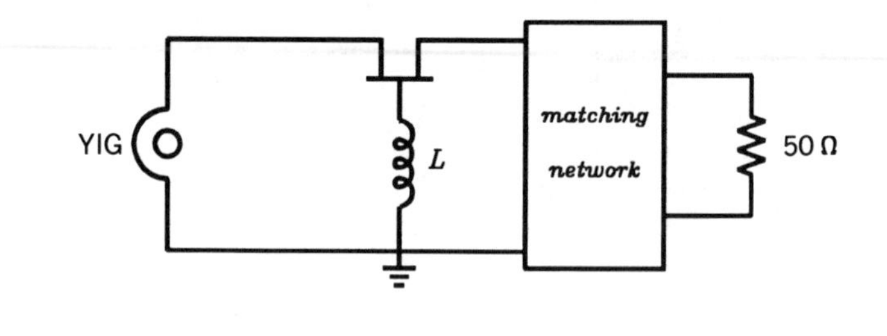

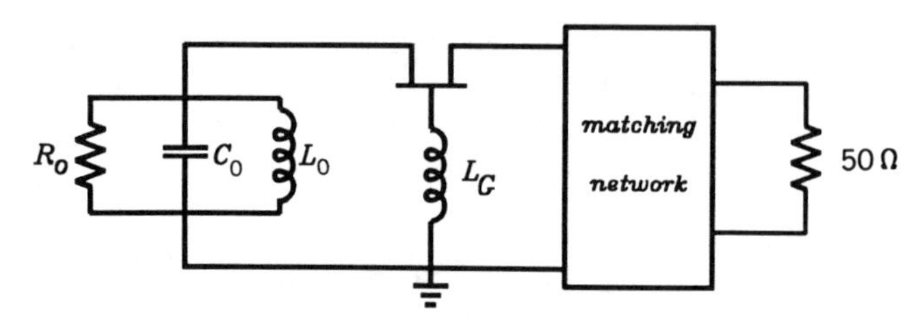

FIGURE 15.9 YIG-tuned oscillator and its equivalent circuit.

strongly to the transmission line and is magnetically saturated. The resonant frequency depends on the applied dc magnetic field according to the following equations:

$$\omega_0 = 2\pi\gamma H_0 \tag{15.14}$$

where γ is the gyromagnetic ratio (2.8 MHz/Oe), H_0 the applied dc magnetic field, and ω_0 the angular resonant frequency. By changing the applied dc current to the YIG, the magnetic field changes and thus the resonant frequency. Therefore, the oscillation frequency can be controlled and tuned by varying the bias current to the YIG.

A varactor diode can be used to electronically tune the oscillation frequency in a manner similar to the Gunn VCO described in Chapter 10. Figures 15.10 and 15.11 show some varactor-tuned transistor oscillators. The frequency tuning can be understood by studying Equation (15.2). Since $X_{out}(V, f)$ is varied by changing the varactor bias voltage, the oscillation frequency, which satisfies Equation (15.2), is changed accordingly. Figure 15.12 shows a low-noise VCO using a bipolar transistor. The oscillator exhibited an 8% tuning bandwidth at 8 GHz [5].

15.7 NOISE IN OSCILLATORS

Since the oscillator is a nonlinear device, the noise voltages and currents generated in an oscillator modulate the signal produced by the oscillator.

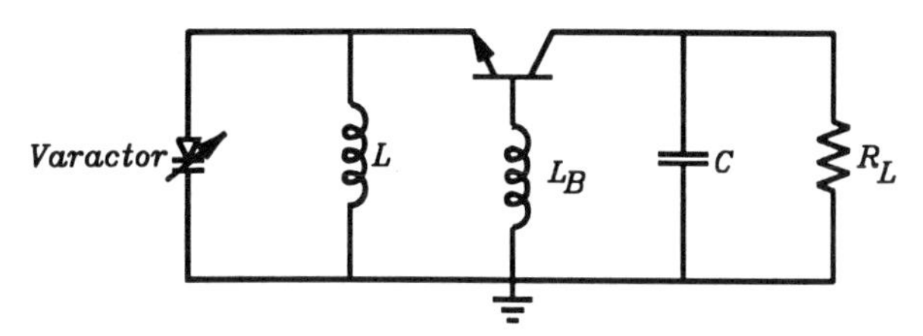

FIGURE 15.10 Varactor-tuned bipolar transistor oscillator. (From Ref. 4 with permission from Wiley.)

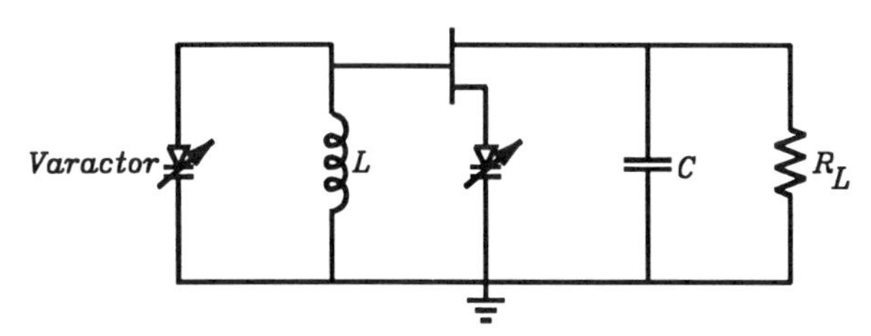

FIGURE 15.11 Varactor-tuned MESFET oscillator. (From Ref. 4 with permission from Wiley.)

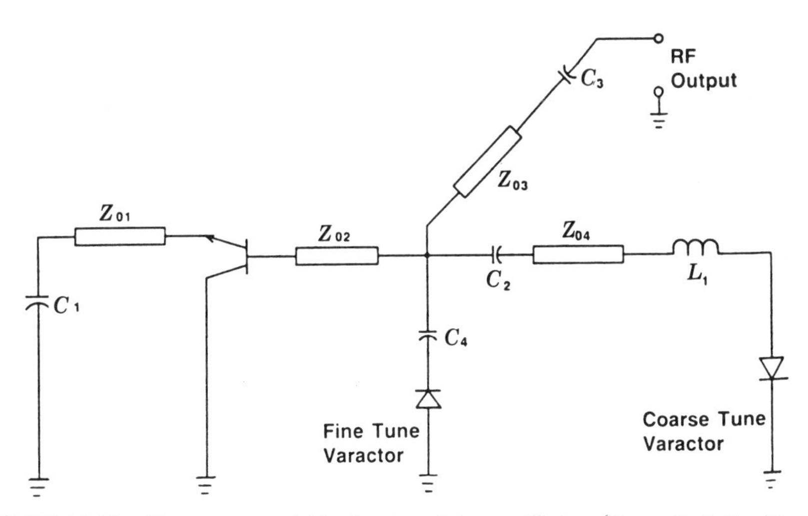

FIGURE 15.12 Varactor-tuned bipolar transistor oscillator. (From Ref. 5 with permission from IEEE.)

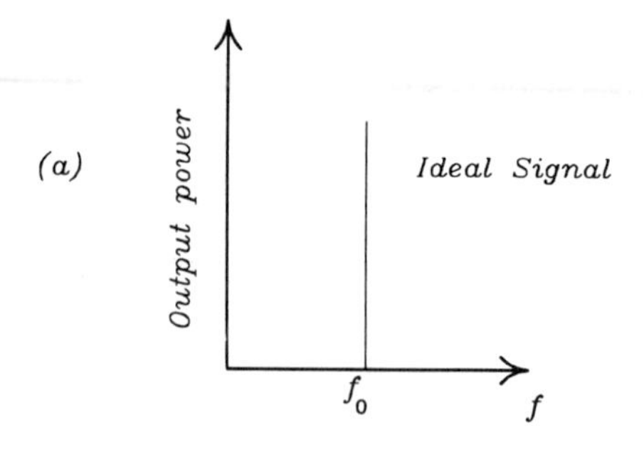

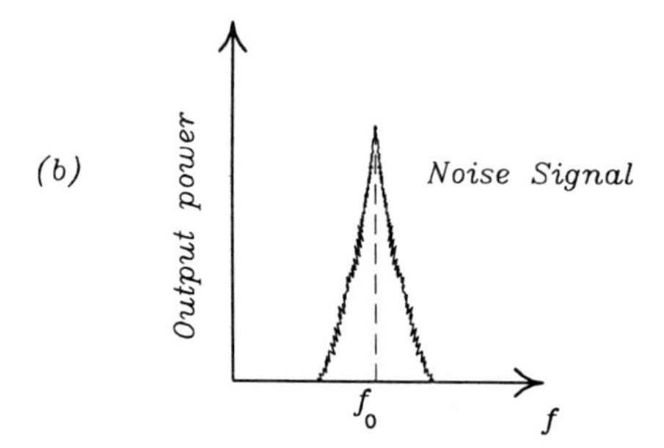

FIGURE 15.13 Ideal signal and noisy signal.

Figure 15.13 shows the ideal signal and the signal modulated by the noise. The noise can be classified as amplitude-modulated (AM) noise, frequency-modulated (FM) noise, or phase noise.

AM noise causes amplitude variations in the output signal. FM noise is indicated in Figure 15.13(*b*) by the spreading of the frequency spectrum. A ratio of single-sideband noise power normalized in a 1-Hz bandwidth to the carrier power is defined as

$$\mathscr{L}(f_m) = \frac{\text{noise power in 1-Hz bandwidth at } f_m \text{ offset from carrier}}{\text{carrier signal power}}$$

$$= \frac{N}{C} \tag{15.15}$$

As shown in Figure 15.14, $\mathscr{L}(f_m)$ is the difference in power between the carrier at f_0 and noise at f_m. The power is plotted on a dB scale. The unit of $\mathscr{L}(f_m)$ is dBc/Hz (dBc means "dB below the carrier power").

It should be mentioned that the bulk of oscillator noise close to the carrier is phase or FM noise. The noise represents the phase jitter or the short-term stability of the oscillator. The oscillator power is not concentrated at a single frequency but rather, is distributed around it. The spectral distributions on

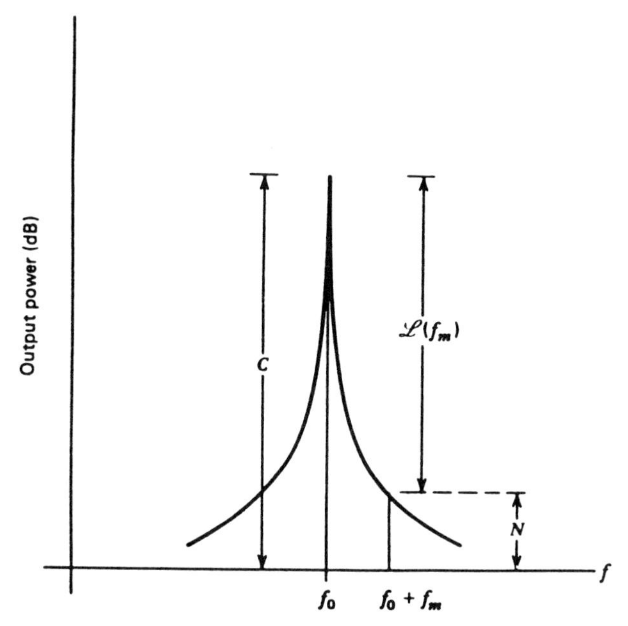

FIGURE 15.14 Oscillator output power spectrum. This spectrum can be seen from the screen of a spectrum analyzer. (From Ref. 2 with permission from Wiley.)

the opposite sides of the carrier are known as noise sidebands. AM noise is much lower than FM noise in a transistor oscillator.

Many methods can be used to measure the FM or phase noise [2, 4, 6]. These methods include the spectrum analyzer method, the two-oscillator method, the single-oscillator method, the delay-line discriminator method, and the cavity discriminator method. Figure 15.15 shows a comparison of noise sideband performance for various oscillators. Shown in the chart are the noise performances of a 10-MHz crystal oscillator, a 40-MHz *LC* lumped-element oscillator, a cavity-tuned oscillator at 500 MHz, an 870- to 990-MHz VCO, and a 2- to 6-GHz YIG oscillator at 6 GHz. To minimize the phase or FM noise, the following methods can be used:

1. Maximize the unloaded Q.
2. Choose an active device with low flicker or $1/f$ noise.
3. Use a high-Q resonator in the circuit. The resonator limits the bandwidth and acts as a filter.
4. Choose an active device with the lowest noise figure.
5. Avoid saturation.
6. Use high-impedance devices such as MESFETs to minimize phase perturbations.

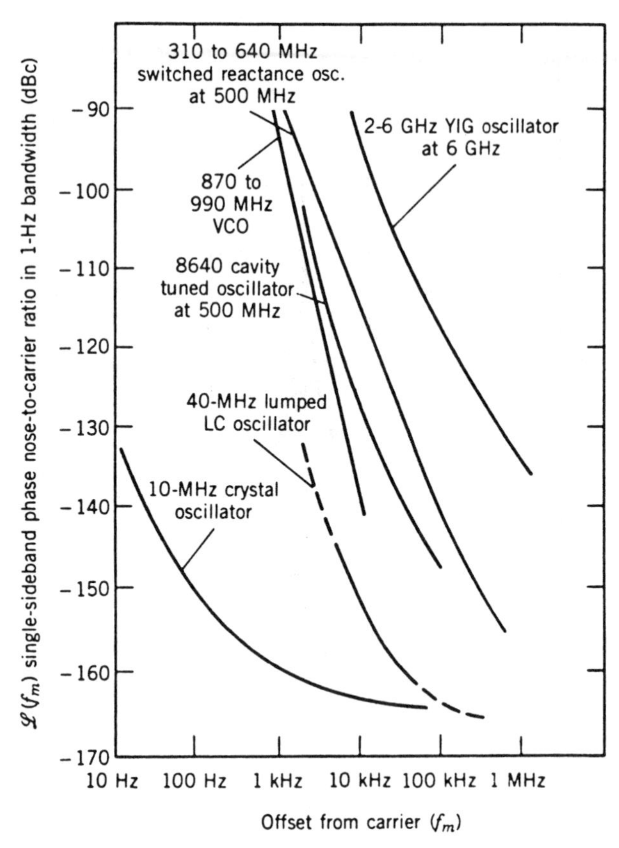

FIGURE 15.15 Comparison of noise performance of various oscillators. (From Ref. 2 with permission from Wiley.)

15.8 PRACTICAL TRANSISTOR OSCILLATOR CIRCUITS

Dielectrically stabilized oscillators, varactor-tuned oscillators, and YIG-tuned oscillators are commercially available at microwave frequencies [7]. Power output for varactor-tuned oscillators is normally 10 or 20 mW, with an output power variation of ± 1.5 dB over the 10 to 20% tuning range. The circuit has the features of fast tuning speed and fast settling time. High stability can be achieved using phase-locked techniques [8].

The dielectrically stabilized oscillator is used for fixed-tuned frequency or local oscillator applications. Narrowband mechanical tuning can be achieved using a tuning screw close to the dielectric resonator. The oscillator provides a signal with very good frequency accuracy (less than 0.1% deviation) and temperature stability ($\pm 0.05\%$) over -54 to $+85°$ C. Typical single-sideband phase noise in a 1-Hz bandwidth is -100 dBc at 10 kHz from the carrier and -120 dBc at 100 kHz from the carrier.

YIG-tuned oscillators provide a very wide tuning range (octave or multioctave) and high power output. For example, a 2- to 8-GHz oscillator with a minimum +20 dBm power output and maximum power variation of 6 dB is available [7]. Compared to varactor-tuned circuits, YIG-tuned circuits are bulky and the tuning speed is slower.

PROBLEMS

P15.1 Prove Equation (15.13).

P15.2 Prove $\Gamma_{out}\Gamma_L = 1$ if $X_{out} + X_L = 0$ and $R_{out} + R_L = 0$.

P15.3 Which of the following two transistors can be used to build an oscillator at 2 GHz? The S parameters at 2 GHz for these two transistors are as follows:

Transistor A	Transistor B
$S_{11} = 0.48 \angle 25°$	$S_{11} = 0.8 \angle 90°$
$S_{12} = 0$	$S_{12} = 0$
$S_{21} = 5.0 \angle 30°$	$S_{21} = 4.0 \angle 65°$
$S_{22} = 0.3 \angle -120°$	$S_{22} = 2.0 \angle 180°$

P15.4 Explain how the circuit shown in Figure 15.9 works.

REFERENCES

1. G. Gonzalez, *Microwave Transistor Amplifiers*, Prentice Hall, Englewood Cliffs, N.J., 1984, Chap. 5.
2. G. D. Vendelin, A. M. Pavio, and U. L. Rohde, *Microwave Circuit Design*, Wiley, New York, 1990, Chap. 6.
3. M. Dydyk, "Cavities and Resonators," in *Handbook of Microwave and Optical Components*, K. Chang, Editor, Wiley, New York, 1989, Vol. 1, Chap. 4.
4. I. Bahl and P. Bhartia, *Microwave Solid State Circuit Design*, Wiley, New York, 1988, Chap. 9.
5. E. Niehenke and R. Hess, "A Microstrip Low Noise X-Band Voltage Controlled Oscillator," *IEEE Transactions on Microwave Theory and Techniques*, Vol. MTT-27, December 1979, pp. 1075–1079.
6. A. L. Lance, "Microwave Measurements," in *Handbook of Microwave and Optical Components*, K. Chang, Editor, Wiley, New York, 1989, Vol. 1, Chap. 9.
7. *Modular and Oscillator Components*, Catalog of Avantek, 1990.
8. K. Chang, K. Louie, A. J. Grote, R. S. Tahim, M. J. Mlinar, G. M. Hayashibara, and C. Sun, "V-Band Low-Noise Integrated Circuit Receiver," *IEEE Transactions on Microwave Theory and Techniques*, Vol. MTT-31, February 1983, pp. 146–154.

CHAPTER 16

Transistor Mixers, Switches, Phase Shifters, and Multipliers

16.1 INTRODUCTION

A transistor is a nonlinear device. Like a diode, it can be used for mixing, switching, multiplying, and many other applications. The use of various diodes for these applications has been around for a long time; however, transistor circuits have gained interest recently due to their insertion gain (instead of loss), potentially good noise performance, compatibility with monolithic integrated circuits, and low cost. In this chapter we discuss briefly the use of transistors for mixers, switches, phase shifters, and multipliers. The fundamental operation principles and use of diodes for these devices were discussed in several earlier chapters.

16.2 TRANSISTOR MIXERS

A transistor mixer could employ MESFETs, HBTs, or HEMTs as nonlinear devices. Figure 16.1 shows the circuit configurations using single-gate and dual-gate MESFETs as mixers. Unlike a diode mixer, a FET mixer can have a conversion gain instead of a conversion loss since the FET is an active device that combines mixing and amplification in one device. A FET mixer normally needs less LO power than the diode mixer, due to the device gain feature. Furthermore, a FET mixer has high IF-to-RF and LO isolation values, due to the device's unilateral feature ($S_{12} \cong 0$). Use of a dual-gate FET gives inherent isolation between RF and LO ports due to the RF and LO being applied to different gates.

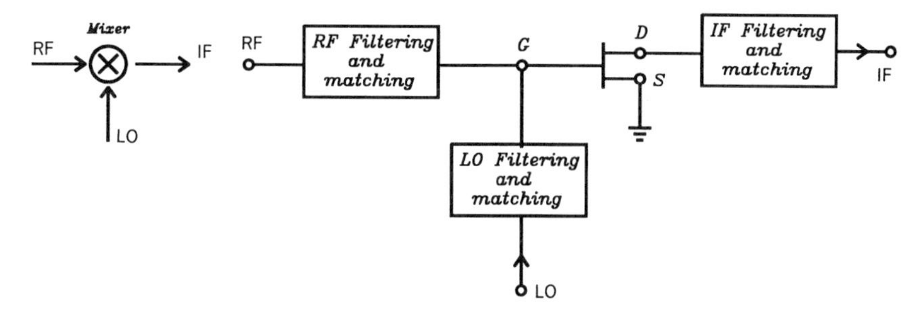

(a) Mixer using a single-gate MESFET

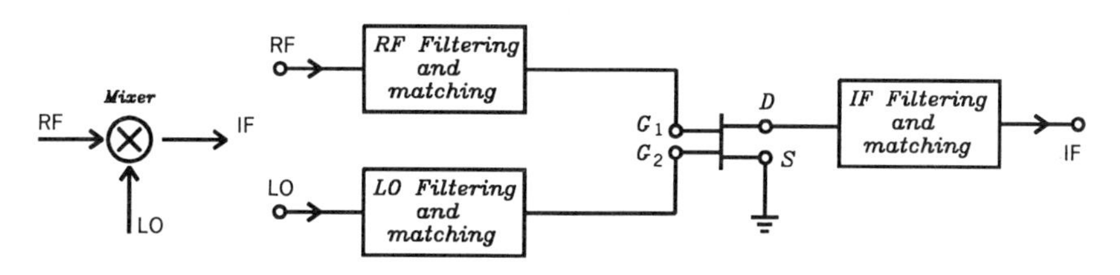

(b) Mixer using a dual-gate MESFET

FIGURE 16.1 Mixer configurations.

One disadvantage of a FET mixer is its higher $1/f$ noise compared to the diode mixer, but a higher IF frequency can be used to avoid this problem. Table 16.1 shows a comparison of a typical diode and FET mixer followed by an IF amplifier with a 2-dB noise figure.

Mixing in FET mixers occurs due to nonlinearities in the FETs. Dominant nonlinear elements for a transistor mixer are C_{gs}, C_{dg}, and I_{ds}. An equivalent circuit is given in Figure 16.2. I_{ds} contributes to two nonlinear elements:

$$g_m = \left. \frac{\partial I_{ds}}{\partial V_{gs}} \right|_{V_{ds}=\text{constant}} \tag{16.1}$$

TABLE 16.1 Comparison of Diode and FET Mixers

Mixer Type	Gain/Loss (dB)	NF (dB)	F_{tot} (dB)
Diode	$L = 6$	5	7.4
FET	$G = 4$	7	7.2

Source: Ref. 1 with permission from Wiley.

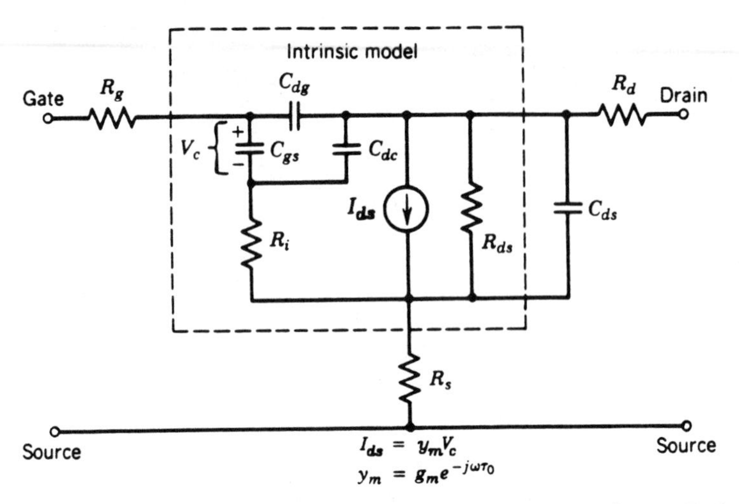

FIGURE 16.2 Small-signal equivalent circuit of an MESFET. (From Ref. 2 with permission from IEEE.)

and

$$g_d = \left.\frac{\partial I_{ds}}{\partial V_{ds}}\right|_{V_{gs}=\text{constant}} \tag{16.2}$$

g_m is the most dominant nonlinear element in mixers. Similar to diode mixers, $V_{\text{RF}} \ll V_{\text{LO}}$ in a normal mixer, and the LO signal modulates the device nonlinearities, causing mixing of LO and RF signals. Similar to the diode mixer, $g_m(t)$ can be expanded in a Fourier series:

$$g_m(t) = \sum_{n=0}^{\infty} g_n \exp(jn\omega_{\text{LO}}t) \tag{16.3}$$

$g_m(t)$ can be solved by nonlinear analysis [3, 4].

Single-gate FET mixers can be arranged as gate mixers or drain mixers [5], although more work has been done with gate mixers. For drain mixers, the LO is applied between the drain and source terminals. The optimum performance of an FET mixer is a compromise between conversion gain and a minimum-noise figure. It seems to be advantageous to make the transconductance dominant in the mixing process and make the influence from the other nonlinear impedances negligible [5]. Figures 16.3 and 16.4 show examples of a gate mixer and a drain mixer. Another example is given in Figure 16.5. The

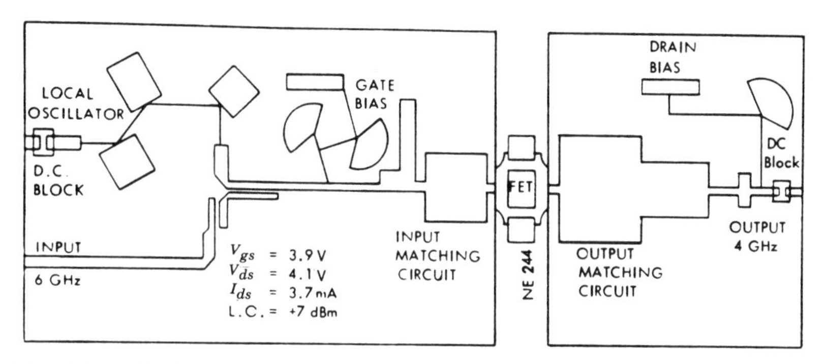

FIGURE 16.3 Design example of a gate mixer. (From Ref. 6 with permission from IEEE.)

mixer operates in the RF frequency range 7.9 to 8.4 GHz, the IF frequency range 0.9 to 1.4 GHz, and at a LO frequency of 7.0 GHz. The optimum conversion gain occurs at +6 dBm LO power and the optimum noise figure occurs at 0 dBm LO power. A low-pass filter in the IF port is used to present a short circuit at the drain to the LO fundamental and harmonics. It is important to short-circuit the IF at the FET gate. A 18-pF capacitor and a quarter-wavelength 100-Ω stub in the gate bias circuit are used to realize the IF short. The LO and RF are applied through a filter diplexer.

Dual-gate FETs can be used to increase the RF/LO isolation since the LO and RF are applied to separate gates. A dual-gate FET can be modeled

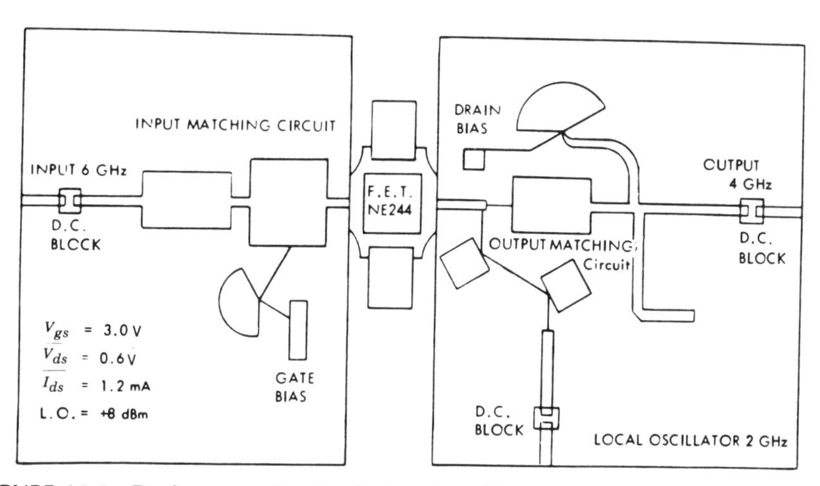

FIGURE 16.4 Design example of a drain mixer. (From Ref. 6 with permission from IEEE.)

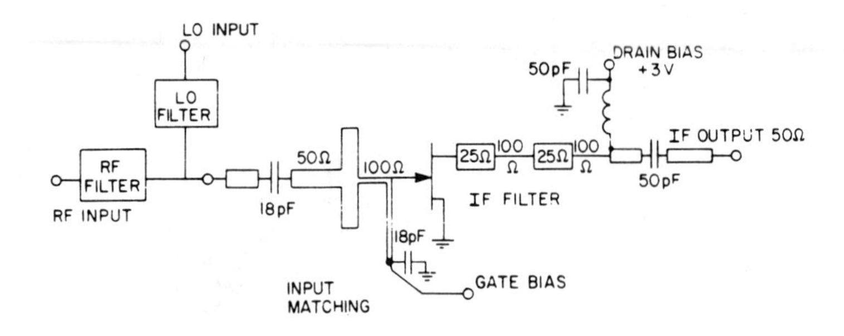

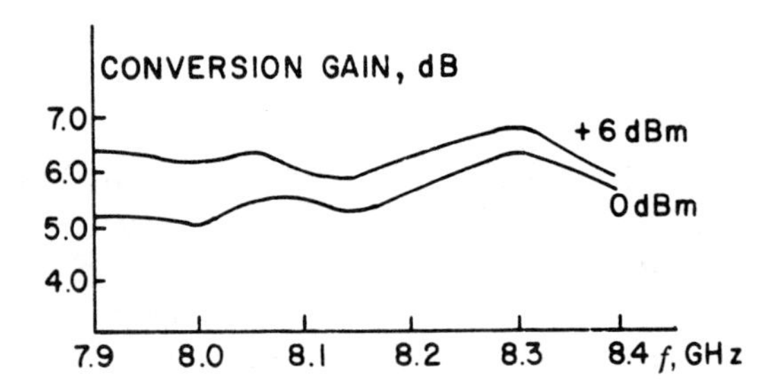

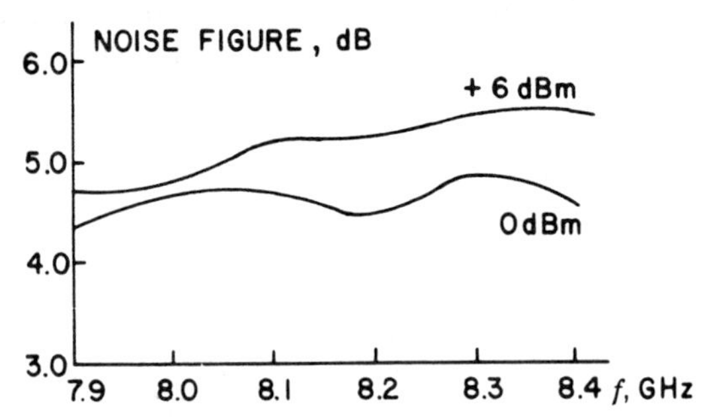

FIGURE 16.5 FET gate mixer and its performance at two LO power levels. (From Ref. 4 with permission from Artech House.)

as two single-gate FETs in series. Like the single-gate FET mixer, the dual-gate FET works best as a transconductance mixer [4]. Therefore, the gates and drain should be short-circuited at all unwanted LO harmonics and mixing frequencies.

Another way to improve LO/RF isolation is to use balanced configurations. Balanced mixers can be realized using 90° or 180° hybrids as shown in Figure 16.6. Balanced mixers also offer better LO noise rejection, better spurious response, and higher dynamic range.

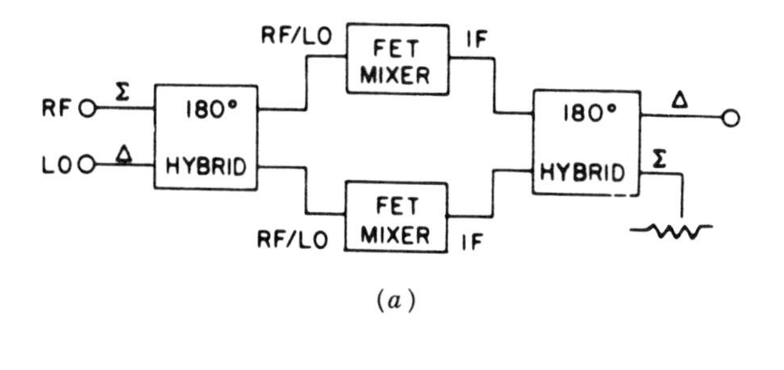

(a)

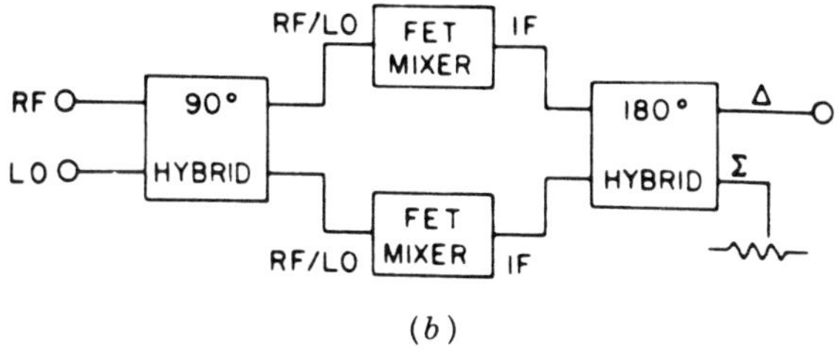

(b)

FIGURE 16.6 Balanced FET mixers using 90° or 180° hybrids. (From Ref. 4 with permission from Artech House.)

HEMTs and HBTs can also be used to build mixers. For millimeter-wave frequencies, HEMTs have the advantages of lower noise and higher electron mobility than those of MESFETs.

16.3 TRANSISTOR SWITCHES AND PHASE SHIFTERS

A MESFET in the passive mode can be used for switching applications. In this case, the gate terminal acts as a port for control of the signal only. The RF connections are made to the drain and source terminals and the gate terminal looks into an open circuit for the RF signal. For switching applications, low-impedance and high-impedance states are obtained by making the gate voltage equal to zero and by using a gate voltage greater than the pinch-off voltage, respectively [7].

A switch using a MESFET in the passive mode is characterized by the low-impedance and high-impedance states. As shown in Figure 16.7(*a*), in the low-impedance (ON) state, the equivalent circuit can be represented by a resistance R_{on} [7]. R_{on} can be calculated from the current path and the cross-sectional area [8]. For a 1-μm X-band MESFET, R_{on} is about 2.5 Ω at 10 GHz.

For the high-impedance (OFF) state shown in Figure 16.7(*b*), the equivalent circuit can be represented by the circuits shown in Figure 16.8 since r_g is

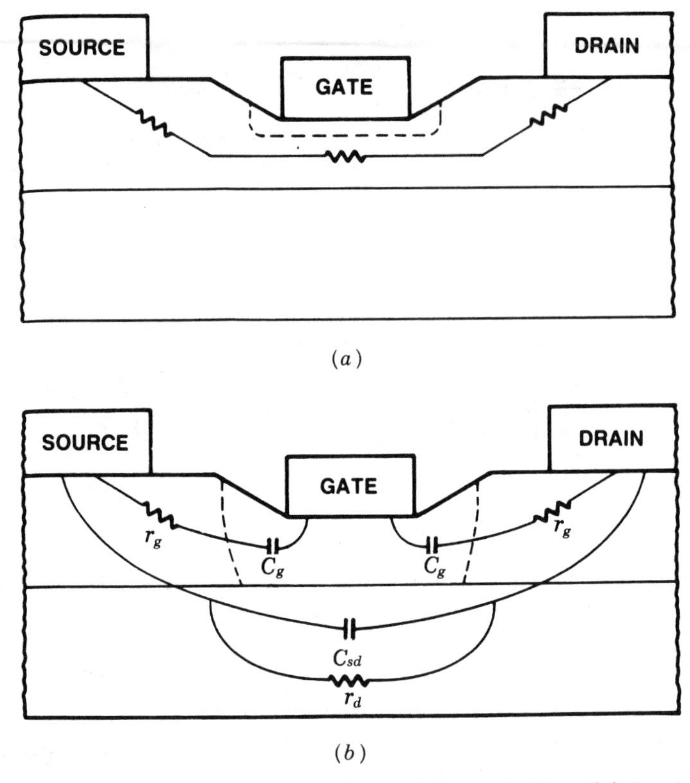

FIGURE 16.7 MESFET in passive mode used for switching: (*a*) low-impedance state; (*b*) high-impedance state. (From Ref. 7 with permission from Wiley.)

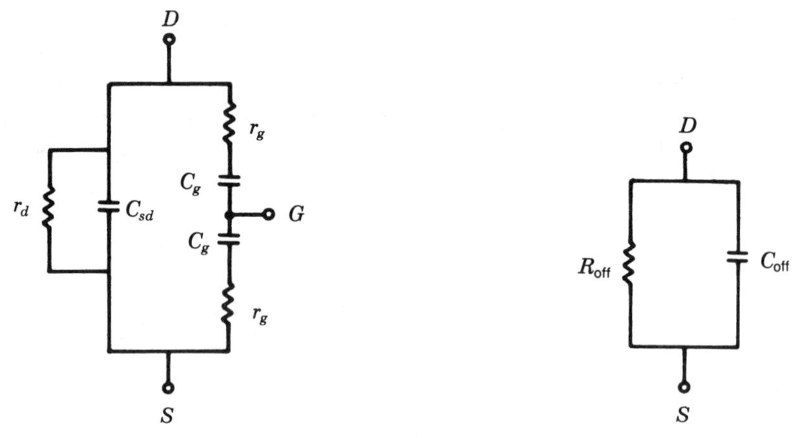

FIGURE 16.8 Equivalent circuits for the high-impedance state in the passive mode. (From Ref. 7 with permission from Wiley.)

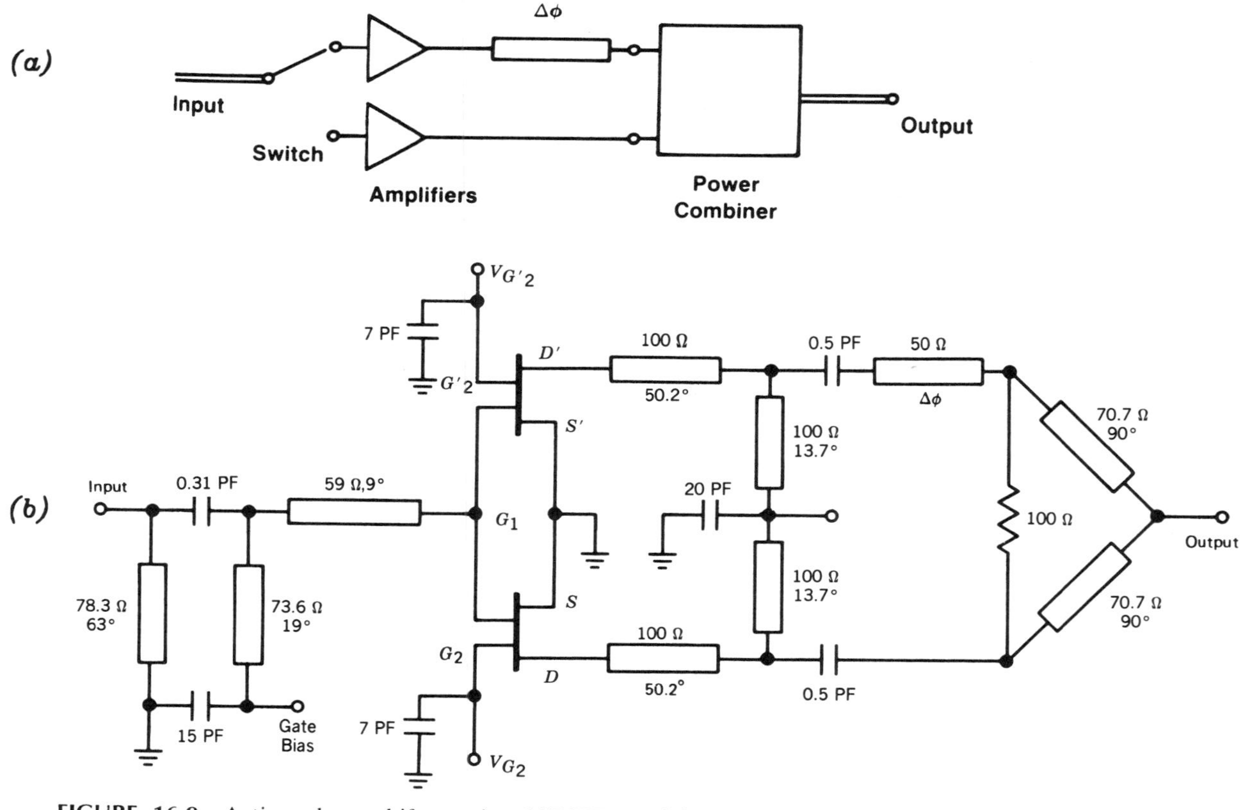

FIGURE 16.9 Active phase shifter using MESFETS: (*a*) block diagram; (*b*) detailed circuits. (From Ref. 9 with permission from IEEE.)

much smaller than $1/\omega C_g$. For a 1-μm 10-GHz MESFET, C_{off} and R_{off} are about 0.2 pF and 2 kΩ [7].

Like the diode switch, the insertion loss and isolation of the MESFET switch can be calculated using the equations given in Chapter 8. Phase shifters can be constructed using hybrid-coupler, loaded-line, or switched-line configurations, as discussed in Chapter 8.

MESFET control devices can also operate in active modes. In the active mode, single-gate or dual-gate MESFETs are used as three-terminal active devices. Signal control and amplification can be accomplished simultaneously. Figure 16.9 shows an active phase shifter using a switched-line configuration [7, 9]. The input signal is switched between two identical amplifiers. At the output of one of the amplifiers, an additional line length is introduced to provide the required phase shift $\Delta\phi$. A Wilkinson-type combiner is used to combine the output. The combiner introduces a 3-dB loss since only one input signal is present at any time, but the gain of the amplifier will offset this loss. Dual-gate MESFETs used for this circuit have an on–off signal ratio of 30 dB and a gain of 9 to 11 dB at X-band. An overall gain of 3 dB has been reported over a 10% bandwidth at X-band [9].

16.4 TRANSISTOR MULTIPLIERS

Active frequency multipliers using MESFETs or other three-terminal devices can offer higher efficiency or even conversion gain than their two-terminal counterparts in broadband operation [10]. Frequency conversion and multiplication are achieved by the nonlinearities of the device. Figure 16.10 shows a frequency-doubler configuration similar to the diode doubler. The input and output matching and filtering networks are used to match the desired frequencies. A practical circuit for a 15- to 30-GHz doubler is shown in

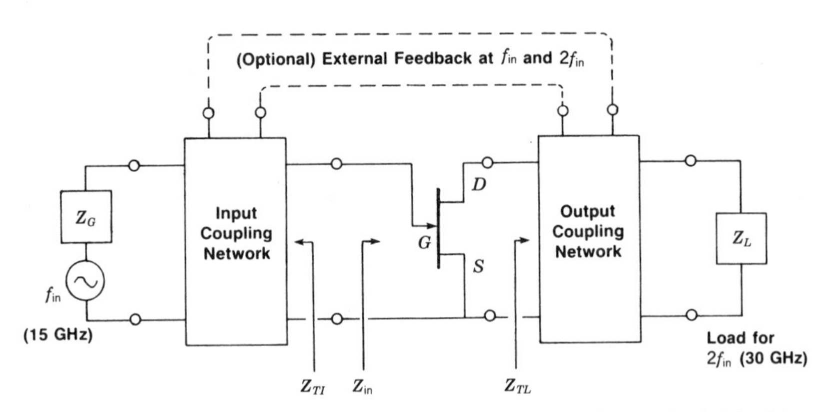

FIGURE 16.10 MESFET frequency-doubler configuration. (From Ref. 11 with permission from IEEE.)

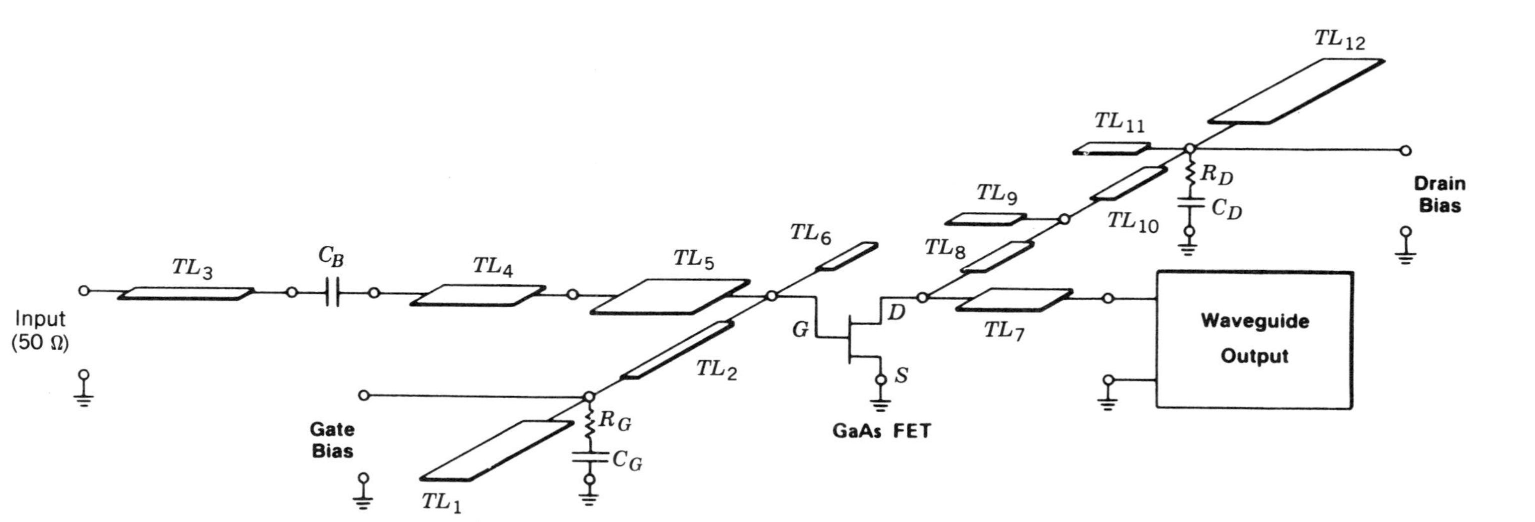

FIGURE 16.11 A 15- to 30-GHz MESFET frequency doubler. The TL elements are microstrip lines. (From Ref. 11 with permission from IEEE.)

FIGURE 16.12 A 15- to 30-GHz MESFET frequency doubler. (From Ref. 11 with permission from IEEE.)

Figure 16.11. TL_3, TL_4, TL_5, TL_6, and C_B comprise the input matching and filtering circuit, which provides a conjugate match at 15 GHz and rejects the 30-GHz signal. TL_6 is an open-circuited stub with a length of $\lambda_g/4$ at 30 GHz. The gate bias circuit consists of R_G, C_G, TL_1, and TL_2. The input circuit uses a 50-Ω coaxial connector. The output circuit consists of a probe coupled to a WR-28 waveguide. The 15-GHz signal is below the cutoff frequency of the waveguide and will not propagate into the output port. This circuit is shown in Figure 16.12, and its performance, in Figure 16.13. It can be seen that a small conversion loss of 0.5 to 3 dB was obtained for this doubler. The loss is much smaller than the diode doubler (typical loss of 5 to 10 dB). Conversion gain can be realized if the MESFET provides higher gain.

Self-oscillating doublers have also been developed to combine the oscillator and doubler in one circuit [12, 13]. The same FET device serves as an oscillator at a fundamental frequency f_0 and simultaneously as a doubler, delivering power to a load at $2f_0$.

In summary, MESFET multipliers offer many advantages over diode multipliers such as conversion gain, broadband operation, low input power, low noise figures, high dynamic range, and isolation between the input and output. The disadvantages are $1/f$ noise near the carrier frequency, lower-frequency operation, and the requirement of two bias voltages.

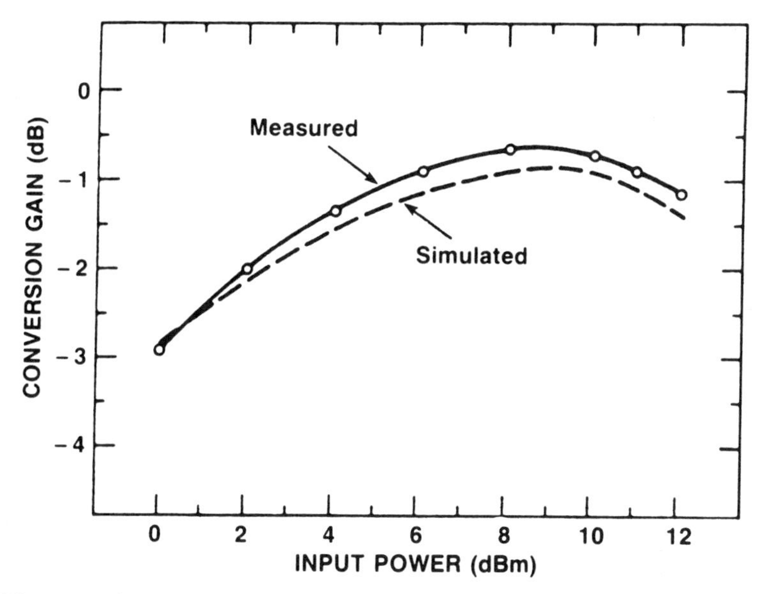

FIGURE 16.13 Measured and predicted performance of a 15- to 30-GHz MESFET frequency doubler. (From Ref. 11 with permission from IEEE.)

REFERENCES

1. I. Bahl and P. Bhartia, *Microwave Solid State Circuit Design*, Wiley, New York, 1988, Chap. 11.
2. C. A. Liechti, "Microwave Field-Effect Transistor—1976," *IEEE Transactions on Microwave Theory and Techniques*, Vol. MTT-24, June 1976, pp. 279–300.
3. D. N. Held and A. R. Kerr, "Conversion Loss and Noise of Microwave and Millimeter-Wave Mixers, Part 1, Theory," *IEEE Transactions on Microwave Theory and Techniques*, Vol. MTT-26, February 1978, pp. 49–55.
4. S. A. Maas, *Microwave Mixers*, Artech House, Norwood, Mass., 1986.
5. E. L. Kollberg, "Mixers and Detectors," in *Handbook of Microwave and Optical Components*, K. Chang, Editor, Wiley, New York, 1990, Vol. 2, Chap. 2.
6. P. Bura and R. Dikshit, "FET Mixers for Communication Satellite Transponders," *IEEE MTT-S International Microwave Symposium Digest*, 1976, pp. 90–92.
7. I. Bahl and P. Bhartia, *Microwave Solid State Circuit Design*, Wiley, New York, 1988, Chap. 12.
8. Y. Ayasli, "Microwave Switching with GaAs FETs," *Microwave Journal*, Vol. 25, November 1982, pp. 61–74.
9. J. L. Vorhaus, R. A. Pucel, and Y. Tajima, "Monolithic Dual-Gate FET Digital Phase Shifters," *IEEE Transactions on Microwave Theory and Techniques*, Vol. MTT-30, July 1982, pp. 982–992.

10. I. Bahl and P. Bhartia, *Microwave Solid State Circuit Design*, Wiley, New York, 1988, Chap. 13.
11. C. Rauscher, "High Frequency Doubler Operation of GaAs Field Effect Transistors," *IEEE Transactions on Microwave Theory and Techniques*, Vol. MTT-31, June 1983, pp. 462–473.
12. T. Saito, M. Iwakuni, T. Sakane, and Y. Tokumitsu, "A 45 GHz GaAs FET MIC Oscillator-Doubler," *IEEE MTT-S International Microwave Conference Technical Digest*, June 1982, pp. 283–285.
13. A. S. Chu and P. T. Chen, "A Osciplier up to K-Band Using Dual-Gate GaAs MESFET," *IEEE MTT-S International Microwave Conference Technical Digest*, June 1980, pp. 383–386.

APPENDIX A

Constants, Units, and Prefixes

TABLE A.1 Physical Constants

Constant	Symbol	Value
Boltzmann constant	k	1.381×10^{-23} J/K
Electron charge	q	1.602×10^{-19} C
Electron mass	m	9.110×10^{-31} kg
Electron volt	eV	1.602×10^{-19} J
Permeability of free space	μ_0	$4\pi \times 10^{-7}$ H/m
Permittivity of free space	ε_0	8.854×10^{-12} F/m
Planck's constant	h	6.626×10^{-34} J · s
Velocity of light in vacuum	c	2.998×10^{8} m/s

TABLE A.2 Conversion Factors and Units

1 ampere (A) = 1 C/s
1 angstrom unit (Å) = 10^{-10} m
1 farad (F) = 1 C/V
1 foot (ft) = 0.305 m
1 hertz (Hz) = 1 cycle/s
1 inch (in.) = 2.54 cm
1 joule (J) = 1 W · s
1 kilogram (kg) = 2.205 lb
1 kilometer (km) = 0.622 mile
1 meter (m) = 39.37 in.
1 micron (μm) = 10^{-6} m
1 mil = 10^{-3} in.
1 mile = 5280 ft
= 1.609 km
1 pound (lb) = 453.6 g
1 volt (V) = 1 W/A
1 weber (Wb) = 1 V · s

TABLE A.3 Prefixes

Prefix	Factor	Symbol
femto	10^{-15}	f
pico	10^{-12}	p
nano	10^{-9}	n
micro	10^{-6}	μ
milli	10^{-3}	m
centi	10^{-2}	c
deci	10^{-1}	d
kilo	10^{3}	k
mega	10^{6}	M
giga	10^{9}	G
tera	10^{12}	T

APPENDIX B

Rectangular Waveguide Properties

EIA WG Designation WR(·)	Recommended Operating Range for TE_{10} Mode (GHz)	Cutoff Frequency for TE_{10} Mode (GHz)	Theoretical CW Power Rating, Lowest to Highest Frequency (MW)	Theoretical Attenuation, Lowest to Highest Frequency (dB/100 ft)	Material Alloy	Inside Dimensions (in.)	JAN WG Designation RG(·)/U
2300	0.32–0.49	0.256	153.0–212.0	0.051–0.031	Alum.	23.000–11.500	
2100	0.35–0.53	0.281	120.0–173.0	0.054–0.034	Alum.	21.000–10.500	
1800	0.41–0.625	0.328	93.4–131.9	0.056–0.038	Alum.	18.000–9.000	201
1500	0.49–0.75	0.393	67.6–93.3	0.069–0.050	Alum.	15.000–7.500	202
1150	0.64–0.96	0.513	35.0–53.8	0.128–0.075	Alum.	11.500–5.750	203
975	0.75–1.12	0.605	27.0–38.5	0.137–0.095	Alum.	9.750–4.875	204
770	0.96–1.45	0.766	17.2–24.1	0.201–0.136	Alum.	7.700–3.850	205
650	1.12–1.70	0.908	11.9–17.2	0.317–0.212	Brass	6.500–3.250	69
				0.269–0.178	Alum.		103
510	1.45–2.20	1.157	7.5–10.7			5.100–2.550	
430	1.70–2.60	1.372	5.2–7.5	0.588–0.385	Brass	4.300–2.150	104
				0.501–0.330	Alum.		105
340	2.20–3.30	1.736	3.1–4.5	0.877–0.572	Brass	3.400–1.700	112
				0.751–0.492	Alum.		113
284	2.60–3.95	2.078	2.2–3.2	1.102–0.752	Brass	2.840–1.340	48
				0.940–0.641	Alum.		75
229	3.30–4.90	2.577	1.6–2.2			2.290–1.145	
187	3.95–5.85	3.152	1.4–2.0	2.08–1.44	Brass	1.872–0.872	49
				1.77–1.12	Alum.		95
159	4.90–7.05	3.711	0.79–1.0			1.590–0.795	
137	5.85–8.20	4.301	0.56–0.71	2.87–2.30	Brass	1.372–0.622	50
				2.45–1.94	Alum.		106
112	7.05–10.00	5.259	0.35–0.46	4.12–3.21	Brass	1.122–0.497	51
				3.50–2.74	Alum.		68
90	8.20–12.40	6.557	0.20–0.29	6.45–4.48	Brass	0.900–0.400	52
				5.49–3.383	Alum.		67
75	10.00–15.00	7.868	0.17–0.23			0.750–0.375	
62	12.4–18.00	9.486	0.12–0.16	9.51–8.31	Brass	0.622–0.311	91
					Alum.		
				6.14–5.36	Silver		107

EIA WG Designation WR(·)	Recommended Operating Range for TE_{10} Mode (GHz)	Cutoff Frequency for TE_{10} Mode (GHz)	Theoretical CW Power Rating, Lowest to Highest Frequency (MW)	Theoretical Attenuation, Lowest to Highest Frequency (dB/100 ft)	Material Alloy	Inside Dimensions (in.)	JAN WG Designation RG(·)/U
51	15.00–22.00	11.574	0.080–0.107			0.510–0.255	
42	18.00–26.50	14.047	0.043–0.058	20.7–14.8	Brass	0.420–0.170	53
				17.6–12.6	Alum.		121
				13.3–9.5	Silver		66
34	22.00–33.00	17.328	0.034–0.048			0.340–0.170	
28	26.50–40.00	21.081	0.022–0.031	—	Brass		
				—	Alum.	0.280–0.140	
				21.9–15.0	Silver		96
22	33.00–50.00	26.342	0.014–0.020	—	Brass	0.224–0.112	
				31.0–20.9	Silver		97
19	40.00–60.00	31.357	0.011–0.015			0.188–0.094	
15	50.00–75.00	39.863	0.0063–0.0090	—	Brass	0.148–0.074	
				52.9–39.1	Silver		98
12	60.00–90.00	48.350	0.0042–0.0060	—	Brass	0.122–0.061	
				93.3–52.2	Silver		99
10	75.00–110.00	59.010	0.0030–0.0041			0.100–0.050	
8	90.00–140.00	73.840	0.0018–0.0026	152–99	Silver	0.080–0.040	138
7	110.00–170.00	90.840	0.0012–0.0017	163–137	Silver	0.065–0.0325	136
5	140.00–220.00	115.750	0.00071–0.00107	308–193	Silver	0.051–0.0255	135
4	170.00–260.00	137.520	0.00052–0.00075	384–254	Silver	0.043–0.0215	137
3	220.00–325.00	173.280	0.00035–0.00047	512–348	Silver	0.034–0.0170	139

Source: M. I. Skolnik, *Radar Handbook* (New York: McGraw-Hill, 1970).

APPENDIX C

Field Distribution for TE and TM Modes in a Rectangular Waveguide

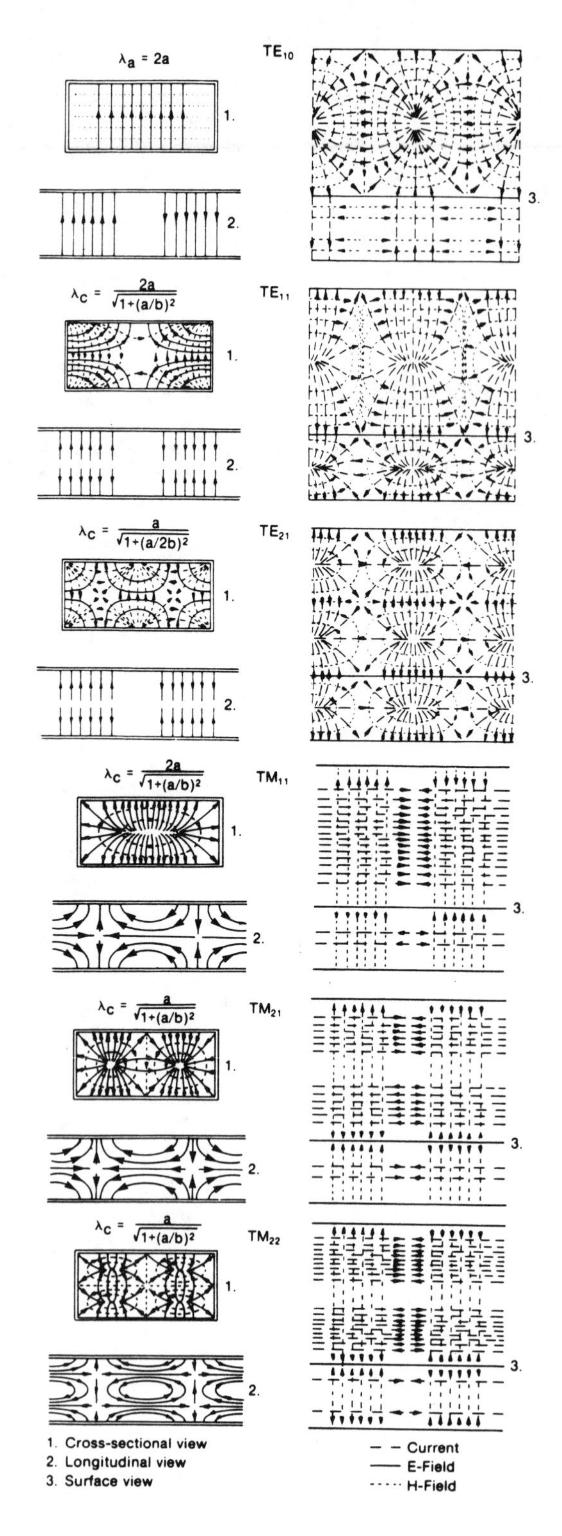

Source: I. J. Bahl "Transmission Lines" in *Handbook of Microwave and Optical Components*, K. Chang, Editor, Vol. 1, Wiley, New York, 1989, pp. 10–11.

APPENDIX D

Microstrip Synthesis Formulas*

Find w/h and ε_{eff} with Z_0 and ε_r given

For narrow strips [i.e., when $Z_0 > (44 - 2\varepsilon_r)\Omega$]:

$$\frac{w}{h} = \left[\frac{\exp(H')}{8} - \frac{1}{4\exp(H')}\right]^{-1} \tag{D.1}$$

where

$$H' = \frac{Z_0\sqrt{2(\varepsilon_r + 1)}}{119.9} + \frac{1}{2}\left(\frac{\varepsilon_r - 1}{\varepsilon_r + 1}\right)\left(\ln\frac{\pi}{2} + \frac{1}{\varepsilon_r}\ln\frac{4}{\pi}\right) \tag{D.2}$$

We may also use, with a slight but significant shift of changeover value to $w/h < 1.3$ [i.e., when $Z_0 > (63 - 2\varepsilon_r)\Omega$]:

$$\varepsilon_{eff} = \frac{\varepsilon_r + 1}{2}\left[1 - \frac{1}{2H'}\left(\frac{\varepsilon_r - 1}{\varepsilon_r + 1}\right)\left(\ln\frac{\pi}{2} + \frac{1}{\varepsilon_r}\ln\frac{4}{\pi}\right)\right]^{-2} \tag{D.3}$$

where H' is given by Equation (D.2) (as a function of Z_0) or, alternatively, as a function of w/h, from Equation (D.1):

$$H' = \ln\left[4\frac{h}{w} + \sqrt{16\left(\frac{h}{w}\right)^2 + 2}\right] \tag{D.4}$$

*The material in this appendix is from T. Edwards, *Foundations for Microstrip Circuit Design*, (Chichester, West Sussex, England: Wiley, 1992).

For wide strips [i.e., when $Z_0 < (44 - 2\varepsilon_r)\Omega$],

$$\frac{w}{h} = \frac{2}{\pi}[(d_\varepsilon - 1) - \ln(2d_\varepsilon - 1)] + \frac{\varepsilon_r - 1}{\pi\varepsilon_r}\left[\ln(d_\varepsilon - 1) + 0.293 - \frac{0.517}{\varepsilon_r}\right] \tag{D.5}$$

where

$$d_\varepsilon = \frac{59.95\pi^2}{Z_0\sqrt{\varepsilon_r}} \tag{D.6}$$

$$\varepsilon_{\text{eff}} = \frac{\varepsilon_r + 1}{2} + \frac{\varepsilon_r - 1}{2}\left(1 + 10\frac{h}{w}\right)^{-0.555} \tag{D.7}$$

Alternatively, where Z_0 is known at first:

$$\varepsilon_{\text{eff}} = \frac{\varepsilon_r}{0.96 + \varepsilon_r(0.109 - 0.004\varepsilon_r)[\log(10 + Z_0) - 1]} \tag{D.8}$$

For microstrip lines on alumina ($\varepsilon_r = 10$) this expression appears to be accurate to $\pm 0.2\%$ over the impedance range

$$8 \leq Z_0 \leq 45\Omega$$

APPENDIX E

Decibel

The decibel (dB) is a dimensionless number that expresses the ratio of two power levels. Specifically,

$$\text{power ratio (in dB)} = 10 \log_{10} \frac{P_2}{P_1} \tag{E.1}$$

where P_1 and P_2 are the two power levels being compared. If power level P_2 is higher than P_1, dB is positive, and vice versa. Since $P = V^2/R$, the voltage definition of dB is given by

$$\text{dB} = 20 \log_{10} \frac{V_2}{V_1} \tag{E.2}$$

The decibel was originally named for Alexander Graham Bell. The unit was used as a measure of attenuation in telephone cable (i.e., the ratio of the power of the signal emerging from a cable to the power of the signal fed in at the other end). It so happened that 1 decibel almost equaled the attenuation of 1 mile of telephone cable.

E.1 CONVERSION FROM POWER RATIOS TO dB, AND VICE VERSA

One can convert any power ratio (P_2/P_1) to decibels, with any desired degree of accuracy, by dividing P_2 by P_1, finding the logarithm of the result, and multiplying it by 10. From Equation (E.1), we can find the power ratio in dB (Table E.1).

As one can see from these results, the use of decibels is very convenient to represent a very large or a very small power ratio. To convert from dB to power ratios, the following equation can be used:

$$\text{power ratio} = 10^{\text{dB}/10} \tag{E.3}$$

TABLE E.1

Power ratio	dB	Power ratio	dB
1	0	1	0
1.26	1	0.794	−1
1.6	2	0.625	−2
2	3	0.5	−3
2.5	4	0.4	−4
3.2	5	0.312	−5
4	6	0.25	−6
5	7	0.2	−7
6.3	8	0.159	−8
8	9	0.125	−9
10	10	0.1	−10
100	20	0.01	−20
1,000	30	0.001	−30
10,000,000	70	1×10^{-7}	−70

E.2 GAIN OR LOSS REPRESENTATIONS

A common use of decibels is in expressing power gains and power losses in circuits. *Gain* is the term for an increase in power level. As shown in Figure E.1, an amplifier is used to amplify an input signal with $P_{\text{in}} = 1$ mW. The output signal is 200 mW. The amplifier has a gain given by

$$\text{gain in ratio} = \frac{\text{output power}}{\text{input power}} = 200 \tag{E.4}$$

$$\text{gain in dB} = 10 \log_{10} \frac{\text{output power}}{\text{input power}} = 23 \text{ dB} \tag{E.5}$$

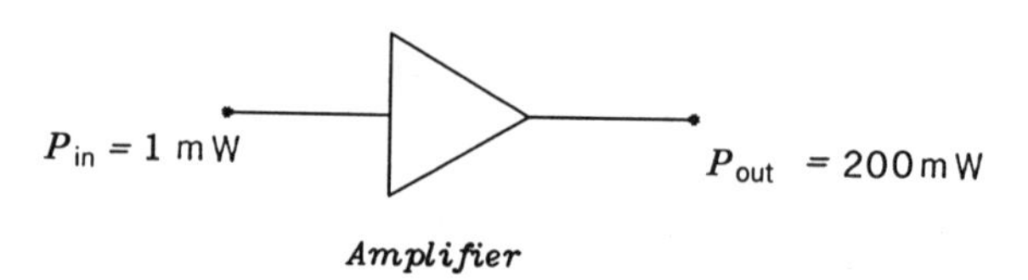

FIGURE E.1 Amplifier circuit

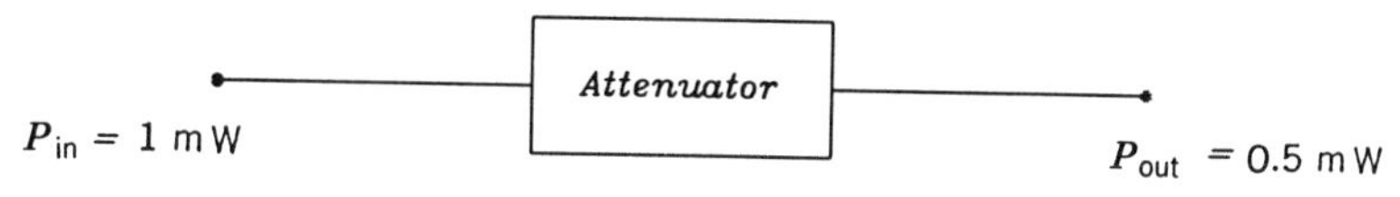

FIGURE E.2 Attenuator circuit.

Now consider an attenuator as shown in Figure E.2. *Loss* is the term for a decrease in power. The attenuator has a loss given by

$$\text{loss in ratio} = \frac{\text{input power}}{\text{output power}} = 2 \tag{E.6}$$

$$\text{loss in dB} = 10 \log_{10} \frac{\text{input power}}{\text{output power}} = 3 \text{ dB} \tag{E.7}$$

The loss described above is called *insertion loss*. Insertion loss occurs in most circuit components, waveguides, and transmission lines. One can consider 3 dB loss to equal −3 dB gain.

For a cascaded circuit, one can add all gains (in dB) together and subtract the losses (in dB). Figure E.3 shows an example. The total gain (or loss) is

$$\text{total gain} = 23 \text{ dB} + 23 \text{ dB} + 23 \text{ dB} - 3 \text{ dB} = 66 \text{ dB}$$

If one uses ratios, the total gain in ratio is

$$\text{total gain} = 200 \times 200 \times 200 \times \frac{1}{2} = 4{,}000{,}000$$

In general, for any number of gains (in dB) and losses (in dB) in a cascaded circuit, the total gain or loss can be found by

$$G_T = (G_1 + G_2 + G_3 + \cdots) - (L_1 + L_2 + L_3 + \cdots) \tag{E.8}$$

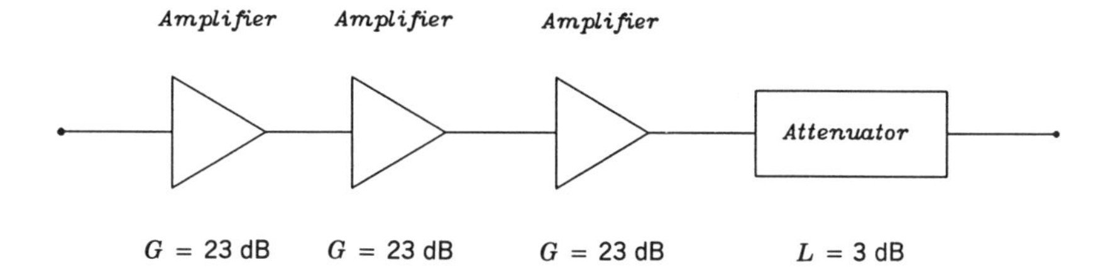

E.3 DECIBELS AS ABSOLUTE UNITS

Decibels can be used to express values of power. All that is necessary is to establish some absolute unit of power as a reference. By relating a given value of power to this unit, the power can be expressed in dB. The reference units used most often are 1 mW and 1 W. If 1 mW is used as a reference, dBm is expressed as dB relative to 1 mW.

$$P\ (\text{in dBm}) = 10 \log P\ (\text{in mW}) \tag{E.9}$$

Therefore, the following results can be written:

$$\begin{aligned} 1\ \text{mW} &= 0\ \text{dBm} \\ 10\ \text{mW} &= 10\ \text{dBm} \\ 1\ \text{W} &= 30\ \text{dBm} \\ 0.1\ \text{mW} &= -10\ \text{dBm} \\ 1 \times 10^{-7}\ \text{mW} &= -70\ \text{dBm} \end{aligned}$$

If 1 W is used as a reference, dBW is expressed as dB relative to 1 W. The conversion equation is given by

$$P\ (\text{in dBW}) = 10 \log P\ (\text{in W}) \tag{E.10}$$

From Equation (E.10) we have

$$\begin{aligned} 1\ \text{W} &= 0\ \text{dBW} \\ 10\ \text{W} &= 10\ \text{dBW} \\ 0.1\ \text{W} &= -10\ \text{dBW} \end{aligned}$$

APPENDIX F

Derivation of Multiplication Factors for Avalanche Breakdown

Consider the p–n junction shown in Figure F.1 and assume that $\alpha_n \neq \alpha_p$. α_n and α_p are the electron and hole ionization rates, respectively. The hole current generated in dx is

$$dI_p = \alpha_p I_p \, dx + \alpha_n I_n \, dx \tag{F.1}$$

Under steady-state conditions, the total current is constant:

$$I = \text{total current} = I_n + I_p = \text{constant} \tag{F.2}$$

Assume that the electric field is high near the junction such that avalanche breakdown occurs. We can define a hole multiplication factor M_p as

$$I_p(w) = M_p I_p(0) \approx I \tag{F.3}$$

The breakdown condition occurs as M_p approaches ∞. From Equations (F.1) and (F.2), we have

$$\frac{dI_p}{dx} = \alpha_p I_p + \alpha_n I_n = \alpha_p I_p + \alpha_n (I - I_p)$$

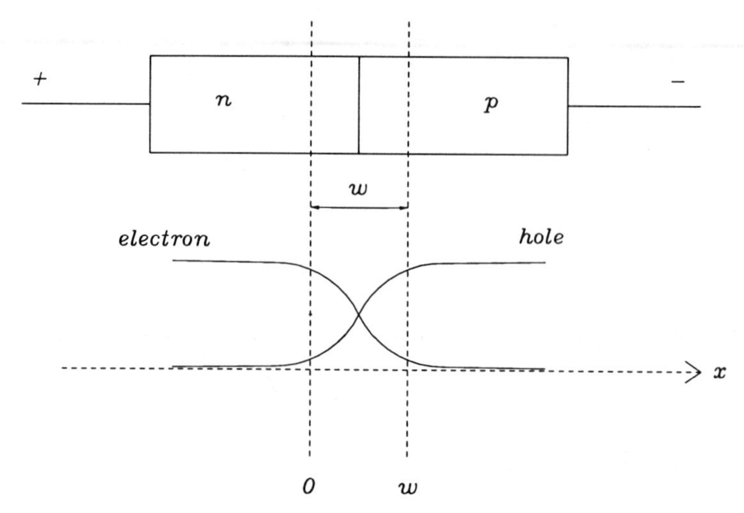

FIGURE F.1 A p–n junction and current distribution.

Therefore,

$$\frac{dI_p}{dx} - (\alpha_p - \alpha_n)I_p = \alpha_n I \tag{F.4}$$

Note that α_p, α_n, and I_p are functions of x, but I is a constant. The solution to Equation (F.4) is

$$I_p(x) = Ce^{\phi} + Ie^{\phi}\int_0^x e^{-\phi}\alpha_n\, dx \tag{F.5}$$

where

$$\phi = \int_0^x (\alpha_p - \alpha_n)\, dx$$

To prove this, one can substitute Equation (F.5) into Equation (F.4). Applying $I_p(x) = I_p(0)$ at $x = 0$, we have $I_p(0) = C$. Therefore, the hole current is given by

$$I_p(x) = I_p(0)\exp\left[\int_0^x (\alpha_p - \alpha_n)\, dx\right] + I\exp\left[\int_0^x (\alpha_p - \alpha_n)\, dx\right] \times \int_0^x \alpha_n \exp\left[-\int_0^{x'} (\alpha_p - \alpha_n)\, dx''\right] dx' \tag{F.6}$$

Substituting $I_p(0) = I/M_p$ into (F.6), we have

$$I_p(x) = I\left\{\frac{1}{M_p} + \int_0^x \alpha_n \exp\left[-\int_0^{x'} (\alpha_p - \alpha_n)\, dx''\right] dx'\right\} \exp\left[\int_0^x (\alpha_p - \alpha_n)\, dx\right] \tag{F.7}$$

Let $x = w$, $I_p(w) \approx I$:

$$I = I\left\{\frac{1}{M_p} + \int_0^w \alpha_n \exp\left[-\int_0^{x'} (\alpha_p - \alpha_n)\, dx''\right] dx'\right\} \exp\left[\int_0^w (\alpha_p - \alpha_n)\, dx\right] \tag{F.8}$$

$1/M_p$ can be expressed as

$$\begin{aligned}\frac{1}{M_p} &= \exp\left[-\int_0^w (\alpha_p - \alpha_n)\, dx\right] - \int_0^w \alpha_n \exp\left[-\int_0^x (\alpha_p - \alpha_n)\, dx'\right] dx \\ &= \exp\left[-\int_0^w (\alpha_p - \alpha_n)\, dx\right] - \int_0^w (\alpha_n - \alpha_p) \exp\left[-\int_0^x (\alpha_p - \alpha_n)\, dx'\right] dx \\ &\quad - \int_0^w \alpha_p \exp\left[-\int_0^x (\alpha_p - \alpha_n)\, dx'\right] dx\end{aligned} \tag{F.9}$$

Using the identity

$$\int_0^w (\alpha_n - \alpha_p) \exp\left[-\int_0^x (\alpha_p - \alpha_n)\, dx'\right] dx = \exp\left[-\int_0^w (\alpha_p - \alpha_n)\, dx\right] - 1 \tag{F.10}$$

we have

$$\frac{1}{M_p} = 1 - \int_0^w \alpha_p \exp\left[-\int_0^x (\alpha_p - \alpha_n)\, dx'\right] dx \tag{F.11}$$

Similarly,

$$\frac{1}{M_n} = 1 - \int_0^w \alpha_n \exp\left[-\int_x^w (\alpha_n - \alpha_p)\, dx'\right] dx \tag{F.12}$$

The identity (F.10) can be proved by letting

$$\phi(x) = \int_0^x (\alpha_p - \alpha_n)\, dx'$$

$$\frac{d\phi(x)}{dx} = \alpha_p - \alpha_n$$

$$\begin{aligned}
\int_0^w (\alpha_n - \alpha_p)\exp\left[-\int_0^x (\alpha_p - \alpha_n)\, dx'\right] dx &= -\int_0^w \frac{d\phi(x)}{dx} e^{-\phi(x)}\, dx \\
&= -\int_0^w e^{-\phi(x)}\, d\phi(x) \\
&= e^{-\phi(x)}\Big|_0^w = e^{-\phi(w)} - e^{-\phi(0)} \\
&= \exp\left[-\int_0^w (\alpha_p - \alpha_n)\, dx'\right] - 1 \\
&= \exp\left[-\int_0^w (\alpha_p - \alpha_n)\, dx\right] - 1
\end{aligned}$$

APPENDIX G

Simplified Small-Signal Model for IMPATT Diode

The simplified small-signal model is useful in obtaining an equivalent circuit for the IMPATT diode. Also, the model gives some physical insight about the existence of negative resistance. In this model the device is divided into three regions: (1) the avalanche region, in which the thickness is very thin, the ionization rate is uniform ($\alpha_n = \alpha_p = \alpha$), and transit-time delay is negligible; (2) the drift region, in which no carriers are generated and all carriers entering from the avalanche region move at saturated velocities ($v_n = v_p = v_s$); and (3) the inactive region, which introduces additional resistance.

G.1 AVALANCHE REGION

In the avalanche region, add Equations (11.10) and (11.11) and integrate from $x = 0$ to $x = x_a$ (where x_a is the width of the avalanche region as shown in Figure G.1, and normally, $x_a \ll w$) to derive the following equation:

$$\frac{dJ}{dt} = \frac{2J}{\tau_a}(\alpha x_a - 1) \tag{G.1}$$

where the boundary conditions at $x = 0$ and $x = x_a$ are used and the transit-time is defined as $\tau_a = x_a/v_s$. Substituting Equations (11.13) through (11.16) into (G.1), we have

$$J_1 = \frac{2\alpha' x_a J_0 E_1}{j\omega\tau_a} \tag{G.2}$$

where J_1 is the avalanche current, which is in phase quadrature with electric

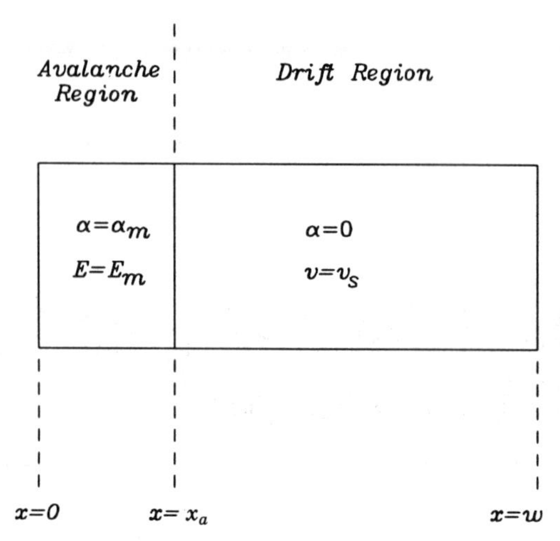

FIGURE G.1 Simplified IMPATT model.

field E_1. The total current is

$$J_T = \frac{2\alpha' x_a J_0 E_1}{j\omega\tau_a} + j\omega\varepsilon_s E_1 \tag{G.3}$$

where the second term on the right-hand side accounts for the displacement current. It can be seen that two components of the total current are in the avalanche region. For a given field, the avalanche current J_1 is imaginary and varies inversely with ω as in an inductor. The other component, $j\omega\varepsilon_s E_1$, is also imaginary and varies proportionally with ω as a capacitor. The equivalent circuit for the avalanche region can be represented by an inductor L_a in parallel with a capacitor C_a as shown in Figure G.2(a), where L_a and C_a are given by

$$L_a = \frac{\tau_a}{2J_0\alpha' A} \tag{G.4}$$

$$C_a = \frac{\varepsilon_s A}{x_a} \tag{G.5}$$

A resonant frequency is given by

$$f_r = \frac{1}{2\pi}\frac{1}{\sqrt{L_a C_a}} = \frac{1}{2\pi}\sqrt{\frac{2\alpha' v_s J_0}{\varepsilon_s}} \tag{G.6}$$

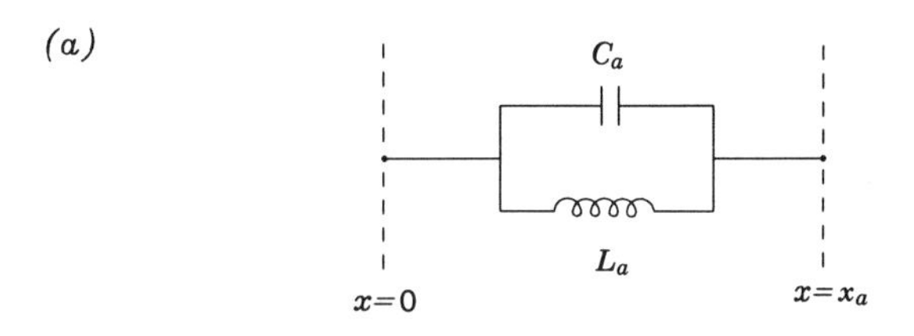

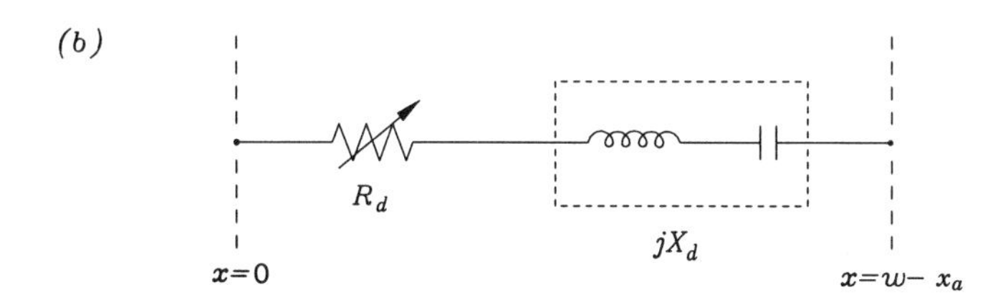

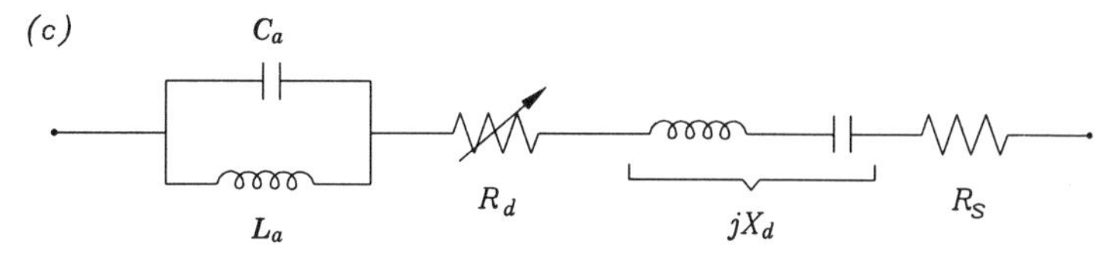

FIGURE G.2 Equivalent circuit of a simplified IMPATT model: (*a*) equivalent circuit of avalanche region; (*b*) equivalent circuit of drift region; (*c*) total equivalent circuit.

and the impedance of the avalanche region is

$$Z_a = L_a \| C_a = \frac{1}{j\omega C_a}\left[\frac{1}{1-(f_r/f)^2}\right] \tag{G.7}$$

G.2 DRIFT REGION

The current injected into the drift region is equal to

$$J(x) = J_T(x) = j\omega\varepsilon_s E_1(x) + J_1 \exp\left(-j\omega\frac{x}{v_s}\right) \tag{G.8}$$

Define the current ratio as

$$r = \frac{J_1}{J} = \frac{1}{1-(f/f_r)^2} \tag{G.9}$$

Then we have

$$E_1(x) = J\frac{1 - re^{-j\omega(x/v_s)}}{j\omega\varepsilon_s} \tag{G.10}$$

The ac voltage across the drift region is

$$V_d = \int_0^{w-x_a} E_1(x)\,dx = \frac{(w-x_a)J}{j\omega\varepsilon_s}\left[1 - \frac{1}{1-(f/f_r)^2}\left(\frac{1-e^{-j\theta_d}}{j\theta_d}\right)\right] \tag{G.11}$$

where

$$\theta_d = \frac{\omega(w-x_a)}{v_s} = \omega\tau_d$$

The impedance of the drift region is

$$Z_d = \frac{V_d}{JA} = R_d + jX_d$$

$$= \frac{1}{\omega C_d}\left[\frac{1}{1-(f/f_r)^2}\left(\frac{1-\cos\theta_d}{\theta_d}\right)\right] + \frac{j}{\omega C_d}\left[-1 + \frac{1}{1-(f/f_r)^2}\left(\frac{\sin\theta_d}{\theta_d}\right)\right] \tag{G.12}$$

where $C_d = A\varepsilon_s/(w - x_a)$. The equivalent circuit is shown in Figure G.2(b). From Equation (G.12), it is seen that $R_d < 0$ for $f > f_r$ except for the nulls at $\theta_d = 2n\pi$, where $R_d = 0$. The imaginary part, jX_d, will be inductive or capacitive depending on the value inside the brackets.

G.3 TOTAL IMPEDANCE

The total impedance of the device is the sum of those of the three regions:

$$Z = Z_a + Z_d + R_s \tag{G.13}$$

The equivalent circuit is shown in Figure G.2(c). R_s represents the series resistance due to the semiconductor.

Index